Thomas Craig

A Treatise on Linear Differential Equations

Vol. 1

Thomas Craig

A Treatise on Linear Differential Equations
Vol. 1

ISBN/EAN: 9783337811525

Printed in Europe, USA, Canada, Australia, Japan

Cover: Foto ©berggeist007 / pixelio.de

More available books at **www.hansebooks.com**

A TREATISE

ON

LINEAR DIFFERENTIAL EQUATIONS

BY

THOMAS CRAIG, PH.D.
ASSOCIATE PROFESSOR IN THE JOHNS HOPKINS UNIVERSITY
ASSOCIATE EDITOR OF THE AMERICAN JOURNAL OF MATHEMATICS

VOLUME I

EQUATIONS WITH UNIFORM COEFFICIENTS

NEW YORK
JOHN WILEY & SONS
15 ASTOR PLACE
1889

[redacted]

DRUMMOND & NEU,
Electrotypers,
1 to 7 Hague Street,
New York.

FERRIS BROS.,
Printers,
326 Pearl Street,
New York.

PREFACE.

THE theory of linear differential equations may almost be said to find its origin in Fuchs's two memoirs published in 1866 and 1868 in volumes 66 and 68 of Crelle's Journal. Previous to this the only class of linear differential equations for which a general method of integration was known was the class of equations with constant coefficients, including of course Legendre's well-known equation which is immediately transformable into one with constant coefficients. After the appearance of Fuchs's second memoir many mathematicians, particularly in France and Germany, including Fuchs himself, took up the subject which, though still in its infancy, now possesses a very large literature.

This literature, however, is so scattered among the different mathematical journals and publications of learned societies that it is extremely difficult for students to read up the subject properly.

I have endeavored in the present treatise to give a by no means complete but, I trust, a sufficient account of the theory as it stands to-day, to meet the needs of students. Full references to original sources are given in every case.

Most of the results in the first two chapters, which deal with the general properties of linear differential equations and with equations having constant coefficients, are of course old, but the presentation of these properties is comparatively new and is due to such mathematicians as Hermite, Jordan, Darboux, and others. All that follows these two chapters is quite new and constitutes the essential part of the modern theory of linear differential equations.

The present volume deals principally with Fuchs's type of equations, *i.e.* equations whose integrals are all regular; a sufficient account has been given, however, of the researches of Frobenius and Thomé on equations whose integrals are not all regular. A pretty full account, due to Jordan, has been given of the application of the theory of substitutions to linear differential equations. This subject will, however, be very much more fully dwelt upon in Volume II, where I intend to take up the question of equations having algebraic integrals and also to give an account of Poincaré's splendid investigations of Fuchsian groups and Fuchsian functions. The theory of the invariants of linear differential equations has been several times touched upon in the present volume, and some of the simpler results of the theory have been employed; but its extended development is necessarily reserved for Volume II, as is also the development of Forsyth's associate equations, about which extremely interesting subject very little is as yet known.

The equation of the second order with the critical points $0, 1, \infty$ has on account of its great importance been very fully treated. In connection with this subject it seemed to me that I could not possibly do better than to reproduce, which has been done in Chapter VII, Goursat's Thesis on equations of the second order satisfied by the hypergeometric series. M. Goursat was kind enough to give me permission to make a translation of his Thesis, which is, I imagine, not very well known among English and American students.

In Chapter XIV I have given only a brief account of equations with doubly-periodic coefficients. I intend, however, to resume this subject in Volume II.

I wish here to tender my thanks to M. Goursat for his kindness in permitting me to make a translation of his most valuable Thesis, and to Dr. Oskar Bolza, Mr. C. H. Chapman, and Dr. J. C. Fields for much valuable assistance.

T. CRAIG.

JOHNS HOPKINS UNIVERSITY,
BALTIMORE, 1889.

CONTENTS.

CHAPTER I.

GENERAL PROPERTIES OF LINEAR DIFFERENTIAL EQUATIONS.

CHAPTER II.

LINEAR DIFFERENTIAL EQUATIONS WITH CONSTANT COEFFICIENTS.

CHAPTER III.

THE INTEGRALS OF THE DIFFERENTIAL EQUATION

$$P = \frac{d^n y}{dx^n} + p_1 \frac{d^{n-1}y}{dx^{n-1}} + p_2 \frac{d^{n-2}y}{dx^{n-2}} + \ldots + p_n y = 0.$$

CHAPTER IV.

FROBENIUS'S METHOD.

CHAPTER V.

LINEAR DIFFERENTIAL EQUATIONS ALL OF WHOSE INTEGRALS ARE REGULAR.

CHAPTER VI.

LINEAR DIFFERENTIAL EQUATIONS OF THE SECOND ORDER, PARTICULARLY THOSE WITH THREE CRITICAL POINTS.

CHAPTER VII.

ON THE LINEAR DIFFERENTIAL EQUATION WHICH ADMITS THE HYPERGEOMETRIC SERIES AS AN INTEGRAL; BY M. EDOUARD GOURSAT.

Part First.

Part Second.

CHAPTER VIII.

IRREDUCIBLE LINEAR DIFFERENTIAL EQUATIONS.

CHAPTER IX.

LINEAR DIFFERENTIAL EQUATIONS SOME OF WHOSE INTEGRALS ARE REGULAR.

CHAPTER X.

DECOMPOSITION OF A LINEAR DIFFERENTIAL EQUATION INTO SYMBOLIC PRIME FACTORS.

CHAPTER XI.

APPLICATION OF THE THEORY OF SUBSTITUTIONS TO LINEAR DIFFERENTIAL EQUATIONS.

CHAPTER XII.

EQUATIONS WHOSE GENERAL INTEGRALS ARE RATIONAL.—HALPHEN'S EQUATIONS.

CHAPTER XIII.

TRANSFORMATION OF A LINEAR DIFFERENTIAL EQUATION.—FORSYTH'S CANONICAL FORM. —ASSOCIATE EQUATIONS.

CHAPTER XIV.

LINEAR DIFFERENTIAL EQUATIONS WITH UNIFORM DOUBLY-PERIODIC COEFFICIENTS.

LINEAR DIFFERENTIAL EQUATIONS.

CHAPTER I.

GENERAL PROPERTIES OF LINEAR DIFFERENTIAL EQUATIONS.

ACCORDING to general usage we will denote by x an independent complex variable; that is, x is the affix of any variable point in the plane. When fixed points, such as the critical points of the various functions we shall encounter, are to be spoken of, we will generally denote their affixes by $a, b, c, \ldots$ Again, y is an unknown function of x, defined by a linear differential equation—that is, a differential equation whose left-hand member is a linear function of y and of the derivatives of y with respect to x; the coefficients in this linear function are arbitrary functions of x; the right-hand member of the linear differential equation is either zero or a function of x alone. The equation has thus the form

$$(1) \qquad \frac{d^n y}{dx^n} + p_1 \frac{d^{n-1} y}{dx^{n-1}} + p_2 \frac{d^{n-2} y}{dx^{n-2}} + \ldots + p_n y = F(x).$$

A first general property of linear differential equations is that *they remain linear when we change the independent variable.* Write, viz., $x = \phi(t)$, and form the new differential equation connecting y and t: it will obviously be an equation of the same form as (1), having new coefficients containing t alone. The verification of this is so simple that it need not be given here.

Another general proposition is (see Jordan's *Cours d'Analyse*, vol. iii. p. 136) as follows: *If $y, z, \ldots, w$ denote n functions of the same independent variable x, and satisfying the system of n linear differential equations,*

$$(2) \qquad E_1 = 0, \qquad E_2 = 0, \; \ldots, \; E_n = 0;$$

and if V is a polynomial in $y, z, \ldots, w$ and their successive derivatives, the different terms of which have for coefficients arbitrary functions of x: then V will satisfy a linear differential equation whose coefficients are rationally expressible in terms of the coefficients of $E_1, \ldots E_m$, V, and of their successive derivatives.

Let $\mu, \mu', \ldots$ denote the highest orders of the derivatives of y, $z, \ldots$, respectively, in equations (2), and let λ denote the degree of the polynomial V. Form now the successive derivatives of V with respect to x; each of these is a polynomial of degree λ in $y, z, \ldots$ and their derivatives; and, further, the derivatives

$$(3) \qquad \frac{d^\mu y}{dx^\mu}, \quad \frac{d^{\mu+1} y}{dx^{\mu+1}}, \ldots, \quad \frac{d^{\mu'} z}{dx^{\mu'}}, \quad \frac{d^{\mu'+1} z}{dx^{\mu'+1}}, \ldots, \ldots$$

are linearly expressible in terms of the derivatives of lower order by means of equations (2) and their derivatives. Substituting these values for the quantities (3), we have

$$(4) \qquad \begin{cases} V = X_1 P_1 + X_2 P_2 + \ldots, \\ \dfrac{dV}{dx} = X_1' P_1 + X_2' P_2 + \ldots, \\ \ldots \ldots \ldots ; \end{cases}$$

where the functions $X, X', \ldots$ are functions of x of the kind mentioned in the above proposition, and $P_1, P_2, \ldots$ are products of the form

$$y^\alpha \left(\frac{dy}{dx}\right)^{\alpha_1}, \ldots, \left(\frac{d^{\mu-1} y}{dx^{\mu-1}}\right)^{\alpha_{\mu-1}}, z^\beta \left(\frac{dz}{dx}\right)^{\beta_1}, \ldots, \left(\frac{d^{\mu'-1} z}{dx^{\mu'-1}}\right)^{\beta_{\mu'-1}}, \ldots,$$

the total number of factors (taking account of their orders of multiplicity) being at most $= \lambda$. The number of these products P is of course limited; let them be denoted by

$$P_1, \quad P_2, \ldots, P_l.$$

Eliminating these quantities between equations (4), we obviously arrive at a linear relation connecting $V, \dfrac{dV}{dx}, \dfrac{d^2V}{dx^2}, \ldots$ From the theory of ordinary differential equations we know that any simultaneous system of such equations can, by the introduction of certain auxiliary functions and the performance of certain simple algebraical operations, be thrown into the *normal form*—that is, can be replaced by a system of equations in each of which there appears only the first derivative of *one* of the (old and new) dependent variables. In case the given system of simultaneous equations is linear, it is obvious that the property of linearity will be retained when the equations are replaced by the normal system. In studying, then, a system of simultaneous linear differential equations, we may at once suppose the system to be replaced by the corresponding normal system. Suppose such a normal system to be

$$(5)\qquad \begin{cases} \dfrac{dy_1}{dx} + a_{11}y_1 + a_{12}y_2 + \ldots + a_{1n}y_n = T_1, \\ \dfrac{dy_2}{dx} + a_{21}y_1 + a_{22}y_2 + \ldots + a_{2n}y_n = T_2, \\ \cdot\quad\cdot\quad\cdot\quad\cdot\quad\cdot\quad\cdot\quad\cdot\quad\cdot\quad\cdot \\ \dfrac{dy_n}{dx} + a_{n1}y_1 + a_{n2}y_2 + \ldots + a_{nn}y_n = T_n; \end{cases}$$

where T_1, T_2, . . . , T_n and the coefficients a_{ik} are functions of x. As a particular case, suppose the given system of simultaneous linear differential equations to reduce to the single one,

$$(6)\qquad \frac{d^ny}{dx^n} + p_1\frac{d^{n-1}y}{dx^{n-1}} + p_2\frac{d^{n-2}y}{dx^{n-2}} + \ldots + p_ny = T.$$

Denoting by y', y'', . . . , y^{n-1} new auxiliary dependent variables, we can replace (6) by the equivalent normal system,

$$(7)\qquad \begin{cases} \dfrac{dy}{dx} - y' = 0, \\ \dfrac{dy'}{dx} - y'' = 0, \\ \cdot\quad\cdot\quad\cdot\quad\cdot\quad\cdot\quad\cdot\quad\cdot\quad\cdot\quad\cdot \\ \dfrac{dy^{n-2}}{dx} - y^{n-1} = 0, \\ \dfrac{dy^{n-1}}{dx} + p_1y^{n-1} + \ldots + p_ny = T. \end{cases}$$

As the reader is supposed to be familiar with the elementary theory of differential equations, and with the general properties of linear differential equations, it will be sufficient here to state some of these properties without proof. First: if $y_1^1, y_2^1, \ldots, y_n^1;\ y_1^2, y_2^2, \ldots, y_n^2;\ \ldots$ are particular solutions of equations (5) when T_1, $T_2, \ldots, T_n$ are all supposed to be zero, then, denoting by C_1, $C_2, \ldots, C_n$ arbitrary constants,

$$(8) \qquad \begin{cases} C_1 y_1^1 + C_2 y_2^1 + \ldots + C_n y_n^1, \\ C_1 y_1^2 + C_2 y_2^2 + \ldots + C_n y_n^2, \\ \cdot \quad \cdot \quad \cdot \quad \cdot \quad \cdot \quad \cdot \quad \cdot \end{cases}$$

are also solutions. This is at once proved by substituting in equations (5) when the right-hand members are all replaced by zero, that is, when equations (5) are of the form

$$(9) \qquad \begin{cases} \dfrac{dy_1}{dx} + a_{11} y_1 + a_{12} y_2 + \ldots + a_{1n} y_n = 0, \\ \dfrac{dy_2}{dx} + a_{21} y_1 + a_{22} y_2 + \ldots + a_{2n} y_n = 0, \\ \cdot \quad \cdot \quad \cdot \quad \cdot \quad \cdot \quad \cdot \quad \cdot \quad \cdot \quad \cdot \\ \dfrac{dy_n}{dx} + a_{n1} y_1 + a_{n2} y_2 + \ldots + a_{nn} y_n = 0. \end{cases}$$

An obvious, but useless, solution of these last equations is $y_1 = y_2 = \ldots = y_n = 0$. Aside from this useless solution, suppose equations (9) to admit the following k particular solutions:

$$(10) \qquad \begin{cases} y_1^1, & y_2^1, \ldots, & y_n^1, \\ y_1^2, & y_2^2, \ldots, & y_n^2, \\ \cdot \quad \cdot & \cdot \quad \cdot \quad \cdot & \cdot \\ y_1^k, & y_2^k, \ldots, & y_n^k. \end{cases}$$

These solutions will be said to be *independent* if one at least of the determinants of order k formed from the different columns of (10) does not vanish. If $k = n$, then the solutions (10) will be independent if the determinant

$$
(11) \qquad \begin{vmatrix} y_1^1, & y_2^1, & \dots, & y_n^1, \\ y_1^2, & y_2^2, & \dots, & y_n^2, \\ \cdot & \cdot & \cdot & \cdot \\ y_1^n, & y_2^n, & \dots, & y_n^n, \end{vmatrix}
$$

does not vanish. The particular application of this to equations (7) and (6) is easily seen, but will be referred to later on.

As an illustration of another well-known general theorem, stated below, consider the system

$$
(12) \qquad \begin{cases} \dfrac{dy}{dx} + ay + bz + cu + dv = 0, \\ \dfrac{dz}{dx} + a_1 y + b_1 z + c_1 u + d_1 v = 0, \\ \dfrac{du}{dx} + a_2 y + b_2 z + c_2 u + d_2 v = 0, \\ \dfrac{dv}{dx} + a_3 y + b_3 z + c_3 u + d_3 v = 0, \end{cases}
$$

and suppose that we know two independent solutions of this system, viz.:

$$
\begin{matrix} y_1, & z_1, & u_1, & v_1; \\ y_2, & z_2, & u_2, & v_2. \end{matrix}
$$

Suppose, for example, that the determinant

$$
\begin{vmatrix} y_1, & z_1, \\ y_2, & z_2, \end{vmatrix}
$$

does not vanish. Let us write now

$$
(13) \qquad \begin{cases} y = C_1 y_1 + C_2 y_2, \\ z = C_1 z_1 + C_2 z_2, \\ u = C_1 u_1 + C_2 u_2 + \xi, \\ v = C_1 v_1 + C_2 v_2 + \eta; \end{cases}
$$

where c_1, c_2, ξ, η, are new functions of x. Substituting the values (13) in (10), and remarking that the terms in c_1 and c_2 vanish (since equations (10) would be satisfied if ξ and η were zero and C_1 and C_2 constants), we have

$$(14)\quad \begin{cases} \dfrac{dC_1}{dx}y_1 + \dfrac{dC_2}{dx}y_2 + c\xi + d\eta = 0, \\ \dfrac{dC_1}{dx}z_1 + \dfrac{dC_2}{dx}z_2 + c_1\xi + d_1\eta = 0, \\ \dfrac{dC_1}{dx}u_1 + \dfrac{dC_2}{dx}u_2 + \dfrac{d\xi}{dx} + c_2\xi + d_2\eta = 0, \\ \dfrac{dC_1}{dx}v_1 + \dfrac{dC_2}{dx}v_2 + \dfrac{d\eta}{dx} + c_3\xi + d_3\eta = 0. \end{cases}$$

Solving these for $\dfrac{dC_1}{dx}$, $\dfrac{dC_2}{dx}$, $\dfrac{d\xi}{dx}$, $\dfrac{d\eta}{dx}$, we derive the system

$$(15)\quad \begin{cases} \dfrac{dC_1}{dx} = A\xi + B\eta, \\ \dfrac{dC_2}{dx} = A_1\xi + B_1\eta, \\ \dfrac{d\xi}{dx} = A_2\xi + B_2\eta, \\ \dfrac{d\eta}{dx} = A_3\xi + B_3\eta. \end{cases}$$

From the last two equations we may suppose ξ and η to be obtained, and then from the first two equations we find C_1 and C_2 by quadratures. Knowing then two independent solutions of the system (12) of four equations, we have made the integration of this system depend on the integration of a similar system of *two* linear differential equations, viz., the last two of equations (15), and on quadratures. We can now show that the new solutions, say ξ' and η', form, with the given solutions

$$\begin{matrix} y_1, & z_1, & u_1, & v_1, \\ y_2, & z_2, & u_2, & v_2, \end{matrix}$$

an independent system. (We of course exclude the obvious solution, $\xi = 0$, $\eta = 0$, of the last two of equations (15)). To fix the ideas, suppose ξ' to be different from zero, and let C_1' and C_2' denote the values of C_1 and C_2 as obtained by the quadratures; then we can show at once that the solution

$$(16)\quad \begin{cases} y_3 = C_1^1 y_1 + C_2^1 y_2, \\ z_3 = C_1^1 z_1 + C_2^1 z_2, \\ u_3 = C_1^1 u_1 + C_2^1 u_2 + \xi^1, \\ v_3 = C_1^1 v_1 + C_2^1 v_2 + \eta^1, \end{cases}$$

is independent of the two known solutions,

$$\begin{matrix} y_1, & z_1, & u_1, & v_1, \\ y_2, & z_2, & u_2, & v_2; \end{matrix}$$

in fact, we see that the determinant

$$\begin{vmatrix} y_1, & z_1, & u_1, \\ y_2, & z_2, & u_2, \\ y_3, & z_3, & u_3, \end{vmatrix}$$

reduces to

$$\xi^1 \begin{vmatrix} y_1, & z_1, \\ y_2, & z_2, \end{vmatrix}$$

and by hypothesis neither of these factors vanishes. The general theorem above referred to, and of which the preceding is a simple illustration, may be stated as follows:

If we know k independent solutions ($k < n$) *of a system of linear differential equations of the first order, the integration of this system can be conducted to that of an analogous system of $n - k$ equations and to quadratures. Each solution of this new system will furnish a solution of the given system which will be independent of those already known.*

Another well-known general theorem which may be stated without proof is: *Every system of linear differential equations of the form* (9) *admits n particular independent solutions,*

$$\begin{matrix} y_1^1, & y_2^1, & \ldots, & y_n^1, \\ y_1^2, & y_2^2, & \ldots, & y_n^2, \\ \cdot & \cdot & \cdot & \cdot \\ y_1^n, & y_2^n, & \ldots, & y_n^n \end{matrix}$$

and its most general solution is

$$\begin{array}{l} c_1 y_1^1 + c_2 y_1^2 + \dots + c_n y_1^n, \\ c_1 y_2^1 + c_2 y_2^2 + \dots + c_n y_2^n, \\ \cdot \quad \cdot \quad \cdot \quad \cdot \quad \cdot \quad \cdot \\ c_1 y_n^1 + c_2 y_n^2 + \dots + c_n y_n^n ; \end{array}$$

where $c_1, c_2, \dots, c_n$ are arbitrary constants.

This is readily verified by substitution in (9), and by aid of the preceding theorem.

Suppose

$$(17) \qquad \left\{ \begin{array}{l} Y_1^1 = c_1^1 y_1^1 + c_2^1 y_2^1 + \dots + c_n^1 y_n^1, \\ Y_1^2 = c_1^1 y_1^2 + c_2^1 y_2^2 + \dots + c_n^1 y_n^2, \\ \cdot \quad \cdot \quad \cdot \quad \cdot \quad \cdot \quad \cdot \quad \cdot \quad \cdot \\ Y_1^n = c_1^1 y_1^n + c_2^1 y_2^n + \dots + c_n^1 y_n^n, \\ \cdot \quad \cdot \quad \cdot \quad \cdot \quad \cdot \quad \cdot \quad \cdot \quad \cdot \\ Y_n^1 = c_1^n y_1^1 + c_2^n y_2^1 + \dots + c_n^n y_n^1, \\ Y_n^2 = c_1^n y_1^2 + c_2^n y_2^2 + \dots + c_n^n y_n^2, \\ \cdot \quad \cdot \quad \cdot \quad \cdot \quad \cdot \quad \cdot \quad \cdot \quad \cdot \\ Y_n^n = c_1^n y_1^n + c_2^n y_2^n + \dots + c_n^n y_n^n, \end{array} \right.$$

to be n arbitrary particular solutions of equations (9). The determinant

$$| Y_i^k | \quad (i, k = 1, 2, \dots n)$$

is obviously equal to the product of the determinants

$$| y_i^k | \quad \text{and} \quad | c_i^k | \quad (i, k = 1, 2, \dots n).$$

Since the determinant $| y_i^k |$ is by hypothesis not zero, it follows that in order that the determinant

$$| Y_i^k |$$

may be different from zero, and consequently that the solutions

(17) may form an independent system, we must have the constants c_i^k so chosen that the determinant

$$| c_i^k |$$

shall not vanish.

To every system of linear differential equations whose right-hand member is zero, such as

$$(18) \quad \begin{cases} \frac{dy_1}{dx} + a_{11}y_1 + a_{12}y_2 + \ldots + a_{1n}y_n = 0, \\ \frac{dy_2}{dx} + a_{21}y_1 + a_{22}y_2 + \ldots + a_{2n}y_n = 0, \\ \cdot \quad \cdot \quad \cdot \quad \cdot \quad \cdot \quad \cdot \quad \cdot \quad \cdot \\ \frac{dy_n}{dx} + a_{n1}y_1 + a_{n2}y_2 + \ldots + a_{nn}y_n = 0, \end{cases}$$

there is associated a system defined by the equations

$$(19) \quad \begin{cases} -\frac{dY_1}{dx} + a_{11}Y_1 + a_{21}Y_2 + \ldots + a_{n1}Y_n = 0, \\ -\frac{dY_2}{dx} + a_{12}Y_1 + a_{22}Y_2 + \ldots + a_{n2}Y_n = 0, \\ \cdot \quad \cdot \quad \cdot \quad \cdot \quad \cdot \quad \cdot \quad \cdot \quad \cdot \quad \cdot \\ -\frac{dY_n}{dx} + a_{1n}Y_1 + a_{2n}Y_2 + \ldots + a_{nn}Y_n = 0. \end{cases}$$

This system (19) will be called the *adjunct* system to (18). A very beautiful relation exists between the solutions of equations (18) and (19). Suppose equations (18) to be multiplied by Y_1, $Y_2, \ldots, Y_n$, respectively, and equations (19) to be similarly multiplied by $-y_1, -y_2, \ldots, -y_n$; add together the results of these two multiplications and we find, after making certain simple reductions,

$$Y_1\frac{dy_1}{dx} + Y_2\frac{dy_2}{dx} + \ldots + Y_n\frac{dy_n}{dx}$$
$$+ y_1\frac{dY_1}{dx} + y_2\frac{dY_2}{dx} + \ldots + y_n\frac{dY_n}{dx} = 0,$$

or

$$\frac{d}{dx}(y_1 Y_1 + y_2 Y_2 + \ldots + y_n Y_n) = 0;$$

and consequently

$$(20) \qquad y_1 Y_1 + y_2 Y_2 + \ldots + y_n Y_n = \text{const.}$$

It is now obvious that the complete solution of either of the systems (18) or (19) will involve the complete solution of the other system. Suppose, in fact, that we know n independent solutions,

$$\begin{array}{llll} y_1^1, & y_2^1, & \ldots, & y_n^1, \\ y_1^2, & y_2^2, & \ldots, & y_n^2, \\ \cdot & \cdot & \cdot & \cdot \\ y_1^n, & y_2^n, & \ldots, & y_n^n, \end{array}$$

of equations (18); then denoting by

$$c_1, \quad c_2, \ldots, c_n,$$

n arbitrary constants, we have

$$(21) \qquad \begin{cases} Y_1 y_1^1 + Y_2 y_2^1 + \ldots + Y_n y_n^1 = c_1, \\ Y_1 y_1^2 + Y_2 y_2^2 + \ldots + Y_n y_n^2 = c_2, \\ \cdot \quad \cdot \quad \cdot \quad \cdot \quad \cdot \quad \cdot \quad \cdot \quad \cdot \\ Y_1 y_1^n + Y_2 y_2^n + \ldots + Y_n y_n^n = c_n. \end{cases}$$

Solving these last equations for $Y_1, Y_2, \ldots, Y_n$, we have the general solution of the system (19) involving, as it should, n arbitrary constants. If we knew only k solutions ($k < n$) of the given system (18), then we would have the k relations

$$(22) \qquad \begin{cases} Y_1 y_1^1 + Y_2 y_2^1 + \ldots + Y_n y_n^1 = c_1, \\ Y_1 y_1^2 + Y_2 y_2^2 + \ldots + Y_n y_n^2 = c_2, \\ \cdot \quad \cdot \quad \cdot \quad \cdot \quad \cdot \quad \cdot \quad \cdot \quad \cdot \\ Y_1 y_1^k + Y_2 y_2^k + \ldots + Y_n y_n^k = c_k. \end{cases}$$

These last relations would enable us to eliminate k of the functions $Y_1, Y_2, \ldots, Y_n$ from the adjunct system (19), and so the problem of the integration of this system would be reduced to that of the integration of a similar system containing only $n - k$ equations. This matter of the adjunct equation will be taken up again in connection with the study of equation (6) and the corresponding system (7), when we will assume for these equations that T, the right-hand member of (6), is equal to zero.

As we shall have but little to do with linear differential equations whose second members are different from zero, and as the *general* properties of such equations are sufficiently well known, we need only state here two fundamental propositions concerning them.

Supposing the linear differential equation whose right-hand member is other than zero to be of the form

$$(\alpha) \qquad \frac{d^n y}{dx^n} + p_1 \frac{d^{n-1} y}{dx^{n-1}} + p_n \frac{d^{n-2} y}{dx^{n-2}} + \ldots + p_n y = \Xi,$$

where Ξ is a function of x; then if Y be a particular solution of this equation, and

$$c_1 y_1 + c_2 y_2 + \ldots + c_n y_n$$

the general solution of the equation when Ξ is replaced by zero, we have as the complete solution of the given equation

$$v = c_1 y_1 + c_2 y_2 + \ldots + c_n y_n + Y.$$

Again, the integration of the equation (α) of order n can be made to depend upon the integration of a similar equation of order $n - k$, provided we know k particular integrals of (α). Proofs of these theorems will be found in Baltzer's *Theorie der Determinanten*, Houël's *Cours d'Analyse*, Forsyth's *Differential Equations*, etc.

We will now go on to the study of the linear differential equation

$$(23) \qquad \frac{d^n y}{dx^n} + p_1 \frac{d^{n-1} y}{dx^{n-1}} + p_2 \frac{d^{n-2} y}{dx^{n-2}} + \ldots + p_n y = 0.$$

We have seen, equation (7), that by introducing new auxiliary variables $n - 1$ in number, say

$$\eta', \quad \eta'', \ldots, \eta^{n-1},$$

we can replace this last equation by the system

$$(24) \quad \begin{cases} \dfrac{d\eta}{dx} - \eta' = 0, \\ \dfrac{d\eta'}{dx} - \eta'' = 0, \\ \cdot \quad \cdot \quad \cdot \quad \cdot \quad \cdot \quad \cdot \quad \cdot \quad \cdot \quad \cdot \quad \cdot \\ \dfrac{d\eta^{n-2}}{dx} - \eta^{n-1} = 0, \\ \dfrac{d\eta^{n-1}}{dx} + p_1\eta^{n-1} + p_2\eta^{n-2} + \ldots + p_n y = 0. \end{cases}$$

Let

$$(25) \quad \begin{cases} y_1, \; y_1', \; y_1'', \ldots, y_1^{n-1}, \\ y_2, \; y_2', \; y_2'', \ldots, y_2^{n-1}, \\ \cdot \quad \cdot \quad \cdot \quad \cdot \quad \cdot \quad \cdot \\ y_n, \; y_n', \; y_n'', \ldots, y_n^{n-1}, \end{cases}$$

denote n independent solutions of this system; then the determinant

$$(26) \quad \begin{vmatrix} y_1, & y_1', & \ldots, & y_1^{n-1}, \\ y_2, & y_2', & \ldots, & y_2^{n-1}, \\ \cdot & \cdot & \cdot & \cdot \\ y_n, & y_n', & \ldots, & y_n^{n-1}, \end{vmatrix}$$

must not vanish. It is perfectly obvious from the form of equations (24) that

$$y_1' = \frac{dy_1}{dx}, \quad y_1'' = \frac{d^2y_1}{dx^2}, \quad \ldots, \quad y_n^{n-1} = \frac{d^{n-1}y_n}{dx^{n-1}};$$

that is, all the integrals y^i are derivatives of the corresponding integrals y, and also that $y_1, y_2, \ldots, y_n$ are independent integrals of (23); the condition then, by hypothesis, that a system of integrals of

(23), viz. $y_1, y_2, \ldots, y_n$, shall be independent, is that the determinant

$$(27)\qquad \begin{vmatrix} \frac{d^{n-1}y_1}{dx^{n-1}}, & \frac{d^{n-2}y_1}{dx^{n-2}}, & \ldots, & y_1, \\ \frac{d^{n-1}y_2}{dx^{n-1}}, & \frac{d^{n-2}y_2}{dx^{n-2}}, & \ldots, & y_2, \\ \cdot & \cdot & \cdot & \cdot \\ \frac{d^{n-1}y_n}{dx^{n-1}}, & \frac{d^{n-2}y_n}{dx^{n-2}}, & \ldots, & y_n, \end{vmatrix}, = D,$$

must not be $= 0$.

The nature of the independence of these integrals must now be shown. The non-vanishing of D does not mean that the functions $y_1, y_2, \ldots, y_n$ are *absolutely* independent, but only that they are *linearly* independent; that is, if D is not zero, then there cannot exist *any* linear relation of the form

$$(28)\qquad c_1y_1 + c_2y_2 + \ldots + c_ny_n = 0;$$

where $c_1, c_2, \ldots, c_n$ are arbitrary constants. On the other hand, if $D = 0$, then some relation of the form (28) exists in which the constants $c_1, c_2, \ldots, c_n$ are not all zero.

The following proof of these theorems is due to Frobenius, and is entirely independent of any considerations involving the linear differential equation:

Suppose $y_1, y_2, \ldots, y_n$ to be defined as series going according to positive integral powers of $x - a$, and all convergent inside a circle of radius ρ, and having its centre at the point $x = a$. The variable x is now of course restricted to the interior of this circle. If among these functions there is a linear relation of the form

$$(29)\qquad c_1y_1 + c_2y_2 + \ldots + c_ny_n = 0,$$

where $c_1, c_2, \ldots, c_n$ are constants, then between this equation and its $n - 1$ derivatives we can of course eliminate the constants c, and as a result have

$$(30) \qquad D = \begin{vmatrix} \frac{d^{n-1}y_1}{dx^{n-1}}, & \frac{d^{n-2}y_1}{dx^{n-2}}, & \cdots, & y_1, \\ \frac{d^{n-1}y_2}{dx^{n-1}}, & \frac{d^{n-2}y_2}{dx^{n-2}}, & \cdots, & y_2, \\ \cdot & \cdot & \cdot & \cdot \\ \frac{d^{n-1}y_n}{dx^{n-1}}, & \frac{d^{n-2}y_n}{dx^{n-2}}, & \cdots, & y_n, \end{vmatrix} = 0.$$

For brevity we will write, as usual, y_i^k instead of $\frac{d^ky_i}{dx^k}$. In order to prove the converse of the theorem, that is, to prove that if $D = 0$ there exists a linear relation of the form (29) between the functions $y_1, y_2, \ldots, y_n$, we will first assume that the minor D_n^{n-1}, corresponding to the element y_n^{n-1} of D, is not identically zero. Consider now the system of $n - 1$ linear equations

$$(31) \qquad \begin{cases} z_1y_1 + z_2y_2 + \ldots + z_ny_n = 0, \\ z_1y_1' + z_2y_2' + \ldots + z_ny_n' = 0, \\ \cdot \quad \cdot \quad \cdot \quad \cdot \quad \cdot \\ z_1y_1^{n-2} + z_2y_2^{n-2} + \ldots + z_ny_n^{n-2} = 0. \end{cases}$$

Solving these for the unknown quantities $z_1, z_2, \ldots, z_n$, we have for the ratios

$$\frac{z_1}{z_n}, \quad \frac{z_2}{z_n}, \quad \ldots, \quad \frac{z_{n-1}}{z_n},$$

perfectly determinate finite functions; but, since by hypothesis $D = 0$, these functions must satisfy the equation

$$(32) \qquad z_1y_1^{n-1} + z_2y_2^{n-1} + \ldots + z_ny_n^{n-1} = 0.$$

Taking this equation into account, we have, as is easily seen, by differentiating equations (31),

$$(33) \qquad \begin{cases} z_1'y_1 + z_2'y_2 + \ldots + z_n'y_n = 0, \\ z_1'y_1' + z_2'y_2' + \ldots + z_n'y_n' = 0, \\ \cdot \quad \cdot \quad \cdot \quad \cdot \quad \cdot \\ z_1'y_1^{n-2} + z_2'y_2^{n-2} + \ldots + z_n'y_n^{n-2} = 0; \end{cases}$$

in which z' denotes $\frac{dz}{dx}$. Now since equations (31) determine definitely the ratios

$$\frac{z_1}{z_n}, \frac{z_2}{z_n}, \ldots, \frac{z^{n-1}}{z_n},$$

equations (33), which are identical in form with equations (31), must equally give perfectly determinate finite values for

$$\frac{z_1'}{z_n'}, \frac{z_2'}{z_n'}, \ldots, \frac{z_{n-1}'}{z_n'},$$

and these values are respectively equal to the values of

$$\frac{z_1}{z_n}, \frac{z_2}{z_n}, \ldots, \frac{z_{n-1}}{z_n};$$

that is, we have

$$\frac{z_k}{z_n} = \frac{z_k'}{z_n'}, \quad \text{or} \quad \frac{d}{dx}\frac{z_k}{z_n} = 0 \; (k = 1, 2, \ldots, n-1),$$

or

$$\frac{z_k}{z_n} = \frac{c_k}{c_n};$$

where $c_1, c_2, \ldots, c_n$ are constants of which c_n is arbitrary but different from zero. From the first of equations (31) we have now

(34) $$c_1y_1 + c_2y_2 + \ldots + c_ny_n = 0.$$

Let us now suppose $D_n^{n-1} = 0$, but the minor of order $n-2$, D_{n-1}^{n-2} of D_n^{n-1}, which corresponds to the element y_{n-1}^{n-2}, to be different from zero; then by proceeding in the same manner we arrive at a relation of the form

(35) $$c_1y_1 + c_2y_2 + \ldots + c_{n-1}y_{n-1} = 0;$$

in which c_{n-1} is not zero. Continuing in this way, we can show that if $D = 0$ there is always a relation of the form (34), in which the constants $c_1, c_2, \ldots, c_n$ are not all zero; remembering, of course, that the equation $y_1 = 0$ is to be included in this form.

We will speak in future of any number of functions being independent if there exists between them no linear homogeneous relation with constant coefficients, and the condition for this independence is expressed in the theorem:

If n functions $y_1, y_2, \ldots, y_n$ are independent, then their determinant (i.e., the determinant D) does not vanish. Conversely, if the functions are not independent, their determinant vanishes.

The theorem has only been proved for the region of the plane inside the circle of radius ρ and centre $x = a$, but by a well-known theorem in the theory of functions the truth of the theorem is established for all parts of the plane.

Referring now to equations (24), let us write them in the form

$$(36)\quad \begin{cases} \dfrac{dy}{dx} + 0y - y' + 0y'' + 0y''' + \ldots + 0y^{n-1} = 0, \\ \dfrac{dy'}{dx} + 0y + 0y' - y'' + 0y''' + \ldots + 0y^{n-1} = 0, \\ \dfrac{dy''}{dx} + 0y + 0y' + 0y'' - y''' + \ldots + 0y^{n-1} = 0, \\ \quad \cdot \quad \cdot \quad \cdot \quad \cdot \quad \cdot \quad \cdot \quad \cdot \\ \dfrac{dy^{n-2}}{dx} + 0y + 0y' + 0y'' + 0y''' + \ldots + 0y^{n-2} - y^{n-1} = 0, \\ \dfrac{dy^{n-1}}{dx} + p_n y + p_{n-1} y' + p_{n-2} y'' + p_{n-3} y''' + \ldots \\ \qquad + p_2 y^{n-2} + p_1 y^{n-1} = 0. \end{cases}$$

The adjunct system to this is:

$$(37)\quad \begin{cases} -\dfrac{dY}{dx} + 0Y + 0Y' + 0Y'' + 0Y''' + \ldots + p_n Y^{n-1} = 0, \\ -\dfrac{dY'}{dx} - Y + 0Y' + 0Y'' + 0Y''' + \ldots + p_{n-1} Y^{n-1} = 0, \\ -\dfrac{dY''}{dx} + 0Y - Y' + 0Y'' + 0Y''' + \ldots + p_{n-2} Y^{n-1} = 0, \\ \quad \cdot \quad \cdot \quad \cdot \quad \cdot \quad \cdot \quad \cdot \quad \cdot \\ -\dfrac{dY^{n-1}}{dx} + 0Y + 0Y' + 0Y'' + 0Y''' + \ldots + p_1 Y^{n-1} = 0. \end{cases}$$

Multiplying equations (36) by Y, Y', . . . , Y^{n-1}, respectively, and equations (37) by $-y$, $-y'$, . . . , $-y^{n-1}$, respectively, and adding, we have

$$\frac{d}{dx}(Yy + Y'y' + \ldots + Y^{n-1}y^{n-1}) = 0,$$

or

$$(38) \qquad Yy + Y'y' + \ldots + Y^{n-1}y^{n-1} = C,$$

where C is a constant. This result has already been obtained in the general case of a system of simultaneous linear differential equations of the first order. If we omit the terms in (36) and (37) whose coefficients are zero, these two systems may be written in the forms:

$$(39) \qquad \begin{cases} \dfrac{dy^{n-1}}{dx} + p_1 y^{n-1} + p_2 y^{n-2} + \ldots + p_n y = 0, \\ \dfrac{dy^{n-2}}{dx} - y^{n-1} = 0, \\ \cdot \quad \cdot \quad \cdot \quad \cdot \quad \cdot \quad \cdot \quad \cdot \quad \cdot \\ \dfrac{dy}{dx} - y' = 0; \end{cases}$$

and, for the adjunct system,

$$(40) \qquad \begin{cases} -\dfrac{dY^{n-1}}{dx} + p_1 Y^{n-1} - Y^{n-2} = 0, \\ -\dfrac{dY^{n-2}}{dx} + p_2 Y^{n-2} - Y^{n-3} = 0, \\ \cdot \quad \cdot \quad \cdot \quad \cdot \quad \cdot \quad \cdot \\ -\dfrac{dY}{dx} + p_n Y^{n-1} = 0. \end{cases}$$

Form now the $(n-1)^{st}$ derivative of the first of these equations, the $(n-2)^{nd}$ derivative of the second equation, etc., and subtract the sum of all the equations of odd order from the sum of those of even order; as a result we have

$$(41)\quad \frac{d^n Y^{n-1}}{dx^n} - \frac{d^{n-1}}{dx^{n-1}}(p_1 Y^{n-1}) + \frac{d^{n-2}}{dx^{n-2}}(p_2 Y^{n-1}) - \ldots + (-1)^n p_n Y^{n-1} = 0.$$

Replacing Y^{n-1} by M, this is

$$(42)\quad \frac{d^n M}{dx^n} - \frac{d^{n-1}}{dx^{n-1}}(p_1 M) + \frac{d^{n-2}}{dx^{n-2}}(p_2 M) - \ldots + (-1)^n p_n M = 0.$$

This equation is the adjunct equation to

$$(43)\quad \frac{d^n y}{dx^n} + p_1 \frac{d^{n-1} y}{dx^{n-1}} + p_2 \frac{d^{n-2} y}{dx^{n-2}} + \ldots + p_n y = 0.$$

The meaning of (42) is easily found. Suppose we multiply (43) by the indeterminate function M and then integrate by parts; we have thus, indicating differential coefficients by accents,

$$\begin{aligned}(44)\quad & My^{n-1} - M'y^{n-2} + M''y^{n-3} - \ldots + \ldots \\ & + p_1 My^{n-2} - (p_1 M)' y^{n-3} + \ldots + p_2 My^{n-3} + \ldots \\ & + (-1)^n \int y \{ M^n - (p_1 M)^{n-1} + (p_2 M)^{n-2} - \ldots \\ & \qquad + (-1)^n p_n M \} \, dx = \text{const.}\end{aligned}$$

If M is a solution of the differential equation

$$(45)\quad M^n - (p_1 M)^{n-1} + (p_2 M)^{n-2} - \ldots + (-1)^n p_n M = 0,$$

the integral in (44) will vanish, and we shall have a linear differential equation of order $n - 1$, containing an arbitrary constant, for the determination of y. If we know k solutions $M_1, M_2, \ldots, M_k$ of (44) we shall obtain, on writing $M = M_1$, $M = M_2, \ldots, M = M_k$, successively, k linear equations of order $n - 1$ in y. Eliminating between these equations the derivatives $y^{n-1}, y^{n-2}, \ldots, y^{n-k+1}$, we shall have for the determination of y an equation of order $n - k$ containing k arbitrary constants. It is clear that if (42) is the adjunct equation to (43), then (43) is the adjunct equation to (42); and, further, that the adjunct equation, say $B = 0$, to a given equation, $A = 0$, is simply one whose integrals are *multipliers* of $A = 0$; that is, supposing $B = 0$ and $A = 0$ to be each of order n, then multiplying A (or B) by an integral of B (or A) and performing a quadrature, the equation obtained for determining the unknown function in A (or B) will be of order $n - 1$, and will involve one arbitrary constant. The subject of adjunct systems of equations will be resumed in another chapter.

The question of the transformation of linear differential equations and the resulting theory of the *invariants* of such equations will not be dealt with in the present volume, but a few remarks may be made here on the subject. Suppose we have given a linear differential equation,

$$P_0 \frac{d^n y}{dx^n} + P_1 \frac{d^{n-1} y}{dx^{n-1}} + P_2 \frac{d^{n-2} y}{dx^{n-2}} + \ldots + P_n y = 0.$$

This equation can be transformed in two different ways, so that after each transformation it shall retain its original form. We may first change the independent variable by a relation of the form $x = f(t)$, and then after effecting this transformation we may change the unknown function y by a relation of the form $y = \phi(t)z$. The different *transforms* of the given equation obtained by giving to $f(t)$ and $\phi(t)$ all possible forms may be considered as belonging to one and the same *class*. Thus all differential equations of the second order form but one class, and they are all reducible to a unique type, say $\frac{d^2\eta}{d\xi^2} + I\eta = 0$; but we do not know how, by means of simple quadratures, to actually make this reduction, nor, having given two equations of the second order, do we know how to find the transformations which will change one into the other. The circumstances are entirely different, however, in the cases of equations of the third or higher orders. We will here only consider briefly the case of an equation of the third order, and show the existence of invariants in this case. Suppose the equation to be

$$\frac{d^3y}{dx^3} + 3P\frac{d^2y}{dx^2} + 3Q\frac{dy}{dx} + Ry = 0. \tag{46}$$

Transform first by making $x = f(t)$; writing

$$(47)\qquad \begin{cases} 3P' = \dfrac{3\dfrac{dt}{dx}\dfrac{d^2t}{dx^2} + 3P\left(\dfrac{dt}{dx}\right)^2}{\left(\dfrac{dt}{dx}\right)^3}, \\[2ex] 3Q' = \dfrac{\dfrac{d^3t}{dx^3} + 3P\dfrac{d^2t}{dx^2} + 3R\dfrac{dt}{dx}}{\left(\dfrac{dt}{dx}\right)^3}, \\[2ex] R' = \dfrac{R}{\left(\dfrac{dt}{dx}\right)^3}, \end{cases}$$

we have

$$(48)\qquad \frac{d^3y}{dt^3} + 3P'\frac{d^2y}{dt^2} + 3Q'\frac{dy}{dt} + R'y = 0.$$

Now make $y = \phi(t)z$, and write

$$(49)\qquad \begin{cases} 3P_0 = \dfrac{3\dfrac{d\phi}{dt} + 3P'\phi}{\phi}, \\[2ex] 3Q_0 = \dfrac{3\dfrac{d^2\phi}{dt^2} + 6P'\dfrac{d\phi}{dt} + 3Q'\phi}{\phi}, \\[2ex] R_0 = \dfrac{\dfrac{d^3\phi}{dt^3} + 3P'\dfrac{d^2\phi}{dt^2} + 3Q'\dfrac{d\phi}{dt} + R'\phi}{\phi}. \end{cases}$$

We have, as the result of this second transformation,

$$(50)\qquad \frac{d^3z}{dt^3} + 3P_0\frac{d^2z}{dt^2} + 3Q_0\frac{dz}{dt} + R_0z = 0.$$

Now, from the first of (47) and the first of (49) we find

(51)
$$P_0 = \frac{\frac{d\phi}{dt}}{\phi} + \frac{\frac{d^2t}{dx^2}}{\left(\frac{dt}{dx}\right)^2} + \frac{P}{\frac{dt}{dx}}$$

and, after easy reductions,

(52)
$$e^{-\int P_0 dt} = \frac{1}{\phi}\frac{dx}{dt}e^{-\int P dx}.$$

The function $e^{-\int P dx}$ is then, relatively to the given differential equation, a true invariant, since, after the transformations, it reproduces itself multiplied by a factor which depends only on the transformations effected. As already mentioned, it is not intended here to go into the subject of these invariants, so we will confine ourselves to the mere enunciation of another invariant of (46). Write

(53)
$$I = 4P^3 + 6P\frac{dP}{dx} + \frac{d^2P}{dx^2} - 6PQ - 3\frac{dQ}{dx} + 2R,$$

and

(54)
$$I_0 = 6P_0^3 + 6P_0\frac{dP_0}{dx} + \frac{d^2P_0}{dx^2} - 6P_0Q_0 - 3\frac{dQ_0}{dx} + 2R_0;$$

from these we can readily find the identity

(55)
$$I_0 = I\left(\frac{dx}{dt}\right)^3.$$

The function I is therefore an invariant of (46.) Combining the two invariants so found, we have the invariant

(56)
$$J = e^{3\int P dx} I,$$

which gives the relation

(57)
$$J_0 = J\phi^3(t).$$

If we consider the adjunct equation

(58)
$$\frac{d^3u}{dx^3} - 3\frac{d^2}{dx^2}(Pu) + 3\frac{d}{dx}(Qu) - Ru = 0,$$

and denote by I and J the two invariants (46), and by I_0 and J_0 the corresponding invariants of the adjunct equation, we have

$$(59) \qquad I_0 = -I, \qquad J_0 = -\frac{I^2}{J}.$$

The preceding results are taken from a paper by Laguerre in the *Comptes Rendus* for 1879. In another volume the investigations of Laguerre, Halphen, and others will be taken up from a more general point of view, and as full an account of their results as is possible will be given. It is, however, impossible to give a really *full* account of the subject within the limits of this treatise, but at least enough will be done to enable the reader to consult with profit the original memoirs.

In what precedes we have assumed that the differential equation possesses an integral, but it is obvious that this fact ought to be proved. The proof of the existence of an integral in the case of a linear differential equation is, however, only a particular case of the proof of the existence of an integral for the general form of an ordinary differential equation or a system of such equations, and it is not within the scope of this treatise to give this general proof, for which the reader is referred to the memoir by Briot and Bouquet in Cahier 36 of the *Journal de l'École Polytechnique* (and also in the treatise *Théorie des Fonctions Elliptiques* by the same authors). Jordan in his *Cours d'Analyse*, vol. iii., also gives this general proof. In connection with Jordan's proof the reader is advised to consult a paper by Picard in the *Bulletin des Sciences Mathématiques* for 1888 entitled *Sur la convergence des séries représentant les intégrales des équations différentielles.*

Fuchs in his first memoir on linear differential equations in vol. 66 of Crelle's Journal gives the special form of Briot and Bouquet's proof which is applicable to the case of linear differential equations, but that will not be reproduced here. Still another proof in the case of the linear differential equations which we are about to study is given by Frobenius in vol. 76 of Crelle; this last proof will be given in the chapter devoted to Frobenius's method for integrating such equations.

CHAPTER II.

LINEAR DIFFERENTIAL EQUATIONS WITH CONSTANT COEFFICIENTS.

THE following investigation of these equations is due principally to Hermite,* and in part to Darboux and Jordan.

The first method of integrating linear differential equations with constant coefficients is due to Euler, and is briefly as follows: Writing the equation in the form

$$(1) \qquad \frac{d^n y}{dx^n} + A_1 \frac{d^{n-1} y}{dx^{n-1}} + A_2 \frac{d^{n-2} y}{dx^{n-2}} + \ldots + A_n y = 0,$$

where $A_1, A_2, \ldots, A_n$ are constants, Euler makes the solution of this equation depend upon the solution of the algebraic equation

$$(2) \qquad F(\alpha), \quad = \alpha^n + A_1 \alpha^{n-1} + A_2 \alpha^{n-2} + \ldots + A_n = 0;$$

in which α^i replaces $\frac{d^i y}{dx^i} (i = 1, 2, \ldots n)$ in equation (1). As is well known, this equation is obtained by replacing y in (1) by $e^{\alpha x}$, and after the substitution dropping the factor $e^{\alpha x}$, which of course does not vanish.

Equation (2), that is $F(\alpha) = 0$, is called by Cauchy the *characteristic equation* of the given differential equation. The details of Euler's method are well known to all students of the elementary theory of linear differential equations; but Cauchy's method, which we now proceed to develop and which is based upon a knowledge of the characteristic equation, is not so well known, at least not to English readers. Let $\Pi(\alpha)$ denote an arbitrary polynomial in α, containing therefore only positive powers of α; we will now consider the integral

$$(3) \qquad y = \int \frac{e^{\alpha x} \Pi(\alpha)}{F(\alpha)} d\alpha,$$

* Équations différentielles linéaires, par M. Ch. Hermite: Bulletin des Sciences Mathématiques. 1879.

where the contour of integration is any closed curve. Suppose first that this contour contains no pole of the function

$$\frac{e^{\alpha x}\Pi(\alpha)}{F(\alpha)},$$

that is, contains no root of the characteristic equation $F(\alpha) = 0$; then by Cauchy's theorem the integral in (3) is zero and we have the known, but useless, integral $y = 0$. In order to obtain effective solutions of the differential equation we must then draw the contour of integration in such a way that it shall contain one or more poles of the function integrated. It is easy to verify that the value of y so obtained is an integral. We have

$$(4)\quad \begin{cases} y = \int \frac{e^{\alpha x}\Pi(\alpha)}{F(\alpha)}d\alpha, \\ \frac{dy}{dx} = \int \frac{e^{\alpha x}\alpha\Pi(\alpha)}{F(\alpha)}d\alpha, \\ \frac{d^2y}{dx^2} = \int \frac{e^{\alpha x}\alpha^2\Pi(\alpha)}{F(\alpha)}d\alpha, \\ \cdot \quad \cdot \quad \cdot \quad \cdot \quad \cdot \\ \frac{d^ny}{dx^n} = \int \frac{e^{\alpha x}\alpha^n\Pi(\alpha)}{F(\alpha)}d\alpha, \end{cases}$$

the integrations in each case being taken round the same closed contour containing one or more of the poles of

$$\frac{e^{\alpha x}\Pi(\alpha)}{F(\alpha)},$$

that is, one or more of the roots of $F(\alpha) = 0$. The expressions (4) substituted in equation (1) give for the first member of this equation the form

$$(5)\quad \int \frac{e^{\alpha x}\Pi(\alpha)}{F(\alpha)}[\alpha^n + A_1\alpha^{n-1} + A_2\alpha^{n-2} + \ldots + A_n]d\alpha,$$

or, from (2),

$$\int e^{\alpha x}\Pi(\alpha)d\alpha;$$

but, since $\Pi(\alpha)$ is a polynomial in α containing only positive powers of α, we have

$$\int e^{\alpha x}\Pi(\alpha)d\alpha = 0$$

for every possible closed path of integration. It follows then that

$$\int \frac{e^{\alpha x}\Pi(\alpha)}{F(\alpha)}d\alpha$$

is always an integral of the given differential equation, no matter what be the contour of integration. A remark must be made here concerning the polynomial $\Pi(\alpha)$, which has been assumed of any degree whatever. This assumption might lead one to think that any number of arbitrary constants might appear in the general integral of the given equation; it is easy to see, however, that the number of such arbitrary constants can never exceed n, the order of the differential equation. The degree of $F(\alpha)$ is of course n. Suppose now that $\Pi(\alpha)$ is of a degree greater than n; then we can write

$$\frac{\Pi(\alpha)}{F(\alpha)} = \Phi(\alpha) + \frac{\Psi(\alpha)}{F(\alpha)};$$

where $\Phi(\alpha)$ is a polynomial and $\Psi(\alpha)$ is also a polynomial, but one whose degree is less than n. Form now the integral

$$\int \frac{e^{\alpha x}\Pi(\alpha)}{F(\alpha)}d\alpha, = \int e^{\alpha x}\Phi(\alpha)d\alpha + \int \frac{e^{\alpha x}\Psi(\alpha)}{F(\alpha)}d\alpha.$$

The first term on the right-hand side of this equation is, by Cauchy's theorem, equal to zero, whatever be the contour of integration, and so the equation reduces to

$$\int \frac{e^{\alpha x}\Pi(\alpha)}{F(\alpha)}d\alpha = \int \frac{e^{\alpha x}\Psi(\alpha)}{F(\alpha)}d\alpha;$$

where $\Psi(\alpha)$ contains at most n arbitrary constants, since its degree is at most $n - 1$. We proceed now to determine the explicit form of the integral

$$y = \int \frac{e^{\alpha x}\Pi(\alpha)}{F(\alpha)}d\alpha.$$

Denote by S the sum of the residues of the function

$$\frac{e^{ax}\Pi(\alpha)}{F(\alpha)}$$

which correspond to the roots of $F(\alpha)$ lying inside the contour of integration. The integral

$$\int \frac{e^{ax}\Pi(\alpha)}{F(\alpha)}d\alpha$$

has now the value $2\pi i S$.

We will first suppose that the characteristic equation $F(\alpha) = 0$ has no multiple roots, and then decompose the function $\frac{\Pi(\alpha)}{F(\alpha)}$ into simple elements. As we have seen, the degree of $\Pi(\alpha)$ may be supposed less than that of $F(\alpha)$, and so we shall have as the result of the decomposition

$$\frac{\Pi(\alpha)}{F(\alpha)} = \frac{C_1}{\alpha - a_1} + \frac{C_2}{\alpha - a_2} + \ldots + \frac{C_n}{\alpha - a_n}. \tag{6}$$

In the function $\frac{e^{ax}\Pi(\alpha)}{F(\alpha)}$ change α into $a_1 + h$; then, since

$$\frac{C_1}{\alpha - a_1}$$

gives only one term containing $\frac{1}{h}$, we have

$$\frac{e^{x(a+h)}\Pi(\alpha+h)}{F(\alpha+h)} = e^{ax}\left[1 + \frac{hx}{1} + \frac{h^2x^2}{1\cdot 2} + \ldots\right] \tag{7}$$

$$\times \left[\frac{C_1}{h} + D_0 + D_1h + D_2h^2 + \ldots\right].$$

The residue in this case is thus $= C_1e^{a_1x}$. We have then as a first solution of the differential equation $2\pi iC_1e^{a_1x}$ obtained by integrating round a contour which contains only the single root a_1 of $F(\alpha) = 0$, that is, only the single pole a_1 of the function

$$\frac{e^{\alpha x}\Pi(\alpha)}{F(\alpha)}.$$

In general the contour of integration contains any number of poles of $\frac{e^{\alpha x}\Pi(\alpha)}{F(\alpha)}$, and consequently the general integral of the equation will be of the form

$$(8) \qquad y = C_1 e^{a_1 x} + C_2 e^{a_2 x} + \ldots + C_n e^{a_n x};$$

where $C_1, C_2, \ldots, C_n$ are constants any or all of which may be zero according to the path of integration, and where $a_1, a_2, \ldots, a_n$ are the roots of the characteristic equation $F(\alpha) = 0$. The factor $2\pi i$ is of course supposed to be contained in the constants $C_1, C_2, \ldots, C_n$.

Suppose now that $F(\alpha) = 0$ has multiple roots, and let

$$(9) \qquad F(\alpha) = (\alpha - a_1)^{\lambda_1 + 1}(\alpha - a_2)^{\lambda_2 + 1} \ldots (\alpha - a_s)^{\lambda_s + 1}.$$

We have now for the formula of decomposition

$$(10) \qquad \begin{aligned} \frac{\Pi(\alpha)}{F(\alpha)} = & \frac{C_1}{(\alpha - a_1)} + \frac{C_2}{(\alpha - a_2)} + \ldots \\ & + \frac{C_1'}{(\alpha - a_1)^2} + \frac{C_2'}{(\alpha - a_2)^2} + \ldots \\ & + \ldots \quad \ldots \\ & + \frac{C_1^{\lambda_1}}{(\alpha - a_1)^{\lambda_1 + 1}} + \frac{C_2^{\lambda_2}}{(\alpha - a_2)^{\lambda_2 + 1}} + \ldots \end{aligned}$$

Changing α into $a_1 + h$ gives, taking account only of the negative powers of h,

$$(11) \qquad \frac{\Pi(a_1 + h)}{F(a_1 + h)} = \frac{C_1}{h} + \frac{C_1'}{h^2} + \ldots + \frac{C_1^{\lambda_1}}{h^{\lambda_1 + 1}};$$

also

(12) $$e^{x(a_1+h)} = e^{a_1 x}\left[1 + \frac{hx}{1} + \frac{h^2x^2}{1,2} + \dots + \frac{h^{\lambda_1}x^{\lambda_1}}{1, 2, \dots, \lambda_1}\right].$$

The residue corresponding to $\alpha = a_1$, that is, the coefficient of $\frac{1}{h}$ in

$$\frac{\Pi(a_1 + h)}{F(a_1 + h)} e^{x(a_1+h)},$$

is found by multiplying together the corresponding terms in the right-hand members of (11) and (12). We find thus for this residue, and consequently for an integral of the differential equation, the expression

$$2\pi i e^{a_1 x}\left[C_1 + \frac{C_1' x}{1} + \dots + \frac{C_1^{\lambda_1} x^{\lambda_1}}{1, 2, \dots, \lambda_1}\right].$$

(Of course the residue alone does not contain the factor $2\pi i$.) The general integral of the given equation is now of the form

(13) $$y = e^{a_1 x}Q_{\lambda_1} + e^{a_2 x}Q_{\lambda_2} + \dots + e^{a_s x}Q_{\lambda_s};$$

where Q_{λ_k} is a polynomial in x of degree λ_k. Since

$$(\lambda_1 + 1) + (\lambda_2 + 1) + \dots + (\lambda_s + 1) = n,$$

we see that the general solution contains n arbitrary constants.

It is often desirable to determine the arbitrary constants in the general integral in such a way that y and its first $n - 1$ derivatives shall have, for a given value of x, certain specified values. In the case where the characteristic equation $F(\alpha) = 0$ has all of its roots different, this determination is easily seen to depend upon the solution of a system of n linear algebraic equations. Suppose, in fact, that

$$y = C_1 e^{a_1 x} + C_2 e^{a_2 x} + \dots + C_n e^{a_n x}$$

is, in the case of unequal roots of $F(\alpha) = 0$, the general integral of the given differential equation, and suppose $x = 0$ to be the particular value of x for which $y, y', y'', \ldots, y^{n-1}$ are to have the values $y_0, y_0', y_0'', \ldots, y_0^{n-1}$; then for the determination of the constants $C_1, C_2, \ldots, C_n$ we have the system of equations

$$(14) \qquad \begin{cases} C_1 + C_2 + \ldots + C_n = y_0, \\ C_1a_1 + C_2a_2 + \ldots + C_na_n = y_0', \\ \cdot \quad \cdot \quad \cdot \quad \cdot \quad \cdot \quad \cdot \quad \cdot \quad \cdot \quad \cdot \\ C_1a_1^{n-1} + C_2a_2^{n-1} + \ldots + C_na_n^{n-1} = y_0^{n-1}. \end{cases}$$

If, however, the characteristic equation has multiple roots, this method cannot be easily applied, as the derivatives of y are more complicated and the different roots do not enter in so simple a manner in the equations to be solved. Cauchy has nevertheless given a very simple method for solving the problem, which applies equally well to the cases of simple and of multiple roots of the characteristic equation. This method we now proceed to develop.

A solution of the differential equation has been given in the form

$$(15) \qquad y = \frac{1}{2\pi i}\int \frac{e^{\alpha x}\Pi(\alpha)}{F(\alpha)}d\alpha$$

In order that this shall be the general integral it is necessary to draw the contour of integration in such a way that it shall contain *all* of the roots of $F(\alpha) = 0$. The contour so drawn can of course be expanded indefinitely without altering the value of the integral (15), and we may therefore make it a circle with centre at the origin and of indefinitely large radius. It is now required to determine the constants in $\Pi(\alpha)$ in such a way that for $x = 0$ the function

$$\frac{1}{2\pi i}\int \frac{e^{\alpha x}\Pi(\alpha)}{F(\alpha)}d\alpha$$

and its first $n - 1$ derivatives shall have given values, say $y_0, y_0', \ldots, y_0^{n-1}$, that is, so that the equations

$$(16)\qquad \left\{\begin{aligned} &\frac{1}{2\pi i}\int\frac{\Pi(\alpha)}{F(\alpha)}d\alpha &&= y_0,\\ &\frac{1}{2\pi i}\int\frac{\alpha\Pi(\alpha)}{F(\alpha)}d\alpha &&= y_0',\\ &\frac{1}{2\pi i}\int\frac{\alpha^2\Pi(\alpha)}{F(\alpha)}d\alpha &&= y_0'',\\ &\quad\cdot\quad\cdot\quad\cdot\quad\cdot\quad\cdot\quad\cdot\\ &\frac{1}{2\pi i}\int\frac{\alpha^{n-1}\Pi(\alpha)}{F(\alpha)}d\alpha &&= y_0^{n-1},\end{aligned}\right.$$

shall be satisfied.

To obtain the values of these integrals we will develop $\frac{\Pi(\alpha)}{F(\alpha)}$ in descending powers of α; and since $\Pi(\alpha)$ is in general of degree $n-1$, the first term of the development will be of degree -1, and we shall have

$$(17)\qquad \frac{\Pi(\alpha)}{F(\alpha)} = \frac{\epsilon_0}{\alpha} + \frac{\epsilon_1}{\alpha_2} + \frac{\epsilon_2}{\alpha_3} + \ldots + \frac{\epsilon_{n-1}}{\alpha^n} + \ldots$$

Substituting this value in equations (16) and integrating round a circle of infinitely great radius, we know that we need only take account in each integral of the term in $\frac{1}{z}$; we have then, by a known formula,

$$(18)\qquad \left\{\begin{aligned} &\epsilon_0 &&= y_0,\\ &\epsilon_1 &&= y_0',\\ &\cdot\quad\cdot &&\cdot\quad\cdot\\ &\epsilon_{n-1} &&= y_0^{n-1}.\end{aligned}\right.$$

We have thus found the coefficients in $\Pi(\alpha)$ of all the terms of degrees equal to or greater than $-n$, and $\Pi(\alpha)$ is thus completely determined since we have identically

$$(19)\qquad \Pi(\alpha) = F(\alpha)\left[\frac{y_0}{\alpha} + \frac{y_0'}{\alpha^2} + \ldots + \frac{y_0^{n-1}}{\alpha^n}\right],$$

and since $\Pi(\alpha)$ is to be an entire polynomial; consequently, since $F(\alpha)$ is of degree n, we see that the first n terms of the series (17) are alone necessary for the determination of the polynomial $\Pi(\alpha)$, and we thus have

$$
\begin{aligned}
(20) \quad \Pi(\alpha) = & \, y_0 \; [A_{n-1} + A_{n-2}\alpha + A_{n-3}\alpha^2 + \ldots + \alpha^{n-1}] \\
& + y_0' \; [A_{n-2} + A_{n-3}\alpha + A_{n-4}\alpha^2 + \ldots + \alpha^{n-2}] \\
& + y_0'' \; [A_{n-3} + A_{n-4}\alpha + A_{n-5}\alpha^2 + \ldots + \alpha^{n-3}] \\
& + \quad \ldots \ldots \ldots \ldots \\
& + y_0^{n-1}.
\end{aligned}
$$

We have thus obtained $\Pi(\alpha)$ by a method equally applicable to the cases of simple and of multiple roots. In order now to obtain the explicit form of y, it is only necessary to calculate by the ordinary elementary algebraic process the residues of the known function

$$\frac{e^{\alpha x}\Pi(\alpha)}{F(\alpha)}.$$

As a simple application of the preceding method consider the well-known differential equation

$$\frac{d^2y}{dx^2} + n^2y = 0.$$

The characteristic equation is obviously

$$\alpha^2 + n^2 = 0,$$

and has for roots $\alpha = \pm in$. Suppose now that for $x = 0$ we wish to have $y = y_0$, $y' = y_0'$; then, by (17), the polynomial $\Pi(\alpha)$ is to be determined by the relation

$$\frac{\Pi(\alpha)}{F(\alpha)} = \frac{y_0}{\alpha} + \frac{y_0'}{\alpha^2} + \frac{y_0''}{\alpha^3} + \ldots,$$

which gives on multiplication by $F(\alpha)$, $= \alpha^2 + n^2$, and disregarding terms of the product containing negative powers of α,

$$\Pi(\alpha) = y_0\alpha + y_0'.$$

We have now to calculate the residues of

$$\frac{e^{\alpha x}\Pi(\alpha)}{F(\alpha)}, \; = \frac{y_0\alpha + y_0'}{\alpha^2 + n^2}e^{\alpha x}.$$

For a root $+\alpha$ the residue is $= \frac{e^{\alpha x}\Pi(\alpha)}{F'(\alpha)}$ or $\frac{1}{2}\left(y_0 + \frac{y_0'}{\alpha}\right)e^{\alpha x}$; for a root $-\alpha$ the residue is $\frac{1}{2}\left(y_0 - \frac{y_0'}{\alpha}\right)e^{-\alpha x}$. The sum of these two residues is

$$\tfrac{1}{2}y_0(e^{\alpha x} + e^{-\alpha x}) + \tfrac{1}{2}y_0'\frac{e^{\alpha x} - e^{-\alpha x}}{\alpha}.$$

Making now $\alpha = in$, we have readily for the general integral sought

$$y = y_0 \cos nx + y_0'\frac{\sin nx}{n}.$$

From the form of the differential equation it is clear that, if $y = \phi(x)$ is a solution, then $y_1 = \phi(x + c)$ will also be a solution, c denoting an arbitrary constant. If then we write

$$y = y_0 \cos n(x - c) + y_0'\frac{\sin n(x - c)}{n},$$

y and y' will take the values y_0 and y_0' for $x = c$. A number of consequences arising from the general integral when x is real might be given; but it is not necessary to give them, as the reader is of course supposed to be familiar with the ordinary elementary theory of differential equations.

We will now briefly consider the case where the right-hand member of the equation is not zero but a function of the independent variable; say the equation is

$$(21)\qquad \frac{d^n y}{dx^n} + A_1\frac{d^{n-1}y}{dx^{n-1}} + A_2\frac{d^{n-2}y}{dx^{n-2}} + \ldots + A_n y = f(x),$$

$f(x)$ denoting an arbitrary function of x. The following method for integrating this equation is due to Cauchy and is given by Darboux in a short paper immediately following Hermite's paper above referred to. We form first a solution of the equation, say

$$y = \Phi(x, t),$$

which, with its first $n - 2$ derivatives, vanishes for $x = t$, and for the same value of x has its $(n - 1)^{st}$ derivative equal to $f(t)$. Such a function being formed, we can easily verify that

$$(22)\qquad y = \int_{x_0}^{x} \Phi(x, t)dt,$$

where x_0 is an arbitrary constant, is a particular solution of the given differential equation. The formation of $\Phi\ (x, t)$ is quite simple. From what precedes we know that the integral

$$\frac{1}{2\pi i}\int \frac{e^{\alpha x}d\alpha}{F(\alpha)}$$

($F(\alpha)$ having the same meaning as before) taken round the circumference of a circle with very great radius is a solution of the differential equation when its second member is zero, and that this solution, together with its first $n - 2$ derivatives, vanishes for $x = 0$, and, finally, that its $n - 1$ derivative is, for this same value of x, equal to unity. If now in this integral we change x into $x - t$ and multiply the integral by $F(t)$, we obviously have the function sought given by the equation

$$(23)\qquad \Phi(x, t) = \frac{f(t)}{2\pi i}\int_R \frac{e^{\alpha(x-t)}d\alpha}{F(\alpha)},$$

the integration extending round the circumference of a circle of indefinitely large radius R. The double integral

$$(24) \qquad Y = \frac{1}{2\pi i}\int_{x_0}^{x} f(t)dt \int_R \frac{e^{\alpha(x-t)}d\alpha}{F(\alpha)}$$

is, as we shall now show, a particular solution of the given differential equation. Differentiating (24), we have

$$(25) \qquad \frac{dY}{dx} = \frac{f(x)}{2\pi i}\int_R \frac{d\alpha}{F(\alpha)} + \frac{1}{2\pi i}\int_{x_0}^{x} f(t)dt \int_R \frac{e^{\alpha(x-t)}\alpha d\alpha}{F(\alpha)};$$

the integral $\int_R \frac{d\alpha}{F(\alpha)}$ is known to be zero, and therefore

$$(26) \qquad \frac{dY}{dx} = \frac{1}{2\pi i}\int_{x_0}^{x} f(t)dt \int_R \frac{e^{\alpha(x-t)}\alpha d\alpha}{F(\alpha)}.$$

Now we know that if $p < n - 1$, all the integrals

$$\int_R \frac{\alpha^p d\alpha}{F(\alpha)}$$

must vanish, and consequently by differentiation of (26) we get

$$(27) \qquad \frac{d^p Y}{dx^p} = \frac{1}{2\pi i}\int_{x_0}^{x} f(t)dt \int_R \frac{e^{\alpha(x-t)}\alpha^p d\alpha}{F(\alpha)}$$

so long as the inequality $p < n$ is satisfied. Suppose $p = n - 1$; then from this last equation we derive

$$(28) \qquad \frac{d^n Y}{dx^n} = \frac{f(x)}{2\pi i}\int_R \frac{\alpha^{n-1} d\alpha}{F(\alpha)} + \frac{1}{2\pi i}\int_{x_0}^{x} f(t)dt \int_R \frac{e^{\alpha(x-t)}\alpha^n d\alpha}{F(\alpha)};$$

and since

$$\int_R \frac{\alpha^{n-1}d\alpha}{F(\alpha)} = 2\pi i,$$

we have finally

$$(29) \qquad \frac{d^n Y}{dx^n} = f(x) + \frac{1}{2\pi i}\int_{x_0}^{x} f(t)dt \int_R \frac{e^{\alpha(x-t)}\alpha^n d\alpha}{F(\alpha)} .$$

Substituting the above values of Y and its derivatives in the differential equation, we see that the equation is satisfied by

$$y = Y = \frac{1}{2\pi i}\int_{x_0}^{x} f(t)dt \int_R \frac{e^{\alpha(x-t)}d\alpha}{F(\alpha)} ,$$

since

$$F(\alpha) = \alpha^n + A_1 \alpha^{n-1} + \ldots + A_n .$$

Referring again to (24), let us write

$$(30) \qquad R(t) = \frac{1}{2\pi i}\int_R \frac{e^{\alpha(x-t)}d\alpha}{F(\alpha)} ;$$

then $R(t)$ is the sum of the residues relative to all the roots of $F(\alpha) = 0$. Suppose $\frac{1}{F(\alpha)}$ to be decomposed into simple fractions, and write

$$\frac{1}{F(\alpha)} = \Sigma\left\{\frac{B_0}{\alpha - a} + \frac{B_1}{(\alpha - a)^2} + \ldots + \frac{B_{p-1}}{(\alpha - a)^p}\right\} ;$$

we have also

$$e^{\alpha(x-t)} = e^{a(x-t)}\left[1 + \frac{(\alpha - a)(x - t)}{1} + \ldots\right],$$

and consequently the residue corresponding to the root a of $F(\alpha) = 0$ is

$$e^{a(x-t)}\left[B_0 + \frac{B_1(x-t)}{1} + \ldots + \frac{B_{p-1}(x-t)^{p-1}}{1.2\ldots p-1}\right].$$

We have therefore for $R(t)$ the value

$$(31) \quad R(t) = \Sigma e^{a(x-t)} \left\{ B_0 + \frac{B_1(x-t)}{1} + \dots + \frac{B_{p-1}(x-t)^{p-1}}{1 \cdot 2 \dots p-1} \right\},$$

and, finally, for Y the value

$$(32) \qquad Y = \int_{x_0}^{x} f(t)R(t)dt.$$

We will now give a brief account, taken from Jordan, of *systems* of linear differential equations with constant coefficients. We have seen that any such system can be reduced to an equivalent system of the first order. Supposing this reduction to have been made, we will assume

$$(33) \quad \begin{cases} \dfrac{dy_1}{dx} + a_{11}y_1 + a_{12}y_2 + \dots + a_{1s}y_s = 0, \\ \dfrac{dy_2}{dx} + a_{21}y_1 + a_{22}y_2 + \dots + a_{2s}y_s = 0, \\ \cdot \quad \cdot \quad \cdot \quad \cdot \quad \cdot \quad \cdot \quad \cdot \\ \dfrac{dy_s}{dx} + a_{s1}y_1 + a_{s2}y_2 + \dots + a_{ss}y_s = 0, \end{cases}$$

as our system of equations. Denote by Δ the *characteristic determinant*

$$\begin{vmatrix} a_{11} + \alpha, & a_{12}, & \dots, & a_{1s} \\ a_{21}, & a_{22} + \alpha, & \dots, & a_{2s} \\ \cdot & \cdot & \cdot & \cdot \\ a_{s1}, & a_{s2}, & \dots, & a_{ss} + \alpha \end{vmatrix}$$

and let $A_{11}, A_{12}, \dots, A_{ss}$ denote the minors

$$\frac{d\Delta}{da_{11}}, \frac{d\Delta}{da_{12}}, \dots, \frac{d\Delta}{da_{ss}}$$

of this determinant. Substitute now in equations (33) the following expressions:

$$(34)\quad \begin{cases} y_1 = \dfrac{1}{2\pi i}\displaystyle\int \dfrac{A_{11}\theta_1 + A_{21}\theta_2 + \ldots + A_{s1}\theta_s}{\Delta} e^{\alpha x} d\alpha, \\ y_2 = \dfrac{1}{2\pi i}\displaystyle\int \dfrac{A_{12}\theta_1 + A_{22}\theta_2 + \ldots + A_{s2}\theta_s}{\Delta} e^{\alpha x} d\alpha, \\ \cdot \quad \cdot \quad \cdot \quad \cdot \quad \cdot \quad \cdot \quad \cdot \quad \cdot \quad \cdot \quad \cdot \\ y_s = \dfrac{1}{2\pi i}\displaystyle\int \dfrac{A_{1s}\theta_1 + A_{2s}\theta_2 + \ldots + A_{ss}\theta_s}{\Delta} e^{\alpha x} d\alpha; \end{cases}$$

where $\theta_1, \theta_2, \ldots, \theta_s$ are functions of α, and where the integration extends around an arbitrary closed contour. From these equations we obtain by differentiation

$$(35)\quad \begin{cases} \dfrac{dy_1}{dx} = \dfrac{1}{2\pi i}\displaystyle\int \alpha\dfrac{A_{11}\theta_1 + A_{21}\theta_2 + \ldots + A_{s1}\theta_s}{\Delta} e^{\alpha x} d\alpha, \\ \dfrac{dy_2}{dx} = \dfrac{1}{2\pi i}\displaystyle\int \alpha\dfrac{A_{12}\theta_1 + A_{22}\theta_2 + \ldots + A_{s2}\theta_s}{\Delta} e^{\alpha x} d\alpha, \\ \cdot \quad \cdot \quad \cdot \quad \cdot \quad \cdot \quad \cdot \quad \cdot \quad \cdot \quad \cdot \quad \cdot \\ \dfrac{dy_s}{dx} = \dfrac{1}{2\pi i}\displaystyle\int \alpha\dfrac{A_{1s}\theta_1 + A_{2s}\theta_2 + \ldots + A_{ss}\theta_s}{\Delta} e^{\alpha x} d\alpha. \end{cases}$$

Substituting in the first of equations (33), and observing that

$$(36)\quad \begin{cases} (a_{11} + \alpha)A_{11} + a_{12}A_{12} + \ldots + a_{1s}A_{1s} = \Delta, \\ (a_{11} + \alpha)A_{21} + a_{12}A_{22} + \ldots + a_{1s}A_{2s} = 0, \\ \cdot \quad \cdot \quad \cdot \quad \cdot \quad \cdot \quad \cdot \quad \cdot \quad \cdot \quad \cdot \\ (a_{11} + \alpha)A_{s1} + a_{12}A_{s2} + \ldots + a_{1s}A_{ss} = 0, \end{cases}$$

we have as the result

$$\frac{1}{2\pi i}\int \theta_1 e^{\alpha x} d\alpha.$$

Making the corresponding substitutions in the remaining equations of the system (33), we have as the result of all the substitutions the expressions

$$(37) \qquad \frac{1}{2\pi i}\int \theta_1 e^{\alpha x} d\alpha, \quad \frac{1}{2\pi i}\int \theta_2 e^{\alpha x} d\alpha, \quad \ldots, \quad \frac{1}{2\pi i}\int \theta_s e^{\alpha x} d\alpha.$$

If we suppose $\theta_1, \theta_2, \ldots, \theta_s$ to be arbitrary constants, the functions $\theta_1 e^{\alpha x}, \ldots, \theta_s e^{\alpha x}$ will be integral functions, and the integrals (37) will all vanish, and therefore the expressions in (34) will be solutions of equations (33), and these solutions will contain s arbitrary constants.

Let us assume that we have chosen a circle of infinite radius as the contour of integration. Now, the initial value of y_1 for $x = 0$ will be

$$(38) \qquad y_1^0 = \frac{1}{2\pi i}\int \frac{A_{11}\theta_1 + A_{21}\theta_2 + \ldots + A_{s1}\theta_s}{\Delta} d\alpha;$$

but we have

$$(39) \qquad \Delta = \alpha^s + B\alpha^{s-1} + \ldots$$

and

$$(40) \quad A_{11}\theta_1 + A_{21}\theta_2 + \ldots + A_{s1}\theta_s = \begin{vmatrix} \theta_1, & a_{12}, & a_{13}, & \ldots, & a_{1s} \\ \theta_2, & a_{22}+\alpha, & a_{23}, & \ldots, & a_{2s} \\ \cdot & \cdot & \cdot & \cdot & \cdot \\ \theta_s, & a_{s2}, & a_{s3}, & \ldots, & a_{ss}+\alpha \end{vmatrix}$$

$$= \theta_1\alpha^{s-1} + D\alpha^{s-2} + \ldots,$$

and consequently

$$(41) \qquad \frac{A_{11}\theta_1 + A_{21}\theta_2 + \ldots + A_{s1}\theta_s}{\Delta} = \frac{\theta_1}{\alpha} + \frac{e}{\alpha^2} + \ldots$$

and

$$(42) \qquad y_1^0 = \frac{1}{2\pi i}\int\left(\frac{\theta_1}{\alpha} + \frac{\epsilon}{\alpha^2} + \ldots\right)d\alpha = \theta_1.$$

We can find in like manner the initial values of $y_2, y_3, \ldots$, and so have finally for these values

$$y_1^0 = \theta_1, \quad y_2^0 = \theta_2, \quad y_3^0 = \theta_3, \quad \ldots, \quad y_s^0 = \theta_s.$$

The solution which we have now found is the general integral, since by properly choosing the constants $\theta_1, \theta_2, \ldots, \theta_s$ we can give $y_1, y_2, \ldots, y_s$ arbitrary initial values. The values of the integrals (34) are easily found; each one is in fact equal to the sum of the residues corresponding to the roots of $\Delta = 0$ of each of the functions which is to be integrated. Consider the integral y_1, and let a_1 denote a root of $\Delta = 0$ whose order of multiplicity is $= \mu$; then

$$(43) \qquad \frac{A_{11}\theta_1 + A_{21}\theta_2 + \ldots + A_{s1}\theta_s}{\Delta} = \frac{F_\mu}{(\alpha - a_1)^\mu} + \ldots + \frac{F_1}{(\alpha - a_1)} + G_0 + \ldots;$$

where $F_\mu, \ldots, F_1$ are linear functions of the constants $\theta_1, \theta_2, \ldots, \theta_s$. Again, we have

$$(44) \qquad e^{\alpha x} = e^{a_1 x}\left[1 + (\alpha - a_1)x + (\alpha - a_1)^2\frac{x^2}{1.2} + \ldots\right].$$

Forming the product of (43) and (44), we have for the sought residue the expression

$$(45) \qquad \left[F_1 + F_2 x + \ldots + F_\mu\frac{x^{\mu-1}}{1.2\ldots(\mu-1)}\right]e^{a_1 x}.$$

That part of the value of y_1 which corresponds to the root a_1 is then of the form $Qe^{a_1 x}$, where Q is a polynomial in x, in general of degree $\mu - 1$. This degree will, however, be lowered to $\mu - \kappa - 1$ if the minors $A_{11}, A_{21}, \ldots, A_{s1}$ are all divisible by $(\alpha - a_1)^\kappa$, since in this case we shall obviously have

$$F_\mu = \ldots = F_{\mu-\kappa+1} = 0.$$

We will now consider briefly the case when the right-hand members of the system of equations are not zero, and write the equations in the form

$$(46)\qquad \begin{cases} \dfrac{dy_1}{dx} + a_{11}y_1 + a_{12}y_2 + \ldots + a_{1s}y_s = f_1(x), \\ \dfrac{dy_2}{dx} + a_{21}y_1 + a_{22}y_2 + \ldots + a_{2s}y_s = f_2(x), \\ \cdot \quad \cdot \quad \cdot \quad \cdot \quad \cdot \quad \cdot \quad \cdot \quad \cdot \quad \cdot \\ \dfrac{dy_s}{dx} + a_{s1}y_1 + a_{s2}y_2 + \ldots + a_{ss}y_s = f_s(x). \end{cases}$$

Formulæ (34) will give a particular solution of this system if we determine the functions $\theta_1, \theta_2, \ldots, \theta_s$ and the contour of integration in such a way that the equations

$$(47)\qquad \begin{cases} \dfrac{1}{2\pi i}\displaystyle\int \theta_1 e^{\alpha x} d\alpha = f_1, \\ \dfrac{1}{2\pi i}\displaystyle\int \theta_2 e^{\alpha x} d\alpha = f_2, \\ \cdot \quad \cdot \quad \cdot \quad \cdot \\ \dfrac{1}{2\pi i}\displaystyle\int \theta_s e^{\alpha x} d\alpha = f_s, \end{cases}$$

shall be satisfied. This determination is easily arrived at if $f_1, f_2, \ldots, f_s$ are of the form $Qe^{\lambda x}$, where Q is a polynomial. Suppose, in fact, that

$$(48)\qquad f_1 = (F_0 + F_1 x + \ldots + F_m x^m)e^{\lambda x};$$

then in order to satisfy the first of equations (47) we have only to make

$$(49)\qquad \theta_1 = \frac{F_0}{\alpha - \lambda} + \ldots + \frac{1 \cdot 2 \ldots m}{(\alpha - \lambda)^{m+1}} F_m,$$

and integrate around a small circle containing the point λ. We see at once from this that if $f_1, f_2, \ldots, f_s$ are polynomials of order m, then $\theta_1, \theta_2, \ldots, \theta_s$ will be sums of simple fractions containing powers of $\alpha - \lambda$ in their denominators up to $(\alpha - \lambda)^{m+1}$.

We have consequently, μ being equal to zero, or, if λ is a root of $\Delta = 0$, μ being the order of multiplicity of this root,

$$(50)\quad \frac{A_{11}\theta_1 + A_{21}\theta_2 + \ldots + A_{s1}\theta_s}{\Delta} = \frac{G}{(\alpha - \lambda)^{m+\mu+1}} + \ldots \\ + \frac{G_{m+\mu}}{\alpha - \lambda} + \ldots,$$

and the corresponding value of y_1, which is equal to the residue of

$$\frac{A_{11}\theta_1 + A_{21}\theta_2 + \ldots + A_{s1}\theta_s}{\Delta} e^{\alpha x}$$

for the point λ, will be of the form

$$Le^{\lambda x},$$

where L is a polynomial in general of degree $m + \mu$, but will be of lower degree if the first coefficients $G, G_1, \ldots$ vanish. Similar results will of course be obtained for $y_2, y_3, \ldots, y_s$.

It is well known that every equation of the form

$$(51)\quad (ax+b)^n \frac{d^n y}{dx^n} + A_1(ax+b)^{n-1}\frac{d^{n-1}y}{dx^{n-1}} + \ldots + A_n y = 0$$

can be thrown into the form of an equation with constant coefficients. Writing, in fact,

$$ax + b = e^t,$$

we have

$$\frac{dy}{dx} = ae^{-t}\frac{dy}{dt},$$

$$\frac{d^2y}{dx^2} = a^2e^{-t}\left[-e^{-t}\frac{dy}{dt} + e^{-t}\frac{d^2y}{dt^2}\right];$$

and, in general,

$$(52) \qquad \frac{d^k y}{dx^k} = a^k e^{-kt} P_k ;$$

where P_k is a linear function, with constant coefficients, of $\frac{dy}{dt}$, $\frac{d^2y}{dt^2}, \ldots, \frac{d^ky}{dt^k}$. Assuming (52) to be true for k ($k = 1, 2, \ldots,$), it is readily seen to be true for $k + 1$; we have in fact

$$\frac{d^{k+1}y}{dx^{k+1}} = a^{k+1}e^{-t}\frac{de^{-kt}P_k}{dt}$$

$$= a^{k+1}e^{-t}\left(-ke^{-kt}P_k + e^{-kt}\frac{dP_k}{dt}\right)$$

$$= a^{(k+1)}e^{-(k+1)t}P_{k+1}.$$

Substituting these values in (51), we have for the determination of y as a function of x a linear differential equation with constant coefficients. If the corresponding characteristic equation has all of its roots unequal, the general integral will be of the form

$$(53) \qquad \begin{aligned} y &= C_1e^{a_1t} + \ldots + C_ne^{a_nt} \\ &= C_1(ax+b)^{a_1} + \ldots + C_n(ax+b)^{a_n}. \end{aligned}$$

If the characteristic equation has multiple roots, then to any one of them, say a_1, there will correspond a solution of the form

$$(54) \qquad e^{a_1t}[C + C_1t + \ldots + C_{\mu-1}t^{\mu-1}]$$
$$= (ax+b)^{a_1}[C + C_1\log(ax+b) + \ldots + C_{\mu-1}\log^{\mu-1}(ax+b)].$$

We will find solutions similar to these in the more general class of linear differential equations which we shall presently study.

In the *Bulletin des Sciences Mathématiques* for August, 1888, M. Ch. Méray gives an investigation of the differential equation with constant coefficients, of which the following is an account:

Consider the k series

$$(a)\quad \begin{cases} U(z) = u_0 + u_1 z + u_2 z^2 + \ldots + u_m z^m + \ldots, \\ V(z) = v_0 + v_1 z + v_2 z^2 + \ldots + v_m z^m + \ldots, \\ \ldots\ldots\ldots\ldots\ldots\ldots\ldots\ldots, \\ T(z) = t_0 + t_1 z + t_2 z^2 + \ldots + t_m z^m + \ldots, \end{cases}$$

depending upon the same variable z; we will say that these series are *co-recurrent* if, for all values of the index m, we have the k relations or *recurrences*

$$(b)\quad \begin{cases} u_{m+1} + a_1 u_m + b_1 v_m + \ldots + h_1 t_m = 0, \\ v_{m+1} + a_2 u_m + b_2 v_m + \ldots + h_2 t_m = 0, \\ \ldots\ldots\ldots\ldots\ldots\ldots\ldots\ldots \\ t_{m+1} + a_k u_m + b_k v_m + \ldots + h_k t_m = 0; \end{cases}$$

of which the coefficients $(a, b, \ldots, h)$ are arbitrarily given constants. The quantities $(u, v, \ldots, t)$ are now all known when we know $(u_0, v_0, \ldots, t_0)$. The summation of these series is easily effected by aid of the theorem: *These series are convergent for values of z with sufficiently small moduli, and writing*

$$(c)\quad \begin{vmatrix} 1 + a_1 z, & b_1 z, & \ldots, & h_1 z \\ a_2 z, & 1 + b_2 z, & \ldots, & h_2 z \\ \vdots & & & \\ a_k z, & b_k z, & \ldots, & 1 + h_k z \end{vmatrix} = F(z),$$

($F(z)$ a polynomial in z of degree k), also writing

$$(d)\quad \begin{cases} A_1(z), & B_1(z), & \ldots, & H_1(z), \\ \ldots & \ldots & \ldots & \ldots \\ A_k(z), & B_k(z), & \ldots, & H_k(z), \end{cases}$$

to denote the minors of F corresponding to the like-placed elements in $F(z)$, we have

$$
(e) \qquad \begin{cases}
U(z) = \dfrac{u_0 A_1(z) + v_0 A_2(z) + \ldots + t_0 A_k(z)}{F(z)}, \\
V(z) = \dfrac{u_0 B_1(z) + v_0 B_2(z) + \ldots + t_0 B_k(z)}{F(z)}, \\
\cdot \quad \cdot \quad \cdot \quad \cdot \quad \cdot \quad \cdot \quad \cdot \quad \cdot \quad \cdot \\
T(z) = \dfrac{u_0 H_1(z) + v_0 H_2(z) + \ldots + t_0 H_k(z)}{F(z)}.
\end{cases}
$$

The numerators in these are obviously polynomials of degree $k - 1$.

Admitting for the moment that the series (a) are convergent, we readily find that their sums are connected by the k simultaneous linear equations

$$
(f) \qquad \begin{cases}
(1 + a_1 z)U + b_1 zV + \ldots + h_1 zT = u_0, \\
a_2 zU + (1 + b_2 z)V + \ldots + h_2 zT = v_0, \\
\cdot \quad \cdot \quad \cdot \quad \cdot \quad \cdot \quad \cdot \quad \cdot \quad \cdot \quad \cdot \\
a_k zU + b_k zV + \ldots + (1 + h_k z)T = t_0.
\end{cases}
$$

For, adding the second members of equations (a) after multiplying them respectively, for example, by

$$1 + a_1 z, \quad b_1 z, \quad \ldots, \quad h_1 z,$$

we see that the term which is independent of z reduces to u_0 and, by the first of recurrences (b), all the other terms reduce to zero when we make m successively $= 0, 1, 2, \ldots$, and the same holds for equations (f) other than the first one. The solution of these equations gives us (e). Now, since we have

$$(g) \qquad F(z), = 1 + p_1 z + p_2 z^2 + \ldots + p_k z^k,$$

a polynomial which does not vanish with z, we know by elementary principles of the theory of functions that the rational fractions (e)

are developable by Maclaurin's theorem for values of z for which mod. z is equal to or less than the least modulus of the roots of $F(z) = 0$. The integral series obtained by these developments cannot differ from the proposed series (a) since they satisfy equations (f), which are equivalent to the recurrences (b). Each of these series considered separately is obviously *recurrent* in the ordinary sense of the word.

If

$$(h) \qquad \frac{\Omega(z)}{F(z)} = w_0 + w_1 z + \ldots + w_m z^m + \ldots$$

is the development in a recurrent series of a rational fraction in z, whose numerator $\Omega(z)$ *is of degree* $< k$, *when k is the degree of the polynomial* $F(z)$, *the integral series*

$$(i) \qquad w_0 + \frac{w_1}{1} z + \frac{w_2}{1\,.\,2} z^2 + \ldots + \frac{w_m}{1\,.\,2\,\ldots\,.\,m} z^m + \ldots$$

is convergent for all the values of z, and, representing its sum by

$$\left\{ \frac{\Omega(z)}{F(z)} \right\}$$

we have for its calculation

$$(j) \qquad \left\{ \frac{\Omega(z)}{F(z)} \right\} = \underset{s}{\mathfrak{L}} \frac{s^{k-1}\Omega\left(\frac{1}{s}\right)}{\left[s^k F\left(\frac{1}{s}\right)\right]} e^{sz}.$$

I. When the rational fraction (h) reduces to the simple fraction

$$(k) \qquad \frac{1}{(1 - s_1 z)^q},$$

s_1 being any constant other than zero, we have evidently

$$\frac{1}{(1-s_1z)^q} = \Sigma\frac{q(q+1)\ldots(q+m-1)}{1\,.\,2\,.\,.\,.\,.\,.\,m}s_1{}^mz^m$$

$$= \frac{1}{1\,.\,2\,.\,.\,.\,.\,.\,(q-1)}\Sigma(m+1)(m+2)\ldots(m+q-1)s_1{}^mz^m.$$

We deduce successively, and without difficulty,

$$\left\{\frac{1}{(1-s_1z)^q}\right\} = \frac{1}{\left.\underline{q-1}\right|}\Sigma(m+1)(m+2)\ldots(m+q-1)\frac{s_1{}^mz^m}{\left.\underline{m}\right|}$$

$$= \frac{1}{\left.\underline{q-1}\right|}\Sigma\left\{\frac{d^{q-1}}{ds^{q-1}}\left[\frac{d^{q-1}}{dz^{q-1}}\frac{(sz)^{m+q-1}}{\left.\underline{m+q-1}\right|}\right]\right\}_{s=s_1}$$

$$= \frac{1}{\left.\underline{q-1}\right|}\left[\frac{d^{q-1}}{ds^{q-1}}\,\frac{d^{q-1}}{dz^{q-1}}e^{sz}\right]_{s=s_1}$$

$$= \frac{1}{\left.\underline{q-1}\right|}\left[\frac{d^{q-1}}{ds^{q-1}}(s^{q-1}e^{sz})\right]_{s=s_1}.$$

Now, by definition, this last expression is simply the residue of the function of s,

$$\frac{s^{q-1}e^{sz}}{(s-s_1)^q},$$

with respect to its infinity s_1 of order of multiplicity q. We have then definitely

$$(l)\qquad \left\{\frac{1}{(1-s_1z)^q}\right\} = \mathop{\mathfrak{L}}_{s=s_1}\cdot\frac{s^{q-1}e^{sz}}{(s-s_1)^q} = \mathop{\mathfrak{L}}_{s=s_1}\cdot\frac{1}{\left(1-s_1\dfrac{1}{s}\right)^q}\frac{e^{sz}}{s},$$

which proves the theorem in this simple case, for under this last $\mathfrak{L}$ the coefficient of e^{sz} is

$$= \frac{s^{q-1} \cdot 1}{s^q\left(1 - s_1\frac{1}{s}\right)^q}.$$

II. In the general case the rational fraction (h) can by decomposition be thrown into the form of a certain linear homogeneous function of simple fractions like (k), and consequently the series (i) into the form of a linear and homogeneous function of partial series which correspond to the simple fractions, or of residues expressing the sums of the partial series, as in equation (l). Collecting all these partial residues into a single residue, the recomposition of the simple fractions analogous to

$$\frac{1}{\left(1 - s_1\frac{1}{s}\right)^q}$$

will evidently put the product

$$e^{sz}\frac{\Omega\left(\frac{1}{s}\right)}{F\left(\frac{1}{s}\right)}, \quad = \frac{s^{k-1}\Omega\left(\frac{1}{s}\right)}{s^kF\left(\frac{1}{s}\right)},$$

under the sign $\mathfrak{L}$, which completes the proof of our theorem.

III. Our reasoning assumes implicitly that $F(z)$ is of the *effective* degree k; that is, that p_k is not zero. For if it were otherwise, the rational fraction (h) would not merely be resolved into simple fractions, but the fractions would be accompanied by integral monomials in z, of which up to this nothing has been said. We can nevertheless show (though this will be left as an exercise for the reader) that the generality of formula (j) will not be infringed upon when we take account of the simple, or multiple, root $s = 0$ which is now possessed by the equation

$$s^kF\left(\frac{1}{s}\right) = 0,$$

and of the evident relation

$$\frac{z^q}{\underline{q|}} = \mathfrak{L}_{s=0} \frac{e^{sz}}{s^{q+1}} = \mathfrak{L}_{s=0} \left(\frac{1}{s}\right)^q \cdot \frac{e^{sz}}{s}.$$

A very few words suffice now for the treatment of our principal question. Consider the system of linear differential equations

$$(m) \qquad \begin{cases} \dfrac{du}{dx} + a_1 u + b_1 v + \ldots + h_1 t = 0, \\ \dfrac{dv}{dx} + a_2 u + b_2 v + \ldots + h_2 t = 0, \\ \cdot \quad \cdot \quad \cdot \quad \cdot \quad \cdot \quad \cdot \\ \dfrac{dt}{dx} + a_k u + b\ v + \ldots + h_k t = 0, \end{cases}$$

the coefficients $(a, b, \ldots, h)$ being arbitrary constants. Let

$$(n) \qquad \begin{vmatrix} s + a_1, & b_1, & \ldots, & h_1 \\ a_2, & s + b_2, & \ldots, & h_2 \\ \cdot & \cdot & \cdot & \cdot \\ a_k, & b_k, & \ldots, & h_k \end{vmatrix} = f(s) = s^k + p_1 s^{k-1} + \ldots + p_k;$$

further, let

$$(o) \qquad \begin{cases} \alpha_1(s), & \beta_1(s), & \ldots, & \eta_1(s), \\ \alpha_2(s), & \beta_2(s), & \ldots, & \eta_2(s), \\ \cdot & \cdot & \cdot & \cdot \\ \alpha_k(s), & \beta_k(s), & \ldots, & \eta_k(s), \end{cases}$$

be the table of principal minors of this determinant. Calling now

$$(p) \qquad u_0, \quad v_0, \quad \ldots, \quad t_0$$

arbitrary constants (k in number), and making

$$\begin{cases} v(s) = u_0\alpha_1(s) + v_0\alpha_2(s) + \ldots + t_0\alpha_k(s), \\ \phi(s) = u_0\beta_1(s) + v_0\beta_2(s) + \ldots + t_0\beta_k(s), \\ \cdot \quad \cdot \quad \cdot \quad \cdot \quad \cdot \quad \cdot \quad \cdot \quad \cdot \\ \tau(s) = u_0\eta_1(s) + v_0\eta_2(s) + \ldots + t_0\eta_k(s); \end{cases}$$

Cauchy's theorem is as follows: *The integrals of the system* (m), *which for* $x = x_0$ *take the initial values* (p), *are given by the formulæ*

$$(q) \qquad \begin{cases} u = \underset{s}{\mathfrak{L}} \dfrac{v(s)e^{s(x-x_0)}}{[f(s)]}, \\ v = \underset{s}{\mathfrak{L}} \dfrac{\phi(s)e^{s(x-x_0)}}{[f(s)]}, \\ \cdot \quad \cdot \quad \cdot \quad \cdot \\ t = \underset{s}{\mathfrak{L}} \dfrac{\tau(s)e^{s(x-x_0)}}{[f(s)]}, \end{cases}$$

which furnish thus the general integrals of the given equations.

Making $x = x_0$ in equations (m) and in all the equations deduced from them by repeated differentiations, we see immediately that the initial values of the integrals and their derivatives of all orders, viz.,

$$\begin{cases} u_0, & u_1, & \ldots, & u_m, & \ldots, \\ v_0, & v_1, & \ldots, & v_m, & \ldots, \\ \cdot & \cdot & \cdot & \cdot & \cdot \\ t_0, & t_1, & \ldots, & t_m, & \ldots, \end{cases}$$

are connected by the recurrences (b).

It follows then that the sums of the co-recurrent series in z which have these initial values as coefficients are given by formulæ (e), and consequently that the series in $x - x_0$ are deduced from these last by replacing z by $x - x_0$ and dividing the terms in $(x - x_0)^m$ by $1 . 2 \ldots . m$; that is, the integrals of (m) having the initial values (p)

are obtained by applying formula (j). These integrals are then the second members of equations (q), for we have evidently

$$f(s) = s^k F\left(\frac{1}{s}\right),$$

and the elements of the table (o) are just what the corresponding elements of (d) become when, after changing z into $\frac{1}{s}$, we multiply each one by s^{k-1}.

The reader is referred to a memoir by J. Collet in the *Annales de l'École Normale Supérieure* for 1888 for another treatment of linear differential equations with constant coefficients.

CHAPTER III.

THE INTEGRALS OF THE DIFFERENTIAL EQUATION

$$P = \frac{d^n y}{dx^n} + p_1 \frac{d^{n-1} y}{dx^{n-1}} + p_2 \frac{d^{n-2} y}{dx^{n-2}} + \ldots + p_n y = 0.$$

THE coefficients $p_1, p_2, \ldots, p_n$ are uniform functions of x, having only poles as critical points. Starting then at any neutral point, say x_0, moving along any path whatever (provided of course it does not pass through a critical point), and returning to x_0, the functions p vary continuously and return at the end of the path to their original values. Let $y_1, \ldots, y_n$ denote a system of fundamental integrals: these integrals may or may not have critical points (in general of course they have), but whatever critical points they have, and of whatever description they are, they must be included among those of the coefficients p. When the variable starts from x_0 and travels by any path back to x_0, the equation resumes its original form, the coefficients p having returned to their original values; the functions $y_1, \ldots, y_n$ vary continuously along this path, remaining always integrals of the equation; we must have then, obviously, at the end of the path that $y_1, \ldots, y_n$ have been changed into linear functions of themselves, *i.e.*, a linear substitution has been imposed on $y_1, \ldots, y_n$. Denote the final values of the integrals by $Sy_1, \ldots, Sy_n$; then we have

$$(1) \qquad \begin{cases} Sy_1 = c_{11} y_1 + \ldots + c_{1n} y_n, \\ Sy_2 = c_{21} y_1 + \ldots + c_{2n} y_n, \\ \cdot \quad \cdot \quad \cdot \quad \cdot \quad \cdot \quad \cdot \\ Sy_n = c_{n1} y_1 + \ldots + c_{nn} y_n; \end{cases}$$

where c_{ik} is a constant. Or we may say that the integrals y_i have submitted to the substitution

$$(2) \qquad S = \left| \begin{matrix} y_1; & c_{11}y_1 + \ldots + c_{1n}y_n \\ \vdots & \cdot \quad \cdot \quad \cdot \quad \cdot \quad \cdot \\ y_n; & c_{n1}y_1 + \ldots + c_{nn}y_n \end{matrix} \right| .$$

Before proceeding farther with the study of the integrals it will be convenient to notice briefly the case of equations having uniform singly or doubly periodic coefficients. Suppose that in the equation

$$P = \frac{d^ny}{dx^n} + p_1\frac{d^{n-1}y}{dx^{n-1}} + \ldots + p_ny = 0$$

the coefficients $p_1, \ldots, p_n$ represent singly periodic functions of x; let ω denote the period; then $p_i(x + \omega) = p_i(x)$. If then we change x into $x + \omega$, the equation retains its original form. In the investigation farther on of the integrals of this equation we will assume them to be uniform functions of x; for the present purpose, however, no such assumption is necessary. Suppose $f_1(x)$, $f_2(x), \ldots,$ $f_n(x)$ a system of fundamental integrals of $P = 0$; *i.e.*, a linearly independent system. Since in travelling from the point x to the point $x + \omega$ the equation resumes its original form, and since the continuously varying functions $f_i(x)$ remain always integrals of the equation, we must have

$$(3) \quad \begin{cases} f_1(x + \omega) = \alpha_{11}f_1(x) + \alpha_{12}f_2(x) + \ldots + \alpha_{1n}f_n(x), \\ f_2(x + \omega) = \alpha_{21}f_1(x) + \alpha_{22}f_2(x) + \ldots + \alpha_{2n}f_n(x), \\ \cdot \quad \cdot \quad \cdot \quad \cdot \quad \cdot \quad \cdot \quad \cdot \quad \cdot \quad \cdot \quad \cdot \\ f_n(x + \omega) = \alpha_{n1}f_1(x) + \alpha_{n2}f_2(x) + \ldots + \alpha_{nn}f_n(x); \end{cases}$$

or the integrals f_i have been submitted to the linear substitution

$$S = \left| \begin{matrix} f_1; & \alpha_{11}f_1 + \ldots + \alpha_{1n}f_n \\ \cdot & \cdot \quad \cdot \quad \cdot \quad \cdot \quad \cdot \\ f_n; & \alpha_{n1}f_1 + \ldots + \alpha_{nn}f_n \end{matrix} \right| .$$

Again, suppose the coefficients p_i to be uniform and doubly periodic functions of x, and let ω and ω' denote the periods; then

$$p_i(x + \omega) = p_i(x), \quad p_i(x + \omega') = p_i(x).$$

Let $f_1(x), \ldots, f_n(x)$ denote a system of fundamental integrals of the equation at the point x. If now we travel by an arbitrary path from x to $x + \omega$, the integrals f_i will obviously submit to the substitution

$$S = \left| \begin{array}{ll} f_1; & \alpha_{11} f_1 + \ldots + \alpha_{1n} f_n \\ \cdot & \cdot \quad \cdot \quad \cdot \quad \cdot \quad \cdot \\ f_n; & \alpha_{n1} f_1 + \ldots + \alpha_{nn} f_n \end{array} \right| .$$

Again, if we travel from x to $x + \omega'$, the integrals will submit to the substitution

$$S' = \left| \begin{array}{ll} f_1; & \alpha'_{11} f_1 + \ldots + \alpha'_{1n} f_n \\ \cdot & \cdot \quad \cdot \quad \cdot \quad \cdot \quad \cdot \\ f_n; & \alpha'_{n1} f_1 + \ldots + \alpha'_{nn} f_n \end{array} \right|$$

Finally, change x into $x + \omega + \omega'$; the effect of this change is clearly the same as first making the substitution S and then making the substitution S', or first making the substitution S' and following it by S. We have therefore the relation

$$SS' = S'S. \tag{4}$$

The order of the substitutions is the order in which they are written; *e.g.*, $S^i S^k$ means that first the substitution S^i is made and then the substitution S^k.

Returning now to our original equation, $P = 0$, let $y_1, \ldots, y_n$ denote a system of fundamental integrals; we have then

$$\frac{d^n y_1}{dx^n} + p_1 \frac{d^{n-1} y_1}{dx^{n-1}} + \ldots + p_n y_1 = 0,$$

$$\frac{d^n y_2}{dx^n} + p_1 \frac{d^{n-1} y_2}{dx^{n-1}} + \ldots + p_n y_2 = 0,$$

$$\cdot \quad \cdot \quad \cdot \quad \cdot \quad \cdot \quad \cdot \quad \cdot \quad \cdot$$

$$\frac{d^n y_n}{dx_n} + p_1 \frac{d^{n-1} y_n}{dx^{n-1}} + \ldots + p_n y_n = 0.$$

Regarding these as forming a system of linear equations in $p_1, p_2, \ldots, p_n$, we have

$$(5)\quad p_i = -\frac{1}{D}\begin{vmatrix} \frac{d^{n-1}y_1}{dx^{n-1}}, \ldots, & \frac{d^{n-i+1}y_1}{dx^{n-i+1}}, & \frac{d^n y_1}{dx^n}, & \frac{d^{n-i-1}y_1}{dx^{n-i-1}}, \ldots, y_1 \\ \frac{d^{n-1}y_2}{dx^{n-1}}, \ldots, & \frac{d^{n-i+1}y_2}{dx^{n-i+1}}, & \frac{d^n y_2}{dx^n}, & \frac{d^{n-i-1}y_2}{dx^{n-i-1}}, \ldots, y_2 \\ \vdots & \vdots & & \\ \frac{d^{n-1}y_n}{dx^{n-1}}, & \frac{d^{n-i+1}y_n}{dx^{n-i+1}}, & \frac{d^n y_n}{dx^n}, & \frac{d^{n-i-1}y_n}{dx^{n-i-1}}, \ldots, y_n \end{vmatrix}$$

where

$$(6)\qquad D = \begin{vmatrix} \frac{d^{n-1}y_1}{dx^{n-1}}, & \frac{d^{n-2}y_1}{dx^{n-2}}, \ldots, & y_1 \\ \frac{d^{n-1}y_2}{dx^{n-1}}, & \frac{d^{n-2}y_2}{dx^{n-2}}, \ldots, & y_2 \\ \vdots & \vdots & \vdots \\ \frac{d^{n-1}y_n}{dx^{n-1}}, & \frac{d^{n-2}y_n}{dx^{n-2}}, \ldots, & y_n \end{vmatrix}.$$

In particular,

$$(7)\quad p_1 = -\begin{vmatrix} \frac{d^n y_1}{dx^n}, & \frac{d^{n-2}y_1}{dx^{n-2}}, \ldots, & y_1 \\ \vdots & \vdots & \vdots \\ \frac{d^n y_n}{dx^n}, & \frac{d^{n-2}y_n}{dx^{n-2}}, \ldots, & y_n \end{vmatrix} \div \begin{vmatrix} \frac{d^{n-1}y_1}{dx^{n-1}}, & \frac{d^{n-2}y_1}{dx^{n-2}}, \ldots, & y_1 \\ \vdots & \vdots & \vdots \\ \frac{d^{n-1}y_n}{dx^{n-1}}, & \frac{d^{n-2}y_n}{dx^{n-2}}, \ldots, & y_n \end{vmatrix}.$$

Then we have

$$(8)\qquad D = Ce^{-\int p_1 dx};$$

where C is a constant necessarily different from zero, since the integrals $y_1, \ldots, y_n$ are linearly independent. The formula for p_i may be written for brevity as

$$p_i = -\frac{D_i}{D};$$

where D_i denotes the determinant into which D is changed when its ith column is replaced by $\frac{d^n y_1}{dx^n}, \ldots, \frac{d^n y_n}{dx^n}$. The result of going round a critical point is to change the integral y_i into Sy_i, and, consequently, to change the determinants D and D_i into ΔD and ΔD_i, where Δ denotes the determinant of the substitution S; viz.,

$$\Delta = \begin{vmatrix} c_{11}, & c_{12}, & \ldots, & c_{1n} \\ c_{21}, & c_{22}, & \ldots, & c_{2n} \\ \cdot & \cdot & \cdot & \cdot \\ \vdots & & & \\ c_{n1}, & c_{n2}, & \ldots, & c_{nn} \end{vmatrix}. \tag{9}$$

Therefore, after going round the critical point, we have, as we should,

$$p_i = -\frac{\Delta D_i}{\Delta D} = -\frac{D_i}{D}.$$

It is of course to be noticed that the determinant Δ is not equal to zero; for if it were, the system Sy_i, and consequently the system y_i, could not form a fundamental system. If $y_1, \ldots, y_n$ denote a system of fundamental integrals, and $z_1, z_2, \ldots, z_\lambda$ linear functions with constant coefficients of $y_1, \ldots, y_\lambda$, given by

$$z_i = \sum_{k=1}^{k=\lambda} C_{ik}\, y_k; \tag{10}$$

then $z_1, \ldots, z_\lambda, y_{\lambda+1}, \ldots, y_n$ form a fundamental system if the determinant

$$\Delta' = \begin{vmatrix} C_{11}, & C_{12}, & \ldots, & C_{1\lambda} \\ C_{21}, & C_{22}, & \ldots, & C_{2\lambda} \\ \vdots & & & \\ C_{\lambda 1}, & C_{\lambda 2}, & \ldots, & C_{\lambda\lambda} \end{vmatrix} \tag{11}$$

is different from zero; in fact, it is obvious that if Δ' be different from zero, there can exist no linear relation with constant coefficients between

$$z_1, \ldots, z_\lambda, y_{\lambda+1}, \ldots, y_n.$$

Suppose a particular integral y_1 (different from zero) to have been found for

$$P = \frac{d^n y}{dx^n} + p_1 \frac{d^{n-1} y}{dx^{n-1}} + \ldots + p_n y = 0;$$

make now

$$y = y_1 \int z dx;$$

then we have for z the equation

$$Q = \frac{d^{n-1} z}{dx^{n-1}} + q_1 \frac{d^{n-2} z}{dx^{n-2}} + \ldots + q_{n-1} z = 0;$$

where

$$(12) \quad \begin{cases} q_1 = \dfrac{n}{y_1} \dfrac{dy_1}{dx} + p_1, \\ q_2 = \dfrac{n(n-1)}{1 \cdot 2 \cdot y_1} \dfrac{d^2 y_1}{dx^2} + \dfrac{np_1}{y_1} \dfrac{dy_1}{dx} + p_2, \\ \cdot \quad \cdot \quad \cdot \quad \cdot \quad \cdot \quad \cdot \quad \cdot \quad \cdot \\ q_r = \dfrac{n(n-1) \ldots (n-r+1)}{1 \cdot 2 \ldots \ldots r \cdot y_1} \dfrac{d^r y_1}{dx^r} \\ \quad + p_1 \dfrac{(n-1)(n-2) \ldots (n-r+1)}{1 \cdot 2 \ldots \ldots (r-1) y_1} \dfrac{d^{r-1} y_1}{dx^{r-1}} \\ \quad + \ldots + p_k \dfrac{(n-k)(n-k-1) \ldots (n-r+1)}{1 \cdot 2 \ldots \ldots (r-k) y_1} \dfrac{d^{r-k} y_1}{dx^{r-k}} + \ldots + p_r, \\ \cdot \quad \cdot \quad \cdot \quad \cdot \quad \cdot \quad \cdot \quad \cdot \quad \cdot \quad \cdot \quad \cdot \quad \cdot \quad \cdot \end{cases}$$

Suppose now that a solution z_1, different from zero, has been found for $Q = 0$; substitute as before

$$z = z_1 \int t dx,$$

and we have for t an equation of order $n - 2$. This process may be continued until we arrive at an equation of the first order in, say, w—an integral of which is w_1. We have now as integrals of $P = 0$ the following:

$$y_1, \quad y_2 = y_1 \int z_1 dx, \quad y_3 = y_1 \int z_1 dx \int t_1 dx, \ \ldots$$
$$y_n = y_1 \int z_1 dx \int t_1 dx \int u_1 dx \ \ldots \int w_1 dx;$$

and these constitute a fundamental system; *i.e.*, we can have no such relation as

$$C_1 y_1 + C_2 y_2 + \ldots + C_n y_n = 0$$

unless all the constant coefficients C are equal to zero. For, suppose this relation does exist; then on substituting the above values of $y_2, \ldots, y_n$ and dividing through by y_1 it becomes

$$0 = C_1 + C_2 \int z_1 dx + C_3 \int z_1 dx \int t_1 dx + \ldots + C_n \int z_1 dx \ldots \int w_1 dx;$$

differentiate and divide by z_1, and we have

$$C_2 + C_3 \int t_1 dx + \ldots + C_n \int t_1 dx \ldots \int w_1 dx = 0;$$

differentiate again and divide by t_1, and so on; we come obviously to the condition $C_n = 0$. Retracing our steps now, and we find

$$C_{n-1} = C_{n-2} = \ldots = C_1 = 0.$$

Denote by D' the determinant formed from a fundamental system of $Q = 0$, in the same way that D was formed from the fundamental integrals of $P = 0$. We found

(13) $$D = Ce^{-\int p_1 dx};$$

consequently

$$D' = C'e^{-\int q_1 dx} = C'e^{-\int p_1 dx - \int \frac{n}{y_1} \frac{dy_1}{dx} dx},$$

or $$D' = C''Dy_1^{-n},$$

and

(14) $$D = C'''D'y_1^{n}.$$

In like manner,

(15) $$D' = C^{iv}D''z^{n-1},$$

and so on. Multiplying together these equations giving D, D', etc., and we have, finally,

(16) $$D = Cy_1^{n} z_1^{n-1} t_1^{n-2} \ldots w_1.$$

All the results given here might have been incorporated in the first chapter, as they only require that the equation shall have uniform coefficients. It seemed, however, better to have them here, as some of them will be of immediate use. We will take up now the equation $P = 0$, as defined at the beginning of this chapter, and see how its integrals behave when we travel round a critical point. For simplicity, let the point be the origin, $x = 0$. As we know, the fundamental system $y_1, \ldots, y_n$ has the linear substitution S imposed upon it when we travel round a critical point. Suppose S to be the substitution peculiar to the point $x = 0$; then on going round this point we have

$$(17) \qquad \begin{cases} Sy_1 = c_{11}y_1 + \ldots + c_{1n}y_n, \\ Sy_2 = c_{21}y_1 + \ldots + c_{2n}y_n, \\ \vdots \qquad \vdots \qquad \vdots \\ Sy_n = c_{n1}y_1 + \ldots + c_{nn}y_n; \end{cases}$$

or

$$(18) \qquad S = \begin{vmatrix} y_1; & c_{11}y_1 + \ldots + c_{1n}y_n \\ y_2; & c_{21}y_1 + \ldots + c_{2n}y_n \\ \vdots & \\ y_n; & c_{n1}y_1 + \ldots + c_{nn}y_n \end{vmatrix}.$$

Let $z_1, \ldots, z_n$ denote another arbitrary system of independent integrals; then

$$(19) \qquad \begin{cases} z_1 = a_{11}y_1 + \ldots + a_{1n}y_n, \\ z_2 = a_{21}y_1 + \ldots + a_{2n}y_n, \\ \vdots \qquad \qquad \qquad \vdots \\ z_n = a_{n1}y_1 + \ldots + a_{nn}y_n. \end{cases}$$

The a_{ij} are arbitrary constants—such, however, that the determinant $| a_{ij} |$ is not zero. Suppose that we travel round $x = 0$; then the y's submit to the substitution S, and consequently the z's are changed into new linear functions of the y's. But by aid of the last equations we can express the y's as linear functions of the z's, and so, after going round $x = 0$, the z's will be changed into linear functions of themselves; *i.e.*, they will have submitted to a linear substitution. As this substitution depends in part upon the arbitrary quantities a_{ij}, we can use this indetermination to affect a simplification of the substitution—viz., we may seek first to determine whether or not there exists an integral which, on travelling round the critical point, changes into itself multiplied by a constant. In the first place, if there *is* any such integral, it must obviously be of the form

$$y = \alpha_1 y_1 + \ldots + \alpha_n y_n.$$

Suppose the effect of going round the point is to change this into sy. Apply the substitution S; *i.e.*, travel round $x = 0$; we must have, by hypothesis,

$$\alpha_1(c_{11}y_1 + \ldots + c_{1n}y_n) + \ldots + \alpha_n(c_{n1}y_1 + \ldots + c_{nn}y_n)$$
$$= s(\alpha_1 y_1 + \ldots + \alpha_n y_n).$$

As no linear relation with constant coefficients can exist among the integrals $y_1, \ldots, y_n$, this must be an identity, and we have therefore, on equating the coefficients of each y, the equations of condition

$$(20) \qquad \begin{cases} (c_{11} - s)\alpha_1 + c_{21}\alpha_2 + \ldots + c_{n1}\alpha_n & = 0, \\ c_{12}\alpha_1 + (c_{22} - s)\alpha_2 + \ldots + c_{n2}\alpha_n & = 0, \\ \ldots \ldots \ldots \ldots \\ c_{1n}\alpha_1 + c_{2n}\alpha_2 \quad + \ldots + (c_{nn} - s)\alpha_n & = 0; \end{cases}$$

giving for the determination of s the algebraic equation of nth order

$$(21) \qquad \Delta = \begin{vmatrix} c_{11} - s, & c_{21}, & \ldots, & c_{n1} \\ c_{12}, & c_{22} - s, & \ldots, & c_{n2} \\ \vdots & & & \\ c_{1n}, & c_{2n}, & \ldots, & c_{nn} - s \end{vmatrix} = 0.$$

This equation will be called the *characteristic equation.*

Suppose $\Delta = 0$ has all of its roots unequal; then if s_i denote one of them, and if this value of s be substituted in the preceding equation, these will obviously serve for the determination of the ratios of $\alpha_1, \alpha_2, \ldots, \alpha_n$, which is all that is necessary. The integral y will therefore be known, and as all the roots of Δ are unequal, we will obviously have n integrals each possessing the required property;

or, in other words, for this set of integrals the substitution S becomes

$$(22) \qquad S = \left| \begin{array}{ll} y_1 ; & s_1 y_1 \\ y_2 ; & s_2 y_2 \\ \vdots & \vdots \\ y_n ; & s_n y_n \end{array} \right|$$

It can be shown in a moment that these integrals constitute a fundamental system; *i.e.*, are not connected by any such linear relation as

$$c_1 y_1 + \ldots + c_n y_n = 0.$$

Suppose such a relation does exist; then on applying the substitution S, that is, turning round the critical point $x = 0$, a similar relation must exist, and so for any number of turns round $x = 0$. Suppose $n - 1$ turns to have been made; then we have

$$\begin{array}{ll} c_1 y_1 + \ldots + c_n y_n & = 0, \\ c_1 s_1 y_1 + \ldots + c_n s_n y_n & = 0, \\ \cdot \quad \cdot \quad \cdot \quad \cdot \quad \cdot \quad \cdot & \cdot \\ c_1 s_1^{n-1} y_1 + \ldots + c_n s_n^{n-1} y_n & = 0. \end{array}$$

As the constants $c_1, \ldots, c_n$ are not zero, we must have

$$\left| \begin{array}{llll} 1, & 1, & \ldots, & 1 \\ s_1, & s_2, & \ldots, & s_n \\ s_1^2, & s_2^2, & \ldots, & s_n^2 \\ \vdots & & & \\ s_1^{n-1}, & s_2^{n-1}, & \ldots, & s_n^{n-1} \end{array} \right| = 0.$$

This is the product of the differences $s_i - s_j$, and cannot therefore vanish unless at least two of the roots s become equal; but by hypothesis the roots are all different; this determinant can therefore not vanish, and hence no such relation as

$$c_1 y_1 + \ldots + c_n y_n = 0$$

can exist; the integrals constitute therefore a fundamental system.

If we had chosen any other set of fundamental integrals, say $v_1, \ldots, v_n$, the substitution S would have the form

$$(23) \qquad S = \left| \begin{array}{l} v_1; \quad \gamma_{11}v_1 + \ldots + \gamma_{1n}v_n \\ \vdots \\ v_n; \quad \gamma_{n1}v_1 + \ldots + \gamma_{nn}v_n \end{array} \right| .$$

The characteristic equation is now

$$(24) \qquad \Delta_1 = \left| \begin{array}{llll} \gamma_{11} - s, & \gamma_{21}, & \ldots, & \gamma_{n1} \\ \gamma_{12}, & \gamma_{22} - s, & \ldots, & \gamma_{n2} \\ \vdots & & & \\ \gamma_{1n}, & \gamma_{2n}, & \ldots, & \gamma_{nn} - s \end{array} \right| = 0.$$

The integrals $y_1, \ldots, y_n$ now become linear functions of $v_1, v_2, \ldots, v_n$ and reproduce themselves, multiplied respectively by $s_1, \ldots, s_n$, and it therefore follows that the equation $\Delta_1 = 0$ is identical with $\Delta = 0$, since it has the same roots. The coefficients of Δ being independent of the choice of fundamental integrals are therefore invariants. A direct proof of this fact, viz., that the characteristic equation is the same whatever set of fundamental integrals is chosen, will now be given. We have

$$(25) \qquad \begin{cases} Sv_1 = \gamma_{11}v_1 + \ldots + \gamma_{1n}v_n, \\ Sv_2 = \gamma_{21}v_1 + \ldots + \gamma_{2n}v_n, \\ \vdots \qquad \vdots \qquad\qquad\quad \vdots \\ Sv_n = \gamma_{n1}v_1 + \ldots + \gamma_{nn}v_n, \end{cases}$$

with the characteristic equation

$$(26) \qquad \Delta_1 = \left| \begin{array}{llll} \gamma_{11} - s, & \gamma_{21}, & \ldots, & \gamma_{n1} \\ \gamma_{12}, & \gamma_{22} - s, & \ldots, & \gamma_{n2} \\ \vdots & & & \\ \gamma_{1n}, & \gamma_{2n}, & \ldots, & \gamma_{nn} - s \end{array} \right| = 0.$$

Again: we must have

$$
(27)\qquad \begin{cases} v_1 = \lambda_{11} y_1 + \dots + \lambda_{1n} y_n, \\ v_2 = \lambda_{21} y_1 + \dots + \lambda_{2n} y_n, \\ \vdots \qquad\qquad\qquad\qquad \vdots \\ v_n = \lambda_{n1} y_1 + \dots + \lambda_{nn} y_n. \end{cases}
$$

The determinant $| \lambda_{ij} |$ is of course not zero. Substitute these values of $v_1, \dots, v_n$ in the right-hand members of equations (25), and we have

$$
(28)\qquad Sv_i = (\gamma_{i1}\lambda_{11} + \gamma_{i2}\lambda_{21} + \dots + \gamma_{in}\lambda_{n1})y_1 + \dots \\ + (\gamma_{i1}\lambda_{1n} + \gamma_{i2}\lambda_{2n} + \dots + \gamma_{in}\lambda_{nn})y_n.
$$

Again: in equations (27) apply the substitution S, and we have

$$
(29)\qquad Sv_i = (c_{11}\lambda_{i1} + c_{21}\lambda_{i2} + \dots + c_{n1}\lambda_{in})y_1 + \dots \\ + (c_{1n}\lambda_{i1} + c_{2n}\lambda_{i2} + \dots + c_{nn}\lambda_{in})y_n.
$$

From (28) and (29) we must then have

$$
\sum_{k=1}^{k=n} c_{kj}\lambda_{ik} = \sum_{k=1}^{k=n} \gamma_{ik}\lambda_{kj}, \; = \delta_{ij}.
$$

Denote by Δ_2 the determinant $| \lambda_{ij} |$; we see then that the two determinants formed from Δ by multiplying it by Δ_1 and Δ_2 respectively, have their elements equal each to each, and that each of these products is equal to

$$
\begin{vmatrix} \delta_{11} - \lambda_{11}s, & \delta_{12} - \lambda_{12}s, & \dots, & \delta_{1n} - \lambda_{1n}s \\ \delta_{21} - \lambda_{21}s, & \delta_{22} - \lambda_{22}s, & \dots, & \delta_{2n} - \lambda_{2n}s \\ \vdots & & & \\ \delta_{n1} - \lambda_{n1}s, & \delta_{n2} - \lambda_{n2}s, & \dots. & \delta_{nn} - \lambda_{nn}s \end{vmatrix} = \Delta_3.
$$

We have then

$$
\Delta_2\Delta = \Delta_2\Delta_1
$$

that is,

$$
\Delta = \Delta_1.
$$

This proof is due to Hamburger.* The following theorem is due to the same author, and is taken from the same paper:

THEOREM.—*If for a given value of s all the minors of order ν in Δ vanish (and consequently all of higher orders), and those of order ν — 1) do not all vanish, then all the minors of Δ_1 of order ν will vanish, and those of order (ν — 1) will not all vanish.*

We have

$$
30)\quad \left\{
\begin{aligned}
\Delta &= \begin{vmatrix} c_{11}-s, & c_{21}, & \ldots, & c_{n1} \\ c_{12}, & c_{22}-s, & \ldots, & c_{n2} \\ \vdots & \vdots & & \vdots \\ c_{1n}, & c_{2n}, & \ldots, & c_{nn}-s \end{vmatrix}, \\
\Delta_1 &= \begin{vmatrix} \gamma_{11}-s, & \gamma_{21}\cdot & \ldots, & \gamma_{n1} \\ \gamma_{12}, & \gamma_{22}-s, & \ldots, & \gamma_{n2} \\ \vdots & \vdots & & \vdots \\ \gamma_{1n}, & \gamma_{2n}, & \ldots, & \gamma_{nn}-s \end{vmatrix}, \\
\Delta_2 &= \begin{vmatrix} \lambda_{11}, & \lambda_{12}, & \ldots, & \lambda_{1n} \\ \lambda_{21}, & \lambda_{22}, & \ldots, & \lambda_{2n} \\ \vdots & \vdots & & \vdots \\ \lambda_{n1}, & \lambda_{n2}, & \ldots, & \lambda_{nn} \end{vmatrix}, \\
\Delta_3 &= \begin{vmatrix} \delta_{11}-\lambda_{11}s, & \delta_{12}-\lambda_{12}s, & \ldots, & \delta_{1n}-\lambda_{1n}s \\ \delta_{21}-\lambda_{21}s, & \delta_{22}-\lambda_{22}s, & \ldots, & \delta_{2n}-\lambda_{2n}s \\ \vdots & \vdots & & \vdots \\ \delta_{n1}-\lambda_{n1}s, & \delta_{n2}-\lambda_{n2}s, & \ldots, & \delta_{nn}-\lambda_{nn}s \end{vmatrix}.
\end{aligned}
\right.
$$

Let A_{ik}, B_{ik}, C_{ik}, D_{ik} denote respectively the minors of order of the determinants Δ, Δ_1, Δ_2, Δ_3; where i and k denote any of the

$$\frac{n(n-1)\ldots(n-\nu+1)}{1.2.3\ldots.\nu}, = \mu,$$

ombinations of ν different numbers formed out of the series 1, 2, . . , n. We have now

$$31)\quad D_{ik} = C_{i1}A_{k1} + \ldots + C_{i\mu}A_{k\mu} = B_{i1}C_{k1} + \ldots + B_{i\mu}C_{k\mu}.$$

* *Journal für reine und angewandte Mathematik*, vol. lxxvi. p. 113.

Since by hypothesis all of the minors A_{rs} are zero, it follows that all of the minors D_{rs} are also zero; we have then the system of equations

(32)
$$\begin{cases} B_{i1} C_{11} + \ldots + B_{i\mu} C_{1\mu} = 0, \\ B_{i1} C_{21} + \ldots + B_{i\mu} C_{2\mu} = 0, \\ \cdot \quad \cdot \quad \cdot \quad \cdot \quad \cdot \quad \cdot \\ B_{i1} C_{\mu 1} + \ldots + B_{i\mu} C_{\mu\mu} = 0. \end{cases}$$

The determinant of this system, viz.,

$$\begin{vmatrix} C_{11}, \ldots, C_{1\mu} \\ C_{21}, \ldots, C_{2\mu} \\ \cdot \quad \cdot \quad \cdot \quad \cdot \\ C_{\mu 1}, \ldots, C_{\mu\mu} \end{vmatrix}.$$

is a power of Δ_2, and therefore does not vanish; consequently we have

(33)
$$B_{i1} = B_{i2} = \ldots = B_{i\mu} = 0;$$

and, as these equations hold for any value of i, it follows that all the minors B_{rs} of order ν of the determinant Δ_1 are equal to zero. If now all of the minors of order $(\nu - 1)$ of Δ_1 vanish, it is clear from the foregoing that all the minors of Δ must also vanish, which is contrary to the hypothesis.

We will take up now the case where the characteristic equation has equal roots.

By employing the results, not yet mentioned, of Hamburger's paper the forms of the integrals can be obtained very readily; but it seems better to first obtain these forms in the same way that Fuchs obtained them, if for no other reason than that of the desirability of developing the subject in historical order.

From what precedes it is clear that there is always at least one integral, say u_1, which is changed into $s_1 u_1$ when the independent variable travels round the critical point $x = 0$. If $y_1, \ldots, y_n$ denote a system of fundamental integrals, we have

(34)
$$u_1 = \alpha_1 y_1 + \ldots + \alpha_n y_n,$$

where the coefficients $\alpha_1, \ldots, \alpha_n$ are not all zero. It is clear now that we can replace one of the integrals $y_1, \ldots, y_n$ by u_1 and still have a fundamental system, provided that the coefficient α corresponding to the replaced y is not zero. Suppose α_1 different from zero; then we can take for a new fundamental system the integrals

$$u_1, \quad y_2, \quad y_3, \quad \ldots, \quad y_n.$$

The result of going round the critical point $x = 0$ is now to change these integrals into the following:

$$(35) \qquad \begin{cases} Su_1 = s_1 u_1, \\ Sy_2 = \beta_{21} u_1 + \beta_{22} y_2 + \ldots + \beta_{2n} y_n, \\ Sy_3 = \beta_{31} u_1 + \beta_{32} y_2 + \ldots + \beta_{3n} y_n, \\ \cdot \quad \cdot \quad \cdot \quad \cdot \quad \cdot \quad \cdot \quad \cdot \quad \cdot \\ Sy_n = \beta_{n1} u_1 + \beta_{n2} y_2 + \ldots + \beta_{nn} y_n. \end{cases}$$

The characteristic equation corresponding to this system is obviously

$$(36) \qquad (s_1 - s) \begin{vmatrix} \beta_{22} - s, & \beta_{32}, & \ldots, & \beta_{n2} \\ \beta_{23}, & \beta_{33} - s, & \ldots & \beta_{n3} \\ \cdot & \cdot & \cdot & \cdot \\ \beta_{2n}, & \beta_{3n}, & \ldots, & \beta_{nn} \end{vmatrix} = 0.$$

As s_1 is a multiple root of the characteristic equation $\Delta = 0$, and as the roots of the characteristic equation are quite independent of the choice of a fundamental system of integrals, it follows that the equation

$$(37) \qquad \begin{vmatrix} \beta_{22} - s, & \ldots, & \beta_{n2} \\ \vdots & & \vdots \\ \beta_{2n}, & \ldots, & \beta_{nn} - s \end{vmatrix} = 0.$$

has at least one root $s = s_1$. This being so, it follows further that there must exist a compatible system of equations, such as

$$(38) \qquad \begin{cases} (\beta_{22} - s_1)A_2 + \beta_{23}A_3 + \ldots + \beta_{2n}A_n = 0, \\ \beta_{32}A_2 + (\beta_{33} - s_1)A_3 + \ldots + \beta_{3n}A_n = 0, \\ \cdot \quad \cdot \quad \cdot \quad \cdot \quad \cdot \quad \cdot \quad \cdot \quad \cdot \quad \cdot \quad \cdot \\ \beta_{n2}A_2 + \beta_{n3}A_3 + \ldots + (\beta_{nn} - s_1)A_n = 0. \end{cases}$$

If now we write

$$u_2 = A_2 y_2 + A_3 y_3 + \ldots + A_n y_n ,$$

and then travel round the critical point $x = 0$, we shall have

$$Su_2 = (\beta_{21} A_2 + \beta_{31} A_3 + \ldots + \beta_{n1} A_n) u_1 + s_1 u_2 ,$$

or

(39) $$Su_2 = s_{21} u_1 + s_1 u_2 ,$$

where s_{21} is a constant. The new function u_2 is obviously an integral, and we can replace by it any one of the integrals $y_2, \ldots , y_n$ of which the coefficient α is different from zero. Suppose that u_2 replaces y_2; then we have a new fundamental system,

$$u_1 , \quad u_2 , \quad y_3 , \quad \ldots , \quad y_n .$$

If s_1 is a triple root of the characteristic equation, we can find a new integral u_3 satisfying the equation

$$Su_3 = s_{31} u_1 + s_{32} u_2 + s_1 u_3 ,$$

where s_{31} and s_{32} are constants. If then s_1 is root of the characteristic equation of multiplicity λ, we can obviously form a group of λ integrals $u_1 , u_2 , \ldots , u_\lambda$, such that

41) $$\begin{cases} Su_1 = s_1 u_1 , \\ Su_2 = s_{21} u_1 + s_1 u_2 , \\ Su_3 = s_{31} u_1 + s_{32} u_2 + s_1 u_3 , \\ \cdot \quad \cdot \quad \cdot \quad \cdot \quad \cdot \quad \cdot \\ Su_{\lambda 1} = s_{\lambda 1} u_1 + s_{\lambda 2} u_2 + \ldots + s_{\lambda, \lambda - 1} u_{\lambda - 1} + s_1 u_\lambda . \end{cases}$$

Finally, if the distinct roots of the characteristic equation $\Delta = 0$ are $s_1 , s_2 , \ldots , s_l$, and if their orders of multiplicity are $\lambda_1 , \lambda_2 , \ldots , \lambda_l$, we can form in the manner above indicated l groups of integrals containing in all $\lambda_1 + \lambda_2 + \ldots + \lambda_l = n$ integrals, and these integrals will, from what precedes, form a fundamental system.

By aid of the above-found properties of a system of fundamental integrals it is easy to find the forms of the integrals.

Write

(42) $$r_1 = \frac{\log s_1}{2\pi i};$$

$\log s_1$ standing for any one of its values. Suppose u_1 an integral such that when the variable turns round the critical point $x = 0$ we have

(43) $$Su_1 = s_1 u_1;$$

then it is clear that in the region of this point the product $x^{-r_1} u_1$ is a uniform function, say $\phi_1(x)$; then

(44) $$u_1 = x^{r_1} \phi_1(x).$$

It follows at once, if all the roots of the characteristic equation are distinct, and if

(45) $$r_k = \frac{\log s_k}{2\pi i},$$

that each integral can be written in the form

(46) $$u_k = x^{r_k} \phi_k(x);$$

where $\phi_k(x)$ is a uniform function of x, and where the exponents r_1, r_2, . . . , r_n do not differ from each other by integers. This last is clear since

$$s_k = e^{2\pi r_k i};$$

if $r_l = r_k + m$, where m is an integer, then we should have $s_k = s_l$, which is contrary to hypothesis. The functions $\phi(x)$ are developable in series containing both positive and negative powers of x, and convergent in the region of $x = 0$.

In the case where s_1 is a multiple root of, say, order λ, the analytical forms of the corresponding integrals may be found as follows: We have already shown that there exists a group of integrals u_1, u_2, . . . , u_λ having the properties expressed in equations (41), and that u_1 is of the form

$$u_1 = x^{r_1} \phi_{11};$$

where ϕ_{11} is uniform in the region of the point $x = 0$, and where

$$r_1 = \frac{1}{2\pi i} \log s_1 .$$

In respect to u_2, equations (41) show that

$$Su_2 = s_{21}u_1 + s_1u_2 .$$

Hence

$$\frac{Su_2}{Su_1} = S\left(\frac{u_2}{u_1}\right) = \frac{s_{21}}{s_1} + \frac{u_2}{u_1}.$$

Hence the substitution S, that is, the result of a circuit of the variable around the point $x = 0$, reproduces $\frac{u_2}{u_1}$ increased by a constant; the point $x = 0$ is therefore a logarithmic critical point for the function $\frac{u_2}{u_1}$, and consequently

$$\frac{u_2}{u_1} - \frac{s_{21}}{s_1 \cdot 2\pi i} \log x$$

is uniform in the region of the point $x = 0$. Designating this function for the moment by $f(x)$, we have

$$\frac{u_2}{u_1} - \frac{s_{21}}{2\pi i \cdot s_1} \log x = f(x),$$

or

$$u_2 = u_1 f(x) + \frac{s_{21}}{2\pi i \cdot s_1} u_1 \log x ;$$

that is, since $u_1 f(x) = x^{r_1}\phi_{11}(x) f(x)$, where $\phi_{11}(x) f(x)$ is necessarily a uniform function,

$$u_2 = x^{r_1}\left[\phi_{11}(x) f(x) + \frac{s_{21}}{2\pi i \cdot s_1} \phi_{11} \log x\right].$$

Making $\phi_{11}(x) f(x) = \phi_{21}(x)$, we have

(47) $$u_2 = x^{r_1}\left[\phi_{21} + \frac{s_{21}}{2\pi i \cdot s_1} \phi_{11} \log x\right].$$

In the same way,

$$S\left(\frac{u_3}{u_1}\right) = \frac{s_{31}}{s_1} + \frac{s_{32}}{s_1} \cdot \frac{u_2}{u_1} + \frac{u_3}{u_1};$$

or, substituting for $\frac{u_2}{u_1}$ its value found above,

$$S\left(\frac{u_3}{u_1}\right) = \frac{s_{31}}{s_1} + \frac{s_{32} \cdot s_{21}}{s_1^2 \cdot 2\pi i} \log x + \frac{u_3}{u_1} + f(x).$$

Now $S(\log^2 x) = \log^2 x + 4\pi i \log x + (2\pi i)^2$, and remembering the form of $f(x)$, it is clear that $\frac{u_3}{u_1}$ behaves like the function

$$c \log^2 x + \psi(x) \log x + \chi(x);$$

where c is a constant, and $\psi(x)$ and $\chi(x)$ are uniform functions. It is therefore plain that we must take

$$\frac{u_3}{u_1} = \chi(x) + \psi(x) \log x + c \log^2 x;$$

or, giving u_1 its above-found value,

(48) $$u_3 = x^{r_1} \{\phi_{31} + \phi_{32} \log x + \phi_{33} \log^2 x\};$$

where ϕ_{31}, ϕ_{32}, and ϕ_{33} are all uniform functions, and, in particular,

$$\phi_{33} = cx^{-r_1}u_1.$$

The law of formation of the successive integrals of this group is now evident. It may be enunciated as follows:

If s_1 is a root of the characteristic equation, of multiplicity λ, then, corresponding to this root, there exists a group of λ integrals,

$$u_1, \quad u_2, \quad \ldots, \quad u_\lambda,$$

having the properties set forth in equations (41), and they can be put into the following form, viz.,

$$
(49) \quad \begin{cases} u_1 = x^{r_1}\phi_{11}, \\ u_2 = x^{r_1}[\phi_{21} + \phi_{22} \log x], \\ u_3 = x^{r_1}[\phi_{31} + \phi_{32} \log x + \phi_{33} \log^2 x], \\ \cdot \quad \cdot \quad \cdot \quad \cdot \quad \cdot \quad \cdot \quad \cdot \quad \cdot \\ u_\lambda = x^{r_1}[\phi_{\lambda 1} + \phi_{\lambda 2} \log x + \ldots + \phi_{\lambda\lambda} \log^{\lambda-1} x]; \end{cases}
$$

where $r_1 = \dfrac{\log s_1}{2\pi i}$, and $\phi_{11}, \phi_{21}, \ldots, \phi_{\lambda\lambda}$ are uniform in the neighborhood of the point $x = 0$.

(I) The quantities ϕ are such that any one of them, ϕ_{ij}, where i is different from j, can be expressed linearly in terms of those whose second subscript is 1.

(II) $\phi_{11}, \phi_{22}, \ldots, \phi_{\lambda\lambda}$ differ from one another only by constant factors.

Assuming the above statements to hold good for the first $k - 1$ of the integrals $u_1, u_2, \ldots, u_\lambda$, they may be shown to hold good for the kth integral, u_k.

For, by equation (41),

$$
(50) \qquad Su_k = s_{k1}u_1 + s_{k2}u_2 + \ldots + s_{k,k-1}u_{k-1} + s_1u_k.
$$

Also, we may always take

$$
(51) \qquad u_k = x^{r_1}\{\phi_{k1} + \phi_{k2} \log x + \ldots + \phi_{kk} \log^{k-1} x\};
$$

$\phi_{k2}, \phi_{k3}, \ldots, \phi_{kk}$ being chosen at will, provided that ϕ_{k1} is properly determined.

Assume then that $\phi_{k2}, \phi_{k3}, \ldots, \phi_{kk}$ are uniform in the region of the point $x = 0$, and let ϕ_{k1} become ϕ'_{k1} when the variable moves round the critical point. Equation (51) thus changes to

$$
Su_k = s_1x^{r_1}\{\phi'_{k1} + \phi_{k2}(\log x + 2\pi i) + \ldots + \phi_{kk}(\log x + 2\pi i)^{k-1}\}.
$$

Expanding and dividing both members by $s_1x^{r_1}$, we have

$$
\frac{Su_k}{s_1x^{r_1}} =
$$

$$\phi_{kk}\log^{k-1}x + \left.\begin{array}{l} 2\pi i(k-1)\phi_{kk} \\ + \qquad \phi_{k,k-1} \end{array}\right| \log^{k-2}x + \left.\begin{array}{l} (2\pi i)^2\dfrac{k-1\,.\,k-2}{2\,!}\phi_{kk} \\ + 2\pi i(k-2)\phi_{k,k-1} \\ + \qquad \phi_{k,k-2} \end{array}\right| \log^{k-3}x$$

$$+ \left.\begin{array}{l} (2\pi i)^3\dfrac{k-1\,.\,k-2\,.\,k-3}{3\,!}\phi_{kk} \\ + (2\pi i)^2\dfrac{k-2\,.\,k-3}{2\,!}\phi_{k,k-1} \\ + 2\pi i(k-3)\phi_{k,k-2} \\ + \qquad \phi_{k,k-3} \end{array}\right| \log^{k-4}x$$

$$+ \ldots + \left.\begin{array}{l} (2\pi i)^{k-4}\dfrac{k-1\,.\,k-2\,.\,k-3}{3\,!}\phi_{kk} \\ + (2\pi i)^{k-5}\dfrac{k-2\,.\,k-3\,.\,k-4}{3\,!}\phi_{k,k-1} \\ + (2\pi i)^{k-6}\dfrac{k-3\,.\,k-4\,.\,k-5}{3\,!}\phi_{k,k-2} \\ + (2\pi i)^{k-7}\dfrac{k-4\,.\,k-5\,.\,k-6}{3\,!}\phi_{k,k-3} \\ \quad . \quad . \quad . \quad . \quad . \quad . \quad . \quad . \\ \quad . \quad . \quad . \quad . \quad . \quad . \quad . \quad . \\ + \qquad \phi_{k,4} \end{array}\right| \log^3 x$$

$$+ \left.\begin{array}{l} (2\pi i)^{k-3}\dfrac{k-1\,.\,k-2}{2\,!}\phi_{kk} \\ + (2\pi i)^{k-4}\dfrac{k-2\,.\,k-3}{2\,!}\phi_{k,k-1} \\ + (2\pi i)^{k-5}\dfrac{k-3\,.\,k-4}{2\,!}\phi_{k,k-2} \\ + (2\pi i)^{k-6}\dfrac{k-4\,.\,k-5}{2\,!}\phi_{k,k-3} \\ + \quad . \quad . \quad . \quad . \quad . \quad . \\ + \quad . \quad . \quad . \quad . \quad . \quad . \\ + (2\pi i) \qquad 3\phi_{k4} \\ + \qquad \phi_{k,3} \end{array}\right| \log^2 x$$

$$
\begin{array}{llr|l}
+ (2\pi i)^{k-2}(k-1)\phi_{kk} & & & \log x + (2\pi i)^{k-1}\phi_{kk} \\
+ (2\pi i)^{k-3}(k-2)\phi_{k,\,k-1} & & & \quad + (2\pi i)^{k-2}\phi_{k,\,k-1} \\
+ (2\pi i)^{k-4}(k-3)\phi_{k,\,k-2} & & & \quad + (2\pi i)^{k-3}\phi_{k,\,k-2} \\
+ (2\pi i)^{k-5}(k-4)\phi_{k,\,k-3} & & & \quad + (2\pi i)^{k-4}\phi_{k,\,k-3} \\
+ \quad . \quad . \quad . \quad . & & & \quad + \quad . \quad . \quad . \\
+ \quad . \quad . \quad . \quad . & & & \quad + \quad . \quad . \quad . \\
+ (2\pi i)^2 & & 3\phi_{k4} & \quad + (2\pi i)^3 \quad \phi_{k4} \\
+ 2\pi i & & 2\phi_{k3} & \quad + (2\pi i)^2 \quad \phi_{k3} \\
+ & & \phi_{k2} & \quad + 2\pi i \quad \phi_{k2} \\
 & & & \quad + \quad . \quad \phi'_{k1} .
\end{array}
$$

But by substituting from equation (50) in equation (51) we have

$$
\frac{Su_k}{x^{r_1}} =
$$

$$
s_1\phi_{kk}\log^{k-1}x + \left.\begin{array}{l} + s_{k,\,k-1}\phi_{k-1,\,k-1} \\ s_1\phi_{k,\,k-1} \end{array}\right| \log^{k-2}x + \left.\begin{array}{l} s_1\phi_{k,\,k-2} \\ + s_{k,\,k-1}\phi_{k-1,\,k-2} \\ + s_{k,\,k-2}\phi_{k-2,\,k-2} \end{array}\right| \log^{k-3}x
$$

$$
\left.\begin{array}{l} + s_1\phi_{k,\,k-3} \\ + s_{k,\,k-1}\phi_{k-1,\,k-3} \\ + s_{k,\,k-2}\phi_{k-2,\,k-3} \\ + s_{k,\,k-3}\phi_{k-3,\,k-3} \end{array}\right| \log^{k-4}x + \quad . \quad . \quad . \quad .
$$

$$
\left.\begin{array}{l} + s_1\phi_{k4} \\ + s_{k,\,k-1}\phi_{k-1,\,4} \\ + s_{k,\,k-2}\phi_{k-2,\,4} \\ + s_{k,\,k-3}\phi_{k-3,\,4} \\ + \quad . \quad . \quad . \\ + s_{k4}\phi_{44} \end{array}\right| \log^3 x + \left.\begin{array}{l} s_1\phi_{k3} \\ + s_{k,\,k-1}\phi_{k-1,\,3} \\ + s_{k,\,k-2}\phi_{k-2,\,3} \\ + s_{k,\,k-3}\phi_{k-3,\,3} \\ + \quad . \quad . \\ + s_{k4}\phi_{43} \\ + s_{k3}\phi_{33} \end{array}\right| \log^2 x
$$

$$
\begin{array}{l|l}
+ s_1\phi_{k2} & \log x + s_1\phi_{k1} \\
+ s_{k,\,k-1}\phi_{k-1,\,2} & \quad + s_{k,\,k-1}\phi_{k-1,\,1} \\
+ s_{k,\,k-2}\phi_{k-2,\,2} & \quad + s_{k,\,k-2}\phi_{k-2,\,1} \\
+ s_{k,\,k-3}\phi_{k-3,\,2} & \quad + s_{k,\,k-3}\phi_{k-3,\,1} \\
+ \ \cdot \quad \cdot \quad \cdot & \quad + \ \cdot \quad \cdot \quad \cdot \\
+ s_{k4}\phi_{42} & \quad + s_{k4}\phi_{41} \\
+ s_{k3}\phi_{32} & \quad + s_{k3}\phi_{31} \\
 & \quad + s_{k2}\phi_{21} \\
 & \quad + s_{k1}\phi_{11}
\end{array}
$$

We thus find two distinct expressions for $\dfrac{Su_k}{x^{r_1}}$ in terms of positive integral powers of $\log x$. Identifying the coefficients of the same powers of $\log x$ in the two expressions, the following equations are obtained:

$$
(52)\quad \left\{
\begin{array}{l}
s_1\phi_{kk} = s_1\phi_{kk}\,, \\
s_1[2\pi i(k-1)\phi_{kk} + \phi_{k,\,k-1}] = s_1\phi_{k,\,k-1} + s_{k,\,k-1}\phi_{k-1,\,k-1}, \\
s_1\left\{ (2\pi i)^2 \dfrac{k-1\,.\,k-2}{2\,!}\phi_{kk} + (2\pi i)(k-2)\phi_{k,\,k-1} + \phi_{k,\,k-2} \right\} \\
\qquad\qquad = s_1\phi_{k,\,k-2} + s_{k,\,k-1}\phi_{k-1,\,k-2} + s_{k,\,k-2}\phi_{k-2,\,k-2}\,. \\
\cdot \quad \cdot \quad \cdot \quad \cdot \quad \cdot \quad \cdot \quad \cdot \quad \cdot \quad \cdot \quad \cdot \quad \cdot \quad \cdot
\end{array}
\right.
$$

The whole number of these equations is k, but the first is only an identity. Observing that the $(k-1)^{\text{st}}$ equation is of the form

$$
s_1\{\,.\,.\,.\, + 2(2\pi i)\phi_{k3} + \phi_{k2}\} = s_1\phi_{k2} + s_{k,\,k-1}\phi_{k-1,\,2} + \cdot\cdot\cdot\,,
$$

the terms in ϕ_{k2} are seen to vanish identically; hence the 2d, 3d, . . . , $(k-1)^{\text{st}}$ equations, in number $k-2$, involve only the $k-2$ unknown quantities

$$
\phi_{kk}, \quad \phi_{k,\,k-1}, \quad \cdot\cdot\cdot\,, \quad \phi_{k3}\,,
$$

which can therefore be expressed in terms of ϕ's of lower index supposed already known.

The determinant of the left-hand members is

$$s_1^{k-2}\left|\begin{array}{cccccc} 2\pi i(k-1), & 0, & 0, & 0, \ldots, & 0, & 0 \\ (2\pi i)^2\dfrac{k-1\,.\,k-2}{2\,!}, & (2\pi i)(k-2,) & 0, & 0, \ldots, & 0, & 0 \\ (2\pi i)^3\dfrac{k-1\,.\,k-2\,.\,k-3}{3\,!}, & (2\pi i)^2\dfrac{k-2\,.\,k-3}{2\,!}, & 2\pi i(k-3), & 0, \ldots, & 0, & 0 \\ \cdot\ \cdot\ \cdot\ \cdot & \cdot\ \cdot & \cdot\ \cdot\ \cdot & & \cdot & \cdot \\ (2\pi i)^{k-4}\dfrac{k-1\,.\,k-2\,.\,k-3}{3\,!}, & (2\pi i)^{k-5}\dfrac{k-2\,.\,k-3\,.\,k-4}{3\,!} & & \ldots, & 0, & 0 \\ (2\pi i)^{k-3}\dfrac{k-1\,.\,k-2}{2\,!}, & (2\pi i)^{k-4}\dfrac{k-2\,.\,k-3}{2\,!} & & \ldots, & 3(2\pi i), & 0 \\ (2\pi i)^{k-2}(k-1), & (2\pi i)^{k-3}(k-2), & & \ldots, & 3(2\pi i)^2, & 2(2\pi i) \end{array}\right|,$$

of which the value is

$$s_1^{k-2}k-1\,.\,k-2\,.\,.\,.\,3\,.\,2\,.\,(2\pi i)^{k-2}=(2\pi i)^{k-2}s_1^{k-2}\,.\,(k-1)\,!$$

Replacing the columns of this determinant successively by

$$s_{k,\,k-1}\phi_{k-1,\,k-1},$$

$$s_{k,\,k-1}\phi_{k-1,\,k-2}+s_{k,\,k-2}\phi_{k-2,\,k-2},$$

$$\cdot\quad\cdot\quad\cdot\quad\cdot\quad\cdot\quad\cdot\quad\cdot\quad\cdot\quad\cdot$$

$$s_{k,\,k-1}\phi_{k-1,\,2}\quad+s_{k,\,k-2}\phi_{k-2,\,2}+\,.\,.\,.\,+s_{k,\,2}\phi_{22},$$

and dividing the results by $(2\pi i)^{k-2}\,.\,s_1^{k-2}\,.\,(k-1)!$, the values of $\phi_{kk},\ldots,\phi_{k,3}$ are obtained in order. In particular it is found that

$$\phi_{kk}=\frac{s_{k,\,k-1}\phi_{k-1,\,k-1}}{2\pi i(k-1)s_1},\tag{53}$$

which establishes proposition (II), since k may have any value from 2 to λ. Also,

(54) $$\phi_{32} = \begin{vmatrix} 2, & \frac{s_{21}}{s_1(2\pi i)}\phi_{22} \\ 2\pi i, & s_{32}\phi_{21} + s_{31}\phi_{11} \end{vmatrix} \div s_1(2\pi i)\,.\,2!\,.$$

But it is already known that $\phi_{22} = \frac{s_{21}\phi_{11}}{s_1(2\pi i)}$; hence ϕ_{32} is expressed in terms of ϕ_{11}, ϕ_{21}. Proceeding in the same manner, it may be seen that proposition (I) holds good throughout.

We have also the equation

$$\begin{aligned} s_1\{(2\pi i)^{k-1}\phi_{kk} + (2\pi i)^{k-2}\phi_{k,k-1} + \dots + (2\pi i)^3\phi_{k4} + (2\pi i)^2\phi_{k3} \\ + (2\pi i)\,\phi_{k2} + \phi'_{k1}\} \\ = s_1\phi_{k1} + s_{k,k-1}\phi_{k-1,1} + \dots + s_{k4}\phi_{41} + s_{k3}\phi_{31} + s_{k2}\phi_{21} + s_{k1}\phi_{11}. \end{aligned}$$

ϕ_{k2} is by hypothesis an arbitrary uniform function; it may therefore be so chosen as to satisfy the following equation:

$$\begin{aligned} s_1\{(2\pi i)^{k-1}\phi_{kk} + \dots + (2\pi i)^2\phi_{k3} + (2\pi i)\phi_{k2}\} \\ = s_{k,k-1}\phi_{k-1,1} + \dots + s_{k2}\phi_{21} + s_{k1}\phi_{11}. \end{aligned}$$

Hence we must have

(55) $$s_1\phi'_{k1} = s_1\phi_{k1};$$

that is, ϕ_{k1} is also a uniform function in the region of the point $x = 0$.

It is thus shown that if the law of formation of the integrals u_1, $u_2, \dots, u_\lambda$, as already stated, holds good for the first $k-1$ of them, it also holds good for the next following. But it was shown by direct investigation to hold for u_2 and u_3; it is therefore true for the whole group.

From the equation

$$\phi_{kk} = \frac{s_{k,k-1}}{2\pi i(k-1)s_1}\phi_{k-1,k-1}$$

we find at once that

(56) $$\phi_{kk} = \frac{s_{k,k-1}s_{k-1,k-2}\dots s_{21}}{(k-1)(k-2)\dots 1\,.\,s_1^{k-1}(2\pi i)^{k-1}}\phi_{11}.$$

In the group of integrals thus obtained, any one of them, u_k, may be replaced by a linear function of the integrals of lower index without altering the form it assumes after undergoing the substitution S; since the integrals involved will all be reproduced increased by linear functions of those of lower index only.*

The system of integrals (49) satisfying the conditions (41) and corresponding to the root s_1 of order of multiplicity λ is called by Fuchs a *group*. We have found that the uniform functions ϕ_{ij} can all be expressed linearly in terms of those whose second suffix is 1, and in particular that the functions $\phi_{11}, \phi_{22}, \ldots, \phi_{\lambda\lambda}$ differ only by constant factors.

Corresponding now to any linear differential equation with uniform coefficients, we have a certain number of distinct groups of integrals corresponding to each critical point. Corresponding to the point $x = 0$, suppose the characteristic equation has k distinct roots $s_1, s_2, \ldots, s_k$, of orders of multiplicity $\lambda_1, \lambda_2, \ldots, \lambda_k$, respectively; then $\lambda_1 + \lambda_2 + \ldots + \lambda_k = n$, and there are k groups of integrals of the form given in equations (49). By aid of the theorem proved above concerning the vanishing of the minors of Δ and Δ_1, Hamburger has shown how each of these groups gives rise to certain sets of *sub-groups* which are independent of one another, and such that the linear relations connecting the uniform functions above denoted by ϕ_{ij} (*i.e.*, the coefficients of the different powers of the logarithm) exist for each sub-group separately. In other words, there is *no* relation connecting any of the uniform functions of one sub-group with those of another sub-group. Hamburger's work (Crelle, vol. 76) is an extension of a process due to Jordan (*Comptes Rendus*, t. lxxiii). Floquet (*Annales d'École Normale Supérieure*, 2e série, t. 12) has also used Hamburger's results in connection with linear differential equations with periodic coefficients. Hamburger's determination of the sub-groups of integrals which immediately follows is taken with but very slight alterations from his paper.

We have $y_1, \ldots, y_n$ for a fundamental system of integrals, and for any other integral

$$(57) \qquad u = \alpha_1 y_1 + \ldots + \alpha_n y_n;$$

---

* The preceding verification of (I) and (II) was worked out by Mr. C. H. Chapman.

for the integrals under consideration we have $Su = s_1 u$, and for the equations determining the ratios of $\alpha_1, \ldots, \alpha_n$,

$$(58)\quad \begin{cases} (c_{11} - s_1)\alpha_1 + c_{21}\alpha_2 + \ldots + c_{n1}\alpha_n = 0, \\ c_{12}\alpha_1 + (c_{22} - s_1)\alpha_2 + \ldots + c_{n2}\alpha_n = 0, \\ \cdot \quad \cdot \quad \cdot \quad \cdot \quad \cdot \quad \cdot \quad \cdot \\ c_{1n}\alpha_1 + c_{2n}\alpha_2 + \ldots + (c_{nn} - s_1)\alpha_n = 0. \end{cases}$$

If we had chosen $v_1, \ldots, v_n$ as our set of fundamental integrals, we should have (equations 23 and 24)

$$(59)\quad \begin{cases} (\gamma_{11} - s_1)\alpha_1 + \gamma_{21}\alpha_2 + \ldots + \gamma_{n1}\alpha_n = 0, \\ \gamma_{12}\alpha_1 + (\gamma_{22} - s_1) + \ldots + \gamma_{n2}\alpha_n = 0, \\ \cdot \quad \cdot \quad \cdot \quad \cdot \quad \cdot \quad \cdot \quad \cdot \\ \gamma_{1n}\alpha_1 + \gamma_{2n}\alpha_2 + \ldots + (\gamma_{nn} - s_1)\alpha_n = 0. \end{cases}$$

The characteristic equations $\Delta = 0$ belonging to the system $y_1, \ldots, y_n$, and $\Delta_1 = 0$ belonging to the system $v_1, \ldots, v_n$, have the same roots. Now from what has been shown above (equations 33 *et seq.*) it is easy to see that there are the same number of independent equations in the system (59) as in the system (58). For, the necessary and sufficient condition that $n - \nu + 1$, and no more, equations of either system are dependent upon the remaining equations of the system is that all the minors of order ν of the corresponding determinant, Δ or Δ_1, shall vanish, while all those of order $\nu - 1$ shall not vanish—a condition which, as we have proved, is simultaneously satisfied by both Δ and Δ_1. Since s_1 is a root of $\Delta = 0$, it follows that at least one of the system of equations (58) is a consequence of the remaining ones. There may, however, be more than one of the equations (58) which are dependent upon the remaining equations of the system. Suppose, for example, that there are $n - \nu$ independent equations. Now, since the choice of the system of fundamental integrals does not at all affect the characteristic equation, it follows that the number of independent equations in (58) and (59) is the same; that is, the number of independent equations is the same whatever set of fundamental integrals is employed. Assuming now $n - \nu$ independent equations in the system (58), we can determine $n - \nu$ of the constants $\alpha_1, \ldots, \alpha_n$ in terms of the remaining ones, or, what is the same thing, we can determine all of the constants $\alpha_1, \ldots, \alpha_n$ as linear homogeneous functions* of ν arbitrary constants. It follows then that all of the integrals which satisfy the relation $Su = s_1 u$ are linearly expressible in terms of ν linearly independent functions which we will denote by u_1, $u_2, \ldots, u_\nu$. We may replace now ν of the fundamental integrals $y_1, \ldots, y_n$ by $u_1, \ldots, u_\nu$—say the new fundamental system is

$$u_1, u_2, \ldots, u_\nu, y_{\nu+1}, \ldots, y_n;$$

and then

(60) $S =$

$$\left| \begin{array}{ll} u_1, u_2, \ldots, u_\nu & ; s_1 u_1, s_1 u_2, \ldots, s_1 u_\nu \\ y_{\nu+1} & ; c_{\nu+1,1} u_1 + \ldots + c'_{\nu+1,\nu} u_\nu + c'_{\nu+1,\nu+1} y_{\nu+1} + \ldots + c'_{\nu+1,n} y_n \\ \vdots & \vdots \\ y_n & ; c'_{n1} u_1 + \ldots + c'_{n\nu} u_\nu + c'_{n,\nu+1} y_{\nu+1} + \ldots + c'_{nn} y_n \end{array} \right|.$$

The characteristic equation becomes now

$$\text{(61)} \qquad \Delta = (s - s_1)^\nu \, \Delta' = 0;$$

where

$$\text{(62)} \qquad \Delta' = \left| \begin{array}{ccc} c'_{\nu+1,\nu+1} - s, & \ldots, & c'_{n,\nu+1} \\ \cdot \quad \cdot & \cdot \quad \cdot & \cdot \quad \cdot \\ \vdots & & \vdots \\ c'_{\nu+1,n}, & \ldots, & c'_{nn} - s \end{array} \right|.$$

As the characteristic equation is independent of the choice of the fundamental system of integrals, it follows that equation (61) has the same roots as the original equation $\Delta = 0$; and if s_1 is a root of multiplicity λ, we must have $\nu \leqq \lambda$. If $\nu = \lambda$, then $u_1, \ldots, u_\nu$ are *all* the integrals associated with the root s_1 and satisfying the relation $Su = s_1 u$. We have in this case $\nu_1 = \lambda$, sub-groups of integrals

* As we have principally to do with *linear homogeneous* functions, it will be convenient to drop the word "homogeneous," so that a "linear function" of any set of quantities will be understood to mean a "linear homogeneous function" of those quantities.

each containing one member. If, however, $\nu < \lambda$, then obviously s_1 is again a root of the equation $\Delta' = 0$. This equation implies the existence of a system of linear equations of the form

$$(63) \quad \begin{cases} \alpha_1'(c'_{\nu+1,\nu+1} - s_1) + \ldots + \alpha'_{n-\nu}c'_{n,\nu+1} & = 0, \\ \cdot \quad \cdot \quad \cdot \quad \cdot \quad \cdot \quad \cdot \quad \cdot \quad \cdot \quad \cdot \quad \cdot & \\ \alpha_1'c_{\nu+1,n} \qquad\qquad + \ldots + \alpha'_{n-\nu}(c'_{nn} - s_1) & = 0, \end{cases}$$

which serve to determine the ratios of the constants $\alpha_1', \ldots, \alpha'_{n-\nu}$. If we multiply the $n - \nu$ equations corresponding to the last $n - \nu$ lines of the substitution (60), we have, by aid of (63), for the new integral sought the equation

$$(64) \qquad v = \alpha_1'y_{\nu+1} + \ldots + \alpha'_{n-\nu}y_n,$$

and this from what precedes is easily seen to satisfy the relation

$$(65) \qquad Sv = s_1v + U;$$

where U is a linear function of $u_1, u_2, \ldots, u_\nu$. If now we find that there are only $n - \nu - \nu'$ independent equations in the system (63), then it follows that we can determine the $n - \nu$ constants $\alpha_1', \ldots, \alpha'_{n-\nu}$ as linear functions of ν' perfectly arbitrary constants, and therefore there will exist ν' linearly independent integrals satisfying the relation (65).

Say these integrals are $v_1, v_2, \ldots, v_{\nu'}$; then

$$(66) \quad Sv_1 = s_1v_1 + U_1, \quad Sv_2 = s_1v_2 + U_2, \quad \ldots, \quad Sv_{\nu'} = s_1v_{\nu'} + U_{\nu'}.$$

where $U_1, U_2, \ldots, U_{\nu'}$ are linear functions of $u_1, u_2, \ldots, u_\nu$. Every other integral satisfying the relation (65) will be a linear function of

$$u_1, \quad u_2, \quad \ldots, \quad u_\nu, \quad v_1, \quad v_2, \quad \ldots, \quad v_{\nu'}.$$

Further, the functions $U_1, U_2, \ldots, U_{\nu'}$ are linearly independent; for, if they were not, it would be possible to form a linear function of $v_1, v_2, \ldots, v_{\nu'}$, say V, which would satisfy the relation $SV = s_1 V$; that is, there would be more than ν linearly independent integrals satisfying this relation, which is contrary to hypothesis. It follows also that ν' cannot be greater than ν, since between any $\nu + 1$ of the functions U_i there necessarily exists a linear relation. Finally, since U_i is a linear function of $u_1, \ldots, u_\nu$, and since each of these functions satisfies the relations $Su_i = s_1 u_i$, we have $SU_i = s_1 U_i$, and writing

$$\overline{U} = \zeta_1 U_1 + \ldots \zeta_{\nu'} U_{\nu'},$$

where $\zeta_1, \ldots, \zeta_{\nu'}$ are constants,

$$S\overline{U} = s_1 \overline{U}.$$

If now $\nu + \nu' = \lambda$, then the λ integrals corresponding to the root s_1 of the characteristic equation are divided into ν' sub-groups of two elements each, and $\nu - \nu'$ sub-groups of one element each; viz., the sub-groups of two elements are

$$(67) \quad \begin{cases} (v_1, \ U_1) \text{ satisfying } Sv_1 = s_1 v_1 + U_1, \ SU_1 = s_1 U_1, \\ \quad \cdot \quad \cdot \quad \cdot \quad \cdot \quad \cdot \quad \cdot \quad \cdot \quad \cdot \\ (v_{\nu'}, \ U_{\nu'}) \quad \ldots \quad Sv_{\nu'} = s_1 v_{\nu'} + U_{\nu'}, \ SU_{\nu'} = s_1 U_{\nu'}; \end{cases}$$

and the sub-groups of one element each are the remaining $\nu - \nu'$ linear functions of $u_1, u_2, \ldots, u_\nu$, which have no linear relation among themselves or with $U_1, \ldots, U_{\nu'}$. There will obviously be no loss of generality if we replace $U_1, \ldots, U_{\nu'}$ simply by $u_1, \ldots, u_{\nu'}$, and the remaining $\nu - \nu'$ functions by $u_{\nu'+1}, \ldots, u_\nu$; denoting then by II the sub-groups of two elements each, and by I the sub-groups of one element each, we have

$$(68) \quad \begin{cases} \text{II} = (u_1, v_1), \quad (u_2, v_2), \quad \ldots, \quad (u_{\nu'}, v_{\nu'}), \\ \text{I} = u_{\nu'+1}, \quad u_{\nu'+2}, \quad \ldots, \quad u_\nu; \end{cases}$$

where $$Sv_i = s_1 v_i + u_i, \quad Su_i = s_1 u_i.$$

Suppose now that $\nu + \nu' < \lambda$; then we can choose for our fundamental system

$$u_1, \ldots, u_\nu, \quad v_1, \ldots, v_{\nu'}, \quad y_{\nu+\nu'+1}, \ldots, y_n;$$

for which

(69) $S =$

$$\left|\begin{matrix} u_1, \ldots, u_\nu ; s_1 u_1, \ldots, s_1 u_\nu \\ v_1, \ldots, v_{\nu'} ; s_1 v_1 + u_1, \ldots, s_1 v_{\nu'} + u_{\nu'} \\ y_{\nu+\nu'+1} ; c''_{\nu+\nu'+1\,1} u_1 + \ldots + c''_{\nu+\nu'+1,\nu} u_\nu + c''_{\nu+\nu'+1,\nu+1} v_1 + \ldots + c''_{\nu+\nu'+1,n} y_n \\ \cdot \quad \cdot \quad \cdot \quad \cdot \quad \cdot \quad \cdot \quad \cdot \quad \cdot \\ y_n ; \quad c''_{n1} u_1 + \ldots + c''_{n\nu} u_\nu + c''_{n,\nu+1} v_1 + \ldots + c''_{nn} y_n \end{matrix}\right| .$$

The characteristic equation is now

(70) $$\Delta = (s - s_1)^{\nu+\nu'} \Delta'' = 0 ;$$

where

(71) $$\Delta'' = \left|\begin{matrix} c''_{\nu+\nu'+1,\nu+\nu'+1} - s_1 , & \ldots , & c''_{n,\nu+\nu'+1} \\ \cdot & \cdot & \cdot \\ c''_{\nu+\nu'+1,n} & \ldots , & c''_{nn} - s \end{matrix}\right| .$$

Since $\nu + \nu' < \lambda$, it follows that s_1 is again a root of $\Delta'' = 0$, and this equation implies the existence of the system

(72) $$\begin{cases} \alpha_1''(c''_{\nu+\nu'+1,\nu+\nu'+1} - s_1) + \ldots + \alpha''_{n-\nu-\nu'} c''_{n,\nu+\nu'+1} = 0, \\ \cdot \quad \cdot \quad \cdot \quad \cdot \quad \cdot \quad \cdot \quad \cdot \quad \cdot \\ \alpha_1'' c_{\nu+\nu'+1,n} + \ldots + \alpha''_{n-\nu-\nu'}(c''_{nn} - s_1) = 0, \end{cases}$$

serving to determine the ratios of $\alpha_1'', \ldots, \alpha''_{n-\nu-\nu'}$. Multiplying the $n - \nu - \nu'$ equations corresponding to the last $n - \nu - \nu'$ rows of (69) by $\alpha_1'', \ldots, \alpha''_{n-\nu-\nu'}$, and adding, we have for the new function

(73) $$w = \alpha_1'' y_{\nu+\nu'+1} + \ldots + \alpha''_{n-\nu-\nu'} y_n ,$$

which is linearly independent of the u's and the v's, and which satisfies the relation

(74) $$Sw = s_1 w + \Gamma ;$$

where Γ is a linear function of $u_1, \ldots, u_\nu, v_1, \ldots, v_{\nu'}$. Suppose now that there are only $n - \nu - \nu' - \nu''$ independent equations in the system (72). Then all of the $n - \nu - \nu'$ constants, α_1'', . . . , $\alpha''_{n-\nu-\nu'}$, are linearly expressible in terms of ν'' arbitrary constants, and consequently there exist ν'' linearly independent functions w_1, . . . , $w_{\nu''}$ which satisfy the relation (74), and every other function satisfying this relation is a linear function of (u, v, w). For these functions w we have

$$(75) \qquad Sw_1 = s_1 w_1 + r_1, \ldots, Sw_{\nu''} = s_1 w_{\nu''} + r_{\nu''}.$$

There can exist no linear relations between the functions r_i of the form

$$(76) \qquad \zeta_1 r_1 + \ldots + \zeta_{\nu''} r_{\nu''} = U;$$

where ζ_1, . . . , $\zeta_{\nu''}$ are constants, and U is a linear function of u_1, . . . , u_ν; for if there could be such a relation, it would also be possible to find a linear relation connecting the functions w_1, . . . , $w_{\nu''}$, say W, alone and satisfying

$$(77) \qquad SW = s_1 W + U';$$

where U' is a linear function of u_1, . . . , u_ν. But by hypothesis the linearly independent functions v_1, . . . , $v_{\nu'}$ are the only ones satisfying such a relation, and consequently we can have no such relation as (76). It follows also that ν'' cannot be greater than ν', and consequently cannot be greater than ν.

Suppose now that $\nu + \nu' + \nu'' = \lambda$; we can obviously without loss of generality replace r_1, . . . , $r_{\nu''}$ by v_1, . . . , $v_{\nu''}$, etc. We find then that the integrals corresponding to the root s_1 of the characteristic equation divide into ν'' sub-groups of three elements each, $\nu' - \nu''$ sub-groups of two elements each, and $\nu - \nu'$ sub-groups of one element each. Denote these by III, II, I, respectively; then

$$(78) \qquad \begin{cases} \text{III} = (u_1, v_1, w_1) & \ldots (u_{\nu''}, v_{\nu''}, w_{\nu''}), \\ \text{II} = (u_{\nu''+1}, v_{\nu''+1}) & \ldots (u_{\nu'}, v_{\nu'}), \\ \text{I} = (u_{\nu'+1}) & \ldots (u_\nu), \end{cases}$$

giving the relations

$$(79)\quad \begin{cases} \text{from III, } Su_1 = s_1u_1,\ Sv_1 = s_1v_1 + u_1,\ Sw_1 = s_1w_1 + v_1, \\ \quad . \quad . \quad . \quad . \quad . \quad . \quad . \quad . \quad . \quad . \\ \qquad Su_{\nu''} = s_1u_{\nu''},\ Sv_{\nu''} = s_1v_{\nu''} + u_{\nu''},\ Sw_{\nu''} = s_1w_{\nu''} + v_{\nu''}; \\ \text{from II, } Su_{\nu''+1} = s_1u_{\nu''+1},\ Sv_{\nu''+1} = s_1v_{\nu''} + u_{\nu''}, \\ \quad . \quad . \quad . \quad . \quad . \quad . \quad . \quad . \quad . \quad . \\ \qquad Su_{\nu'} = s_1u_{\nu'},\ Sv_{\nu'} = s_1v_{\nu'} + u_{\nu'}; \\ \text{from I, } Su_{\nu'+1} = s_1u_{\nu'+1},\ .\ .\ .\ ,\ Su_\nu = s_1u_\nu. \end{cases}$$

By a continuation of this process we find a series of numbers, ν, ν', ν'', . . . , $\nu^{(k)}$, such that

$$\nu + \nu' + \ldots + \nu^{(k)} = \lambda,$$

and where no ν is greater than the preceding one. For $\nu^{(\lambda)}$ we shall have

$\nu^{(k)}$	sub-groups of $k + 1$ elements each,	
$\nu^{(k-1)} - \nu^{(k)}$	" " k elements each,	
⋮		
$\nu' - \nu''$	" " 2 " "	
$\nu - \nu'$	" " 1 element each.	

The elements of a group containing m elements are, say, y_1, y_2, . . . , y_m; then

$$Sy_1 = s_1y_1,\quad Sy_2 = s_1y_2 + y_1,\quad \ldots,\quad Sy_m = s_1y_m + y_{m-1}.$$

The substitution S corresponding to the point $x = 0$ now takes the form

$$(80)\quad S = \left| \begin{matrix} y_1, y_2, \ldots, y_a; & s_1y_1, s_1y_2 + y_1, & \ldots, s_1y_a + y_{a-1} \\ y_1', y_2', \ldots, y'_{a'}; & s_1y_1', s_1y_2' + y_1', & \ldots, s_1y'_{a'} + y'_{a'-1} \\ . \quad . \quad . \quad . & . \quad . \quad . \quad . & . \quad . \quad . \quad . \\ y_1^{(i)}, y_2^{(i)} \ldots, y^{(i)}_{a^i}; & s_1y_1^{(i)}, s_1y_2^{(i)} + y_1^{(i)}, & \ldots, s_1y^{(i)}_{a^i} + y^{(i)}_{a^i-1} \\ z_1, z_2, \ldots, z_\beta; & s_2z_1, s_2z_2 + z_1, & \ldots, s_2z_\beta + z_{\beta-1} \\ . \quad . \quad . \quad . & . \quad . \quad . \quad . & . \quad . \quad . \quad . \end{matrix} \right|$$

Jordan in his *Cours d'Analyse* (vol. iii. p. 175) gives S a slightly different form and one which is a trifle more convenient. Jordan's form is derived simply from (79), viz.: Write

$$\begin{array}{llll} Y_1 = y_1, & Y_2 = s_1 y_2, & Y_3 = s_1^2 y_3, \ldots, & Y_\alpha = s_1^{\alpha-1} y_\alpha, \\ Y_1' = y_1', & Y_2' = s_1 y_2', & Y_3' = s_1 y_3', \ldots, & Y_{\alpha'} = s_1^{\alpha'-1} y_{\alpha'}, \\ \cdot & \cdot & \cdot & \cdot \\ Z_1 = z_1, & Z_2 = s_2 z_2, & Z_3 = s_2^2 z_3, \ldots, & Z_\beta = s_2^{\beta-1} z_\beta, \\ \cdot & \cdot & \cdot & \cdot \end{array}$$

We have now

$$(80') \quad S = \left| \begin{array}{l} Y_1, Y_2, \ldots; s_1 Y_1, s_1(Y_2 + Y_1), \ldots, s_1(Y_\alpha + Y_{\alpha-1}) \\ Y_1', Y_2', \ldots; s_1 Y_1', s_1(Y_2' + Y_1'), \ldots, s_1(Y_{\alpha'} + Y_{\alpha'-1}) \\ \cdot \quad \cdot \quad \cdot \quad \cdot \quad \cdot \quad \cdot \\ Z_1, Z_2, \ldots; s_2 Z_1, s_2(Z_2 + Z_1), \ldots, s_2(Z_\beta + Z_{\beta-1}) \\ \cdot \quad \cdot \quad \cdot \quad \cdot \quad \cdot \quad \cdot \end{array} \right| .$$

Before going on with this subject it will be convenient to take up the case of differential equations with uniform doubly periodic coefficients, and give Jordan's method for reducing the substitutions S and S' above referred to to their respective canonical forms. Assuming $f_1(x), \ldots, f_n(x)$ as a system of fundamental integrals and letting S and S' denote the substitutions arising from the change of x into $x + \omega$ and $x + \omega'$ respectively, we have seen that

$$(a) \qquad S = \left| \begin{array}{l} f_1;\ \alpha_{11} f_1 + \ldots + \alpha_{1n} f_n \\ \cdot \quad \cdot \quad \cdot \quad \cdot \quad \cdot \\ f_n;\ \alpha_{n1} f_1 + \ldots + \alpha_{nn} f_n \end{array} \right| ,$$

and

$$(b) \qquad S' = \left| \begin{array}{l} f_1;\ \alpha'_{11} f_1 + \ldots + \alpha'_{1n} f_n \\ \cdot \quad \cdot \quad \cdot \\ f_n;\ \alpha'_{n1} f_1 + \ldots + \alpha'_{nn} f_n \end{array} \right| ,$$

and also

$$(c) \qquad SS' = S'S.$$

We can by a method entirely similar to the preceding one reduce the substitutions S and S' to their canonical forms. Let s denote a root of the characteristic equation corresponding to S, and let $y_1, y_2, \ldots$ denote those independent integrals of the differential equation which are multiplied by s when the substitution S is made, *i.e.* when x is changed into $x + \omega$. The general form of the integrals possessing this property is then

$$\alpha_1 y_1 + \alpha_2 y_2 + \ldots.$$

Suppose that on applying the substitution S' to y_1 we change y_1 into Y_1; now the substitution SS' changes y_1 into sY_1, and $S'S$ must produce the same result; but S' changes y_1 into Y_1, and so S should change Y_1 into sY_1; it follows then that Y must have the form

$$\alpha_1 y_1 + \alpha_2 y_2 + \ldots.$$

The substitution S' thus replacing each of the integrals $y_1, y_2, \ldots$ by linear functions of these same integrals, there must exist at least one linear function, say u, of these integrals which S' changes into $s'u$. We have thus shown that there exists at least one integral, u, which the substitutions S and S' change into su and $s'u$ respectively.

We proceed now to show that it is always possible to find a set of fundamental integrals, say

$$\begin{matrix} y_{11}, \ldots, y_{1l_1}, & y_{21}, \ldots, y_{2l_2}, & \ldots, y_{\lambda 1}, \ldots, y_{\lambda l_\lambda}, \\ z_{11}, \ldots, z_{1m_1}, & z_{21}, \ldots, z_{2m_2}, & \ldots, z_{\mu 1}, \ldots, z_{\mu l_\mu}, \\ \cdot \quad \cdot \quad \cdot & \cdot \quad \cdot \quad \cdot & \cdot \quad \cdot \quad \cdot \end{matrix}$$

such that the substitutions S and S' shall take the forms

$$(d)\ S = \left| \begin{matrix} y_{1k}, \ldots, y_{ik}, \ldots; & s_1 y_{1k}, \ldots, s_1(y_{ik} + Y_{ik}), \ldots \\ z_{1k}, \ldots, z_{ik}, \ldots; & s_2 z_{1k}, \ldots, s_2(z_{ik} + Z_{ik}), \ldots \\ \cdot \quad \cdot \quad \cdot & \cdot \quad \cdot \quad \cdot \end{matrix} \right|,$$

$$(e)\quad S' = \begin{vmatrix} y_{1k}, & \ldots, & y_{ik}, & \ldots; & s_1'y_{1k}, & \ldots, & s_1'(y_{ik}+Y'_{ik}), \ldots \\ z_{1k}, & \ldots, & z_{ik}, & \ldots; & s_2'z_{1k}, & \ldots, & s_2'(z_{ik}+Z'_{ik}), \ldots \\ \cdot & \cdot & \cdot & \cdot & \cdot & \cdot & \cdot \end{vmatrix};$$

where (s_1, s_1'), (s_2, s_2'), . . . are pairs of different constants, *i.e.*, s_1 is never equal to s_2, s_1' is never equal to s_2', etc.; and where Y_{ik}, Y'_{ik} are linear functions of the integrals y whose first suffix is less than i; Z_{ik}, Z'_{ik} are linear functions of the integrals z, whose first suffix is less than i; etc. Assuming this proposition true for substitutions containing less than n variables, we will prove it to be true of the substitutions S and S' containing n variables. We have seen that there always exists an integral u which is changed into su and $s'u$ by the substitutions S and S' respectively. In the system of fundamental integrals $f_1, f_2, \ldots, f_n$ let us replace any one, say f_n, by the integral u. The substitutions S and S' then take the forms

$$(f)\quad S = |f_1, \ldots, f_{n-1}, u; F_1 + a_1 u, \ldots, F_{n-1} + a_{n-1}u, su|,$$

$$(g)\quad S' = |f_1, \ldots, f_{n-1}, u; F'_1 + a'_1 u, \ldots, F'_{n-1} + a'_{n-1}u, s'u|,$$

the functions F_i and F_i' denoting linear functions of $f_1, \ldots, f_{n-1}$.

Consider now the substitutions

$$(h)\qquad \Sigma = |f_1, \ldots, f_{n-1}; \ F_1, \ldots, F_{n-1}|,$$

$$(i)\qquad \Sigma' = |f_1, \ldots, f_{n-1}; \ F'_1, \ldots, F'_{n-1}|.$$

of $n - 1$ variables. From the relation

$$SS' = S'S$$

follows at once

$$(j)\qquad \Sigma\Sigma' = \Sigma'\Sigma.$$

Applying now the above-stated theorem to these substitutions, they may be written in the forms (d) and (e), $n - 1$ variables appearing instead of n. This same change of independent integrals will put S and S' in the forms

$$(k)\quad S = \begin{vmatrix} y_{1k}, \ldots, y_{ik}, \ldots; s_1y_{1k}+c_{1k}u, \ldots, s_1(y_{ik}+Y_{ik})+c_{ik}u, \ldots \\ z_{1k}, \ldots, z_{ik}, \ldots; s_2z_{1k}+d_{1k}u, \ldots, s_2(z_{ik}+Z_{ik})+d_{ik}u, \ldots \\ \cdot \quad \cdot \quad \cdot \quad \cdot \quad \cdot \quad \cdot \quad \cdot \quad \cdot \quad \cdot \quad \cdot \quad \cdot \\ u \qquad ; su \end{vmatrix},$$

$$(l)\quad S' = \begin{vmatrix} y_{1k}, \ldots, y_{ik}, \ldots; s'_1y_{1k}+c'_{1k}u, \ldots, s'_1(y_{ik}+Y'_{ik})+c'_{ik}u, \ldots \\ z_{1k}, \ldots, z_{ik}. \ldots; s'_2z_{1k}+d'_{1k}u, \ldots, s'_2(z_{ik}+Z'_{ik})+d'_{ik}u, \ldots \\ \cdot \quad \cdot \quad \cdot \quad \cdot \quad \cdot \quad \cdot \quad \cdot \quad \cdot \quad \cdot \quad \cdot \quad \cdot \\ u \qquad ; s'u \end{vmatrix}.$$

Suppose now we replace the independent integrals y_{ik} by the following:

$$(m)\qquad y'_{ik} = y_{ik} + \alpha_{ik}u;$$

the substitutions S and S' will retain their forms; but the constants c_{ik} and c'_{ik} will be changed into $[c_{ik}]$ and $[c'_{ik}]$, where

$$(n)\qquad [c_{ik}] = (s - s_1)\alpha_{ik} - s_1H_{ik} + c_{ik},$$

$$(o)\qquad [c'_{ik}] = (s - s'_1)\alpha_{ik} - s'_1H'_{ik} + c'_{ik};$$

H_{ik} and H'_{ik} denoting what Y_{ik} and Y'_{ik} respectively become when in them we replace the functions y by the corresponding constants α.

We will now first assume that s is not equal to s_1; in this case we can clearly assign such values to the constants α that all the new coefficients $[c_{ik}]$ shall vanish. It is easy to see, by aid of the relation $SS' = S'S$, that the vanishing of these constants will involve the vanishing of the constants $[c'_{ik}]$. Equating the coefficients of u in the expressions

$$SS'y_{ik} \qquad \text{and} \qquad S'Sy_{ik},$$

and denoting by Γ_{ik} and Γ'_{ik} what Y_{ik} and Y'_{ik} become when the integrals y are replaced by the corresponding constants c and c', we have

$$(p)\qquad s_1(c'_{ik} + \Gamma_{ik}) + s'c_{ik} = s_1'(c_{ik} + \Gamma_{ik}) + sc'_{ik}.$$

If now the constants c_{ik} are all zero these relations (p) reduce to the form

$$(q) \qquad (s_1 - s)c'_{ik} + s_1\Gamma'_{ik} = 0.$$

These equations (q) are linear and homogeneous in the quantities c'_{ik}, and their determinant is a power of $s_1 - s$; but since by hypothesis s_1 is not equal to s, this determinant cannot vanish, and therefore, in order that equations (q) may be satisfied, we must have all of the quantities c'_{ik} equal zero. In the same way, if s' is not equal to s_1' we can make all the constants c'_{ik} vanish, and their vanishing will also involve the vanishing of all the constants c_{ik}. Continuing this process, suppose that none of the relations

$$s,\ s' = s_1\,,\ s_1',\quad s,\ s' = s_2\,,\ s_2',\ \ldots$$

are satisfied; then we can make all of the constants c_{ik}, c'_{ik}; d_{ik}, d'_{ik}; . . . , disappear, and so the substitutions S and S' will be in the canonical form, and to the different classes of integrals y, z, . . . we have added the class composed of one integral only, viz., u.

Suppose now that $s = s_1$, $s' = s_1'$; as before we can make all the coefficients d_{ik}, d'_{ik}; . . . , vanish. If now we write

$$c_{ik} = s_1\,\gamma_{ik},\quad c'_{ik} = s_1'\gamma'_{ik},$$

we will again have the substitutions S and S' in the normal form, the new integral u entering now into the category of integrals y_{1k} belonging to the class of integrals y, which have unity for their first suffix, *i.e.*, the class which the substitution S multiplies by s, and which the substitution S' multiplies by s'.

We resume now our original problem of determining the forms of the integrals of the linear differential equation with uniform coefficients. Starting from equation (80′), Hamburger proceeds to determine the forms of the integrals; but Jordan gives a briefer and rather more elegant shape to Hamburger's method, so we shall employ it. Write as before

$$\frac{\log s}{2\pi i} = r$$

(dropping the subscripts for convenience). The substitution S being in the canonical form (80′), let $y_0, y_1, \ldots, y_k$ denote a group of integrals to which S applies, and let s denote the corresponding root of $\Delta = 0$. Write

$$y_0 = x^r z_0, \quad y_1 = x^r z_1, \quad \ldots, \quad y_k = x^r z_k.$$

Since after once turning round $x = 0$, x^r reproduces itself multiplied by $e^{2\pi i r}$, $= s$, it follows that the functions z submit to the substitution

$$\Sigma = | z_0, \ldots, z_k; \quad z_0, \ldots, z_k + z_{k-1} | ; \tag{81}$$

since

$$Sy_0 = sy_0, \quad Sy_1 = s(y_1 + y_0), \quad \ldots, \quad Sy_k = s(y_k + y_{k-1}).$$

It is only necessary now to find the forms of $z_0, z_1, \ldots, z_k$. The first, z_0, is obviously a uniform function, since $Sz_0 = z_0$. To get the forms of $z_1, \ldots, z_k$ we introduce a new function, θ_1, defined by

$$\theta_1 = \frac{\log x}{2\pi i}. \tag{82}$$

The effect of turning once round $x = 0$, that is, of applying the substitution S, is

$$S\theta_1 = \theta_1 + 1. \tag{83}$$

Introduce now the series of functions $\theta_1, \theta_2, \ldots, \theta_k$, defined by the equations

$$\theta_1 = \frac{\log x}{2\pi i}, \quad \ldots, \quad \theta_k = \frac{\theta_1(\theta_1 - 1) \ldots (\theta_1 - k + 1)}{1 \cdot 2 \ldots k}; \tag{84}$$

for these we have, as the result of turning once round $x = 0$,

$$S\theta_1 = \theta_1 + 1, \quad \ldots, \quad S\theta_k = \frac{(\theta_1 + 1)\,\theta_1(\theta_1 - 1) \ldots (\theta_1 - k + 2)}{1 \cdot 2 \ldots k}. \tag{85}$$

Add and subtract θ_k in $S\theta_k$; then for the right-hand member we have

$$(86)\quad \theta_k + \frac{(\theta_1+1)\theta_1(\theta_1-1)\dots(\theta_1-k+2)}{1.2\dots k} - \frac{\theta_1(\theta_1-1)\dots(\theta_1-k+1)}{1.2\dots k}$$

$$= \theta_k + [(\theta_1+1)-(\theta_1-k+1)]\frac{\theta_1(\theta_1-1)\dots(\theta_1-k+2)}{1.2\dots k}$$

$$= \theta_k + \frac{\theta_1(\theta_1-1)\dots(\theta_1-k+2)}{1.2\dots k-1} = \theta_k + \theta_{k-1}.$$

We have then finally for the functions θ,

$$(87)\qquad S\theta_1 = \theta_1 + 1,\quad S\theta_2 = \theta_2 + \theta_1,\quad \dots,\quad S\theta_k = \theta_k + \theta_{k-1}.$$

The result of λ turns round $x = 0$ gives

$$(88)\qquad S^\lambda\theta_1 = \theta_1 + \lambda,\quad S^\lambda\theta_2 = \theta_2 + \lambda\theta_1 + \frac{\lambda(\lambda-1)}{1.2}$$

$$S^\lambda\theta_k = \theta_k + \lambda\theta_{k-1} + \frac{\lambda(\lambda-1)}{1.2}\theta_{k-2} + \dots + \theta_{k-\lambda}.$$

If $\lambda = k$, this last is

$$(89)\qquad S^k\theta_k = \theta_k + k\theta_{k-1} + \frac{k(k-1)}{1.2}\theta_{k-2} + \dots + 1.$$

If $\lambda = k + l$, the coefficients in this equation change into the binomial coefficients corresponding to the exponent $(k+l)$; as there are no functions such as θ_{-i} (i a positive integer), the last term is simply the binomial coefficient

$$\frac{(l+k)(l+k-1)\dots(l+1)}{\underline{k|}}.$$

If now we choose a system of uniform functions $u_0, u_1, \dots, u_k$, we can write

$$(90)\qquad \begin{cases} z_0 = u_0, \\ z_1 = \theta_1 u_0 + u_1, \\ \cdot \quad \cdot \quad \cdot \quad \cdot \\ z_k = \theta_k u_0 + \theta_{k-1}u_1 + \dots + u_k. \end{cases}$$

And these values obviously satisfy the condition (81).

We have now for the integrals $y_0, \ldots, y_k$ the same forms as those in equations (49); retaining, however, Jordan's notation, we have

$$(91)\quad \begin{cases} y_0 = x^r M_0, \\ y_1 = x^r[M_1 \log x + N_1], \\ \cdot \quad \cdot \quad \cdot \quad \cdot \quad \cdot \quad \cdot \\ y_k = x^r[M^r \log^k x + N_k \log^{k-1} x + \ldots]. \end{cases}$$

The uniform functions $M_0, \ldots, M_k, N_1, \ldots, N_k \ldots$ are obviously linear functions of the $k + 1$ independent functions $u_0, u_1, \ldots, u_k$ of equations (90), and in particular $M_0, M_1, \ldots, M_k$ differ only by constant factors from each other and from u_0. It follows from this that the functions

$$x^r M_1, \quad x^k M_2, \quad \ldots, \quad x^r M_k,$$

which differ only from $x^r M_0$ by constant factors, are integrals of the differential equation. These functions M and N are so far perfectly arbitrary uniform functions; they will, however, in particular cases be seen to divide themselves into two classes—one containing only a finite number of negative powers of x (or of $x - a$, if a be the critical point considered) and one containing an infinite number of such powers.

When *all* of the functions M, N entering into any one of the integrals of the equation contain only a finite number of negative powers of the variable x, the integral is said to be *regular* in the region of the point $x = 0$. A very important class of equations, first investigated by Fuchs, is that class in which all of the integrals in the region of a critical point are regular. The investigation of this class will be given in the following chapters.

The substitution S which we have been considering is of course only one of a number, finite or infinite, which belongs to the linear differential equation with uniform coefficients. If the variable be made to describe all possible paths enclosing one or more of the critical points $a, b, c \ldots$ of the equation, we shall have a certain substitution corresponding to each of the paths; the aggregate of all these substitutions is called the *group* of the equation. We will denote this group by the letter G. It is clear that G will assume different forms according to the choice of the system of fundamental, *i.e.*, linearly independent, integrals. This notion of the group of a linear differential equation is of the highest importance in the theory, and will be treated of more fully in another chapter of this volume, and still more fully in Volume II. From a knowledge of the group of an equation we can derive all the essential properties of the equation. Suppose, for example, $P = 0$ is an equation having all of its integrals regular: it is obvious that among equations of this type are included all equations having only algebraic integrals. What, then, are the conditions which $P = 0$ must satisfy in order that all of its integrals may be algebraic? It is obviously necessary that the different functions into which the substitutions of the group G change the chosen system, say $y_1, y_2, \ldots, y_n$, of fundamental integrals must be limited in number, and consequently it is sufficient that the group G contain only a finite number of substitutions. The case of these equations will also be returned to later on. Among the substitutions which enter into a group there are, since we consider only the case of a finite number of critical points, only a finite number of independent ones. Suppose, for example, that S is a substitution corresponding to a given closed contour K, and that S_1 is another belonging to a given closed contour K_1; then, if we describe successively the two contours K and K_1, we shall arrive at a substitution SS_1 which is the resultant of the two substitutions S and S_1. All of the substitutions in a group therefore result from the combinations which can be made among the substitutions belonging to each of the critical points taken separately.

CHAPTER IV.

FROBENIUS'S METHOD.

WE will consider, as before, the region of the critical point $x = 0$ and let $P_0(x), P_1(x), \ldots, P_n(x)$ denote convergent series proceeding according to positive integral powers of x, and further assume that $P_0(x)$ does not vanish for $x = 0$. If all the integrals of the given differential equation, $P(y) = 0$, are regular in the region of $x = 0$, the equation, as will subsequently be seen, can be put in the form

$$(1) \qquad P_0(x)x^n \frac{d^n y}{dx^n} + P_1(x)x^{n-1}\frac{d^{n-1}y}{dx^{n-1}} + \ldots + P_n(x)y = 0.$$

We will now seek to determine the form of the integrals of this equation by an extremely elegant and ingenious method due to Frobenius.* For simplicity we will assume $P_0(x) = 1$. In equation (1) make the substitution

$$(2) \qquad y = g(x, r) = \sum_{\nu=0}^{\nu=\infty} g_\nu\, x^{r+\nu}.$$

The limits $\nu = 0$ and $\nu = \infty$ will hereafter be omitted from the summation sign, but will be always understood. We see at once that this substitution gives rise to the equation

$$(3) \qquad P(\Sigma g_\nu x^{r+\nu}) = \Sigma g_\nu P(x^{r+\nu}).$$

---

* *Ueber die Integration der linearen Differentialgleichungen durch Reihen.* (Von Herrn G. Frobenius.) Crelle, vol. 76.

If now we write

$$(4)\qquad f(x, r) = r(r-1)\ldots(r-n+1)P_0(x) + r(r-1)\ldots(r-n+2)P_1(x) + \ldots + P_n(x),$$

we have

$$(5)\qquad P(x^r) = x^r f(x, r),$$

and consequently

$$(6)\qquad P[g(x, r)] = \Sigma g_\nu f(x, r+\nu)x^{r+\nu}.$$

Since the functions $P_i(x)$ are developable in convergent series going according to positive integral powers of x, it follows that the series

$$(7)\qquad f(x, r) = \Sigma f_\nu(r)x^\nu$$

is also convergent, and that the coefficients of the different powers of x are integral functions of r of the degree n at most. The substitution $y = g(x, r)$ being made in (1) gives us now

$$(8)\qquad \Sigma\left[g_\nu f(r+\nu) + g_{\nu-1}f_1(r+\nu-1) + \ldots + g_1 f_{\nu-1}(r+1) + g_0 f_\nu(r)\right]x^{r+\nu} = 0.$$

In order, then, that $y = g(x, r)$ shall be an integral of (1), we must have

$$(9)\qquad \begin{cases} g_0 f(r) = 0, \\ g_1 f(r+1) + g f_1(r) = 0, \\ \cdot\quad\cdot\quad\cdot\quad\cdot\quad\cdot\quad\cdot\quad\cdot\quad\cdot\quad\cdot\quad\cdot \\ g_\nu f(r+\nu) + g_{\nu-1}f_1(r+\nu-1) + \ldots + g_1 f_{\nu-1}(r+1) + g_0 f_\nu(r) = 0. \end{cases}$$

If we suppose now that $g_0 x^r$ is the first term in the series

$$g(x, r) = \Sigma g_\nu x^{r+\nu},$$

then g_0 cannot be zero, and consequently r must be a root of the equation $f(r) = 0$, which is of the nth degree in r. In what follows we will consider r as a variable parameter, and g_0 (or simply g for convenience) as an arbitrary function of r. Neglecting the first of equations (9), we can at once determine g_1, g_2, . . . as functions of r. Write

$$(10)\quad (-1)^\nu h_\nu(r) = \begin{vmatrix} f_1(r+\nu-1), & f_2(r+\nu-2), & \ldots, & f_{\nu-1}(r+1), & f_\nu(r) \\ f(r+\nu-1), & f_1(r+\nu-2), & \ldots, & f_{\nu-2}(r+1), & f_{\nu-1}(r) \\ 0, & f(r+\nu-2), & \ldots, & f_{\nu-3}(r+1), & f_{\nu-2}(r) \\ \cdot & \cdot & \cdot & \cdot & \cdot \\ 0, & 0, & \ldots, & f(r+1), & f_1(r) \end{vmatrix};$$

now from (9) and (10) we have

$$(11)\qquad g_\nu(r) = \frac{g(r)h_\nu(r)}{f(r+1)f(r+2)\ldots f(r+\nu)}.$$

The variable parameter r will be so restricted that all of its values shall be found in the regions of the roots of the equation $f(r) = 0$. Since the roots of this equation have their moduli all less than a certain determinate finite quantity, it is easy to see that these regions can be chosen so small that the denominator of the rational function $g_\nu(r)$ shall only vanish for the roots of the equation $f(r) = 0$. This vanishing of the denominator of $g_\nu(r)$ would, however, make $g_\nu(r)$ infinite in general; this difficulty can nevertheless be avoided by a proper choice of the arbitrary function $g(r)$. We will suppose that the roots of $f(r) = 0$ are arranged in groups in the manner described in the last chapter, and suppose further that ϵ (necessarily an integer) is the maximum difference of two roots in any of the groups. Writing now

$$(12)\qquad g(r) = f(r+1)f(r+2)\ldots f(r+\epsilon)C(r),$$

where $C(r)$ is an arbitrary function of r, it follows that for all the values of r under consideration the functions $g_\nu(r)$ are finite, and consequently that if the series $y = g(x, r)$ is convergent, then $y = g(x, r)$ is an integral of the differential equation

$$(13)\qquad P(y) = f(r)g(r)x^r.$$

We have now first to investigate the conditions for convergence of the series

$$g(x, r) = \Sigma g_\nu x^{r+\nu}.$$

If we assume $\nu > \epsilon$, then, recalling the definition of ϵ, it is clear that $f(r + \nu + 1)$ cannot vanish for any of the values of r to which we are restricted, and so, from equations (9), we have

$$(14) \qquad g_{\nu+1} = -\frac{1}{f(r+\nu+1)}\Big[g_\nu f_1(r+\nu) + g_{\nu-1} f_2(r+\nu-1) + \ldots + g f_{\nu+1}(r)\Big].$$

Denoting now by $F_\nu(r)$ and $G_\nu(r)$ the moduli of $f_\nu(r)$ and $g_\nu(r)$, we have at once the inequality

$$(15) \qquad G_{\nu+1} \leqq \frac{1}{F(r+\nu+1)}\Big[G_\nu F_1(r+\nu) + G_{\nu-1} F_2(r+\nu-1) + \ldots + G F_{\nu+1}(r)\Big].$$

We will now suppose a circle of radius K drawn with the point $x = 0$ as centre, and where K is as little less as we please than the radius of a circle inside of which the functions $P_1(x)$, $P_2(x)$, . . . , $P_n(x)$ are *all* convergent; also let $f'(x, r)$ denote the derivative of $f(x, r)$ with respect to x. The series

$$f(x, r), \quad = \Sigma f_\nu(r) x^\nu,$$

and

$$f'(x, r), \quad = \Sigma(\nu + 1) f_{\nu+1}(r) x^\nu,$$

are both convergent so long as the inequality

$$\text{mod. } x < K$$

is satisfied; *i.e.*, so long as the point x remains inside the circle of radius K, or, we may say for brevity, so long as x remains inside the circle K. Let now $M(r)$ denote the maximum value which the modulus of $f'(x, r)$ takes on the circumference of the circle K; then by a well-known theorem we have

$$(16) \qquad F_{\nu+1}(r) < \frac{1}{\nu+1} M(r) K^{-\nu} < M(r) K^{-\nu},$$

and consequently, from (15),

$$(17)\qquad G_{\nu+1} < \frac{1}{F(r+\nu+1)}[G_\nu M(r+\nu) + G_{\nu-1} M(r+\nu-1)K^{-1} + \ldots + GM(r)K^{-\nu}].$$

Denoting the right-hand member of this inequality by $a_{\nu+1}$, we have

$$(18)\qquad a_{\nu+1} = \frac{G_\nu M(r+\nu)}{F(r+\nu+1)} + \frac{a_\nu F(r+\nu)}{KF(r+\nu+1)};$$

or, since $G_\nu < a_\nu$,

$$(19)\qquad a_{\nu+1} < a_\nu\left[\frac{M(r+\nu)}{F(r+\nu+1)} + \frac{F(r+\nu)}{KF(r+\nu+1)}\right].$$

Still assuming $\nu > \epsilon$, we will define certain quantities b_ν by the formula

$$(20)\qquad b_{\nu+1} = b_\nu\left[\frac{M(r+\nu)}{F(r+\nu+1)} + \frac{F(r+\nu)}{KF(r+\nu+1)}\right];$$

also assume b_ϵ, as we obviously can, so that we shall have the inequalities

$$(21)\qquad G_\nu < a_\nu < b_\nu.$$

We know that the integral function $f(r)$ is of degree n in r, and therefore when ν increases indefinitely the quotient

$$\frac{f(r+\nu)}{f(r+\nu+1)},$$

and of course its modulus,

$$\frac{F(r+\nu)}{F(r+\nu+1)},$$

tends to the limit unity. Further: we have assumed $P_0(x) = 1$, and consequently $f'(x, r)$ is an integral function of r of degree at most equal to $n - 1$; also, $M(r)$ denotes the maximum value which this function can have when mod. $x = K$. It is now easy to see (the rigorous proof will be given immediately) that if ν increases indefinitely we have

$$(22) \qquad \lim. \frac{M(r+\nu)}{F(r+\nu+1)} = 0.$$

It follows now from (20) that

$$(23) \qquad \lim. \frac{b_{\nu+1}}{b_\nu} = \frac{1}{K},$$

and therefore that the series $\Sigma b_\nu x^\nu$, and therefore also the series

$$(24) \qquad g(x, r), \quad = \Sigma g_\nu(r) x^{r+\nu},$$

is convergent inside the circle K.

It is necessary now to show, by aid of the results already obtained, that the series (24) is uniformly convergent for each of the values of r under consideration. If we denote by δ a given arbitrary small quantity, we must show that for all of the considered values of r it is possible to find a finite number, say j, such that the modulus of the sum

$$\sum_{\nu=j}^{\nu=\infty} g_\nu(r) x^{r+\nu}$$

shall be less than δ. In establishing this we will first prove the truth of equation (22). Let s denote the modulus of r; *i.e.*,

$$s = \text{mod. } r = |\, r \,|.$$

(The symbol $|\, X \,|$, due to Weierstrass, will be used when convenient to denote the modulus of the quantity X, whatever X may stand for.) Also, let $M_1, M_2, \ldots, M_n$ denote the maximum values of the moduli of $P_1'(x), P_2'(x), \ldots, P_n'(x)$ on the circumference of the circle K; if then we write *

$$(25) \quad \psi(s) = s(s+1)\ldots(s+n-2)M_1 \\ + s(s+1)\ldots(s+n-3)M_2 + \ldots + M_n,$$

* For the moment we will allow r to vary indefinitely until it is necessary to re-introduce $r+\nu$ as the indefinitely-increasing argument.

we have

$$M(r) < \psi(s). \tag{26}$$

The function $f(r)$ is of degree n in r, and we may thus write

$$f(r) = r^n + [f(r) - r^n],$$

and so obtain the inequality

$$F(r) \geqq s^n - | f(r) - r^n |;$$

again,

$$f(r) = r(r-1)\ldots(r-n+1)$$
$$+ P_1(0)r(r-1)\ldots(r-n+2) + \ldots + P_n(0).$$

Let $\Pi_1, \ldots, \Pi_n$ denote the moduli of $P_1(0), \ldots, P_n(0)$; then, if we write

$$\phi(s) = s(s+1)\ldots(s+n-1) + \Pi_1 s(s+1)\ldots(s+n-2)$$
$$+ \ldots + \Pi_n - s^n,$$

we have

$$| f(r) - r^n | < \phi(s), \tag{27}$$

and consequently

$$F(r) > s^n - \phi(s), \tag{28}$$

provided only we choose s so large that the right-hand side of this inequality shall be positive; this can, of course, always be done, since $\phi(s)$ is only of degree $n-1$. From the inequalities obtained above we derive at once

$$\frac{M(r+\nu)}{F(r+\nu+1)} < \frac{\psi|r+\nu|}{|r+\nu+1|^n - \phi|r+\nu+1|}; \tag{29}$$

but

$$|r+\nu| < \nu+s \qquad \text{and} \qquad |r+\nu+1| > \nu-s;$$

and since the positive functions $\psi(s)$ and $s^n - \phi(s)$ always increase after a certain value of the argument s, we have, by giving ν a sufficiently large value,

$$(30)\qquad \frac{M(r+\nu)}{F(r+\nu+1)} < \frac{\psi(\nu+s)}{(\nu-s)^n - \phi(\nu-s)}.$$

The numerator of the right-hand member of this inequality is of lower degree than the denominator, and so this member tends to zero as ν increases indefinitely; that is, we have, as above stated,

$$(31)\qquad \lim. \frac{M(r+\nu)}{F(r+\nu+1)} = 0.$$

For the completion of the proof of the convergence of the series $g(x, r)$, we notice that since all of the roots of $f(r) = 0$ lie in a finite region, and since the parameter r can only vary in the regions of these roots, then s must be always less than a certain determinate quantity, say t. If now we take ν sufficiently large, we have obviously

$$(32)\qquad \frac{M(r+\nu)}{F(r+\nu+1)} < \frac{\psi(\nu+t)}{(\nu-t)^n - \phi(\nu-t)}$$

and

$$(33)\qquad \frac{F(r+\nu)}{F(r+\nu+1)} < \frac{(\nu+t)^n + \phi(\nu+t)}{(\nu-t)^n - \phi(\nu-t)}.$$

Suppose $\nu = \mu$ is the value of ν for which these inequalities first hold; they will, *à fortiori*, continue to hold for all values of ν greater than μ. Suppose now quantities c_ν to be defined, for all values of ν equal to and greater than μ, by the formula

$$(34)\qquad c_{\nu+1} = c_\nu\left[\frac{\psi(\nu+t)}{(\nu-t)^n - \phi(\nu-t)} + \frac{1}{K}\,\frac{(\nu+t)^n + \phi(\nu+t)}{(\nu-t)^n - \phi(\nu-t)}\right];$$

then if we choose $c_\mu > b_\mu$, as we obviously may, we have generally $c_\nu > b_\nu$. Now if k be chosen as little smaller as we please than K, then, since $\lim. \dfrac{c_{\nu+1}}{c_\nu} = \dfrac{1}{K}$, the series $\Sigma c_\nu k^\nu$ is convergent. Beginning with the first term of this series, and counting forward, we can cut off a finite number of terms such that the sum of the remaining ones, say $\sum_{\nu=j}^{\nu=\infty} c_\nu k^\nu$, shall be less than an arbitrarily chosen quantity δk^{-s}. Now, since we can of course choose j greater than μ, it follows that we have

$$\left|\sum_{\nu=j}^{\nu=\infty} g_\nu(r) x^{r+\nu}\right| < \delta$$

for all values of r inside the regions of the roots of $f(r) = 0$, and for all values of x inside the circle of radius k and having the point $x = 0$ as its centre. The series

$$g(x, r) = \Sigma g_\nu x^{r+\nu}$$

is therefore uniformly convergent and can be differentiated with respect to r, and its differential coefficient so formed will be equal to the sum of the differential coefficients of its successive terms.

Having established now the convergence of the series $g(x, r)$, we proceed to investigate the forms of the integrals of the equation $P(y) = 0$.

We will consider the group of $\mu + 1$ roots $r_0, r_1, \ldots, r_\mu$ of the equation $f(r) = 0$. These roots are so arranged (as already described) that for $\alpha < \beta$ the difference $r_\alpha - r_\beta$ is a positive integer. As certain of the roots of this group may be equal, we will further assume that $r_0, r_\alpha, r_\beta, r_\gamma, \ldots$ are the distinct ones; then

$r_0 = r_1 = \ldots = r_{\alpha-1}$ will be an α-tuple root of $f(r) = 0$;
$r_\alpha = r_{\alpha+1} = \ldots = r_{\beta-1}$ will be a $(\beta - \alpha)$-tuple root of $f(r) = 0$;
$r_\beta = r_{\beta+1} = \ldots = r_{\gamma-1}$ will be a $(\gamma - \beta)$-tuple root of $f(r) = 0$;
etc. etc.

We have assumed for $g(r)$ the form

$$g(r) = f(r+1) f(r+2) \ldots f(r+\epsilon)\, C(r);$$

where $C(r)$ is an arbitrary function of r. Recalling now the definition of ϵ, we have the inequality $\epsilon \geqq r_0 - r_\mu$, and therefore $g(r)$ cannot vanish

for $r = r_0 = r_1 = \ldots = r_{\alpha-1}$; but
for $r = r_\alpha = r_{\alpha+1} = \ldots = r_{\beta-1}$, $g(r)$ is zero of the order α,
for $r = r_\beta = r_{\beta+1} = \ldots = r_{\gamma-1}$, $g(r)$ is zero of the order β, etc.;

and generally for $r = r_k$, $g(r)$ is zero at most of the order k. Again: the expression $f(r)g(r)x^r$ is zero

$$\text{of the order } \alpha \text{ for } r = r_0 = r_1 = \ldots = r_{\alpha-1},$$
$$\text{of the order } \beta \text{ for } r = r_\alpha = r_{\alpha+1} = \ldots = r_{\beta-1}, \text{ etc.};$$

and generally $f(r)\, g(r)x^r$ is zero of at least the order $k+1$ for $r = r_k$, and consequently its kth derivative with respect to r must vanish for $r = r_k$. Write now equation (13) in the identical form

$$P[g(x, r)] = f(r)g(r)x^r,$$

and differentiate this k times with respect to r, and in the result write r_k for r; we have as the result of this operation

$$P[g^k(x, r_k)] = 0; \tag{35}$$

where

$$g^k(x, r) = \frac{d^k g(x, r)}{dr^k}.$$

It follows at once that

$$y = g^k(x, r_k) \tag{36}$$

is an integral of the equation $P(y) = 0$. Since the series

$$g(x, r), = x^r \Sigma g_\nu(r) x^\nu,$$

is a uniformly convergent series, the same is true of the series

$$g^k(x, r^k), = x^{r_k} \Sigma \Big\{ g_\nu^k(r_k) + k g_\nu^{k-1}(r_k) \log x \tag{37}$$
$$+ \frac{k(k-1)}{1 \cdot 2} g_\nu^{k-2}(r_k) \log^2 x + \ldots + g_\nu(r_k) \log^k x \Big\} x^\nu,$$

which is an integral of $P(y) = 0$. Since $g(r)$ vanishes for $r = r_k$ at most of the order k, it follows that

$$g(r_k), \quad g'(r_k), \quad \ldots, \quad g^k(r_k)$$

cannot all be zero, and so, by Fuchs's definition, the integral (37) *belongs to the exponent* r_k. It is obvious that (37) is of the same form as

$$x^{r_k} \Sigma \{ \phi_0 + \phi_1 \log x + \ldots + \phi_k \log^k x \}; \tag{38}$$

where the ϕ's are uniform and continuous functions of x, and are not all zero for $x = 0$. Suppose $k < \alpha$; then from (37) we find, as the coefficient of $\log^k x$, the series

$$(39) \qquad x^{r_k}\Sigma g_r(r_k)x_\nu \,.$$

Since $k < \alpha$, we have $g(r_k) = g(r_0)$, and so this series cannot vanish identically, and as a consequence we have that k is the exponent of the highest power of $\log x$ which can appear in the integral $g^k(x, r_k)$. It is easy to extend this result, and so observe that in general the integrals which belong to *equal* roots of $f(r) = 0$ have *different exponents* for the highest powers of $\log x$ that enter, and consequently that these integrals are linearly independent. In particular, if r is an n-tuple root of $f(r) = 0$, then the n corresponding integrals of $P(y) = 0$, viz., $g^k(x, r_k)$, $(k = 1, 2, \ldots, n)$, are linearly independent. From what has been said it is clear that the *form* of the integral $g^k(x, r_k)$ is unaltered if we add to it a linear function (with arbitrary constant coefficients) of $g^{k-1}(x, r_{k-1}), \ldots, g(x, r_0)$; we derive from this fact the conclusion that the most general value of the integral belonging to the exponent k contains $k + 1$ arbitrary constants. Let us suppose (which we may do without any loss of generality) that when the arbitrary function $C(r)$, which enters as a factor in all the coefficients of the series $\Sigma g_\nu(r)x^{r+\nu}$, becomes unity, the function $g(x, r)$ becomes $h(x, r)$; then $g(x, r) = C(r)h(x, r)$, and consequently

$$(40) \quad g^k(x, r) = Ch^k(x, r) + kC'h^{k-1}(x, r) + \ldots + C_k h(x, r),$$

where g^α, C^α, h^α denote derivatives of g, C, h with respect to r. The functions

$$h^k(x, r_k), \quad h^{k-1}(x, r_k), \quad \ldots, \quad h(x, r_k)$$

are linearly independent, and consequently the integral $g^k(x, r_k)$ contains the $k + 1$ arbitrary constants $C, C', \ldots, C^k$. We have thus formed the general integral belonging to the exponent k of the differential equation without the aid of the integrals

$$g^{k-1}(x, r_{k-1}), \quad \ldots, \quad g(x, r_0).$$

In the particular case of the root r_0 the integral $g(x, r_0)$ contains only one arbitrary constant. It follows at once from what precedes that one integral of the differential equation $P(y) = 0$ is, to an arbitrary constant près, completely determined by the following condition: If the integral is divided by x^r the quotient must be uniform, and must also be finite for $x = 0$, r denoting a root of the algebraic equation $f(r) = 0$ which does not exceed any other root of this equation by a positive integer.

We have above made an assumption concerning the form of the arbitrary function $g(r)$, in order to prevent the coefficients of the series $g(x, r)$ from becoming infinite: viz., we have assumed

$$g(r) = f(r + 1)f(r + 2) \ . \ . \ . \ f(r + \epsilon)C(r);$$

where $C(r)$ is an arbitrary function of r, and where ϵ denotes the greatest (integral) difference between two roots of any group of roots of $f(r) = 0$. It is obvious that ϵ may take any other value greater than this, and the foregoing results will still hold. The value of this liberty of choice of ϵ is seen if we wish to actually compute the value of the series $g(x, r)$ up to a given power of x, say x^{r+k}. In this case assume $\epsilon = k$; then all of the assumed coefficients $g_\nu(r)$ will be integral functions of r, and so the inconvenience of differentiating fractional forms will be evaded. Of course it is understood that the arbitrary function $C(r)$ must be so chosen as not to become infinite for any of the considered values of r. For further remarks on this point the reader is referred to Frobenius's memoir.

We have seen both by Fuchs's method and that of Frobenius that logarithms generally appear in the integrals of a *group*. The *groups* of integrals arise, it is to be remembered, from the fact that certain roots of the indicial equation differ from each other by integers, including zero. The separation of these roots into groups, as already described, gives rise to the corresponding groups of integrals. It may be, however, that logarithms will not appear in a group of integrals. Fuchs investigated the conditions necessary to be satisfied in order that no logarithms should appear in a given group. The reader is referred to Fuchs's memoir in Crelle (vol. 68) for his investigation of this subject, and also to one by Cayley in Crelle, vol. 100. The test to be applied in order to ascertain whether or not logarithms exist is obtained by Frobenius in a very simple manner in his memoir which we are now considering. Before taking up Frobenius's remarks on this subject, we can show (Tannery, p. 167) that if in a group two roots are *equal*, say $r_1 = r_2$, then logarithms necessarily appear in this group. The integrals belonging to r_1 and r_2 are in general of the form (in the region of $x = 0$)

$$y_1 = x^{r_1}\phi_{11},$$

$$y_2 = x^{r_2}[\phi_{21} + \phi_{22} \log x];$$

where the functions ϕ are uniform and continuous in the region of $x = 0$. If now in $P(y) = 0$ we make the transformation

$$y = y_1 \int z dx,$$

we know that the integral z_1 of the linear differential equation of order $n - 1$ in z belongs to the exponent $r_2 - r_1 - 1$; but since $r_2 = r_1$, we have in this case $r_2 - r_1 - 1 = -1$ as the exponent to which z_1 belongs, and therefore z_1 is of the form

$$z_1 = \frac{A}{x} + \theta(x);$$

where the constant A does not vanish, and where $\theta(x)$ is, in the region of $x = 0$, a uniform and continuous function. From this we see that

$$\int z_1 dx$$

is of the form

$$A \log x + \psi(x),$$

and consequently y_2 is of the form

$$x^{r_1}[\phi_{21}(x) + \phi_{22} \log x].$$

It is clear from this simple illustration that if there are equal roots in a group, logarithms must necessarily appear in the corresponding group of integrals.

By aid of equation (37) the Frobenius method leads us to the same conclusion; we have seen, in fact, that in the integrals belonging to an α-tuple root of $f(r) = 0$ there are, in general, at least $\alpha - 1$ logarithms, and so in order that no logarithms exist in the group we must have the roots $r_0, r_1, \ldots, r_\mu$, among which the r of equation (37) is to be found, all different. We can now quickly find by Frobenius's method the conditions to be satisfied in order that logarithms may not appear in a given group of integrals. We have already seen that the series $g^k(x, r_k)$ of equation (37) may represent, without the aid of the series $g^{k-1}(x, r_{k-1}), \ldots, g(x, r_0)$, the most general integral of the differential equation $P(y) = 0$ belonging to the exponent r_k. It follows now at once that in order that the integral $g^k(x, r_k)$ may contain no logarithms we must have all of the functions $g_\nu(r)$ zero of order k for $r = r_k$. Now,

$$(41) \qquad g_\nu(r) = \frac{g(r)h_\nu(r)}{f(r+1)f(r+2)\ldots f(r+\nu)};$$

in which $g(r)$ is zero of order k for $r = r_k$. The function

$$(42) \qquad H_\nu(r), \quad = \frac{h_\nu(r)}{f(r+1)f(r+2)\ldots f(r+\nu)},$$

must, therefore, not become infinite for $r = r_k$. Since the functions $g_a(r)$ are connected by the formula (see equations (9))

$$(43) \quad g_\nu f(r+\nu) + g_{\nu-1}f_1(r+\nu-1) + \ldots + gf_\nu(r) = 0,$$

and since

$$(44) \qquad H_\nu(r) = \frac{g_\nu(r)}{g(r)},$$

we have for the functions H the formula

$$(45) \quad H_\nu f(r+\nu) + H_{\nu-1}f_1(r+\nu-1) + \ldots + Hf_\nu(r) = 0.$$

If now $H_{\nu-1}, H_{\nu-2}, \ldots, H$ are all finite for $r = r_k$, the same must also be true for $H_\nu f(r+\nu)$.

Now from (44) we have $H = 1$, and consequently $H_\nu(r_k)$ will be finite for all values of ν, provided only it does not become infinite for such values of ν as make $r_k + \nu$ a root of the equation $f(r) = 0$; that is, for values of ν equal to some one of the quantities

$$r_{k-1} - r_k, \quad r_{k-2} - r_k, \quad \ldots, \quad r_0 - r_k.$$

It is easy to see now that in order that $H_\nu(r_k)$ shall be finite for $\nu = r_{k-1} - r_k$ it is necessary and sufficient that $h_\nu(r)$ shall be zero of the first order for $r = r_k$. For $\nu = r_{k-2} - r_k$ and $r = r_k$ we have now

$$(45)\quad H_\nu(r)f(r+\nu) = \frac{h_\nu(r)}{f(r+1)f(r+2)\ldots f(r+\nu-1)}$$

and $h_\nu(r)$ is zero of order one. In order that $H_\nu(r)$ shall be finite it is now necessary that $h_\nu'(r_k)$ shall vanish. Continuing this process, we readily find the following conditions which must be satisfied in order that the integral belonging to the root r_k of the indicial equation $f(r) = 0$ may contain no logarithms. Defining the function $h_\nu(r)$ by the equation

$$(46)\quad (-1)^\nu h_\nu(r) = \begin{vmatrix} f_1(r+\nu-1), & f_2(r+\nu-2), & \ldots, & f_{\nu-1}(r+1), & f_\nu(r) \\ f(r+\nu-1), & f_1(r+\nu-2), & \ldots, & f_{\nu-2}(r+1), & f_{\nu-1}(r) \\ 0, & f(r+\nu-2), & \ldots, & f_{\nu-3}(r+1), & f_{\nu-2}(r) \\ \cdot & \cdot & \cdot & \cdot & \cdot \\ 0, & 0, & \ldots, & f(r+1), & f_1(r) \end{vmatrix},$$

then for $r = r_k$ the following equations must be satisfied:

$$(47)\quad \begin{cases} h_\nu(r) = 0 & \text{for} \quad \nu = r_{k-1} - r_k, \\ h_\nu'(r) = 0 & \text{for} \quad \nu = r_{k-2} - r_k, \\ \cdot \quad \cdot \quad \cdot & \cdot \quad \cdot \quad \cdot \\ h_\nu^{k-1}(r) = 0 & \text{for} \quad \nu = r_0 - r_k. \end{cases}$$

CHAPTER V.

LINEAR DIFFERENTIAL EQUATIONS ALL OF WHOSE INTEGRALS ARE REGULAR.

IN what follows the critical point under consideration will be taken, as in the preceding chapters, as $x = 0$. The results obtained for this point can be changed into the corresponding results for any other critical point x_1 by changing x into $x - x_1$; the functions ϕ_{ij} or (M, N) will of course be different for each critical point, but will always be uniform. It has been seen that functions of the form

$$(1)\qquad F = x^r[\phi_0 + \phi_1 \log x + \ldots + \phi_k \log^k x],$$

or, for a critical point $x = x_l$,

$$F = (x - x_l)^r [\phi_0 + \phi_1 \log(x - x_l) + \ldots + \phi_k \log^k (x - x_l)],$$

play an important part in the theory. Since the uniform functions ϕ contain (for the present) only a finite number of negative values of x, or $x - x_l$, and since for p positive the products

$$x^p \log^q x, \quad (x - x_l)^p \log^q (x - x_l)$$

are zero for $x = 0$ and $x = x_l$ respectively, it follows that we can always find a number r such that the product

$$x^{-r}F \quad \text{or} \quad (x - x_l)^{-r}F$$

shall be different from zero, and shall become infinite in the same way as the function

$$(2)\qquad \alpha + \beta \log x + \ldots + \lambda \log^k x,$$

or

$$\alpha + \beta \log (x - x_l) + \ldots + \lambda \log^k (x - x_l),$$

where α, β, . . . , λ are constants.

We will use now only the critical point $x = 0$, as the necessary changes for any other point are obvious.

The function F of (1) is now said to *belong to the exponent r.*

If two functions F and F_1 belong to exponents r and r_1 respectively, the product FF_1 will belong to the exponent $r + r_1$; also,

$$SFF_1 = SF \times SF_1, \quad S\left[\frac{F}{F_1}\right] = \frac{SF}{SF_1}, \text{ etc.}$$

If F belongs to the exponent r, then in general $\dfrac{dF}{dx}$ will belong to the exponent $r - 1$. A case of exception arises, however, when $r = 0$, and when at the same time F does not become infinite for $x = 0$. Suppose

$$F = x^r[\phi_0 + \phi_1 \log x + \ldots + \phi^k \log^k x],$$

r being the exponent above described, it follows that the uniform functions ϕ contain only positive integer powers of x in their development, and are not all zero for $x = 0$. We have now

$$\frac{dF}{dx} = \ldots + x^{r-1}\left[r\phi_p + (p + 1)\phi_{p+1} + x\frac{d\phi_p}{dx}\right]\log^p x + \ldots$$
$$+ x^{r-1}\left[r\phi_k + x\frac{d\phi_k}{dx}\right]\log^k x.$$

The coefficients of the different powers of $\log x$ cannot all vanish for $x = 0$; for if they could we should have the system of equations

$$(3) \qquad \begin{cases} r\phi_p + (p + 1)\phi_{p+1} = 0, \\ \quad \cdot \qquad \cdot \qquad \cdot \qquad \cdot \\ r\phi_k \qquad\qquad\qquad = 0; \end{cases}$$

but this system gives us

$$\phi_0 = \phi_1 = \ldots = \phi_k = 0$$

for $x = 0$, which is contrary to hypothesis. If, however, $r = 0$, it is only necessary, in order that all the coefficients of $\log x$ vanish in $\dfrac{dF}{dx}$, *i.e.*, that equations (3) be satisfied, that we have

$$\phi_k = \phi_{k-1} = \ldots = \phi_1 = 0$$

for $x = 0$. In this case F reduces to the finite and non-vanishing value of ϕ_0, and $\dfrac{dF}{dx}$ belongs to the exponent zero, or, in certain cases, to a positive integral exponent.

It is easy to see, if F belongs to the exponent r, that by properly choosing the constant of integration the function

$$\int F dx$$

will be of the same character as F, and will belong to the exponent $r + 1$. The preceding results (taken from Tannery's memoir) may be briefly expressed as follows:

1. Every regular function

$$F = x^r[\phi_k \log^k x + \phi_{k-1} \log^{k-1} x + \dots + \phi_0]$$

has a derivative

$$x^r \left\{ \frac{r}{x} [\phi_k \log^k x + \dots + \phi_0] + \frac{k\phi_k \log^{k-1} x}{x} + \frac{d\phi_k}{dx} \log^k x + \dots \right\}, \tag{4}$$

which is also regular.

2. Every product or quotient of regular functions is also regular. The following derivation of the form of the differential equation all of whose integrals are regular in the region of a critical point, is taken from Jordan. Denote by $y_1, \dots, y_n$ a system of regular integrals of the equation in the region of the critical point $x = 0$, and let y_l^a denote the differential coefficient $\dfrac{d^a y_l}{dx^a}$; the equation can be written in the form

$$\begin{vmatrix} y, & y_1, & \dots, & y_n \\ y', & y_1', & \dots, & y_n' \\ \vdots & \cdot & \cdot & \cdot \\ y^n, & y_1^n, & \dots, & y_n^n \end{vmatrix} = 0. \tag{5}$$

The coefficients of $y, y', \ldots, y^n$ are the principal minors of the determinant corresponding in order to these quantities, and they are obviously sums of regular functions, such as

$$(6) \qquad x^r[\psi_i \log^i x + \ldots] + x^{r_1}[\psi_{i_1}' \log^{i_1} x + \ldots] + \ldots .$$

Now, when x turns round the critical point $x = 0$, the integrals

$$y_1, \quad \ldots, \quad y_n,$$

and their derivatives,

$$y_1^a, \quad \ldots, \quad y_n^a, \quad \alpha = 1, 2, \ldots, n,$$

submit to the same linear substitution; the coefficients of $y, y', \ldots, y^n$ therefore reproduce themselves each multiplied by the determinant, say δ, of the substitution. [If we divide the equation through by the coefficient of y^n, the coefficients, say $p_1, p_2, \ldots, p_n$ of $y', y'', \ldots, y^n$, will therefore reproduce themselves exactly after the substitution; in other words, they are, as by the original hypothesis, uniform functions.] In order that an expression of the form (6) may reproduce itself multiplied by the determinant δ after x turns round $x = 0$, it is clear that the logarithms must all disappear, and that the exponents $r, r_1, \ldots$ can only differ by integers from the quantity

$$\frac{\log \delta}{2\pi i}, \quad = \text{, say, } \beta.$$

The coefficients of the differential equation (5) are therefore of the form $x^\beta P$, where P, like the functions ϕ, is a uniform function of x, having $x = 0$ only as an ordinary point or a pole.

It follows also from the remark in brackets that $p_1, p_2, \ldots, p_n$ can have $x = 0$ only as an ordinary point or a pole.

Fuchs's theorem regarding equations which have a system of independent integrals all of which are regular may be stated as follows:

In order that the linear differential equation

$$\frac{d^n y}{dx^n} + p_1 \frac{d^{n-1} y}{dx^{n-1}} + \ldots + p_n y = 0$$

may have a system of linearly independent regular integrals in the region of the point $x = 0$, it is necessary and sufficient that each coefficient, say p_i, of the equation shall have the point $x = 0$ for an ordinary point or a pole; in case this point is a pole, its order of multiplicity must not be greater than i.

The first part of this theorem has already been established. The groups of integrals which have been formed corresponding to the distinct roots $s_1, s_2, \ldots$ of the characteristic equation have belonging to each of them a certain exponent r defined by

$$\frac{\log s_i}{2\pi i} = r_i;$$

the values $r_1, r_2, \ldots$ derived from $s_1, s_2, \ldots$, and associated with each group of integrals, must then differ from each other by quantities other than integers. As a change of r_i into $r_i + m$, where m is an integer, would leave s_i unchanged, it is clear that the factor x^{r_i} in the group associated with s_i might have different values of r_i in the different integrals of the group, provided those values differed only by integers. In establishing the second part of Fuchs's theorem we shall, however, arrive at an algebraic equation (the equation $f(r) = 0$ of the last chapter) determining the exponent r for each group without ambiguity. Jordan proves this second part of the theorem in a very brief and elegant manner. Fuchs's proof, which is rather longer, will be given later. If we transform the equation

$$\frac{d^n y}{dx^n} + p_1 \frac{d^{n-1} y}{dx^{n-1}} + \ldots + p_n y = 0$$

by the substitution

$$y = X\eta,$$

where η is the new dependent variable, and X is of the form

$$x^\lambda \sum_{i=0}^{i=\infty} c_i x^i$$

and therefore

(7)
$$\frac{1}{X} = x^{-\lambda} \sum_{i=0}^{i=\infty} c_i' x^i,$$

we have the equation

(8) $$X\frac{d^n\eta}{dx^n}+q_1'\frac{d^{n-1}\eta}{dx^{n-1}}+\ldots+q_n'\eta=0.$$

The values of $q_1', \ldots, q_n'$ are, from (12), Chap. III,

$$q_1'=n\frac{dX'}{dx}+p_1X,$$

$$q_2'=\frac{n(n-1)}{1.2}\frac{d^2X}{dx^2}+np_1\frac{dX}{dx}+p_2X,$$

$$\cdot \quad \cdot \quad \cdot \qquad \qquad \cdot \quad \cdot$$

$$q_n'=\frac{d^nX}{dx^n}+p_1\frac{d^{n-1}X}{dx^{n-1}}+\ldots+p_nX;$$

or, dividing by X, calling the new coefficients $q_1, q_2, \ldots, q_n$, and writing $X^{(i)}$ for $\frac{d^iX}{dx^i}$,

(9) $$\begin{cases} q_1=n\frac{X'}{X}+p_1, \\ q_2=\frac{n(n-1)X''}{1.2\quad X}+np_1\frac{X'}{X}+p_2, \\ \cdot \quad \cdot \quad \cdot \quad \cdot \quad \cdot \quad \cdot \quad \cdot \\ q_n=\frac{1}{X}[X^{(n)}+p_1X^{(n-1)}+\ldots+p_nX]. \end{cases}$$

Now $\frac{X'}{X}$ has obviously the point $x=0$ as a pole of order 1; $\frac{X''}{X}$ has $x=0$ as a pole of order 2; and in general $\frac{X^{(i)}}{X}$ has $x=0$ as a pole of order i.

If then $p_1, p_2, \ldots, p_n$ have $x=0$ as a pole of at most the order 1, 2, . . . , n, respectively, it follows that the coefficients $q_1, q_2, \ldots, q_n$ in the equation

(10) $$\frac{d^n\eta}{dx^n}+q_1\frac{d^{n-1}\eta}{dx^{n-1}}+\ldots+q_n\eta=0$$

have $x = 0$ as a pole of at most the orders 1, 2, . . . , n, respectively. (It is easy to see that the order of multiplicity of the pole $x = 0$ in q_i may be less than i, but under the hypothesis as to the coefficients p_i it can never be greater than i).

If now the equation in y with coefficients p has all of its integrals regular in the region of the critical point $x = 0$, it follows, from the relation

$$\eta = \frac{1}{X} y = yx^{-\lambda} \sum_{i=0}^{i=\infty} c_i x^i ,$$

that all the integrals of the transformed equation (10) are regular; conversely, if all the integrals of (10) are regular and its coefficients possess the property in question, then all the integrals of

$$(11) \qquad \frac{d^n y}{dx^n} + p_1 \frac{d^{n-1} y}{dx^{n-1}} + \ldots + p_n y = 0 \qquad .$$

will be regular, and the coefficients p will possess the same property, since (11) is derived from (10) by the transformation

$$\eta = \frac{1}{X} y.$$

We have seen that the equation (11) with uniform coefficients always admits of at least *one* integral of the form

$$y_0 = x^r \phi_0 ,$$

where ϕ_0 is a uniform function containing only positive powers of x, and different from zero when $x = 0$. If we replace the function X by y_0, $= x^r \phi_0$, equation (10) retains its form, but the coefficient q_n becomes

$$\frac{1}{y_0}\left[\frac{d^n y_0}{dx^n} + p_1 \frac{d^{n-1} y_0}{dx^{n-1}} + \ldots + p_n y_0\right],$$

which vanishes, and therefore (10) must be written in the form

$$(12) \qquad \frac{d^n \eta}{dx^n} + q_1 \frac{d^{n-1} \eta}{dx^{n-1}} + \ldots + q_{n-1} \frac{d\eta}{dx} = 0;$$

a differential equation of the nth order having 1, or any constant, as an integral; such an integral is of course regular.

Consider the differential equation of the first order

$$\frac{dy}{dx} + Py = 0, \tag{13}$$

which admits a regular integral,

$$y = x^r \Sigma c_i x^i.$$

The quotient,

$$\frac{1}{y}\frac{dy}{dx},$$

admits obviously the point $x = 0$ as a pole of order 1 at most, and therefore P must have $x = 0$ for a pole of order 1 at most; for if P had $x = 0$ for a pole of order, say, $\alpha > 1$, then there would be terms in P which could not cancel with any terms in $\frac{1}{y}\frac{dy}{dx}$.

Again, take the differential equation of the second order

$$\frac{d^2y}{dx^2} + P\frac{dy}{dx} + Qy = 0, \tag{14}$$

having all its integrals regular in the region of the point $x = 0$. Suppose $Y, = x^r\phi_0$, to be one of them; then writing

$$y = Yz,$$

it is clear that z must be regular; substituting, we have

$$\frac{d^2z}{dx^2} + \left[\frac{2}{Y}\frac{dY}{dx} + P\right]\frac{dz}{dx} + \frac{1}{Y}\left[\frac{d^2Y}{dx^2} + P\frac{dY}{dx} + QY\right] = 0,$$

or, making $\frac{dz}{dx} = v$ (v is therefore regular),

$$\frac{dv}{dx} + \left[\frac{2}{Y}\frac{dY}{dx} + P\right]v = 0. \tag{15}$$

As already seen, the coefficient of v here has $x = 0$ as a pole of order 1 at most, and since $\frac{1}{Y}\frac{dY}{dx}$ possesses this property, it follows that P has $x = 0$ as a pole of order 1 at most.

Write

$$w = e^{-\frac{1}{2}\int P dx}, \quad y = wv;$$

then substituting in (14), we have

$$(16) \qquad \begin{cases} \dfrac{d^2v}{dx^2} + Iv = 0, \\ I = Q - \dfrac{1}{2}\dfrac{dP}{dx} - \dfrac{1}{4}P^2. \end{cases}$$

If y_1 and y_2 are two independent regular integrals of the given equation, and corresponding to them are v_1 and v_2, then

$$(17) \qquad v_1 = y_1 e^{\frac{1}{2}\int P dx}, \quad v_2 = y_2 e^{\frac{1}{2}\int P dx}.$$

From what has been found concerning P it follows that

$$e^{\frac{1}{2}\int P dx}$$

is regular, and therefore that v_1 and v_2 are regular. The equation

$$(18) \qquad \frac{d^2v}{dx^2} + Iv = 0$$

has then its integrals regular, and consequently

$$\frac{1}{v}\frac{d^2v}{dx^2}$$

admits $x = 0$ as a pole of order 2 at most, and consequently I, and therefore Q, admits $x = 0$ as a pole of order 2 at most. (It is easy to see from (18) that P and Q have the required property.) A similar process might be employed for the differential equation of the third order, but it is rather long, and besides unnecessary. These two illustrations show the truth of the theorem for $n = 1, 2$.

Returning now to equation (12), write $\frac{d\eta}{dx} = \eta'$, and the equation becomes

$$(19) \qquad \frac{d^{n-1}\eta'}{dx^{n-1}} + q_1 \frac{d^{n-2}\eta'}{dx^{n-1}} + \ldots + q_{n-1}\eta' = 0.$$

The integrals of this are the derivatives of the integrals of (12); but these latter are regular, therefore the integrals of (19) are regular. Suppose now the theorem to hold true for (19); then q_i admits $x=0$ as a pole of order i at most. The theorem is then true for equation (12); and since (12) is derived from (11), by the transformation

$$(20) \qquad \eta = \frac{1}{X} y$$

it is also true for (11). We have then for the region of $x = 0$ the equations

$$(21) \qquad p_1 = \frac{P_1}{x}, \quad p_2 = \frac{P_2}{x^2}, \quad \ldots, \quad p_n = \frac{P_n}{x^n};$$

where $P_1, P_2, \ldots, P_n$ are in the region of $x=0$ uniform continuous functions, such as

$$(22) \qquad \begin{cases} P_1 = a_0 + a_1 x + a_2 x^2 + \ldots \\ P_2 = b_0 + b_1 x + b_2 x^2 + \ldots \\ \cdot \quad \cdot \quad \cdot \quad \cdot \quad \cdot \quad \cdot \\ P_n = l_0 + l_1 x + l_2 x^2 + \ldots \end{cases}$$

If we know the radius of the common circle of convergence of these series, say ρ, then the values of the coefficients $a, b, \ldots$ will be limited by the inequalities

$$\text{mod. } a_m \overset{=}{<} \frac{M}{\rho^m}, \quad \text{mod. } b_m \overset{=}{<} \frac{M}{\rho^m}, \ldots$$

where M is a properly chosen constant. If we substitute in (11) for y the value

$$(23) \qquad y = x^r,$$

we have as the result

$$(24)\qquad F(r)x^r+\phi_1(r)x^{r+1}+\phi_2(r)x^{r+2}+\ldots, \quad = x^r\Phi(x,r).$$

where

$$(25)\quad F(r)=r(r-1)\ldots(r-n+1)+a_0r(r-1)\ldots(r-n+2) \\ +\ldots+b_0r(r-1)\ldots(r-n+3)+\ldots,$$

$$(26)\quad \phi_m(r)=a_mr(r-1)\ldots(r-n+2) \\ +b_mr(r-1)\ldots(r-n+3)+\ldots$$

The equation $F(r) = 0$ is called by Fuchs the "*determinirende Fundamentalgleichung;*" following Cayley (*Quart. Journ. of Math.*, 1886) we will call it the *Indicial Equation*.

We will now give Fuchs's investigation concerning the forms of the coefficients in the case where the differential equation has all of its integrals regular. Let $0, x_1, x_2, \ldots, x_\rho$ denote the critical points of the integrals, and write

$$\psi(x)=x(x-x_1)\ldots(x-x_\rho);$$

then the differential equation under consideration has the form

$$(27)\qquad \frac{d^ny}{dx^n}+\frac{P_\rho(x)}{\psi(x)}\frac{d^{n-1}y}{dx^2}+\frac{P_{2\rho}(x)}{\psi^2(x)}\frac{d^{n-2}y}{dx^{n-2}} \\ +\ldots+\frac{P_{n\rho}(x)}{\psi^n(x)}y=0;$$

where $P_{k\rho}$ denotes a polynomial in x of the degree $k\rho$, or of a lower degree.

We have seen in Chapter III that a system of fundamental integrals of the equation

$$(28)\qquad \frac{d^ny}{dx^n}+p_1\frac{d^{n-1}y}{dx^{n-1}}+\ldots+p_ny=0$$

can be obtained of the form

$$(29)\qquad y_1,\quad y_2=y_1\int z_1dx,\quad y_3=y_1\int z_1dx\int t_1dx,\ \ldots$$

where z_1, t_1, . . . are integrals of linear differential equations of orders $n - 1$, $n - 2$, . . . respectively. It is obvious that these auxiliary functions can be so chosen that the fundamental integrals $y_1, \ldots, y_n$ shall submit to the substitution (80) (Chapter III) when the variable turns round the point $x = 0$. The integrals of the equation of the nth order are

$$y_1, \quad y_2, \quad \ldots, \quad y_n;$$

those of the equation of order $n - 1$ are

$$z_1, \quad z_2, \quad \ldots, \quad z_{n-1};$$

those of the equation of order $n - 2$ are

$$t_1, \quad t_2, \quad \ldots, \quad t_{n-2};$$

etc. It is easy to see that the integrals y_1, z_1, t_1, . . . are all of the form $x^r\phi(x)$, where $\phi(x)$ is a uniform function different from zero for $x = 0$, and that they belong respectively to the exponents

$$r_1, \quad r_2 - r_1 - 1, \quad r_3 - r_2 - 1 \ \ldots$$

The case where the difference of two exponents, say r_i and r_k, is zero gives rise to the "case of exception" mentioned above; the difficulty arising here can, however, be removed by replacing one of the two corresponding integrals by a proper linear function of both. For example, take the differential equation

$$\frac{d^2y}{dx^2} - \frac{x-4}{2x(x-2)}\frac{dy}{dx} + \frac{x-3}{2x^2(x-2)}y = 0;$$

this admits as solutions the two independent functions

$$x^{\frac{1}{2}} \quad \text{and} \quad (x^2 - 2x)^{\frac{1}{2}},$$

which both belong, in the region of $x = 0$, to the exponent $\frac{1}{2}$; we can, however, replace them by the independent functions

$$x^{\frac{1}{2}}, \quad (x^2 - 2x)^{\frac{1}{2}} - \sqrt{-2}\,.\,x^{\frac{1}{2}},$$

or

$$x^{\frac{1}{2}}, \quad \frac{x^{\frac{3}{2}}}{(x-2)^{\frac{1}{2}} + \sqrt{-2}\,.\,x^{\frac{1}{2}}},$$

belonging to the exponents $\frac{1}{2}$ and $\frac{3}{2}$ respectively. This "case of exception" being borne in mind, we need not refer to it again unless it is absolutely necessary in some particular case.

Suppose now that y_1 is known and is of the form

$$(30) \qquad y_1 = x^{r_1}\phi(x),$$

and satisfying from (80′) the relation

$$Sy_1 = s_1 y_1;$$

suppose further that we write, as we may,

$$(31) \qquad y_2 = y_1 \int z_1 dx;$$

where $Sy_2 = s_1 y_2 + s_1 y_1$, and where y_2 belongs to the exponent r_2: we have

$$(32) \qquad z_1 = \frac{d}{dx}\frac{y_2}{y_1},$$

from which follows

$$(33) \qquad Sz_1 = \frac{d}{dx} S \frac{y_2}{y_1} = \frac{d}{dx}\frac{s_1 y_2 + s_1 y_1}{s_1 y_1} = \frac{d}{dx}\frac{y_2}{y_1}.$$

From this it is clear that z_1 is a uniform function belonging to the exponent $r_2 - r_1 - 1$. It is easy to show that the remaining integrals $z_2, \ldots, z_{n-1}$, which with z_1 form a fundamental system for the equation of the $(n-1)^{st}$ order belong to the exponents

$$r_3 - r_1 - 1, \quad r_4 - r_1 - 1, \quad \ldots;$$

and also that in

$$y_3 = y_1 \int z_1 dx \int t_1 dx$$

the function t_1 is uniform and belongs to the exponent $r_3 - r_2 - 1$.

By continuing in the manner indicated, we find finally that the functions

$$y_1, \quad y_1 z_1, \quad y_1 z_1 t_1, \quad \ldots, \quad y_1 z_1 t_1 \ldots w,$$

belong respectively to the exponents

$$r_1, \quad r_2 - 1, \quad r_3 - 2, \quad \ldots, \quad r_n - (n-1).$$

From equation (16), Chapter III, we have,

$$(34) \qquad D = Cy_1^n z_1^{n-1} \ldots w_1,$$

where C is a non-vanishing constant. It follows then on substituting for y_1 its value, and remembering that $z_1, t_1, \ldots, w_1$ are all uniform functions, that

$$(35) \qquad D = x^{r_1 + r_2 + \ldots + r_n - \frac{n(n-1)}{2}} \psi(x);$$

where $\psi(x)$ is uniform and continuous in the region $x = 0$, and does not vanish when $x = 0$. A similar form is at once obtained for the point infinity: it is only necessary to change the variable by the relation

$$x = \frac{1}{t},$$

and, as before, investigate the point $t = 0$. We shall return to this point in a moment. We have

$$(36) \qquad p_i = -\frac{D_i}{D};$$

denote, as above, by δ the determinant of the substitution arising from travelling round $x = 0$, and write

$$(37) \qquad \frac{\log \delta}{2\pi i} = \beta.$$

It follows now at once that

$$(38) \qquad \begin{cases} D = x^{m+\beta} \psi(x), \\ D_i = x^{m'+\beta} \psi'(x); \end{cases}$$

m and m' being integers, and $\psi(x)$, $\psi'(x)$ uniform and continuous functions of x in the region of $x = 0$. In developing these determinants no logarithms will appear. These results have already been obtained. Suppose now we multiply each element in D_i and D by x raised to the power denoted by the negative of the exponent to which the corresponding element belongs. The determinant D_i will then be multiplied by

$$A, \quad = x^{-\left\{\sum_1^n r_i - \frac{n(n-1)}{2} - i\right\}},$$

and D by

$$B_i \quad = x^{-\left\{\sum_1^n r_i - \frac{n(n-1)}{2}\right\}}.$$

The quotient,

$$\frac{AD_i}{BD},$$

differs from $\frac{D_i}{D}$, *i.e.*, $-p_i$, by the factor $\frac{1}{x^i}$. It follows now at once that $p_i x^i$ is a uniform and continuous function of x which may become zero for $x = 0$, but may not become infinite for this value of x. We have then, finally,

$$(39) \qquad p_i = \frac{P_i}{x^i},$$

where P_i is a uniform and continuous function in the region of $x = 0$; and if the integrals are regular in the regions of all the critical points $0, x_1, \ldots, x_\rho$, then

$$(40) \qquad p_i = \frac{P_i}{\psi^i};$$

where $\psi = x(x - x_1) \ldots (x - x_\rho)$, and P_i now denotes a uniform and continuous function in all the plane. The equation has now the form

$$(41) \qquad \frac{d^ny}{dx^n} + \frac{P_1}{\psi}\frac{d^{n-1}y}{dx^{n-1}} + \frac{P_2}{\psi^2}\frac{d^{n-2}y}{dx^{n-2}} + \ldots + \frac{P_n}{\psi^n}y = 0.$$

To study the point ∞, write $x = \frac{1}{t}$. Transforming the equation, we find without difficulty

$$(42) \qquad \frac{d^ny}{dt^n} + \left.\begin{array}{l} a_{n,n-1}\frac{1}{t} \\ -\frac{P_1}{\psi}\frac{1}{t^2} \\ \\ \end{array}\right| \frac{d^{n-1}y}{dt^{n-1}} + \left.\begin{array}{r} a_{n,n-2}\frac{1}{t^2} \\ -a_{n-1,n-2}\frac{P_1}{\psi}\frac{1}{t^3} \\ +\frac{P_2}{\psi^2}\frac{1}{t^4} \end{array}\right| \frac{d^{n-2}y}{dt^{n-2}} + \ldots = 0;$$

where $x = \frac{1}{t}$ has to be substituted in $\psi, P_1, \ldots, P_n$. The quantities $a_{k,k-1}, a_{k,k-2}, \ldots$ are integers, the first of which $= k(k-1)$. The point $t = 0$ is now a pole (or ordinary point) of order at most $= 1, 2, \ldots, n$, for each of the coefficients of

$$\frac{d^{n-1}y}{dt^{n-1}}, \quad \frac{d^{n-2}y}{dt^{n-2}}, \quad \ldots, \quad y;$$

it is therefore necessary that $\frac{P_1}{\psi}, \frac{P_2}{\psi^2}, \ldots$ be developable in positive ascending powers of t, and that the first term in each of these developments shall be of the degree $1, 2, \ldots, n$, respectively. Now,

$$\frac{1}{\psi} = \frac{1}{\frac{1}{t}\left(\frac{1}{t} - x_1\right) \ldots \left(\frac{1}{t} - x_\rho\right)} = t^{\rho+1} + c_1 t^\rho + \ldots;$$

we must therefore have

$$(43) \quad \begin{cases} P_1 = \frac{d_1}{t^\rho} + \frac{d_2}{t^{\rho-1}} + \ldots = d_1 t^\rho + d_2 t^{\rho-1} + \ldots \\ P_2 = \frac{c_1}{t^{2\rho}} + \frac{c_2}{t^{2\rho-1}} + \ldots = c_1 t^{2\rho} + c_2 t^{2\rho-1} + \ldots \\ \ldots \quad \ldots \quad \ldots \end{cases}$$

That is, $P_1, P_2, \ldots, P_n$ are polynomials of the degrees

$$\rho, \quad 2\rho, \quad 3\rho, \quad \ldots, \quad n\rho;$$

where $\rho + 1$ is the number of finite critical points.

We shall now give Fuchs's proof (which has only so far been faintly indicated) of the converse of the theorem just proved; viz., we shall prove that *every linear differential equation of the form*

$$(44) \qquad \frac{d^n y}{dx^n} + \frac{M_1}{x}\frac{d^{n-1}y}{dx^{n-1}} + \frac{M_2}{x^2}\frac{d^{n-2}y}{dx^{n-2}} + \ldots + \frac{M_n}{x^n} y = 0,$$

where $M_1, M_2, \ldots, M_n$ are uniform and continuous functions of x in the region of the critical point $x = 0$, admits in the region of this point a system of fundamental integrals all of which are regular.

The method employed in Fuchs's proof is the same as that used by Weierstrass in his general proof of the existence of an integral of an algebraic differential equation. Instead, however, of referring directly to Fuchs's memoir, we shall follow Tannery's exposition of the same.

From what has been shown we know that (44) admits at least one integral of the form

$$y = x^r\phi(x), \tag{45}$$

where $\phi(x)$ is a uniform and continuous function of x, and is not zero for $x = 0$. If now we substitute

$$y = x^r\eta \tag{46}$$

in (44), we shall have

$$\frac{d^n\eta}{dx^n} + \frac{M_1'}{x}\frac{d^{n-1}\eta}{dx^{n-1}} + \frac{M_2'}{x^2}\frac{d^{n-2}\eta}{dx^{n-2}} + \ldots + \frac{M_n'}{x^n}\eta = 0; \tag{47}$$

where

$$\left\{\begin{aligned} M_k' = {} & \frac{n(n-1)\ldots(n-k+1)}{\underline{k|}} r(r-1)\ldots(r-k+1) \\ & + \frac{(n-1)(n-2)\ldots(n-k+1)}{\underline{k-1|}} r(r-1)\ldots(r-k+2)M_1(x) \\ & + \ldots + (n-k+1)rM_{k-1}(x) + M_k(x); \\ M_n' = {} & r(r-1)\ldots(r-n+1) + r(r-1)\ldots(r-n+2)M_1(x) \\ & + \ldots + rM_{n-1}(x) + M_n(x). \end{aligned}\right. \tag{48}$$

Now if (44) admits as solution the value of y in (45), then (47) must admit a solution of the form

$$\eta = C_0 + C_1x + C_2x^2 + \ldots; \tag{49}$$

where C_0 is necessarily different from zero. Before going farther take, for example, the case $n = 2$, viz.:

$$\frac{d^2y}{dx^2} + \frac{M_1}{x}\frac{dy}{dx} + \frac{M_2}{x^2}y = 0. \tag{50}$$

Substitute here $y = x^r\eta$, and we have

$$(51)\quad \frac{d^2\eta}{dx^2} + \frac{2r(r-1) + M_1(x)}{x}\frac{d\eta}{dx} + \frac{r(r-1) + rM_1(x) + M_2(x)}{x^2}\eta = 0.$$

Now this must have an integral of the form

$$(52)\qquad \eta = C_0 + C_1x + C_2x^2 + \ldots;$$

where C_0 cannot vanish. This value of η, substituted in (51), must give a result which is identically zero; *i.e.*, the coefficient of every power of x must vanish. Bearing in mind now that M_1 and M_2 are uniform, and, in the region of $x = 0$, continuous functions, it is clear that x^{-2} is the highest negative power of x that will appear after making the substitution; the coefficient of x^{-2} is obviously the quantity

$$C_0[r(r-1) + rM_1(0) + M_2(0)];$$

that this may vanish we must have, since C_0 is not $= 0$,

$$(53)\qquad r(r-1) + rM_1(0) + M_2(0) = 0.$$

In the same way we see that if (49) is a solution of (47) we must have that the coefficient of x^{-n} vanishes; that is,

$$M_n'(0) = 0,$$

or

$$(54)\quad \begin{aligned} r(r-1)\ldots(r-n+1) &+ r(r-1)\ldots(r-n+2)M_1(0) \\ &+ \ldots + r(r-1)\ldots(r-n+3)M_2(0) \\ &+ \ldots + rM_{n-1}(0) + M_n(0) = 0. \end{aligned}$$

This is the same as equation (25), and will be called the indicial equation. If r_i is a root of the indicial equation, and s_i the corresponding root of the characteristic equation, we have

$$r_i = \frac{\log s_i}{2\pi i},$$

a relation which has already been shown.

Notice that if $M_n(x) = 0$, and of course then $M_n(0) = 0$, the indicial equation is divisible by r. If after division by r we change r into $r + 1$, we obtain the indicial equation corresponding to the differential equation of order $n - 1$, which is formed by taking $\frac{dy}{dx}$ for the new unknown function.

Suppose a system of fundamental integrals of (44) to be $y_1, y_2, \ldots, y_n$, belonging respectively to the roots $r_1, r_2, \ldots, r_n$ of the indicial equation. We will further suppose that the roots $r_1, \ldots, r_n$ are so arranged that none of the differences

$$r_2 - r_1 - 1, \quad r_3 - r_1 - 1, \quad \ldots, \quad r_n - r_1 - 1$$

is zero or a positive integer: this arrangement will be made if the real part of none of the roots written in the order $r_1, r_2, \ldots, r_n$ is greater than the real part of the preceding one.

We have already seen that if a regular function, say F, belongs to an exponent r, its derivative belongs to the exponent $r - 1$; also, if two regular functions, F and G, belong to exponents r and ρ respectively, their quotient $\frac{F}{G}$ belongs to the exponent $r - \rho$.

We can write the integrals of equation (44) in the form

$$y_1, \quad y_2 = y_1 \int z_1 dx, \quad y_3 = y_1 \int z_2 dx, \quad \ldots;$$

where z satisfies a linear differential equation of order $n - 1$. If now $y_1, y_2, \ldots, y_n$ belong to exponents $r_1, r_2, \ldots, r_n$, and if in (44) we make the substitution

$$y = y_1 \int z dx,$$

then the integrals (properly chosen) of the equation of order $n - 1$ in z will belong to the exponents

$$r_2 - r_1 - 1, \quad r_3 - r_1 - 1, \quad \ldots, \quad r_n - r_1 - 1;$$

or, in other words,

$$r_2 - r_1 - 1, \quad r_3 - r_1 - 1, \quad \ldots, \quad r_n - r_1 - 1$$

are the roots of the indicial equation corresponding to the differential equation in z. Again: if we write

$$z = \int t dx,$$

the roots of the indicial equation corresponding to the differential equation of the order $n - 2$ in t will be

$$r_3 - r_2 - 1, \quad r_4 - r_2 - 1, \quad \ldots, \quad r_n - r_2 - 1,$$

and so on. Suppose now that the roots of the indicial equation (54) or (25) are arranged in the prescribed order, viz., such that the differences

$$(55) \qquad r_2 - r_1 - 1, \quad r_3 - r_1 - 1, \quad \ldots, \quad r_n - r_1 - 1$$

are neither zero nor positive integers; then, *The differential equation* (44) *admits in the region of* $x = 0$ *a regular integral of the form*

$$y = x^{r_1} \phi(x);$$

i.e., $\phi(x)$ is in the region of $x = 0$ a uniform and continuous function of x, and does not vanish for $x = 0$.

In order to prove this, make in (44) the substitution

$$(56) \qquad y = x^{r_1} \eta,$$

and we have

$$(57) \qquad \frac{d^n\eta}{dx^n} + \frac{N_1}{x}\frac{d^{n-1}\eta}{dx^{n-1}} + \frac{N_2}{x^2}\frac{d^{n-2}\eta}{dx^{n-2}} + \ldots + \frac{N_n}{x^{n-1}}\eta = 0.$$

The functions $N_1, N_2, \ldots, N_{n-1}$ differ only from $M_1', M_2', \ldots, M'_{n-1}$ of equations (46) by the replacing of r by r_1, and by making

$$(58) \qquad N_n(x) = \frac{M_n'(x)}{x}.$$

If we write, for brevity,

$$N_1'(x) = \frac{N_1(x) - N_1(0)}{x} \quad \ldots N'_{n-1}(x) = \frac{N_{n-1}(x) - N_{n-1}(0)}{x},$$

equation (57) can be thrown into the form

$$(59)\qquad x^{n-1}\frac{d^n\eta}{dx^n} + N_1(0)x^{n-2}\frac{d^{n-2}\eta}{dx^{n-2}} + \ldots + N_{n-1}(0)\frac{d\eta}{dx}$$
$$= N_1'(x)x^{n-1}\frac{d^{n-1}\eta}{dx^{n-1}} + N_2'(x)x^{n-2}\frac{d^{n-2}\eta}{dx^{n-2}}$$
$$+ \ldots + N'_{n-1}(x)x\frac{d\eta}{dx} + N_n(x)\eta.$$

We have to show that this equation admits in the region of $x = 0$ an integral of the form

$$(60)\qquad \eta = C_0 + C_1x + C_2x^2 + \ldots;$$

substituting this value of η in (56), and equating the coefficients of x^k where k is an integer, we have

$$(61)\ [(k+1)k(k-1)\ldots(k-n+2) + N_1(0)(k+1)k(k-1)\ldots(k-n+3)$$
$$+ N_2(0)(k+1)k(k-1)\ldots(k-n+4) + \ldots + N_{n-1}(0)(k+1)]C_{k+1}$$
$$= A_0C_k + A_1C_{k-1} + \ldots + A_kC_0;$$

where the coefficients $A_0, A_1, \ldots, A_k$ are made up from mere numerical quantities and from the coefficients of the different powers of x in the development of the functions

$$N_1'(x),\quad N_2'(x),\quad \ldots,\quad N'_{n-1}(x),\quad N_n(x).$$

The coefficient of C_{k+1} equated to zero gives

$$(62)\quad (k+1)k(k-1)\ldots(k-n+2)+N_1(0)(k+1)k(k-1)\ldots(k-n+3)$$
$$+ \ldots + N_{n-1}(0)(k+1) = 0.$$

This is simply the indicial equation corresponding to equation (57), divided through by r, and then r changed into $k+1$. Now the roots of this equation have been shown to be $r_2 - r_1 - 1, \ldots, r_n - r_1 - 1$, none of which are either zero or positive integers; therefore, since k is a positive integer, no such equation as (62) can exist, and consequently the coefficient of C_{k+1} can never vanish for any positive integer value of k, nor for $k = 0$.

From (61) we can now obviously determine each coefficient C in the form

$$(63)\qquad C_a = \mathfrak{A}_a c_0.$$

We have now to establish the convergence of this series

$$\eta = C_0 + C_1 x + C_2 x^2 + \ldots$$

in the region of $x = 0$; for this purpose we compare it with another series, the convergence of which is readily ascertained.

Let $\Omega_1, \Omega_2, \ldots, \Omega_{n-1}, \Omega_n$ denote the maximum moduli of $N_1'(x), \ldots, N'_{n-1}(x), N_n(x)$; then, by a known theorem in the theory of functions,

$$(64) \quad \begin{cases} \text{mod.} \left[\dfrac{d^\alpha N_1'(x)}{dx^\alpha}\right]_{x=0} < \underline{\alpha|}\dfrac{\Omega_1}{\rho^\alpha}, \\ \text{mod.} \left[\dfrac{d^\alpha N_2'(x)}{dx^\alpha}\right]_{x=0} < \underline{\alpha|}\dfrac{\Omega_2}{\rho^\alpha}, \\ \cdot \quad \cdot \quad \cdot \quad \cdot \quad \cdot \quad \cdot \\ \text{mod.} \left[\dfrac{d^\alpha N_n(x)}{dx^\alpha}\right]_{x=0} < \underline{\alpha|}\dfrac{\Omega_n}{\rho^\alpha}; \end{cases}$$

where ρ denotes the distance from the origin to the next nearest critical point. We form now the auxiliary differential equation

$$(65) \quad \gamma_1 x^{n-2}\frac{d^{n-1}v}{dx^{n-1}} + \gamma_2 x^{n-3}\frac{d^{n-2}v}{dx^{n-2}} + \ldots + \gamma_{n-1}\frac{dv}{dx}$$
$$= \frac{\Omega_1}{1-\frac{x}{\rho}}x^{n-1}\frac{d^{n-1}v}{dx^{n-1}} + \frac{\Omega_2}{1-\frac{x}{\rho}}x^{n-2}\frac{d^{n-2}v}{dx^{n-2}} + \ldots + \frac{\Omega_{n-1}}{1-\frac{x}{\rho}}x\frac{dv}{dx}$$
$$+ \frac{\Omega_n}{1-\frac{x}{\rho}}v;$$

where the quantities $\gamma_1, \ldots, \gamma_{n-1}$ are arbitrary positive quantities subject merely to the condition that no root of the equation

$$(66) \quad \gamma_1 w(w-1)\ldots(w-n+3) + \gamma_2 w(w-1)\ldots(w-n+4)$$
$$+ \ldots + \gamma_{n-1} = 0$$

shall be either zero or a positive integer. We seek now to satisfy this equation by a series

$$(67) \qquad v = \sum_0^\infty g_\alpha x^\alpha,$$

where g_0 is not zero. As above, substitute this value in (65), and equate coefficients of x^k: we have without difficulty

$$(68)\quad [\gamma_1(k+1)k\ldots(k-n+3)+\gamma_2(k+1)k\ldots(k-n+4)\ldots$$
$$+\gamma_{n-1}(k+1)]g_{k+1}=B_0g_k+B_1g_{k-1}+\ldots+B_kg_0;$$

where the coefficients B are formed in the same way as the coefficients A in (61); viz., the B's are formed out of the coefficients in the developments of

$$\frac{\Omega_1}{1-\frac{x}{\rho}}, \ldots, \frac{\Omega_n}{1-\frac{x}{\rho}},$$

according to ascending powers of x, and out of certain numerical quantities, just as the A's are formed from the coefficients in the developments of

$$N_1'(x),\quad N_2'(x),\quad \ldots,\quad N_n(x),$$

and the same numerical quantities. Since the B's are obviously all positive it follows from (64) that

$$(69)\qquad B_\alpha > \text{mod. } A_\alpha,\quad \alpha = 1, 2, \ldots, k.$$

Equation (68) gives now

$$(70)\qquad g_\alpha = \mathfrak{B}_\alpha g_0,$$

where $\mathfrak{B}_\alpha$ is a positive quantity; if then we take g_0 positive, as we obviously may, all the coefficients $g_1, g_2, \ldots$ will be positive. It is obvious that there must exist a finite limit for k, say $k = t$, such that for $k \geqq t$ we have always

$$(71)\quad \text{mod.}\left[k(k-1)\ldots(k-n+2)+N_1(0)k(k-1)\ldots(k-n+3)\right.$$
$$\left.+\ldots+N_{n-1}(0)\right]$$
$$>\gamma_1k(k-1)\ldots(k-n+2)+\gamma_2k(k-1)\ldots(k-n+3)+\ldots\gamma_{n-1}.$$

Now from (61), (68), (69), and (71) it is clear that we shall have

$$g_k > \text{mod. } C_k,$$

if this can be shown to hold for $k \leqq t$.

Let **A** denote the largest modulus in the series

$$1, \quad \mathfrak{A}_1, \quad \mathfrak{A}_2, \quad \ldots, \quad \mathfrak{A}_t,$$

and let **B** denote the least of the quantities

$$1, \quad \mathfrak{B}_1, \quad \mathfrak{B}_2, \quad \ldots, \quad \mathfrak{B}_t,$$

and suppose we choose C_0 so that

(72) $$\mathbf{A} \text{ mod. } C_0 < \mathbf{B} g_0 .$$

Since **A**, **B**, g_0 are different from zero, this inequality is of course always possible, and gives for C_0 a value different from zero. It is clear now from (63), (70), and (72), remembering the hypothesis as to **A** and **B**, that for $k \leqq t$, and consequently for all values k, that we have always

$$g_k > \text{mod. } C_k .$$

If then the series

(73) $$v = g_0 + g_1 x + g_2 x^2 + \ldots$$

is convergent, it follows *à fortiori* that

(74) $$\eta = C_0 + C_1 x + C_2 x^2 + \ldots$$

is convergent. To establish the convergence of (73) in the region of $x = 0$, we proceed as follows: Multiply out (65) by $1 - \frac{x}{\rho}$, and then substitute the above value of v; this gives, on equating the coefficients of x^k,

(75) $$[(k+1)k(k-1)\ldots(k-n+3)\gamma_1 + \ldots + (k+1)\gamma_{n-1}]g_{k+1}$$
$$= \Big[k(k-1)\ldots(k-n+2)\Big(\frac{\gamma_1}{\rho}\Big) + \Omega_1\Big)$$
$$+ k(k-1)\ldots(k-n+3)\Big(\frac{\gamma_2}{\rho} + \Omega_2\Big)$$
$$+ \ldots k\Big(\frac{\gamma_{n-1}}{\rho} + \Omega_{n-1}\Big) + \Omega_n\Big]g_k.$$

We have from this, for the limit of the ratio of two consecutive coefficients g_{k+1}, g_k,

$$\lim_{k=\infty} \frac{g_{k+1}}{g_k} = \frac{\gamma_1 + \rho\Omega_1}{\rho\gamma_1},$$

and for the limit of the ratio of the two corresponding terms in the series (73)

$$\lim_{k=\infty} \frac{g_{k+1}x^{k+1}}{g_k x^k} = \frac{\gamma_1 + \rho\Omega_1}{\rho\gamma_1} x.$$

If

$$\text{mod.} \frac{\gamma_1 + \rho\Omega_1}{\rho\gamma_1} x, \quad = \frac{\gamma_1 + \rho\Omega_1}{\rho\gamma_1} \text{mod.}\, x, < 1,$$

then the series is convergent; that this may be, we need only limit x by the inequality

$$\text{mod.}\, x < \frac{\rho\gamma_1}{\gamma_1 + \rho\Omega_1}.$$

Since γ_1 and ρ are positive non-vanishing quantities it is obvious that (73) has a circle of convergence having $x = 0$ as its centre. It follows at once that (74) is convergent inside this same circle. As γ_1 can be taken as large as we please, we have

$$\lim_{\gamma_1=\infty} \frac{\rho\gamma_1}{\gamma_1 + \rho\Omega_1} = \rho,$$

so that it is clear that the series (74) is convergent in the entire region of $x = 0$. This can also be shown by a well-known process in the theory of functions. We have shown now that (59) admits as a solution the uniform and continuous function

$$\eta = C_0 + C_1 x + C_2 x^2 + \ldots,$$

and it therefore follows that (44) admits the solution

$$y_1 = x^{r_1}\eta, \qquad \text{or, say,} \qquad y_1 = x^{r_1}\phi(x);$$

where $\phi(x)$ is a uniform and continuous function of x in the region of $x = 0$. We have now shown that the given equation, (44), possesses one regular integral; the remainder of the theorem, viz., that *all* of the integrals of (44) are regular, is quickly proved. If in (44) we write

$$y = y_1 \int z dx,$$

we find, as already shown, for z an equation of order $n - 1$, which in form is the same as (44). The roots of the indicial equation corresponding to the z differential equation are

$$r_2 - r_1 - 1, \quad r_3 - r_1 - 1, \quad \ldots, \quad r_n - r_1 - 1;$$

denote these for a moment by

$$\rho_2, \quad \rho_3, \quad \ldots, \quad \rho_n;$$

then, since $r_1, r_2, \ldots, r_n$ are so arranged that for $\beta > \alpha$ the difference $r_\beta - r_\alpha$ is never a positive integer, it follows that the same property exists for the roots $\rho_2, \rho_3, \ldots, \rho_n$; viz., $\rho_\beta - \rho_\alpha$ is for $\beta > \alpha$ never a positive integer. The differential equation in z therefore admits a regular integral of the form

$$z_1 = x^{r_2 - r_1 - 1}\psi(x),$$

$\psi(x)$ being a uniform and continuous function of x in the region of $x = 0$. Corresponding to z_1 we have

$$y_2 = y_1 \int z_1 dx,$$

an integral of (44) belonging to the exponent

$$r_2 - r_1 - 1 + 1 + r_1, \quad = r_2.$$

By the same process we can find a third regular integral of (44),

$$y_3 = y_1 \int z_1 dx \int t_1 dx,$$

where t_1 is a regular integral of a differential equation of order $n - 2$, and belongs to the exponent $r_3 - r_2 - 1$. The integral y_3 therefore belongs to the exponent

$$r_3 - r_2 - 1 + 1 + r_2 - r_1 - 1 + 1 + r_1, \quad = r_3.$$

The integrals z_1, t_1, . . . , w_1 of the auxiliary differential equations are always of the form

$$x^r \theta(x),$$

where $\theta(x)$ is a uniform and continuous function of x. The logarithms which appear in the integrals of (44) can then only enter through the different integrations which have to be performed.

We have thus established the existence of a fundamental system of regular integrals of equation (44), the elements of which system belong respectively to the exponents r_1, r_2, . . . , r_n, which are the roots of the indicial equation. It is also obvious that these exponents are respectively equal to

$$\frac{\log s_1}{2\pi i}, \quad \frac{\log s_2}{2\pi i}, \quad . \; . \; . \; , \quad \frac{\log s_n}{2\pi i};$$

where s_1, s_2, . . . , s_n are the roots of the characteristic equation—a result arrived at in what precedes.

Resume for a moment equation (41), viz.,

$$\frac{d^n y}{dx^n} + \frac{P_1}{\psi}\frac{d^{n-1} y}{dx^{n-1}} + \dots + \frac{P_n}{\psi^n} y = 0;$$

where $\psi = x(x - x_1)(x - x_2) \dots (x - x_\rho)$, and consider the critical point $x = x_i$. We can show that the sum of the roots of the indicial equations for the critical points is equal to $\rho \dfrac{n(n-1)}{2}$. Denote by $\psi'(x_i)$ the derivative $\left(\dfrac{d\psi}{dx}\right)_{x = x_i}$; the indicial equation relative to this point is then

$$(76) \quad r(r-1)\dots(r-n+1) + \frac{P_1(x_i)}{\psi'(x_i)} r(r-1)\dots(r-n+2)$$
$$+ \frac{P_2(x_i)}{[\psi'(x_i)]^2} r(r-1)\dots(r-n+3) + \dots = 0.$$

The sum of the roots of this equation is

$$\frac{n(n-1)}{2} - \frac{P_1(x_i)}{\psi'(x_i)}.$$

Again: the indicial equation corresponding to the critical point $x = \infty$ is

$$r(r-1)\ldots(r-n+1)+(a_{n,n-1}-d_1)r(r-1)\ldots(r-n+2)+\ldots=0.$$

The sum of the roots of this is

$$= \frac{n(n-1)}{2} - a_{n,n-1} + d_1 = -\frac{n(n-1)}{2} + d_1,$$

since $a_{n,n-1} = n(n-1)$.

The sum total of the roots of all the indicial equations is

$$\rho\frac{n(n-1)}{2} + d_1 - \Sigma\frac{P_1(x_i)}{\psi'(x_i)}.$$

Now by a well-known formula for the decomposition of rational fractions we have

$$\Sigma\frac{P_1(x_i)}{\psi'(x_i)}\frac{1}{x-x_i} = \frac{P_1(x)}{\psi(x)};$$

multiplying this by x and then making $x = \infty$, we have

$$\Sigma\frac{P_1(x_i)}{\psi'(x_i)} = d_1;$$

therefore, finally, we have for the sum of all the roots the value

$$\rho\frac{n(n-1)}{2}.$$

In concluding this chapter we will give a few illustrations of the development of the integrals in series, and the method of obtaining the coefficients. We shall use for convenience only the point $x = 0$, and begin with an equation of the third order, viz.,

(1) $$\frac{d^3y}{dx^3} + \frac{Q_1(x)}{x(x-a_1)(x-a_2)\ldots(x-a_\mu)}\frac{d^2y}{dx^2} + \frac{Q_2(x)}{x^2\{(x-a_1)\ldots(x-a_\mu)\}^2}\frac{dy}{dx} + \frac{Q_3(x)}{x^3\{(x-a_1)\ldots(x-a_\mu)\}^3}y = 0;$$

where the critical points of the coefficients are

$$x = 0, \quad x = a_1, \quad \ldots, \quad x = a_\mu$$

in number $\mu + 1$; and $Q_1(x)$, $Q_2(x)$, ... are polynomials in x of degrees μ, 2μ, ... respectively.

Equation (1) may be put into the following form:

$$(2) \qquad \frac{d^3y}{dx^3} + \frac{P_1(x)}{x}\frac{d^2y}{dx^2} + \frac{P_2(x)}{x^2}\frac{dy}{dx} + \frac{P_3(x)}{x^3}y = 0;$$

$P_1(x)$, $P_2(x)$, $P_3(x)$ being rational functions of x for which $x = 0$ is not a pole.

The indicial equation is

$$r(r-1)(r-2) + P_1(0)r(r-1) + P_2(0)r + P_3(0) = 0,$$

of which the roots (a, b, c) will first be supposed all different, and the difference between no pair of them an integer. Equation (2) will then have, in the region of the point $x = 0$, three integrals of the form

$$x^a\Sigma c_i x^i, \quad x^b\Sigma c_i' x^i, \quad x^c\Sigma c_i'' x^i, \quad (i = 0, 1, \ldots \infty.)$$

Equation (2) may be conveniently written

$$(4) \qquad x^3\frac{d^3y}{dx^3} + P_1 x^2\frac{d^2y}{dx^2} + P_2 x\frac{dy}{dx} + P_3 y = 0.$$

Substituting in this equation the series

$$y = \sum_{i=0}^{i=\infty} c_i x^{i+a},$$

and equating to zero the coefficient of $x^{\mu+a}$, we find

$$(5) \quad c_\mu F(\mu+a) + c_{\mu-1}\phi_1(\mu+a-1) + c_{\mu-2}\phi_2(\mu+a-2) + \ldots = 0.$$

The symbols F and ϕ_1, ϕ_2, ... have here the following meanings: The point $x = 0$ being an ordinary point for P_1, P_2, P_3, we may write

$$P_1 = a_0 + a_1 x + a_2 x^2 + \ldots,$$
$$P_2 = b_0 + b_1 x + b_2 x^2 + \ldots,$$
$$P_3 = d_0 + d_1 x + d_2 x^2 + \ldots;$$

if now in (4) we substitute $y = x^r$, then

$$F(r) = r(r-1)(r-2) + a_0 r(r-1) + b_0 r + d_0,$$
$$\phi_1(r) = a_1 r(r-1) + b_1 r + d_1,$$
$$\phi_2(r) = a_2 r(r-1) + b_2 r + d_2,$$
$$\cdots\cdots\cdots$$
$$\phi_\nu(r) = a_\nu r(r-1) + b_\nu r + d_\nu.$$

In equation (5), making $\mu = 0$, the quantities $c_{\mu-1}, \ldots$ all vanish identically, leaving only $c_0 F(a) = 0$; but $F(a) = 0$, since a is a root of $F(r) = 0$; hence c_0 is indeterminate.

Making $\mu = 1$, we have

$$c_1 F(1+a) + c_0 \phi_1(a) = 0.$$

If $\mu = 2$,

(7) $$c_2 F(a+2) + c_1 \phi_1(a+1) + c_0 \phi_2(a) = 0.$$

If $\mu = 3$,

$$c_3 F(a+3) + c_2 \phi_1(a+2) + c_1 \phi_2(a+1) + c_0 \phi_3(a) = 0,$$
$$\cdots\cdots\cdots\cdots$$

Hence

$$c_1 = -\frac{c_0 \phi_1(a)}{F(1+a)},$$

$$c_2 = \frac{+c_0}{F(a+2)} \left\{ \frac{\phi_1(a)\phi_1(a+1)}{F(1+a)} - \phi_2(a) \right\},$$
$$\cdots\cdots\cdots$$

Thus every c of index > 0 can, as seen above, be expressed as a product of c_0, which cannot $= 0$, into a function of the root a, and so the series $y = \sum_{i=0}^{i=\infty} c_i x^{\mu+a}$ is completely known when any value arbitrarily chosen has been assigned to c_0. In a manner precisely similar the series corresponding to the roots b and c may be calculated.

Suppose one of the roots, as a, to be zero: the corresponding integral will be $y = \sum_{i=0}^{i=\infty} c_i x^i$, which on being substituted in equation (4) gives for the coefficient of x^μ

$$c_\mu F(\mu) + c_{\mu-1}\phi_1(\mu - 1) + c_{\mu-2}\phi_2(\mu - 2) + \ldots = 0.$$

Making $\mu = 0$, we find

$$c_0 F(0) = 0, \quad \text{but} \quad F(0) = 0;$$

hence c_0 is arbitrary, since it cannot vanish. It may be noted that P_3 must be divisible by x in this case, or $d_0 = 0$.
For $\mu = 1$,

$$c_1 F(1) + c_0\phi_1(0) = 0.$$

For $\mu = 2$,

$$c_2 F(2) + c_1\phi_1(1) + c_0\phi_2(0) = 0.$$

$$\cdot \quad \cdot \quad \cdot \quad \cdot \quad \cdot \quad \cdot \quad \cdot$$

Therefore the coefficients c are to be obtained just as before. The case of all three roots of the indicial equation distinct is thus comparatively simple; but a numerical example may not be found entirely useless in this connection.

Suppose the coefficients P_1, P_2, P_3, when developed by Taylor's series in ascending powers of x, to have the following values:

$$P_1 = \tfrac{7}{6} + a_1x + a_2x^2 + a_3x^3 + \ldots,$$
$$P_2 = \tfrac{1}{6} + b_1x + b_2x^2 + b_3x^3 + \ldots,$$
$$P_3 = -\tfrac{1}{6} + d_1x + d_2x^2 + d_3x^3 + \ldots.$$

Thus $a_0 = \frac{7}{6}$, $b_0 = \frac{1}{6}$, $d_0 = -\frac{1}{6}$, and the indicial equation for $x = 0$ becomes

$$F(r) = r(r-1)(r-2) + \tfrac{7}{6}r(r-1) + \tfrac{1}{6}r - \tfrac{1}{6} = 0, \text{ or}$$

(6) $$r^3 - \tfrac{11}{6}r^2 + r - \tfrac{1}{6} = 0.$$

The roots of this cubic are 1, $\frac{1}{2}$, and $\frac{1}{3}$. Since no two of them differ by an integer, there are no logarithms in the integrals, and we may obtain them by putting 1, $\frac{1}{2}$, $\frac{1}{3}$ successively for a in equation (5). We have then, first,

$$c_1 F(2) + c_0 \phi_1(1) = 0. \tag{7}$$

But

$$F(2) = 2^3 - \tfrac{11}{6} \times 2^2 + 2 - \tfrac{1}{6} = 2\tfrac{1}{2},$$
$$\phi_1(1) = b_1 + d_1;$$

equation (7) therefore becomes

$$2\tfrac{1}{2}\, c_1 + (b_1 + d_1)\, c_0 = 0, \quad \text{or} \quad c_1 = -\frac{2(b_1 + d_1)}{5} c_0 .$$

Again,

$$F(a + 2) = F(3) = 3^3 - \tfrac{11}{6} \times 3^2 + 3 - \tfrac{1}{6} = 13\tfrac{1}{3},$$
$$\phi_1(a + 1) = \phi_1(2) = 2a_1 + 2b_1 + d_1,$$
$$\phi_2(1) = b_2 + d_2;$$

whence equation (7) becomes

$$13\tfrac{1}{2} c_2 - (2a_1 + 2b_1 + d_1)\frac{2(b_1 + d_1)}{5} c_0 + c_0 (b_2 + d_2) = 0;$$

that is,

$$c_2 = \frac{2c_0}{135}\{2(b_1 + d_1)(2a_1 + 2b_1 + d_1) - 5(b_2 + d_2)\}.$$

In the same way, as many of the coefficients may be calculated as are desirable. It will often happen that c_0, which is a common factor to all the terms, may be so chosen as to simplify the series more or less.

The roots $\frac{1}{3}$ and $\frac{1}{2}$ may now be treated in the same way, and each will give rise to a convergent infinite series all of whose coefficients except the first are determinate.

Returning to the supposition that a, b, c are the roots of $F(r) = 0$, let two of them, as a and b, be equal, and corresponding to the double root a we shall have two integrals; the first being the convergent infinite series

$$y_1 = \Sigma\, b_i x^{a+i}, \quad (i = 1, 2, \ldots, \infty,)$$

and the second of the form

$$y_2 = x^a\{\Sigma c_i x^i + \Sigma c_i' x^i \,.\, \log x\}.$$

Substituting y_2 in equation (4), the following results are obtained for the coefficients of $x^{a+\mu}$ and $x^{a+\mu}\log x$, which must separately vanish.

Coefficient of $x^{a+\mu}$:

$$\begin{aligned}(8)\quad & F(a+\mu)\,c_\mu + \phi_1(a+\mu-1)c_{\mu-1} + \phi_2(a+\mu-2)c_{\mu-2} + \dots \\ & + F'(a+\mu)c_\mu' + \phi_1'(a+\mu-1)c'_{\mu-1} + \phi_2'(a+\mu-2)c'_{\mu-2} + \dots = 0.\end{aligned}$$

Here $F'(a+\mu)$ denotes the result obtained by differentiating $F(r)$ with respect to r, and substituting $a+\mu$ for r in the result; and similarly for the other functions, ϕ_1', ϕ_2', . . . It may also be noted that if $\phi(x, r)x^r$ be the result of substituting x^r for y in any linear differential equation, the result of substituting $x^r \log^\lambda x$ is $\dfrac{d^\lambda}{dr^\lambda}[\phi(x, r)x^r]$. For $x^r \log^\lambda x = \dfrac{d^\lambda}{dr^\lambda} x^r$, and it is easy to see that it makes no difference whether we perform the differentiation λ times upon x^r and substitute the result, or substitute x^r and differentiate the result λ times with respect to r. In fact, the result of substituting x^r is

$$A\frac{d^n x^r}{dx^n} + B\frac{d^{n-1}x^r}{dx^{n-1}} + \dots = \phi(x, r)x^r,$$

where the coefficients A, B, . . . do not contain r, and since

$$\frac{d^\lambda}{dr^\lambda}\frac{d^n}{dx^n}(x^r) = \frac{d^n}{dx^n}\frac{d^\lambda}{dr^\lambda}(x^r).$$

Hence, knowing the result obtained by substituting x^r, that for $x^r \log^\lambda x$ is given by the following identity:

$$\begin{aligned}\frac{d^\lambda}{dr^\lambda}[\phi(x, r)x^r] &= \phi(x, r)x^r \log^\lambda x^r + \lambda\frac{d\phi(x, r)}{dr}x^r \log^{\lambda-1} x \\ & + \frac{\lambda(\lambda-1)}{2!}\frac{d^2\phi(x, r)}{dr^2}x^r \log^{\lambda-2} x + \dots + \frac{d^\lambda \phi(x, r)}{dr^\lambda}x^r.\end{aligned}$$

If $\lambda = 1$, as in the example now under consideration, the formula ends with the second term.

The coefficient of $x^{a+\mu}\log x$ is therefore

(9) $F(a+\mu)c_\mu' + \phi_1(a+\mu-1)c'_{\mu-1} + \phi_2(a+\mu-2)c'_{\mu-2} + \ldots$

Making $\mu = 0$ in equation (8), we find

$$c_0 F(a) + c_0' F'(a) = 0;$$

but since $F(a)$ and $F'(a)$ are both equal to 0, c_0 and c_0' are both arbitrary.

In the same equation, making $\mu = 1, 2, 3, \ldots$, we find

$$c_1 F(a+1) + c_0\phi_1(a) + c_1' F'(a+1) + c_0'\phi_1'(a) = 0;$$
$$c_2 F(a+2) + c_1\phi_1(a+1) + c_0\phi_2(a) + c_2' F'(a+2)$$
$$+ c_1'\phi_1'(a+1) + c_0'\phi_2'(a) = 0;$$
$$\cdot \quad \cdot \quad \cdot \quad \cdot \quad \cdot \quad \cdot \quad \cdot \quad \cdot \quad \cdot \quad \cdot \quad \cdot$$

From equation (9), in like manner,

$$c_0' F(a) = 0;$$
$$c_1' F(a+1) + c_0'\phi_1(a) = 0;$$
$$\cdot \quad \cdot \quad \cdot \quad \cdot \quad \cdot$$

From the latter set of equations the coefficients c_1', c_2', . . . are all obtained in terms of c_0', which is still undetermined.

We know by what has been previously shown that the series $\sum_{i=0}^{i=\infty} c_i x^{a+i}$, the coefficient of log x, can differ only by a constant factor from $y_1 = \Sigma b_i x^{a+i}$ $(i = 0, 1, \ldots, \infty)$. The root b of the equation $F(r) = 0$ corresponds to a series entirely similar to y_1, which may be found in the manner already amply illustrated.

Let us now suppose that the three roots of the indicial equation are all equal, say the common value is a.

In this case the three integrals in the region of the point $x = 0$ are of the following forms:

$y_1 = x^a \Sigma c_i'' x^i$, $(i = 0, 1, \ldots, \infty;)$

$y_2 = x^a\{\Sigma c_i''' x^i + \Sigma c_i'' x^i \,.\, \log x\}$, $(i = 0, 1, \ldots, \infty;)$

$y_3 = x^a\{\Sigma c_i x^i + \Sigma c_i' x^i \,.\, \log x + \Sigma c_i'' x^i \,.\, \log^2 x\}, (i = 0, 1, \ldots, \infty.)$

The manner of finding the integrals y_1 and y_2 will now present no difficulty; as to y_3, remembering the useful formula given above for substituting $x^r \log^\lambda x$ in the differential equation, which is mentioned by Jordan (*Cours d'Analyse*, iii. p. 82), we find for the coefficients of $x^{a+\mu} \log^2 x$, $x^{a+\mu} \log x$, $x^{a+\mu}$ the following expressions, all equal to zero:

$$(10)\quad c''_\mu F(a+\mu) + c''_{\mu-1}\phi_1(a+\mu-1) + c''_\mu\phi_2(a+\mu-2) + \ldots = 0;$$

$$(11)\quad c'_\mu F(a+\mu) + c'_{\mu-1}\phi_1(a+\mu-1) + c'_{\mu-2}\phi_2(a+\mu-2) + \ldots$$
$$+ 2[c''_\mu F'(a+\mu) + c''_{\mu-1}\phi_1'(a+\mu-1)$$
$$+ c''_{\mu-2}\phi_2'(a+\mu-2) + \ldots] = 0;$$

$$(12)\quad c_\mu F(a+\mu) + c_{\mu-1}\phi_1(a+\mu-1) + c_{\mu-2}\phi_2(a+\mu-2) + \ldots$$
$$+ c'_\mu F'(a+\mu) + c'_{\mu-1}\phi_1'(a+\mu-1)$$
$$+ c'_{\mu-2}\phi_2'(a+\mu-2) + \ldots + c''_\mu F''(a+\mu)$$
$$+ c''_{\mu-1}\phi_1''(a+\mu-1) + c''_{\mu-2}\phi_2''(a+\mu-2) + \ldots = 0.$$

Making $\mu = 0$, we obtain the following equations:

$$c_0''F(a) = 0,$$
$$c_0' F(a) + 2c_0''F'(a) = 0,$$
$$c_0F(a) + c_0'F'(a) + c_0''F''(a) = 0;$$

but as a is three times a root of $F(r) = 0$, $F(a)$, $F'(a)$, $F''(a)$ all vanish; and since c_0, c_0', c_0'' cannot vanish, they may have any values whatever; the integral y_3 therefore contains three arbitrary constants, the maximum number for an integral of an equation of the third order. Equation (10) enables us to determine all the remaining constants c_1'', c_2'', . . . ; by aid of them equation (11) gives c_1', c_2', . . . ; and knowing these, equation (12) gives the constants c_1, c_2, . . .

The integral y_1 is thus found incidentally while calculating y_3 (a result readily foreseen from the general theory). It is proper to remark that upon substituting either y_1 or y_2 in the differential equation, we obtain precisely the same set of equations for determining the constants c_1'', c_2'', . . . as already found by substituting y_3.

From the preceding principles and examples the reader will deduce with ease the following *Rule for finding the integrals of a linear differential equation of the mth order when they are all regular:*

Take for the origin that critical point in the region of which the integrals are to be determined. Multiply the equation by x^m; the origin will then no longer be a critical point for the products px, $p_2x^2, \ldots, p_mx^m$. Develop these products by Taylor's theorem, or otherwise, in the following forms:

$$p_1x = a_0 + a_1x + a_2x^2 + \ldots$$
$$p_2x^2 = b_0 + b_1x + b_2x^2 + \ldots$$
$$\cdot \quad \cdot \quad \cdot \quad \cdot \quad \cdot$$
$$p_mx^m = m_0 + m_1x + m_2x^2 + \ldots$$

Form the indicial equation,

$$F(r) = r(r-1)\ldots(r-n+1) + a_0r(r-1)\ldots(r-n+2) + \ldots + m_0 = 0,$$

and obtain its roots. If no two roots are equal, or differ by an integer, the equation has m integrals of the form

$$\Sigma c_i x^{a+i}, \quad (i = 0, 1, \ldots, \infty,)$$

where a is one of the roots. If, however, two or more roots are equal, or differ by integers, some of the integrals will in general contain logarithms which occur in a manner already explained. Let $y_1, y_2, \ldots, y_\lambda$ be the integrals corresponding to the root a of multiplicity λ. Substitute y_2 in the equation, and equate to 0 the coefficients of every term in x and every term in log x separately. From the system of equations thus formed the constants of the series may be successively determined, with the exception of two which will remain arbitrary; and the series multiplying log x will be the integral y_1. In the same way the constant coefficient in the integral y may be determined, and the series multiplying $\log^\lambda x$ will again be the integral y_1.

Proceeding in a similar manner with each root of the indicial equation, the integrals corresponding to the region of the point $x = 0$ may all be found.

As a further illustration, let us obtain the integrals of the equation

$$\frac{d^4y}{dx^4} + \frac{4x + 2x^2 + x^3}{8x(x+1)(x-1)(x-2)} \frac{d^3y}{dx^3}$$
$$+ \frac{64 + 16x^2 + 4x^4 + x^6}{32x^2(x+1)^2(x-1)^2(x-2)^2} \frac{d^2y}{dx^2}$$
$$- \frac{512 + 64x^3 + 8x^6 + x^9}{64x^3\{(x+1)(x-1)(x-2)\}^3} \frac{dy}{dx}$$
$$+ \frac{51200 + 3200x^4 + 25x^{12}}{2048x^4\{(x+1)(x-1)(x-2)\}^4} y = 0$$

in the region of the point $x = 0$. This equation is seen upon examination to satisfy the conditions that all of its integrals shall be regular, and each numerator is of the maximum degree in x.

We have

$$p_1(x) = \frac{4x + 2x^2 + x^3}{8\{(x+1)(x-1)(x-2)\}}$$
$$= \tfrac{1}{4}\{x + x^2 + \tfrac{7}{4}x^3 + \tfrac{11}{8}x^4 + \tfrac{31}{16}x^5 + \tfrac{47}{32}x^6 + \tfrac{127}{64}x^7 + \ldots\},$$

$$p_2x^2 = \frac{64 + 16x^2 + 4x^4 + x^6}{32\{(x+1)(x-1)(x-2)\}^2}$$
$$= \tfrac{1}{2}\{1 + x + 3x^2 + \tfrac{22}{8}x^3 + \tfrac{89}{16}x^4 + \tfrac{156}{32}x^5 + \tfrac{536}{64}x^6 + \tfrac{916}{128}x^7 + \ldots\},$$

$$p_3x^3 = -\frac{512 + 64x^3 + 8x^6 + x^9}{64\{(x+1)(x-1)(x-2)\}^3}$$
$$= -\{1 + \tfrac{3}{2}x + \tfrac{18}{4}x^2 + \tfrac{47}{8}x^3 + \tfrac{186}{16}x^4 + \tfrac{447}{32}x^5 + \tfrac{1471}{64}x^6 + \tfrac{3354}{128}x^7 + \ldots\},$$

$$p_4x^4 = \frac{51200 + 3200x^4 + 25x^{12}}{2048\{(x+1)(x-1)(x-2)\}^4}$$
$$= \tfrac{25}{16}\{1 + 2x + \tfrac{26}{4}x^2 + \tfrac{84}{8}x^3 + \tfrac{356}{16}x^4 + \tfrac{1020}{32}x^5 + \tfrac{3550}{64}x^6 + \tfrac{6340}{128}x^7 + \ldots\}.$$

Hence

$$a_0 = 0, \quad b_0 = \tfrac{1}{2}, \quad c_0 = -1, \quad d_0 = \tfrac{25}{16}.$$

The indicial equation corresponding to $x = 0$ for an equation of the fourth order is

$$r(r-1)(r-2)(r-3) + a_0r(r-1)(r-2) + b_0r(r-1) + c_0r + d_0 = 0,$$

which upon reduction becomes

$$r^4 + (a_0 - 6)r^3 + (b_0 - 3a_0 + 11)r^2 + (c_0 - b_0 + 2a_0 - 6)r + d_0 = 0.$$

Substituting in this the above values for a_0, b_0, c_0, d_0, it becomes

$$r^4 - 6r^3 + 11\tfrac{1}{2}r^2 - 7\tfrac{1}{2}r + \tfrac{25}{16} = 0,$$

of which the roots are found to be $\frac{1}{2}$, $\frac{1}{2}$, $\frac{5}{2}$, $\frac{5}{2}$, the second pair differing from the first by the integer 2. We can now obtain an integral of the following form:

$$y = \Sigma_0{}' x^{\frac{1}{2}+\mu}(c_\mu + c'_\mu \log x)$$
$$+ \Sigma_1{}^{\infty} x^{\frac{1}{2}+\mu}(c_\mu + c'_\mu \log x + c''_\mu \log^2 x + c'''_\mu \log^3 x).$$

For, substituting this expression in the differential equation, and placing equal to zero the coefficients of

$$x^{\frac{1}{2}+\mu}\log^3 x, \quad x^{\frac{1}{2}+\mu}\log^2 x, \quad x^{\frac{1}{2}+\mu}\log x, \quad x^{\frac{1}{2}+\mu},$$

we obtain the following equations:

$$F(\tfrac{1}{2}+\mu)c'''_\mu + \phi_1(\tfrac{1}{2}+\mu-1)c'''_{\mu-1} + \phi_2(\tfrac{1}{2}+\mu-2)c'''_{\mu-2} + \ldots = 0; \quad \text{(A)}$$

$$F(\tfrac{1}{2}+\mu)c''_\mu + \phi_1(\tfrac{1}{2}+\mu-1)c''_{\mu-1} + \phi_2(\tfrac{1}{2}+\mu-2)c''_{\mu-2} + \ldots$$
$$+ 3[F'(\tfrac{1}{2}+\mu)c'''_\mu + \phi_1'(\tfrac{1}{2}+\mu-1)c'''_{\mu-1} + \phi_2'(\tfrac{1}{2}+\mu-2)c'''_{\mu-2}$$
$$+ \ldots] = 0; \quad \text{(B)}$$

$$F(\tfrac{1}{2}+\mu)c'_\mu + \phi_1(\tfrac{1}{2}+\mu-1)c'_{\mu-1} + \phi_2(\tfrac{1}{2}+\mu-2)c'_{\mu-2} + \ldots$$
$$+ 2[F'(\tfrac{1}{2}+\mu)c''_\mu + \phi_1'(\tfrac{1}{2}+\mu-1)c''_{\mu-1} + \phi_2'(\tfrac{1}{2}+\mu-2)c''_{\mu-2} + \ldots]$$
$$+ 3[F''(\tfrac{1}{2}+\mu)c'''_\mu + \phi_1''(\tfrac{1}{2}+\mu-1)c'''_{\mu-1} + \phi_2''(\tfrac{1}{2}+\mu-2)c'''_{\mu-2}$$
$$+ \ldots] = 0; \quad \text{(C)}$$

$$F(\tfrac{1}{2}+\mu)c_\mu + \phi_1(\tfrac{1}{2}+\mu-1)c_{\mu-1} + \phi_2(\tfrac{1}{2}+\mu-2)c_{\mu-2} + \ldots$$
$$+ F'(\tfrac{1}{2}+\mu)c'_\mu + \phi_1'(\tfrac{1}{2}+\mu-1)c'_{\mu-1} + \phi_2'(\tfrac{1}{2}+\mu-2)c'_{\mu-2} + \ldots$$
$$+ F''(\tfrac{1}{2}+\mu)c''_\mu + \phi_1''(\tfrac{1}{2}+\mu-1)c''_{\mu-1} + \phi_2''(\tfrac{1}{2}+\mu-2)c''_{\mu-2} + \ldots$$
$$+ F'''(\tfrac{1}{2}+\mu)c'''_\mu + \phi_1'''(\tfrac{1}{2}+\mu-1)c'''_{\mu-1} + \phi_2'''(\tfrac{1}{2}+\mu-2)c'''_{\mu-2}$$
$$+ \ldots = 0. \quad \text{(D)}$$

The meaning of the symbols F, F', ..., ϕ_1, ϕ_1', ... has already been explained; it is now, however, convenient to write them down in full. We have, viz.,

$$F(r) = r^4 - 6r^3 + \tfrac{23}{2}r^2 - \tfrac{15}{2}r + \tfrac{25}{16},$$

$$F'(r) = 4r^3 - 18r^2 + 23r - \tfrac{15}{2},$$

$$F''(r) = 12r^2 - 36r + 23,$$

$$F'''(r) = 24r - 36;$$

$$\begin{aligned}\phi_1(r) &= \tfrac{1}{4}r(r-1)(r-2) + \tfrac{1}{2}r(r-1) - \tfrac{3}{2}r + \tfrac{25}{8}\\ &= \tfrac{1}{4}r^3 - \tfrac{1}{4}r^2 - \tfrac{3}{2}r + \tfrac{25}{8},\end{aligned}$$

$$\phi_1'(r) = \tfrac{3}{4}r^2 - \tfrac{1}{2}r - \tfrac{3}{2},$$

$$\phi_1''(r) = \tfrac{3}{2}r - \tfrac{1}{2},$$

$$\phi_1'''(r) = \tfrac{3}{2}.$$

In fact, as before mentioned,

$$\begin{aligned}\phi_m(r) &= a_m r(r-1)(r-2) + b_m r(r-1) + c_m r + d_m\\ &= a_m r^3 + (b_m - 3a_m)r^2 + (c_m - b_m + 2a_m)r + d_m,\end{aligned}$$

$$\phi'_m(r) = 3a_m r^2 + 2(b_m - 3a_m)r + (c_m + b_m - 2a_m),$$

$$\phi_m''(r) = 6a_m r + 2(b_m - 3a_m),$$

$$\phi_m'''(r) = 6a_m;$$

a_m, b_m, c_m, d_m being the coefficients of x^m in the expressions of

$$p_1x, \quad p_2x^2, \quad p_3x^3, \quad p_4x^4.$$

Directing our attention to equation (A) and making $\mu = 0$, we obtain only the identity $F(\frac{1}{2})c_0''' = 0$; making $\mu = 1$, we find

$$F(1\tfrac{1}{2})c_1''' + \phi_1(\tfrac{1}{2})c_0''' = 0.$$

Now $\phi_1(\frac{1}{2}) = \frac{1}{32} - \frac{1}{16} - \frac{3}{4} + \frac{25}{8} = \frac{75}{32}$; $F(1\frac{1}{2}) = 11\frac{7}{8}$. Therefore

$$11\tfrac{7}{8}c_1''' + \tfrac{75}{32}c_0''' = 0,$$

which will not alone suffice to determine c_1''' and c_0'''. However, making $\mu = 0$ in equation (C), the result is $3F''(\frac{1}{2})c_0''' = 0$. Since

$\frac{1}{2}$ is a root of multiplicity 2 only, $F''(\frac{1}{2})$ is not $= 0$; hence $c_0''' = 0$, and consequently $c_1''' = 0$. By making $\mu = 2$ in equation (A), we find $F(2\frac{1}{2})\, c_2''' = 0$, whence it may be concluded that c_2''' is arbitrary.

In equation (B), making $\mu = 1$, we obtain $11\frac{7}{8}c_1'' + \frac{75}{32}c_0'' = 0$, and making $\mu = 0$ in equation (D), $F''(\frac{1}{2})c_0'' = 0$; hence $c_0'' = 0$ and $c_1'' = 0$. Assigning the value 2 to μ in equation (B), the result is, since $F'(2\frac{1}{2}) = 0$, $F(2\frac{1}{2})c_2'' = 0$, showing that c_2'' is also arbitrary. The only conditions by which to determine c_0 and c_0' are the following, derived from (C) and (B):

$$F(\tfrac{1}{2})c_0' = 0 \qquad \text{and} \qquad F(\tfrac{1}{2})c_0 + F'(\tfrac{1}{2})c_0' = 0,$$

both of which are identities, so that c_0 and c_0' are also arbitrary.

Returning to equation (A), let $\mu = 3$. Then

$$F(3\tfrac{1}{2})c_3''' + \phi_1(2\tfrac{1}{2})c_2''' = 0$$

whence

$$c_3''' = -\frac{\phi_1(2\frac{1}{2})}{F(3\frac{1}{2})}c_2'''.$$

In like manner, by making $\mu = 4, 5, \ldots$, the coefficients c_4''', c_5''', ... may all be found from equation (A). Aided by these values, c_3'', c_4'', ... are to be obtained in like manner from equation (B), c_1', ... from (C), and c_1, c_2, ... from equation (D). Thus we have an integral of the form announced containing four arbitrary constants; it is therefore the general integral of the given equation in the region of the point $x = 0$.*

Among functions of the kind considered is obviously the function y, defined by the irreducible algebraic equation $f(x, y) = 0$ of, say, the nth degree in y. If the n branches of the function y (*i.e.*, the n roots of $f = 0$) so defined are linearly independent, then y satisfies a linear differential equation of the above form of the nth order; if, however, there are only m ($m < n$) linearly independent branches, it is clear that y will satisfy an equation of the mth order. If a differential equation is satisfied by a particular root of the irreducible algebraic equation $f(x, y) = 0$, it must be satisfied by all the roots. Since the remaining roots are branches of the one func-

* The preceding illustrations of the general theory are due to Mr. C. H. Chapman.

tion y obtained by travelling round certain critical points, and if y_1 were the chosen integral, it will remain an integral during all the motion of the variable, though the branch y_1 changes into, say, y_2, etc. Suppose the given differential equation to be of order n, and suppose that among the n branches of the function y there are only m $(m < n)$ linearly independent; then the functions $y_1, \ldots, y_m$ satisfy a differential equation of order m, and so this equation has *only algebraic integrals*, and the given equation has these same integrals and some additional ones. The first equation, having all the integrals of the second for integrals, is said to be a *reducible* equation. We have here the first notion of *reducibility* and *irreducibility* in differential equations, the notion being entirely analogous to that of reducibility and irreducibility in algebraic equations. This subject will be taken up later on for a fuller discussion, but it is convenient to give here a few theorems in connection with the notion of reducible equations. Suppose the linear differential equation $P = 0$ has among its integrals all of the integrals of $Q = 0$, where Q is of a lower order than P. Since among the integrals of $P = 0$ there are functions which do not satisfy $Q = 0$, it is clear that the equation $P = 0$ may have critical points which do not belong to $Q = 0$. Conversely, in spite of the fact that the integrals of $Q = 0$ are *all* integrals of $P = 0$, it may happen that $Q = 0$ has critical points which do not belong to $P = 0$. It is easy to see what the character of the indicial equation is for these points, and consequently the character of the integrals in the region of the points. Suppose a one of the critical points of $Q = 0$ which does not belong to $P = 0$; now, remembering that the only critical points the integrals of a differential equation can have are those of the equation itself, it follows that in the region of $x = a$ the integrals of $Q = 0$ must be uniform and continuous functions, since otherwise, as these integrals are also integrals of $P = 0$, this last equation must have $x = a$ as a critical point, which is contrary to hypothesis.

In the region of a non-critical (or neutral) point, say a, of $Q = 0$ (or of any other linear differential equation), the m integrals will be developed in series of powers of $x - a$, whose first terms are respectively

$$(x-a)^0, \quad (x-a)^1, \quad (x-a)^2, \quad \ldots, \quad (x-a)^{m-1};$$

if a is a critical point of $Q = 0$ of the kind just mentioned, the developments of the integrals in positive integral powers of $x - a$ will begin respectively with

$$(x-a)^{r_1}, \quad (x-a)^{r_2}, \quad \ldots, \quad (x-a)^{r_m};$$

where the positive integers $r_1, r_2, \ldots, r_m$ are all different, and do not coincide with the numbers

$$0, \quad 1, \quad 2, \quad \ldots, \quad m-1,$$

and can consequently not all be less than m. It is also clear that if any one of these roots $r_1, r_2, \ldots, r_m$ is greater than $n - 1$, the point $x = a$ is a critical point of $P = 0$, and that the integrals in the region of this point will be developed in positive ascending powers of $x - a$, the first terms of the developments being as above, $(x-a)^{r_1}, (x-a)^{r_2}, \ldots$ Suppose for a moment we call these points *quasi-critical* points; we have then for $m < n$ the theorem: *If a linear differential equation of order n has among its integrals all the integrals of a linear differential equation of order m, then all the critical points of the latter equation which are not critical points of the first are quasi-critical points, and the roots of the indicial equations belonging to such points are all integers each of which is less than n, the order of the given equation.*

Among the different branches of an integral of a given differential equation (uniform coefficients always understood) of order n there can, of course, exist at most only n which are linearly independent; if, however, there are only $m (m < n)$ (it will be assumed hereafter, unless something is said to the contrary, that m is always $< n$) linearly independent branches of the function, it will satisfy an equation of order m, and consequently the given equation is reducible. It follows conversely from this that the number of linearly independent branches of a function which is an integral of an irreducible linear differential equation is exactly equal to the order of the equation. It is also evident that among the integrals of a reducible equation there are always some the number of whose linearly independent branches is less than the order of the equation.

Again, if a function y is an integral of a given equation, then all of its branches are integrals of the same equation. Suppose the given equation to be an irreducible one of order m; then among the branches of y there are just m linearly independent ones, viz., $y_1, y_2, \ldots, y_m$. If now y satisfies another equation of the same kind but of order n, then $y_1, y_2, \ldots, y_m$ satisfy it, and consequently, denoting by $c_1, c_2, \ldots, c_m$ arbitrary constants,

$$Y = c_1 y_1 + c_2 y_2 + \ldots + c_m y_m$$

also satisfies it. But Y is the general integral of the irreducible equation; and so it follows that if a given equation has for an integral *one* of the integrals of an irreducible equation, it has among its integrals all of the integrals of the irreducible equation. Suppose now that we have a reducible linear differential equation of order n; it must have an integral y in common with an equation of lower order m: and suppose that among the branches of the function y there are l which are linearly independent; then l can be at most equal m. From what has been said, we see that y satisfies a differential equation of order l, of which the general integral is

$$Y = c_1 y_1 + c_2 y_2 + \ldots + c_l y_l.$$

Now every differential equation which is satisfied by y must be satisfied by $y_1, y_2, \ldots, y_l$, and consequently Y satisfies the reducible differential equation of order n. If, therefore, a given linear differential equation is reducible, there exists a linear differential equation of lower order all of whose integrals are integrals of the given equation. As this last equation may be again reducible, we have that if a given linear differential equation is reducible there must, as is easily seen, be one or more irreducible equations all of whose integrals are integrals of the given equation.

We have seen that if an integral of a linear differential equation changes into itself, multiplied by a constant s when the independent variable turns round a critical point, s is a root of a certain algebraic equation, viz., the characteristic equation corresponding to the critical point. If s is a simple root of this equation, there is only one integral satisfying the condition

$$Sy = sy;$$

but if s is a multiple root of the characteristic equation, say of order $k + 1$, then there are corresponding to it $k + 1$ independent integrals.

If these are regular integrals, we know from what precedes that their general form is, in the region of $x = a$,

$$y = \phi_0 + \phi_1 \log (x - a) + \phi_2 \log^2 (x - a) + \ldots + \phi_\alpha \log^\alpha (x - a);$$

where $\phi_0, \phi_1, \ldots$ are functions which may vanish for $x = a$, and which when x turns round the point $x = a$ change into themselves each multiplied by s. The highest exponent of $\log (x - a)$ which can enter into this group of integrals cannot be greater than k; if this highest exponent is equal to k, then there can exist but a single integral corresponding to s which contains no logarithm. This has already been seen, but it is convenient here to prove it in another way.*

Suppose y_α to be an integral corresponding to the root s of the characteristic equation for the point $x = a$; then y_α is of the form

$$y_\alpha = \phi_{\alpha, 0} + \phi_{\alpha, 1} \log (x - a) + \ldots + \phi_{\alpha, \alpha} \log^\alpha (x - a);$$

where $\phi_{\alpha, \alpha}$ is of course not zero. If we apply the substitution S, *i.e.*, make x turn round the point $x = a$, the result, Sy_α, is again an integral of our equation, and is of the form

$$y_\alpha + 2\pi i \alpha y_{\alpha - 1},$$

where

$$y_{\alpha - 1} = \phi_{\alpha - 1, 0} + \phi_{\alpha - 1, 1} \log (x - a) + \ldots + \phi_{\alpha - 1, \alpha - 1} \log^{\alpha - 1} (x - a),$$

and where

$$\phi_{\alpha - 1, \alpha - 1} = \phi_{\alpha, \alpha};$$

and consequently $\phi_{\alpha - 1, \alpha - 1}$ is not zero.

Now since $y_\alpha + 2\pi i \alpha y_{\alpha - 1}$ is an integral of the given equation, it is clear that $y_{\alpha - 1}$ is also an integral. It follows now that if to the root s of the characteristic equation for the point $x = a$ there is an integral in which the highest power of $\log (x - a)$ is $\log^\alpha (x - a)$, there exist integrals in which β is the exponent of the highest power of $\log (x - a)$, where β is any one of the numbers $0, 1, 2, \ldots, \alpha - 1$.

---

* What immediately precedes and follows is taken directly from an important memoir by Frobenius—*Ueber den Begriff der Irreductibilität in der Theorie der linearen Differentialgleichungen* (Crelle, vol. 76). This memoir will again be referred to later on.

We see now that if, corresponding to the root s of the characteristic equation, there is an integral in which k is the exponent of the highest power of $\log(x-a)$, there exists a group of $k+1$ linearly independent integrals having as exponents of the highest power of this logarithm $0, 1, 2, \ldots, k$; say these integrals are $y_0, y_1, \ldots, y_k$. Now since to a $k+1$-fold root of the characteristic equation there can exist only $k+1$ linearly independent integrals, it follows that if x turns round the point $x=a$, any one of these integrals, say y_0, will become

$$Sy_0 = c_0y_0 + c_1y_1 + \ldots + c_ky_k;$$

equating the coefficients of the different powers of $\log(x-a)$, we have

$$c_k = c_{k-1} = \ldots = c_1 = 0,$$

and consequently

$$Sy_0 = c_0y_0.$$

We will say that two integrals whose ratio is a constant do not differ from each other; *e.g.*, y_0 and Sy_0, $= c_0y_0$, do not differ from each other.

Suppose now, conversely, that the given equation has only one integral in the region of $x=a$ which, when the substitution S is applied, changes into itself multiplied by the $(k+1)$-fold root s of the corresponding characteristic equation: we must have then that k is the exponent of the highest power of $\log(x-a)$ which can enter into any of the integrals belonging to the root s. Suppose y_0 to be the integral considered; then $Sy_0 = sy_0$. If $k>0$, the equation has other integrals than y_0 which correspond to the root s; and since y_0 is the only integral satisfying the relation

$$Sy_0 = sy_0,$$

these other integrals must involve logarithms, and one of them, say y_1, will be of the form

$$y_1 = \phi_{1,0} + \phi_{1,1}\log(x-a).$$

Now we know that $\phi_{1,1}$ can only differ from y_0 by a constant factor, and of course by choosing y_1 properly this factor may be made unity, and so we have $\phi_{1,1} = y_0$, and consequently

$$y_1 = \phi_{1,0} + y_0\log(x-a).$$

Suppose now there exists a second integral in which $\log(x-a)$ enters to the first power, say y_1'; it must have the form

$$y_1' = \phi'_{1,0} + y_0 \log(x-a).$$

We have, therefore,

$$y_1' - y_1 = \phi'_{1,0} - \phi_{1,0};$$

but the difference, $y_1' - y_1$, is an integral of the equation, and, as it contains no logarithms, we must have

$$y_1' - y_1 = c_0 y_0, \quad \text{or} \quad y_1' = c_0 y_0 + y_1;$$

where c_0 is a constant. It follows then that every integral which only contains the first power of $\log(x-a)$ is a linear function of y_0 and y_1. If $k > 1$, then corresponding to the $(k+1)$-fold root s there must be more than two linearly independent integrals, and among them there must be one, say y_2, of the form

$$y_2 = \phi_{2,0} + \phi_{2,1} \log(x-a) + \phi_{2,2} \log^2(x-a).$$

It can be shown just as before that all integrals which contain $\log(x-a)$ only to the second power are linearly expressible in terms of y_0, y_1, and y_2. By continuing this process we see that if to the root s of order of multiplicity $k+1$ there corresponds only one integral free from logarithms, then k is the exponent of the highest power of $\log(x-a)$ which can appear in the group of integrals belonging to the root s. We arrive thus at the theorem: *In order that a linear differential equation shall possess only a finite number of integrals which satisfy the condition $Sy = sy$ (where S refers to any critical point, and s is any root of the corresponding characteristic equation), it is necessary and sufficient that the roots of the characteristic equation corresponding to the critical point in question shall be all distinct, or, in the case of a root of order of multiplicity $k+1$, that k shall be the exponent of the highest power of the logarithm in the integrals of the group belonging to this root.*

CHAPTER VI.

LINEAR DIFFERENTIAL EQUATIONS OF THE SECOND ORDER, PARTICULARLY THOSE WITH THREE CRITICAL POINTS.

THE linear differential equation of the second order with three critical points is the one which has been studied more than any other, and is of the highest importance, as its integrals include many of the most important functions with which mathematicians are, as yet, thoroughly familiar. Before taking up the case of three critical points it will be convenient to prove a theorem due to Fuchs* concerning a transformation of differential equations of the second order—a transformation which does not apply to equations of a higher order. As shown in the last chapter, the most general form of a differential equation of the second order with ρ finite critical points and all of whose integrals are regular, is

$$(1)\qquad \frac{d^2z}{dt^2}+\frac{P_1(t)}{\psi(t)}\frac{dz}{dt}+\frac{P_2(t)}{\psi^2(t)}z=0;$$

where

$$\psi(t)=(t-t_1)(t-t_2)\ .\ .\ .\ (t-t_\rho),$$

and where $P_1(t)$, $P_2(t)$ are polynomials of the degrees $\rho-1$ and $2\rho-2$ respectively. Fuchs proves that such an equation can by a change of the dependent variable z be thrown into the form

$$(2)\qquad \psi(t)\frac{d^2y}{dt^2}+Q_1(t)\frac{dy}{dt}+Q_2(t)y=0;$$

where the degrees of ψ, Q_1, and Q_2 are ρ, $\rho-1$, and $\rho-2$ respectively.

* Heffter: *Inaugural-dissertation*. Berlin, 1886.

Suppose r_i a root of the indicial equation corresponding to the critical point t_i; also suppose that r_i is not zero. The indicial equation is

$$(3) \qquad r(r-1)+r\frac{P_1(t_1)}{\psi'(t_1)}+\frac{P_2(t_1)}{\psi'^2(t_1)}=0.$$

Make the substitution

$$(4) \qquad z=\Pi(t-t_i)^{s_i}y; \quad (i=1, 2, \ldots, \rho;)$$

equation (1) becomes now

$$(5) \qquad \Pi\psi^2\frac{d^2y}{dt^2}+\left[2\Pi\psi^2\sum_1^\rho\frac{r_k}{t-t_k}+\Pi\psi P_1(t)\right]\frac{dy}{dt}$$
$$+\left\{\Pi\psi^2\left[\sum_1^\rho\frac{r_k(r_k-1)}{(t-t_k)^2}+\sum_{\substack{i,k=1,2,\ldots,\rho\\ i<k}}\frac{2r_ir_k}{(t-t_i)(t-t_k)}\right]\right.$$
$$\left.+\Pi\psi P_1(t)\sum_1^\rho\frac{r_k}{t-t_k}+\Pi P_2(t)\right\}y=0.$$

Dividing through by $\psi\Pi(t-t_1)$ we see at once that the coefficients of $\frac{d^2y}{dt^2}$ and $\frac{dy}{dt}$ have the required form. That the coefficient of y has the required form can be seen by aid of (3), or directly as follows: If we divide first by $\Pi(t-t_i)$ the coefficient of y is seen to be an integral function of t of the degree $2\rho-2$. Denote this for the moment by $P_2'(t)$. In consequence of the transformation (4) there will be one root of each of the indicial equations corresponding to $t_1, \ldots, t_\rho$ which is zero, and therefore

$$\frac{P_2'(t_i)}{\psi'^2(t_1)}=0; \quad (i=1, 2, \ldots, \rho;)$$

and so $P_2'(t)$ is divisible by $\psi(t)$, and the quotient is a polynomial of degree $\rho-2$. This transformation cannot be made when all the roots r_i are zero; but in this case the differential equation has already the required form.

The differential equation of the second order may now be written in the simpler form

(6)
$$\frac{d^2y}{dt^2} + \frac{P_1(t)}{\psi(t)}\frac{dy}{dt} + \frac{P_2(t)}{\psi(t)}y = 0;$$

where $P_1(t)$ and $P_2(t)$ are polynomials of the degrees $\rho - 1$ and $\rho - 2$ respectively, and $\psi(t) = (t - t_1) \ldots (t - t_\rho)$. We will take up now the case of only three critical points; it will be convenient to obtain directly the result of the above theorem for this particular case (Jordan, *Cours d'Analyse*, vol. iii. p. 220).

Suppose the given differential equation, say $P = 0$ (where z is the dependent and t the independent variable) has three critical points, and denote by t the independent variable; then the critical points may be denoted by t_1, t_2, t_3. We will first transform the equation by the relation

(7)
$$t = \frac{\alpha x + \beta}{\gamma x + \delta}.$$

There are here, of course, only three arbitrary constants, viz., the ratios of any three of the arbitrary constants $\alpha, \beta, \gamma, \delta$ to the fourth one. Let τ and ξ denote two corresponding values of t and x; we have obviously in the region of these values a relation of the form

$$t - \tau = a_1(x - \xi) + a_2(x - \xi)^2 + \ldots,$$

in which, when τ or ξ become infinite, we must replace $t - \tau$ or $x - \xi$ by $\frac{1}{t}$ or $\frac{1}{x}$. This value of $t - \tau$ being substituted in an integral function of $t - \tau$ will obviously give an integral function of $x - \xi$. Again, if it be substituted in a *regular* function of $t - \tau$, such as

$$(t-\tau)^r[T_0 + T_1 \log(t-\tau) + \ldots + T_\lambda \log^\lambda (t-\tau)],$$

where $T_0, T_1, \ldots, T_\lambda$ are integral functions of $t - \tau$, we shall obviously have a regular function of $x - \xi$ of the form

$$(x - \xi)^r[X_0 + X_1 \log(x - \xi) + \ldots + X_\lambda \log^\lambda (x - \xi)],$$

where $X_0, X_1, \ldots, X_\lambda$ are integral functions of $x - \xi$. It follows then from what has been shown in Chapter V that the transformed equation in z and x admits as critical points the three points ξ_1, ξ_2, ξ_3 corresponding to t_1, t_2, t_3, and that its integrals are all regular in the regions of these points. Since (7) involves three perfectly arbitrary constants, we can choose for the new critical points any three points we please: suppose we choose the points 0, 1, ∞. This can be done in six different ways, as shown in the following table. t_1, t_2, t_3 correspond to ξ_1, ξ_2, ξ_3, and by a proper determination of the coefficients $\alpha, \beta, \gamma, \delta$ (or rather their ratios) we can make any three values of ξ we choose correspond to the three given values of t; we have then

	t_1	t_2	t_3
	ξ_1	ξ_1	ξ_3
1	0	1	∞
2	0	∞	1
3	1	0	∞
4	1	∞	0
5	∞	0	1
6	∞	1	0

Denote by (λ, λ'), (μ, μ'), (ν, ν') the roots of the indicial equations corresponding to the critical points 0, 1, ∞; then if the differences $\lambda - \lambda'$, $\mu - \mu'$, $\nu - \nu'$ are not integers, the general forms of the integrals in the region of these points will be:

$$(\alpha)\quad \begin{cases} \text{For } x = 0, & c x^{\lambda} U + c' x^{\lambda'} U'; \\ \text{For } x = 1, & c(x-1)^{\mu} V + c'(x-1)^{\mu'} V'; \\ \text{For } x = \infty, & c \dfrac{1}{x^{\nu}} W + c' \dfrac{1}{x^{\nu'}} W'; \end{cases}$$

where c and c' are arbitrary constants, U and U' are integral functions of x, V and V' are integral functions of $x - 1$, and W and W' are integral functions of $\dfrac{1}{x}$; that is.

$$U = C_0 + C_1 x + C_2 x^2 + \ldots;$$
$$U' = C_0' + C_1' x + C_2' x^2 + \ldots;$$
$$V = D_0 + D_1(x-1) + D_2(x-1)^2 + \ldots;$$
$$V' = D_0' + D_1'(x-1) + D_2'(x-1)^2 + \ldots;$$
$$W = E_0 + E_1 \frac{1}{x} + E_2 \frac{1}{x^2} + \ldots;$$
$$W' = E_0' + E_1' \frac{1}{x} + E_2' \frac{1}{x^2} + \ldots$$

The differential equation $P = 0$ has now been transformed into one with only the two finite critical points $x = 0$ and $x = 1$; supposing then λ and μ not equal zero, the transformation (4) becomes

$$z = x^{\lambda}(x-1)^{\mu} y, \tag{8}$$

and the corresponding equation between y and x has for integrals:

$$(\alpha')\begin{cases} \text{In the region of } x = 0, & cU_1 + c' x^{\lambda' - \lambda} U_1'; \\ \text{In the region of } x = 1, & cV_1 + c'(x-1)^{\mu' - \mu} V_1'; \\ \text{In the region of } x = \infty, & c\dfrac{1}{x^{\lambda+\mu+\nu}} W_1 + c' \dfrac{1}{x^{\lambda+\mu+\nu'}} W_1'; \end{cases}$$

where c and c' are arbitrary constants, and

$$U_1 = (1-x)^{-\mu} U, \qquad U_1' = (1-x)^{-\mu} U',$$
$$V_1 = (-1)^{\mu} x^{-\lambda} V, \qquad V_1' = (-1)^{\mu} x^{-\lambda} V',$$
$$W_1 = \left(\frac{1}{x} - 1\right)^{-\mu} W, \qquad W_1' = \left(\frac{1}{x} - 1\right)^{-\mu} W';$$

where (U_1, U_1'), (V_1, V_1'), (W_1, W_1') are respectively developable in positive integral powers of x, $x - 1$, and $\frac{1}{x}$.

The new equation, say $H = 0$, in y and x has thus all of its integrals regular, and its indicial equations relative to the critical points $x = 0$ and $x = 1$ have for roots

$$x = 0; \quad r_1 = 0, \quad r_2 = \lambda' - \lambda.$$
$$x = 1; \quad r_1 = 0, \quad r_2 = \mu' - \mu.$$

There exist four ways in which this result can be arrived at; because we can take for λ, in (8), either of the roots of the indicial equation corresponding to $x = 0$, and for μ either of the roots corresponding to $x = 1$. Now, unless these indicial equations have λ and λ', μ and μ', as conjugate imaginary roots, there will always be one of the roots $\lambda' - \lambda$, $\mu' - \mu$, which will have the real part positive for both. The equation $H = 0$ is now of the form

$$\frac{d^2y}{dx^2} + \frac{P_1(x)}{x(x-1)}\frac{dy}{dx} + \frac{P_2(x)}{x^2(x-1)^2}y = 0; \tag{9}$$

where (Chapter V) $P_1(x)$ and $P_2(x)$ are polynomials of the degrees 1 and 2 respectively. The indicial equations relative to the critical points 0 and 1 are respectively

$$\begin{cases} r(r-1) - rP_1(0) + P_2(0) = 0; \\ r(r-1) + rP_1(1) + P_2(1) = 0. \end{cases} \tag{10}$$

As each of these has zero for a root, it follows that $P_2(x)$ must vanish for $x = 0$ and for $x = 1$, and consequently is divisible by $x(x-1)$; as $P_2(x)$ is of the degree 2, this division can only have a constant as a quotient, and the transformed differential equation is

$$\frac{d^2y}{dx^2} + \frac{Ax + B}{x(x-1)}\frac{dy}{dx} + \frac{C}{x(x-1)}y = 0, \tag{11}$$

a particular case of equation (6). (By Fuchs's theorem we might at once have written the differential equation in this form.) The three constants A, B, C are perfectly arbitrary, and may be replaced by the following,

$$A = \alpha + \beta + 1, \quad B = -\gamma, \quad C = \alpha\beta,$$

so that (11) becomes

$$\frac{d^2y}{dx^2} + \frac{(\alpha + \beta + 1)x - \gamma}{x(x-1)}\frac{dy}{dx} + \frac{\alpha\beta}{x(x-1)}y = 0, \tag{12}$$

the well-known differential equation for the hypergeometric series. If in (6) we assumed only two critical points, 0 and 1, then we should at once have (11) for $\rho - 1 = 1$, $\rho - 2 = 0$, $\psi = x(x-1)$. [An interesting property of (12) arises from differentiation, viz.,

$$x(1-x)\frac{d^3y}{dx^3} + [\gamma+1-(\alpha+\beta+3)x]\frac{d^2y}{dx^2} - (\alpha+1)(\beta+1)\frac{dy}{dx} = 0;$$

or, if $\dfrac{dy}{dx} = \eta$,

$$(13) \qquad x(1-x)\frac{d^2\eta}{dx^2} + [\gamma+1-(\alpha+1+\beta+1+1)x]\frac{d\eta}{dx} - (\alpha+1)(\beta+1)\eta = 0,$$

which is of the same form as (12), and differs from it only in having $\alpha+1$, $\beta+1$, and $\gamma+1$ in place of α, β, and γ. It is obvious that when $\dfrac{d^my}{dx^m} = \eta$ we shall have

$$(14) \qquad x(1-x)\frac{d^2\eta}{dx^2} + [\gamma+m+(\alpha+m+\beta+m+1)]\frac{d\eta}{dx} - (\alpha+m)(\beta+m)\eta = 0,$$

differing from (12) by having $\alpha+m$, $\beta+m$, and $\gamma+m$ in place of α, β, and γ.]

In order to get the indicial equation corresponding to the point $x = \infty$ in (12), transform by writing $x = \dfrac{1}{t}$; we have then

$$(15) \qquad \frac{d^2y}{dt^2} + \frac{(\alpha+\beta-1)-(\gamma-2)t}{t(t-1)}\frac{dy}{dt} - \frac{\alpha\beta}{t^2(t-1)}y = 0.$$

The three indicial equations and their roots are now, from (12) and (15):

$$(16) \qquad \begin{cases} \text{For } x = 0, \begin{cases} r(r-1) + \gamma r = 0, \\ r_1 = 0, \quad r_2 = 1-\gamma; \end{cases} \\ \text{For } x = 1, \begin{cases} r(r-1) - r(\gamma-\alpha-\beta-1) = 0, \\ r_1 = 0, \quad r_2 = \gamma-\alpha-\beta; \end{cases} \\ \text{For } x = \infty, \begin{cases} r(r-1) - r(\alpha+\beta-1) + \alpha\beta = 0, \\ r_1 = \alpha, \quad r_2 = \beta. \end{cases} \end{cases}$$

In the region of $x = 0$ we have corresponding to the root $r_1 = 0$ an integral of the form

$$y_1 = C_0 + C_1 x + C_2 x^2 + \ldots;$$

substituting this in (12) and equating to zero the coefficient of x^k, we have

$$[k(k-1) + k(\alpha+\beta+1) + \alpha\beta]C_k = [k(k+1) + \gamma(k+1)]C_{k+1},$$

giving

$$C_{k+1} = \frac{(\alpha+k)(\beta+k)}{(1+k)(\gamma+k)} C_k; \tag{17}$$

and so, making the arbitrary constant C_0 equal to unity, we have for y_1 the value

$$y_1 = 1 + \frac{\alpha \cdot \beta}{1 \cdot \gamma} x + \frac{\alpha(\alpha+1)\beta(\beta+1)}{1 \cdot 2 \cdot \gamma(\gamma+1)} x^2 + \ldots; \tag{18}$$

or, employing the usual notation for hypergeometric series,

$$y_1 = F(\alpha, \beta, \gamma, x).$$

We have seen above that there are six ways of transforming the differential equation in question into an analogous one when the independent variable is changed by the relation

$$x = \frac{\alpha u + \beta}{\gamma u + \delta}, \tag{19}$$

and corresponding to each of these there are four transformations of the dependent variable of the form

$$y = x^\lambda (x-1)^\mu z; \tag{20}$$

where λ and μ are respectively roots of the indicial equations corresponding to $x = 0$ and $x = 1$.

The critical points of (12) are 0, 1, ∞, and the six different forms of (19) are given by the following table:

(21)

	$x = 0,$	$1,$	∞	
$x = u$	$u = 0,$	$1,$	∞	i
$x = \frac{u}{u-1}$	$u = 0,$	$\infty,$	1	ii
$x = 1 - u$	$u = 1,$	$0,$	∞	iii
$x = \frac{u-1}{u}$	$u = 1,$	$\infty,$	0	iv
$x = \frac{1}{1-u}$	$u = \infty,$	$0,$	1	v
$x = \frac{1}{u}$	$u = \infty,$	$1,$	0	vi

As an illustration of these transformations take the case of $x = \frac{u-1}{u}$. The critical points corresponding to the original ones $x = 0$, $x = 1$, $x = \infty$ are respectively $u = 1$, $u = \infty$, $u = 0$, and the roots of the indicial equations are:

$$(22)\quad \begin{cases} \text{For } u = 1, & r_1 = 0, \quad r_2 = 1 - \gamma; \\ \text{For } u = \infty, & r_1 = 0, \quad r_2 = \gamma - \alpha - \beta; \\ \text{For } u = 0, & r_1 = \alpha, \quad r_2 = \beta. \end{cases}$$

These values are written at once by aid of what precedes. The transformed equation—for which we have, however, no use—is

$$(23)\quad \frac{d^2y}{du^2} + \frac{\alpha+\beta-1-(\alpha+\beta-\gamma-1)u}{u(u-1)}\frac{dy}{du} - \frac{\alpha\beta}{u^2(u-1)}y = 0.$$

Make now the transformation

$$(24)\qquad y = u^{\lambda}(1-u)^{\mu}z,$$

λ and μ being roots respectively of the indicial equations corresponding to $u = 0$ and $u = 1$. Referring to the formulæ (α), we have

$$\lambda = \alpha \text{ or } \beta, \quad \mu = 0 \text{ or } 1 - \gamma.$$

Suppose now we choose

$$\lambda = \alpha, \quad \mu = 1 - \gamma;$$

the formula of transformation is then

$$y = u^{\alpha}(1 - u)^{1-\gamma} z. \tag{25}$$

From (α') we have at once, for the roots of the indicial equations belonging to the (z, u) equation,

$$\left\{\begin{array}{llll} u = 0; & r_1 = 0, & r_2 = \beta - \alpha; \\ u = 1; & r_1 = 0, & r_2 = \gamma - 1; \\ u = \infty; & r_1 = \alpha + 1 - \gamma, & r_2 = 1 - \beta. \end{array}\right. \tag{26}$$

The (z, u) equation is

$$\frac{d^2z}{du^2} + \frac{(\beta-\alpha-1)-(\beta+\gamma-\alpha-3)u}{u(u-1)}\frac{dz}{du} - \frac{(\beta-1)(\alpha-\gamma+1)}{u(u-1)}z = 0, \tag{27}$$

giving at once the roots in equations (26). Write

$$\alpha' = \alpha - \gamma + 1, \quad \beta' = 1 - \beta, \quad \gamma' = \alpha - \beta + 1;$$

then this equation has in the region of $u = 0$ the integral

$$z = F(\alpha', \beta', \gamma', u).$$

Transforming back to the variables y and x, we have, on neglecting a constant factor which is merely a power of -1,

$$y = x^{-\alpha}\left(1 - \frac{1}{x}\right)^{\gamma-\alpha-1} F\left(\alpha - \gamma + 1, 1 - \beta, \alpha - \beta + 1, \frac{1}{1-x}\right). \tag{28}$$

Proceeding in this way, it is clear that we can form 24 particular integrals of (12).

A full discussion of these functions is contained in the following chapter, which is a translation of a Thesis by M. E. Goursat; the reader is also referred to another memoir by the same author in the *Annales de l'École Normale* for 1883. In the following table of the 24 functions we must make some supposition as to the values of the different powers of x, $1-x$, $1-\frac{1}{x}$ which appear as factors: suppose we choose the values such that the different powers x, $1-x$, $1-\frac{1}{x}$ shall reduce to unity when we make respectively $x=1$, $x=0$, $x=\infty$.

I. $F(\alpha, \beta, \gamma, x)$.

II. $(1-x)^{\gamma-\alpha-\beta}F(\gamma-\alpha, \gamma-\beta, \gamma, x)$.

III. $(1-x)^{-\alpha}F\left(\alpha, \gamma-\beta, \gamma, \frac{x}{x-1}\right)$.

IV. $(1-x)^{-\beta}F\left(\beta, \gamma-\alpha, \gamma, \frac{x}{x-1}\right)$.

V. $x^{1-\gamma}F(\alpha-\gamma+1, \beta-\gamma+1, 2-\gamma, x)$.

VI. $x^{1-\gamma}(1-x)^{\gamma-\alpha-\beta}F(1-\alpha, 1-\beta, 2-\gamma, x)$.

VII. $x^{1-\gamma}(1-x)^{\gamma-\alpha-1}F\left(\alpha-\gamma+1, 1-\beta, 2-\gamma, \frac{x}{x-1}\right)$.

VIII. $x^{1-\gamma}(1-x)^{\gamma-\beta-1}F\left(\beta-\gamma+1, 1-\alpha, 2-\gamma, \frac{x}{x-1}\right)$.

IX. $F(\alpha, \beta, \alpha+\beta-\gamma+1, 1-x)$.

X. $x^{1-\gamma}F(\alpha-\gamma+1, \alpha+\beta-\gamma+1, 1-x)$.

XI. $x^{-\alpha}F\left(\alpha, \alpha-\gamma+1, \alpha+\beta-\gamma+1, \frac{x-1}{x}\right)$.

XII. $x^{-\beta}F\left(\beta, \beta-\gamma+1, \alpha+\beta-\gamma+1, \frac{x-1}{x}\right)$.

XIII. $(1-x)^{\gamma-\alpha-\beta}F(\gamma-\alpha, \gamma-\beta, \gamma-\alpha-\beta+1, 1-x)$.

XIV. $(1-x)^{\gamma-\alpha-\beta}x^{1-\gamma}F(1-\alpha, 1-\beta, \gamma-\alpha-\beta+1, 1-x)$.

XV. $(1-x)^{\gamma-\alpha-\beta}x^{\alpha-\gamma}F\left(1-\alpha,\ \gamma-\alpha,\ \gamma-\alpha-\beta+1,\ \frac{1-x}{x}\right).$

XVI. $(1-x)^{\gamma-\alpha-\beta}x^{\beta-\gamma}F\left(1-\beta,\ \gamma-\beta,\ \gamma-\alpha-\beta+1,\ \frac{1-x}{x}\right).$

XVII. $x^{-\alpha}F\left(\alpha,\ \alpha-\gamma+1,\ \alpha-\beta+1,\ \frac{1}{x}\right).$

XVIII. $x^{-\alpha}\left(1-\frac{1}{x}\right)^{\gamma-\alpha-\beta}F\left(1-\beta,\ \gamma-\beta,\ \alpha-\beta+1,\ \frac{1}{x}\right).$

XIX. $x^{-\alpha}\left(1-\frac{1}{x}\right)^{-\alpha}F\left(\alpha,\ \gamma-\beta,\ \alpha-\beta+1,\ \frac{1}{1-x}\right).$

XX. $x^{-\alpha}\left(1-\frac{1}{x}\right)^{\gamma-\alpha-1}F\left(\alpha-\gamma+1,\ 1-\beta,\ \alpha-\beta+1,\ \frac{1}{1-x}\right).$

XXI. $x^{-\beta}F\left(\beta,\ \beta-\gamma+1,\ \beta-\alpha+1,\ \frac{1}{x}\right).$

XXII. $x^{-\beta}\left(1-\frac{1}{x}\right)^{\gamma-\alpha-\beta}F\left(1-\alpha,\ \gamma-\alpha,\ \beta-\alpha+1,\ \frac{1}{x}\right).$

XXIII. $x^{-\beta}\left(1-\frac{1}{x}\right)^{-\beta}F\left(\beta,\ \gamma-\alpha,\ \beta-\alpha+1,\ \frac{1}{1-x}\right).$

XXIV. $x^{-\beta}\left(1-\frac{1}{x}\right)^{\gamma-\beta-1}F\left(\beta-\gamma+1,\ 1-\alpha,\ \beta-\alpha+1,\ \frac{1}{1-x}\right).$

The general condition for the convergence of the series $F(\alpha, \beta, \gamma, \xi)$ is mod. $\xi < 1$; from this we can see at once what are the regions of convergence for the above series of functions, viz.—(1): for the argument x the region of convergence is the interior of the circle of radius unity, whose centre is at $x=0$; (2): for the argument $\frac{1}{x}$ the region of convergence is the entire plane outside of this circle; (3): for the argument $1-x$ the region of convergence is the interior of the circle whose centre is at the point $x=1$ and whose radius is unity; (4): for the argument $\frac{1}{1-x}$ the region of convergence is all of the plane outside of this circle; (5): for the argument $\frac{x}{x-1}$ suppose $x = \xi + i\eta$, and make $\xi = \frac{1}{2}$, then

$$\text{mod.}\, \frac{x}{x-1} = \text{mod.}\, \frac{\frac{1}{2} + i\eta}{-\frac{1}{2} + i\eta} = 1,$$

and so the region of convergence is all of the plane to the left of the line $\xi = \frac{1}{2}$; (6): for $\frac{x-1}{x}$ the required region is obviously all of the plane to the right of the same line.

The dissection of the plane which is necessary in order that the above functions may remain uniform while the variable travels all over the plane without crossing any of the cuts is obviously affected by drawing a cut from 0 to $-\infty$, and another from $+1$ to $+\infty$.

It is now easy to see that the above 24 functions divide into six groups of four each. In the region of the point $x = 0$ the indicial equation has for roots 0 and $1 - \gamma$; now the functions I to IV inclusive are uniform and convergent in the region of the point $x = 0$, and they reduce to unity for $x = 0$; there can, however, be but one integral in this region possessing this property, viz. the integral

$$y_1 = F(\alpha, \beta, \gamma, x)$$

belonging to the exponent zero; therefore each of the integrals I, II, III, IV represents the same function. Again, the functions V to VIII inclusive are the products of $x^{1-\gamma}$ into expressions which are uniform and convergent for the same region $x = 0$, and which reduce to unity for $x = 0$; they represent therefore the one regular integral, say y_2, corresponding to the root $1 - \gamma$ of the indicial equation for $x = 0$. The expressions IX to XII, and XIII to XVI, represent in the same way the two regular integrals, say y_3 and y_4, corresponding to the point $x = 1$; and XVII to XX, and XXI to XXIV, represent the two regular integrals, say y_5 and y_6, belonging to the point $x = \infty$.

It will be convenient to insert here a direct determination of the integrals of (12) which belong to the critical points 0, 1, and ∞ respectively, and to show what conditions must in each case be satisfied by the constants α, β, γ, in order that the integrals may be of the forms sought. Writing equation (12) in the form

$$(29)\qquad (x^2 - x)\frac{d^2y}{dx^2} - [\gamma - (\alpha + \beta + 1)x]\frac{dy}{dx} + \alpha\beta y = 0,$$

we have as above, for the roots of the indicial equations:

$$\begin{aligned} &\text{For } x = 0, \quad r_1 = 0, \quad r_2 = 1 - \gamma; \\ &\text{For } x = 1, \quad r_1 = 0, \quad r_2 = \gamma - \alpha - \beta; \\ &\text{For } x = \infty, \quad r_1 = \alpha, \quad r_2 = \beta. \end{aligned}$$

When γ is not a negative integer we have in the region of $x = 0$ the regular integral

$$y_1 = F(\alpha, \beta, \gamma, x),$$

where $F(\alpha, \beta, \gamma, x)$ is a convergent series inside the circle whose centre is $x = 0$ and whose radius is unity. Inside this same circle there must exist another integral belonging to the exponent $1 - \gamma$; to find it, write

$$y = x^{1-\gamma} Y,$$

and we have for Y the equation

$$(30)\qquad (x^2 - x)\frac{d^2Y}{dx^2} - [2 - \gamma - (\alpha + \beta - 2\gamma + 3)x]\frac{dY}{dx} - (\alpha - \gamma + 1)(\beta - \gamma + 1)Y = 0;$$

which differs from (29) only by having α, β, γ replaced by $\alpha - \gamma + 1$, $\beta - \gamma + 1$, $2 - \gamma$. When $2 - \gamma$ is not a negative integer, equation (30) admits the solution

$$(31)\qquad Y = F(\alpha - \gamma + 1, \beta - \gamma + 1, 2 - \gamma, x),$$

and consequently (29) has the solution

$$(32) \qquad y_2 = x^{1-\gamma} F(\alpha - \gamma + 1, \beta - \gamma + 1, 2 - \gamma, x).$$

It follows then that if γ is not an integer, the general integral of (29) is, in the region of $x = 0$,

$$(33) \qquad y = Cy_1 + C'y_2,$$

where C and C' are arbitrary constants. For the region of $x = 1$ make $x = 1 - \xi$; equation (29) becomes now

$$(34) \quad \xi(\xi - 1)\frac{d^2y}{d\xi^2} - [\alpha + \beta - \gamma + 1 - (\alpha + \beta + 1)\xi]\frac{dy}{d\xi} + \alpha\beta y = 0,$$

which differs from (29) only by the change of γ into $\alpha + \beta - \gamma + 1$. If then $\alpha + \beta - \gamma + 1$ is not an integer, (29) admits in the region of the point $x = 1$ the solutions

$$(35) \quad \begin{cases} y_3 = F(\alpha, \beta, \alpha + \beta - \gamma + 1, 1 - x); \\ y_4 = x^{\gamma - \alpha - \beta} F(\gamma - \beta, \gamma - \alpha, 1 + \gamma - \alpha - \beta, 1 - x); \end{cases}$$

and the complete solution in the region of this point will be

$$y = Cy_3 + C'y_4,$$

where C and C' are arbitrary constants. For the region of $x = \infty$ it is only necessary to write $x = \frac{1}{t}$, and find the indicial equation for $t = 0$. We find readily that if $\alpha - \beta + 1$ is neither zero nor a negative integer,

$$(36) \qquad y_5 = x^{-\alpha} F\left(\alpha, \alpha - \gamma + 1, \alpha - \beta + 1, \frac{1}{x}\right)$$

is an integral of (29); and so if $\beta - \alpha + 1$ is neither zero nor a negative integer,

$$(36) \qquad y_6 = x^{-\beta} F\left(\beta, \beta - \gamma + 1, \beta - \alpha + 1, \frac{1}{x}\right).$$

These are the six integrals spoken of above; as to their regions of convergence, it is clear that for the pair (y_1, y_2) we have the circle of radius unity whose centre is at the origin; for the pair (y_5, y_6) we have the entire plane outside of this circle; and for (y_3, y_4) the circle whose centre is at $x = 1$ and whose radius is unity.

We see now that the regions of $x = 0$ and $x = 1$ have an area in common, as have also the regions of $x = 1$ and $x = \infty$. If $1 - \gamma$ and $\gamma - \alpha - \beta$ have their real parts positive or zero, the series $y_1, \ldots, y_6$ will be convergent at each point on their respective circles (a known property of the convergence of these series which need not be gone into here); it follows therefore that in the area common, say, to $x = 0$ and $x = 1$ the functions y_3 and y_4 are linearly expressible in terms of y_1 and y_2. If now in this common area we have expressed y_3 and y_4 in terms of y_1 and y_2, we can by a known process in the theory of functions obtain uniform and convergent developments for y_3 and y_4 in the whole region of $x = 0$. Similar remarks apply to the linear relations connecting the integrals y_5 and y_6 with y_3 and y_4, and consequently with y_1 and y_2. Following Jordan, write for subsequent convenience, instead of $y_1, \ldots, y_6$,

$$c_1 y_1, \quad c_2 y_2, \quad c_3 y_3, \quad c_4 y_4, \quad c_5 y_5, \quad c_6 y_6;$$

then it is clear that we must have the relations

$$(37) \left\{ \begin{array}{l} c_3 y_3 = A c_1 y_1 + B c_2 y_2; \\ c_4 y_4 = C c_1 y_1 + D c_2 y_2; \end{array} \right. \qquad (38) \left\{ \begin{array}{l} c_5 y_5 = E c_1 y_1 + F c_2 y_2; \\ c_6 y_6 = G c_1 y_1 + H c_2 y_2; \end{array} \right.$$

where the constants $A, \ldots, H$ have now to be determined.

Suppose x to describe a circle of radius unity around the origin; the integral y_1 will not change, but y_2, y_5, y_6 will be multiplied respectively by $e^{2\pi i(1-\gamma)}$, $e^{-2\pi i\alpha}$, $e^{-2\pi i\beta}$, and equations (38) become

$$(39) \qquad \left\{ \begin{array}{l} e^{-2\pi i\alpha} c_5 y_5 = E c_1 y_1 + F e^{2\pi i(1-\gamma)} c_2 y_2; \\ e^{-2\pi i\beta} c_6 y_6 = G c_1 y_1 + H e^{2\pi i(1-\gamma)} c_2 y_2. \end{array} \right.$$

Now in equations (37) make $x = 0$; also in (37), (38), (39) make $x = 1$, and we have the following eight equations for the determination of the constants $A, \ldots, H$:

$$
(40)\quad \begin{cases}
c_3(y_3)_0 = Ac_1(y_1)_0 + Bc_2(y_2)_0; \quad c_4(y_4)_0 = Cc_1(y_1)_0 + Dc_2(y_2)_0; \\
c_3(y_3)_1 = Ac_1(y_1)_1 + Bc_2(y_2)_1; \quad c_4(y_4)_1 = Cc_1(y_1)_1 + Dc_2(y_2)_1; \\
c_5(y_5)_1 = Ec_1(y_1)_1 + Fc_2(y_2)_1; \quad c_6(y_6)_1 = Gc_1(y_1)_1 + Hc_2(y_2)_1; \\
\qquad e^{-2\pi i\alpha} c_6(y_5)_1 = Ec_1(y_1)_1 + Fe^{2\pi i(1-\gamma)} c_2(y_2)_1; \\
\qquad e^{-2\pi i\beta} c_6(y_6)_1 = Gc_1(y_1)_1 + He^{2\pi i(1-\gamma)} c_2(y_2)_1.
\end{cases}
$$

The values of $(y_1)_0, \ldots, (y_2)_1$ are known to be

$$
(41)\quad \begin{cases}
(y_1)_0 = 1; & (y_2)_0 = 0; \\
(y_3)_0 = \dfrac{\Gamma(\alpha+\beta-\gamma+1)\Gamma(1-\gamma)}{\Gamma(\beta-\gamma+1)\Gamma(\alpha-\gamma+1)}; & (y_4)_0 = \dfrac{\Gamma(\gamma-\alpha-\beta+1)\Gamma(1-\gamma)}{\Gamma(1-\alpha)\Gamma(1-\beta)}; \\
(y_1)_1 = \dfrac{\Gamma(\gamma)\Gamma(\gamma-\alpha-\beta)}{\Gamma(\gamma-\alpha)\Gamma(\gamma-\beta)}; & (y_2)_1 = \dfrac{\Gamma(2-\gamma)\Gamma(\gamma-\alpha-\beta)}{\Gamma(1-\alpha)\Gamma(1-\beta)}; \\
(y_3)_1 = 1; & (y_4)_1 = 0; \\
(y_5)_1 = \dfrac{\Gamma(\alpha-\beta+1)\Gamma(\gamma-\alpha-\beta)}{\Gamma(1-\beta)\Gamma(\gamma-\beta)}; & (y_6)_1 = \dfrac{\Gamma(\beta-\alpha+1)\Gamma(\gamma-\alpha-\beta)}{\Gamma(1-\alpha)\Gamma(\gamma-\alpha)}.
\end{cases}
$$

These values introduced in (40) give us the means of determining the constants $A, \ldots, H$ in terms of the constants $c_1, \ldots, c_6$, exponentials and Γ-functions. A choice of values must now be made for the constants $c_1, \ldots, c_6$; if we make them each equal to unity, then $A, \ldots, H$ are determined as functions of the exponentials and the Γ-functions. A still simpler result is obtained by writing

$$
(42)\quad \begin{cases}
c_1 = \dfrac{\Gamma(\alpha)\Gamma(\gamma-\alpha)}{\Gamma(\gamma)}; & c_2 = \dfrac{\Gamma(1-\beta)\Gamma(\beta-\gamma+1)}{\Gamma(2-\gamma)}; \\
c_3 = \dfrac{\Gamma(\alpha)\Gamma(\beta-\gamma+1)}{\Gamma(\alpha+\beta-\gamma+1)}; & c_4 = \dfrac{\Gamma(1-\beta)\Gamma(\gamma-\alpha)}{\Gamma(\gamma-\alpha-\beta+1)}; \\
c_5 = \dfrac{\Gamma(\alpha)\Gamma(1-\beta)}{\Gamma(\alpha-\beta+1)}; & c_6 = \dfrac{\Gamma(\beta-\gamma+1)\Gamma(\gamma-\alpha)}{\Gamma(\beta-\alpha+1)}
\end{cases}
$$

Equations (40) are now

(43)
$$\begin{cases} \dfrac{\Gamma(\alpha)\Gamma(1-\gamma)}{\Gamma(\alpha-\gamma+1)} = A\dfrac{\Gamma(\alpha)\Gamma(\gamma-\alpha)}{\Gamma(\gamma)}; \\ \dfrac{\Gamma(1-\gamma)\Gamma(\gamma-\alpha)}{\Gamma(1-\alpha)} = C\dfrac{\Gamma(\alpha)\Gamma(\gamma-\alpha)}{\Gamma(\gamma)}; \\ \dfrac{\Gamma(\alpha)\Gamma(\beta-\gamma+1)}{\Gamma(\alpha+\beta-\gamma+1)} = A\dfrac{\Gamma(\gamma-\alpha-\beta)\Gamma(\alpha)}{\Gamma(\gamma-\beta)} \\ \qquad + B\dfrac{\Gamma(\gamma-\alpha-\beta)\Gamma(\beta-\gamma+1)}{\Gamma(1-\alpha)}; \\ 0 = C\dfrac{\Gamma(\gamma-\alpha-\beta)\Gamma(\alpha)}{\Gamma(\gamma-\beta)} \\ \qquad + D\dfrac{\Gamma(\gamma-\alpha-\beta)\Gamma(\beta-\gamma+1)}{\Gamma(1-\alpha)}; \\ \dfrac{\Gamma(\gamma-\alpha-\beta)\Gamma(\alpha)}{\Gamma(\gamma-\beta)} = E\dfrac{\Gamma(\gamma-\alpha-\beta)\Gamma(\alpha)}{\Gamma(\gamma-\beta)} \\ \qquad + F\dfrac{\Gamma(\gamma-\alpha-\beta)\Gamma(\beta-\gamma+1)}{\Gamma(1-\alpha)}; \\ \cdots\cdots\cdots\cdots \end{cases}$$

If we divide each of these equations by the product of the Γ-functions which appear in each numerator and then employ the relation

$$\Gamma(p)\Gamma(1-p) = \frac{\pi}{\sin p\pi},$$

we have finally

(44)
$$\begin{cases} \sin(\gamma-\alpha)\pi = A\sin\gamma\pi; \quad \sin\alpha\pi = C\sin\gamma\pi; \\ \sin(\gamma-\alpha-\beta)\pi = A\sin(\gamma-\beta)\pi + B\sin\alpha\pi; \\ 0 = C\sin(\gamma-\beta)\pi + D\sin\alpha\pi; \\ \sin(\gamma-\beta)\pi = E\sin(\gamma-\beta)\pi + F\sin\alpha\pi; \\ \sin\alpha\pi = G\sin(\gamma-\beta)\pi + H\sin\alpha\pi; \\ e^{-2\pi i\alpha}\sin(\gamma-\beta)\pi = E\sin(\gamma-\beta)\pi + Fe^{2\pi i(1-\gamma)}\sin\alpha\pi; \\ e^{-2\pi i\beta}\sin\alpha\pi = G\sin(\gamma-\beta)\pi + He^{2\pi i(1-\gamma)}\sin\alpha\pi. \end{cases}$$

These give $A, \ldots, H$ in terms of sines and exponentials only, viz.:

$$
(45)\quad \left\{
\begin{aligned}
A &= \frac{\sin(\gamma-\alpha)\pi}{\sin\gamma\pi};\\
B &= \frac{1}{\sin\alpha\pi}\left[\sin(\gamma-\alpha-\beta)\pi - \frac{\sin(\gamma-\alpha)\pi\sin(\gamma-\beta)\pi}{\sin\gamma\pi}\right];\\
C &= \frac{\sin\alpha\pi}{\sin\gamma\pi};\\
D &= -\frac{\sin\alpha\pi\sin(\gamma-\beta)\pi}{\sin\alpha\pi\sin\gamma\pi};\\
E &= \frac{e^{2\pi i(1-\gamma)} - e^{-2\pi i\alpha}}{e^{2\pi i(1-\gamma)} - 1}.\\
F &= \frac{e^{-2\pi i\alpha} - 1}{e^{2\pi i(1-\gamma)} - 1}\cdot\frac{\sin(\beta-\gamma)\pi}{\sin\alpha\pi};\\
G &= \frac{1}{\sin(\gamma-\beta)\pi} - \frac{\sin\alpha\pi}{\sin(\gamma-\beta)\pi}\cdot\frac{e^{-2\pi i\beta}-1}{e^{2\pi i(1-\gamma)}-1};\\
H &= \frac{e^{-2\pi i\beta} - 1}{e^{2\pi i(1-\gamma)} - 1}.
\end{aligned}
\right.
$$

These forms have been taken from Jordan; they are to be found in a number of places, *e.g.*, Forsyth's *Differential Equations* (here in a somewhat modified form) and Goursat's Thesis, above referred to.

If γ (and therefore also $1-\gamma$) is an integer, then one of the integrals in the region of $x = 0$ *may* contain a logarithm; so if $\gamma - \alpha - \beta$ is an integer, or if $\alpha - \beta$ is an integer, a logarithm may appear in one of the integrals belonging to $x = 1$ and to $x = \infty$. It has been shown in general that logarithms enter in such cases as these; they *may*, however, fall out even when these conditions are satisfied. This question will be entered into later.

A single illustration of the entrance of logarithms will only be given. We have already seen that if the two roots of an indicial equation are equal, then a logarithm must necessarily appear. Suppose now that we have $\gamma = 1$, and consider the region of $x = 0$. The roots of the indicial equation are now each zero, and the differential equation itself is

$$(46) \qquad x(x-1)\frac{d^2y}{dx^2}+[(\alpha+\beta+1)x-1]\frac{dy}{dx}+\alpha\beta y=0$$

one integral of which is

$$F(\alpha,\ \beta,\ 1,\ x), \quad = \text{say}, A_0 + A_1x + A_2x^2 + \ldots,$$

when for brevity we have written

$$A_k = \frac{\alpha(\alpha+1)\ldots(\alpha+k-1)\beta(\beta+1)\ldots(\beta+k-1)}{\underline{k}|^2}.$$

For convenience write

$$\Phi = F(\alpha,\ \beta,\ 1,\ x);$$

then a second integral of (46) is of the form

$$y = \phi + \Phi \log x,$$

where ϕ is a uniform and continuous function of x in the region of $x=0$. Substituting this value of y in (46), and equating to zero the aggregate of terms which do not contain the logarithm as factor, we have for ϕ the equation

$$(47) \quad x(x-1)\frac{d^2\phi}{dx^2}+[(\alpha+\beta+1)x-1]\frac{d\phi}{dx}+\alpha\beta\phi+2(x-1)\frac{d\Phi}{dx} + (\alpha+\beta)\Phi = 0.$$

Assume now

$$(48) \qquad \phi = A_0a_0 + A_1a_1x + A_2a_2x^2 + \ldots,$$

and replace in (47) Φ and ϕ by their development in series; we have, on equating to zero the coefficient of x^k,

$$(49) \quad (\alpha+k)(\beta+k)A_ka_k - (k+1)^2A_{k+1}a_{k+1} + (\alpha+\beta+2k)A_k - 2(k+1)A_{k+1} = 0;$$

giving, since

$$\frac{A_{k+1}}{A_k} = \frac{(\alpha+k)(\beta+k)}{(k+1)^2},$$

$$(50) \qquad a_{k+1} - a_k = \frac{-2}{k+1} + \frac{1}{\alpha+k} + \frac{1}{\beta+k}.$$

From this last we have

$$(51) \qquad a_k = a_0 + \frac{1}{\alpha} + \frac{1}{\alpha+1} + \ldots + \frac{1}{\alpha+k-1}$$
$$+ \frac{1}{\beta} + \frac{1}{\beta+1} + \ldots + \frac{1}{\beta+k-1}$$
$$- 2\left(1 + \frac{1}{2} + \frac{1}{3} + \ldots + \frac{1}{k}\right).$$

We may make $a_0 = 0$ without loss of generality, since this only amounts to subtracting $a_0\Phi$ from ϕ; the series giving ϕ is thus determined by its general term $A_k a_k x^k$.

It is not difficult to see from the form of this general term that the solution

$$\phi + \Phi \log x$$

thus obtained can be written in the form

$$\Phi \log x + \frac{d\Phi}{d\alpha} + \frac{d\Phi}{d\beta} + \frac{d\Phi}{d\gamma};$$

where $\frac{d\Phi}{d\alpha}, \frac{d\Phi}{d\beta}, \frac{d\Phi}{d\gamma}$ denote the values of the derivatives of the function

$$\Phi = F(\alpha, \beta, \gamma, x)$$

when we have made $\gamma = 1$.

It is not possible to give any adequate account of the many investigations that have been made upon the hypergeometric series, but the results obtained by M. André Markoff in two interesting notes in the *Mathematische Annalen* may be here stated. In his first note* M. Markoff proposes to find all the cases where the product of two values of y satisfying the differential equation

$$x(1-x)\frac{d^2y}{dx^2} + [\gamma - (\alpha+\beta+1)x]\frac{dy}{dx} - \alpha\beta y = 0$$

---

* *Math. Annalen*, vol. xxviii. p. 586.

reduces to an integral function of x. Writing $z = y_1 y_2$, the differential equation of the third order satisfied by z is found to be

$$x^2(1-x)^2\frac{d^3z}{dx^3} + 3x(1-x)(ax+b)\frac{d^2z}{dx^2}$$
$$+ (cx^2 + dx + e)\frac{dz}{dx} + (fx+g)z = 0;$$

where

$$\begin{aligned}
a &= -(\alpha + \beta + 1),\\
b &= \gamma,\\
c &= 2\alpha^2 + 8\alpha\beta + 2\beta^2 + 3\alpha + 3\beta + 1,\\
d &= -2\gamma(2\alpha + 2\beta + 1) - 4\alpha\beta,\\
e &= 2\gamma^2 - \gamma,\\
f &= 4\alpha\beta(\alpha + \beta),\\
g &= -2\alpha\beta(2\gamma - 1).
\end{aligned}$$

It remains now to find all the cases where this last differential equation admits the solution

$$z = \text{integral function of } x.$$

If n denote the degree of the function, the conditions are found to be as follows:

(A) n even:

(1) $\alpha = -\frac{n}{2}$,

(2) $\beta = -\frac{n}{2}$,

(3) $\alpha + \beta = -n$, $\gamma = \frac{1}{2}, \ -\frac{1}{2}, \ -\frac{3}{2}, \ \ldots, \ -\frac{2n-1}{2}$;

(B) n odd:

$$(1)\ \alpha = -\frac{n}{2},\quad \gamma = \frac{1}{2},\ -\frac{1}{2},\ -\frac{3}{2},\ \ldots,\ -\frac{n-2}{2},$$
$$\beta,\ \beta - 1,\ \ldots,\ \beta - \frac{n-1}{2},$$

$$(2)\ \beta = -\frac{n}{2},\quad \gamma = \frac{1}{2},\ -\frac{1}{2},\ -\frac{3}{2},\ \ldots,\ -\frac{n-2}{2},$$
$$\alpha,\ \alpha - 1,\ \ldots,\ \alpha - \frac{n-1}{2},$$

$$(3)\ \alpha + \beta = -n,\quad \gamma = \frac{1}{2},\ -\frac{1}{2},\ -\frac{3}{2},\ \ldots,\ -\frac{n-2}{2},$$
$$-\frac{n}{2},\ \ldots,\ -\frac{2n-1}{2}.$$

In his second note* (closely allied to certain investigations by Schwarz) Markoff seeks to find all the cases where the same differential equation of the second order admits an integral of the form

$$Xy'^2 + Yy'y + Zy^2,\quad \left(y' = \frac{dy}{dx}\right)$$

where X, Y, and Z are rational integral functions of x. The conditions found in this case are as follows (n an integer):

(1) $\alpha + \beta = -n,\quad \gamma = \frac{1}{2},\ -\frac{1}{2},\ -\frac{3}{2},\ \ldots,\ -n + \frac{1}{2},$

(2) $\alpha + \beta = n,\quad \gamma = \frac{3}{2},\ \frac{5}{2},\ \ldots,\ n - \frac{1}{2},$

(3) $\alpha + \beta - 2\gamma = -n,\quad \gamma = \frac{3}{2},\ \frac{5}{2},\ \ldots,\ n - \frac{1}{2},$

(4) $\alpha + \beta - 2\gamma = n,\quad \gamma = \frac{1}{2},\ -\frac{1}{2},\ -\frac{3}{2},\ \ldots,\ -n + \frac{1}{2},$

(5) $\left.\begin{matrix} 2\alpha - \gamma \\ \text{or} \\ 2\beta - \gamma \end{matrix}\right\} = -n,\quad \alpha + \beta - \gamma = -\frac{1}{2},\ -\frac{3}{2},\ \ldots,\ -n + \frac{1}{2},$

(6) $\left.\begin{matrix} 2\alpha - \gamma \\ \text{or} \\ 2\beta - \gamma \end{matrix}\right\} = n,\quad \alpha + \beta - \gamma = \frac{1}{2},\ \frac{3}{2},\ \ldots,\ n - \frac{1}{2},$

---

* *Math. Annalen*, vol. xxix. p. 247.

$$(7)\qquad \left.\begin{matrix}\alpha-\beta+\gamma\\ \text{or}\\ \beta-\alpha+\gamma\end{matrix}\right\} = -n,\quad \gamma=\tfrac{1}{2},\quad -\tfrac{1}{2},\quad -\tfrac{3}{2},\quad \ldots,\quad -n+\tfrac{1}{2};$$

$$(8)\qquad \left.\begin{matrix}\alpha-\beta+\gamma\\ \text{or}\\ \beta-\alpha+\gamma\end{matrix}\right\} = n,\quad \gamma=\tfrac{3}{2},\quad \tfrac{5}{2},\quad \ldots,\quad n-\tfrac{1}{2}.$$

An interesting case of the differential equation of the second order has been studied by Dr. Heun in the *Mathematische Annalen.**

If $\xi_1, \xi_2, \ldots, \xi_\rho$ are the finite critical points of the equation, and if $\psi(x) = (x-\xi_1), (x-\xi_2), \ldots, (x-\xi_\rho)$, we have seen, equation (2), that the differential equation can be put in the form

$$(52)\qquad \psi(x)\frac{d^2y}{dx^2} + Q_1(x)\frac{dy}{dx} + Q_2(x)y = 0;$$

where $Q_1(x)$ and $Q_2(x)$ are polynomials of the degrees $\rho + 1$ and $\rho - 2$ respectively.

The conditions† that the points ξ shall be critical points determine $\rho + 1$ of the $2\rho - 1$ constants in (52), and there remain therefore (in $Q_2(x)$) $\rho - 2$ independent constants, which we will call the *characteristic parameters* of (52). Equation (52) written out fully is of the form

$$(53)\qquad \psi(x)\frac{d^2y}{dx^2} + (k_1x^{\rho-1} + k_2x^{\rho-2} + \ldots + k_\rho)\frac{dy}{dx} + (k_2'x^{\rho-2} + k_3'x^{\rho-3} + \ldots + k_\rho')y = 0.$$

Denoting by $(x, 1)^n$ a rational integral function of x of degree n, and writing

$$y_1 = (x, 1)^n,$$

then in order that y_1 may satisfy (53) we must have

$$n(n-1) + k_1 n + k_2' = 0.$$

---

* "Integration regulärer Differentialgleichungen zweiter Ordnung durch die Kettenbruchentwicklung von ganzen Abel'schen Integralen dritter Gattung." Von Karl Heun in München. *Math. Annalen*, vol. xxx. p. 553.

† *American Journal of Mathematics*, vol. x. p. 205 *et seq.*

This condition is manifestly not sufficient, as the characteristic parameters k_3', k_4', . . . , k_ρ' must satisfy certain algebraic equations found by substituting y_1 in (53). These conditions serve for the complete determination of the characteristic parameters.* The number of integral functions y_1, $= (x, 1)^n$, obtained in this way, is equal

$$\frac{(n+1)(n+2)\ \ldots\ (n+\rho-2)}{1\ .\ 2\ \ldots\ (\rho-2)}.$$

From these rational solutions of (53) the second (and transcendental) integral is obtained by the known formula

$$(54) \qquad y_2 = y_1 \int \frac{e^{-\int \frac{Q_1(x)}{\psi(x)} dx}}{y_1^2} dx.$$

In the special case where $Q_1(x) = \frac{1}{2}\frac{d}{dx}\psi(x)$ Heine has shown† that we have

$$(55) \qquad \frac{y_2}{\psi(x)} = y_1 \sum_{\nu=1}^{\nu=\rho-1} C_\nu \int_{\xi_\nu}^{\xi_{\nu+1}} \frac{dz}{(x-z)\psi(z)} - Z;$$

where Z is a rational integral function of x, and C_1, C_2, . . . , $C_{\rho-1}$ are constants determined by aid of a certain system of algebraic equations. This last equation can be written in the form

$$(56) \qquad y_2 = Vy_1 + E,$$

where V and E are functions of x satisfying the relation

$$(57) \qquad y_1^2\left[\frac{dV}{dx} + \frac{d}{dx}\left(\frac{E}{y_1}\right)\right] = e^{-\int \frac{Q_1(x)}{\psi(x)} dx};$$

---

* Heine: *Handbuch der Kugelfunctionen*, Bd. I, § 136.

† Crelle, vol. 61.

viz., from (56) we have

(58) $$\frac{d}{dx}\left(\frac{y_2}{y_1}\right) = \frac{dV}{dx} + \frac{d}{dx}\left(\frac{E}{y_1}\right),$$

and from (54)

(59) $$\frac{d}{dx}\left(\frac{y_2}{y_1}\right) = -\frac{e^{-\int\frac{Q_1(x)}{\psi(x)}dx}}{y_1^2};$$

equating the right-hand members of (58) and (59), we have (57).

This result of Heine's is, of course, restricted by the condition imposed upon $Q_1(x)$, viz.,

$$Q_1(x) = \frac{1}{2}\frac{d}{dx}\psi(x).$$

Starting from equation (57) we can, however, determine V and E in the general case. Write

(60) $$\gamma_\nu = \frac{Q_1(\xi_\nu)}{\psi'(\xi_\nu)}, \quad \nu = 1, \quad 2, \ldots, \quad \rho.$$

Since the polynomials $Q_1(x)$ and $\psi(x)$ are respectively of the degrees $\rho - 1$ and ρ, we have

(61) $$\frac{Q_1(x)}{\psi(x)} = \frac{\gamma_1}{x-\xi_1} + \frac{\gamma_2}{x-\xi_2} + \ldots + \frac{\gamma_\rho}{x-\xi_\rho}.$$

Write, again,

(62) $$(x-\xi_1)^{\gamma_1}(x-\xi_2)^{\gamma_2}\ldots(x-\xi_\rho)^{\gamma_\rho} = \theta(x);$$

now in (57) we can make

(63) $$V = \int\frac{\phi(x)}{\theta(x)}dx, \qquad E = -Z\frac{\psi(x)}{\theta(x)};$$

where Z is a rational integral function, as yet to be determined, whose degree is $n - 1$, and $\phi(x)$ is an equally undetermined rational integral function whose degree cannot exceed $\rho - 2$.

Equation (56) gives now

(64) $$y_2 = y_1 \int \frac{\phi(x)}{\theta(x)}\, dx - \frac{\psi(x)}{\theta(x)} Z,$$

or

(65) $$\frac{\theta(x)}{\psi(x)} y_2 = y_1 \frac{\theta(x)}{\psi(x)} \int \frac{\phi(x)}{\theta(x)}\, dx - Z.$$

Denoting now by W a transcendental function of x, we have

(66) $$R^{(n)} = N^{(n)} W - Z^{(n)};$$

where $Z^{(n)} \div N^{(n)}$ is the nth convergent of the development of W in continued fractions, and $R^{(n)}$ is the corresponding remainder. We can now write

(67) $$W = \frac{\theta(x)}{\psi(x)} \int \frac{\phi(x)}{\theta(x)}\, dx, \quad Z^{(n)} = Z, \quad N^{(n)} = y_1, \quad R^{(n)} = \frac{\theta(x)}{\psi(x)} y_2,$$

if we can establish the following conditions:

I. That W is developable in a continued fraction of the form

$$W = \cfrac{C_0}{a_0 + b_0 x + \cfrac{C_1}{a_1 + b_1 x + \cfrac{C_2}{a_2 + b_2 x + \cdots}}}$$

which, for a given value of n, is *regular* up to, and including, the nth partial denominator.

II. If there exist as many values of $\phi(x)$ as there are rational solutions y_1 of the differential equation, viz.,

$$\frac{(n+1)(n+2)\ldots(n+\rho-2)}{1\,.\,2\ldots(\rho-2)}.$$

III. If the function $R^{(n)} \dfrac{\psi(x)}{\theta(x)}$ is a *second* solution of the differential equation.

To show that W is developable in a continued fraction of the form in (I), it is necessary first to show that W is developable in a convergent series containing only negative powers of x. Write $\frac{\psi(x)}{\theta(x)} = \Omega(x)$; then, by a known theorem of Abel's,*

$$(68)\quad \Omega(x)\int\frac{dz}{(x-z)\Omega(z)} - \theta(z)\int\frac{dx}{(z-x)\theta(x)}$$

$$= \sum_{m=0}^{m=\rho-2}\sum_{n=0}^{n=\rho-2}\left\{\frac{m+1}{m+n+2}b^{(m+n+1)} - \frac{n+1}{m+n+2}c^{(m+n+1)}\right\}J_{m,n};$$

where

$$J_{m,n} = \int\frac{x^m dx}{\theta(x)}\cdot\int\frac{z^n dz}{\Omega(z)},$$

$$\theta(x)\frac{d\Omega(x)}{dx} = b^{(0)} + b^{(1)}x + \ldots + b^{(\rho-1)}x^{\rho-1},$$

$$\Omega(x)\frac{d\theta(x)}{dx} = c^{(0)} + c^{(1)}x + \ldots + c^{(\rho-1)}x^{\rho-1}.$$

The roots of the equation $\theta(x) = 0$ are $\xi_1, \xi_2, \ldots, \xi_\rho$. In (68) integrate from one of these roots, say ξ_ν, to the following, $\xi_{\nu+1}$; we thus obtain expressions of the form

$$(69)\quad \phi(x)\int_{\xi_\nu}^{\xi_{\nu+1}}\frac{dz}{(x-z)\Omega(z)}$$

$$= \sum_{m=0}^{m=\rho-2}\Gamma_\rho^{(m)}\int\frac{x^m dx}{\theta(x)},\quad \nu = 1, 2, \ldots, \rho-1,$$

where the quantities Γ are known constants. Multiply now each of equations (69) by the corresponding constant of the series

$$C_1,\quad C_2,\quad \ldots,\quad C_{\rho-1},$$

and add the results; we have then

$$(70)\qquad \frac{\theta(x)}{\psi(x)}\sum_\nu\sum_m\Gamma_\nu^{(m)}C_\nu\int\frac{x^m dx}{\theta(x)} = \sum_\nu C\int_{\xi_\nu}^{\xi_{\nu+1}}\frac{\theta(z)}{(x-z)\psi(z)}dz.$$

* Jacobi: *Gesammelte Werke*, Bd. II. pp. 125–127.

Letting δ denote a constant, we have, from what precedes, for $\phi(x)$ the form

$$(71)\qquad \phi(x) = \delta_0 + \delta_1 x + \delta_2 x^2 + \ldots + \delta_{\rho-2} x^{\rho-2}.$$

From this it follows very easily that the integral on the left-hand side of (70) is identical with the above-given value of W if we have

$$(72)\qquad \begin{cases} \Gamma_1^{(0)} C_1 + \ldots + \Gamma_{\rho-1}^{(0)} C_\rho = \delta_0, \\ \Gamma_1^{(1)} C_1 + \ldots + \Gamma_{\rho-1}^{(1)} C_\rho = \delta_1, \\ \cdot \quad \cdot \quad \cdot \quad \cdot \quad \cdot \quad \cdot \quad \cdot \quad \cdot \quad \cdot \\ \Gamma_1^{(\rho-2)} C_1 + \ldots + \Gamma_{\rho-1}^{(\rho-2)} C_\rho = \delta_{\rho-2}. \end{cases}$$

W now takes the form

$$(73)\qquad W = \sum_{\nu=1}^{\nu=\rho-1} C_\nu \int_{\xi_\nu}^{\xi_{\nu+1}} \frac{\theta(z)}{(x-z)\psi(z)}\, dz.$$

From this equation it immediately follows that W is developable in a convergent series containing only negative powers of x.

In what follows we will assume that $\theta(z)$ has the form

$$(74)\qquad \theta(z) = (z-\xi_1)^{\frac{\lambda_1}{\mu_1}} (z-\xi_2)^{\frac{\lambda_2}{\mu_2}} \ldots (z-\xi_\rho)^{\frac{\lambda_\rho}{\mu_\rho}},$$

where λ_k and μ_k are positive integers satisfying the condition

$$\lambda_k < \mu_k.$$

Under this hypothesis we have

$$(75)\qquad Q_1(x) = \frac{\lambda_1}{\mu_1} \cdot \frac{(x-\xi_2)(x-\xi_3)\ldots(x-\xi_\rho)}{(\xi_1-\xi_2)(\xi_1-\xi_3)\ldots(\xi_1-\xi_\rho)} \psi'(\xi_1) + \ldots$$
$$+ \frac{\lambda_\rho}{\mu_\rho} \cdot \frac{(x-\xi_1)(x-\xi_2)\ldots(x-\xi_{\rho-1})}{(\xi_\rho-\xi_1)(\xi_\rho-\xi_2)\ldots(\xi_\rho-\xi_{\rho-1})} \psi'(\xi_\rho),$$

and consequently the function W is a sum of entire Abelian integrals of the third kind.

We proceed now to the complete determination of the function $\phi(x)$. (One coefficient only in $\phi(x)$ will remain indeterminate.) From equation (56) and the following ones we have

$$y_2 = y_1 \int \frac{\phi(x)}{\theta(x)} dx - \frac{\psi(x)}{\theta(x)} Z,$$

$$y_2 = \frac{\psi(x)}{\theta(x)} R^{(n)}, \quad Z = Z^{(n)}.$$

Dropping the x when it is not necessary to write it, we have from these last equations

$$\frac{\psi}{\theta} R^{(n)} = y_1 \int \frac{\phi}{\theta} dx - \frac{\psi}{\theta} Z^{(n)},$$

or

$$\frac{\psi}{\theta} \cdot \frac{R^{(n)}}{y_1} = \int \frac{\phi}{\theta} dx - \frac{\psi}{\theta} \cdot \frac{Z^{(n)}}{y_1},$$

giving

$$\frac{d}{dx}\left[\frac{\psi}{\theta} \cdot \frac{R^{(n)}}{y_1}\right] = \frac{\phi}{\theta} - \frac{d}{dx}\left[\frac{\psi}{\theta} \cdot \frac{Z^{(n)}}{y_1}\right],$$

or

(76) $$\theta y_1^2 \frac{d}{dx}\left[\frac{\psi}{\theta} \cdot \frac{R^{(n)}}{y_1}\right] = y_1^2 \left\{ \phi - \frac{d}{dx}\left[\frac{\psi}{\theta} \cdot \frac{Z^{(n)}}{y_1}\right] . \theta \right\}.$$

The right-hand member of this equation is equal to a constant, and consequently the left-hand member must also be a constant. This last, however, cannot be true if the order* of the remainder $R^{(n)}$ is $= -(n+1)$. If the $(n+1)^{st}$ partial denominator of the development of W in a continued fraction is then not linear (as the preceding ones are), but of the degree $\rho - 1$, then $R^{(n)}$ will be of degree $-(n+\rho-1)$, and consequently

(77) $$\theta(x)\, y_1^2 \frac{d}{dx}\left[\frac{\psi(x)}{\theta(x)} \cdot \frac{R^{(n)}}{y_1}\right] = \text{const.} = 1.$$

---

* By order of the transcendental function $R^{(n)}$ we mean, with Heine, the *least negative* exponent in the development of $R^{(n)}$ in series.

It follows from this that $R^{(n)}\dfrac{\psi(x)}{\theta(x)}$ actually represents a *second* solution of the given differential equation.

The fact of $R^{(n)}$ being of degree $-(n+\rho-1)$ gives sufficient data for the determination of the coefficients of $\phi(x)$; for in the product $y_1 W$ the terms in $x^{-1}, x^{-2}, \ldots, x^{-(n+\rho-2)}$ must drop out. Writing now

$$y_1 = h_0 x^n + h_1 x^{n-1} + \ldots + h_{n-1}x + h_n,$$

$$W = \frac{a_1}{x} + \frac{a_2}{x^2} + \ldots + \frac{a_r}{x^r} + \ldots,$$

we have at once the following equations of condition:

$$(78)\quad \begin{cases} a_1 h_n + a_2 h_{n-1} + \ldots + a_{n+1} h_0 = 0; \\ a_2 h_n + a_3 h_{n-1} + \ldots + a_{n+2} h_0 = 0; \\ \cdot \quad \cdot \quad \cdot \quad \cdot \quad \cdot \quad \cdot \quad \cdot \\ a_{n+\rho-2} h_n + a_{n+\rho-1} h_{n-1} + \ldots + a_{2n+\rho-2} h_0 = 0. \end{cases}$$

The quantities a have the form

$$a_r = \alpha^{(1)} C_1 + \alpha^{(2)} C_2 + \ldots + \alpha^{(\rho-1)} C_{\rho-1};$$

where the coefficients α are again linearly expressible in terms of the primitive moduli of periodicity of the Abelian integrals which enter into the problem. From the first n of equations (78) we can now express the coefficients $h_0, h_1, \ldots, h_n$ as functions of these moduli, and of the ratios

$$\frac{C_1}{C_{\rho-1}}, \quad \frac{C_2}{C_{\rho-1}}, \quad \ldots, \quad \frac{C_{\rho-2}}{C_{\rho-1}},$$

and consequently also as functions of the ratios

$$\frac{\delta_0}{\delta_{\rho-2}}, \quad \frac{\delta_1}{\delta_{\rho-2}}, \quad \ldots, \quad \frac{\delta_{\rho-3}}{\delta_{\rho-2}}.$$

Introducing the found values of $h_0, h_1, \ldots, h_n$ in the remaining $\rho-2$ equations of (78), we have a system of equations giving

$$\frac{(n+1)(n+2)\ldots(n+\rho-2)}{1.2.3\ldots(\rho-2)}$$

groups of values for the ratios

$$\frac{\delta_0}{\delta_{\rho-2}}, \quad \frac{\delta_1}{\delta_{\rho-2}}, \quad \ldots, \quad \frac{\delta_{\rho-3}}{\delta_{\rho-2}};$$

and, from equations (72), the same number of groups of values for the ratios of the constants C.

The results of the investigation may be summed up as follows:

If the regular differential equation

$$\psi(x)\frac{d^2y}{dx^2} + Q_1(x)\frac{dy}{dx} + Q_2(x)y = 0$$

has a rational integral function of degree n as its *first* solution, then this function is the denominator of the nth convergent in the development of the function

$$W = \sum_{i=1}^{i=\rho-1} C_i \int_{\xi_i}^{\xi_{i+1}} \frac{\theta(z)}{(x-z)\psi(z)}\, dz$$

in a continued fraction, where

$$\psi(\xi_i) = 0, \quad \theta(z) = (z-\xi_1)^{\frac{Q_1(\xi_1)}{\psi'(\xi_1)}}(z-\xi_2)^{\frac{Q_1(\xi_2)}{\psi'(\xi_2)}} \ldots (z-\xi_\rho)^{\frac{Q_1(\xi_\rho)}{\psi'(\xi_\rho)}},$$

and where the constants C are so determined that the first n partial denominators of the continued fraction are linear, while the $(n+1)^{\text{st}}$ is of degree $\rho - 1$. Further, the remainder multiplied by $\dfrac{\psi(x)}{\theta(x)}$ is a *second* solution of the same differential equation.

In particular, if the quantities $\dfrac{Q_1(\xi_i)}{\psi'(\xi_i)}$ are proper fractions, then the second solution, y_2, contains no higher transcendents than entire Abelian integrals of the third kind.

It is not the intention here to go into any extended account of the differential equations which give rise to Spherical Harmonics and the other allied functions of Analysis, but it is interesting to show how these equations are connected with the differential equation satisfied by Riemann's P-function. (We will use this form of the letter **P** to denote Riemann's function, in order to avoid confusion when the P of ordinary zonal harmonics has also to be employed.)

Riemann's definition of the **P**-function is as follows:

Denote by

$$\mathbf{P}\left\{\begin{matrix} a & b & c & \\ \alpha & \beta & \gamma & x \\ \alpha' & \beta' & \gamma' & \end{matrix}\right\}$$

a function of x satisfying the following conditions:

(I) For all values of x other than a, b, c, the function is uniform and continuous.

(II) Between any three branches of the function, say $\mathbf{P}'$, $\mathbf{P}''$, $\mathbf{P}'''$, there exists a linear relation with constant coefficients, viz.:

$$c'\mathbf{P}' + c''\mathbf{P}'' + c'''\mathbf{P}''' = 0.$$

(III) The function can be put in the form

$$c_\tau \mathbf{P}^\tau + c_{\tau'}\mathbf{P}^{\tau'} \quad (\tau = \alpha, \alpha', \beta, \beta', \gamma, \gamma')$$

where the products

$$\mathbf{P}^\tau(x-\sigma)^{-\tau}, \quad \mathbf{P}^{\tau'}(x-\sigma)^{-\tau'} \quad (\sigma = a, b, c)$$

are uniform for $x = \sigma$, and are not equal either zero or infinity.

With reference to the six real quantities α, α', β, β', γ, γ', we assume that none of the differences $\alpha - \alpha'$, $\beta - \beta'$, $\gamma - \gamma'$ are integers, and, further, that $\alpha + \alpha' + \beta + \beta' + \gamma + \gamma' = 1$. It may be left as an exercise to the student to show from the point of view of the linear differential equation satisfied by **P** that under the above conditions α, α', β, β', γ, γ' *must* be real.

The first columns of the function

$$\mathbf{P}\left\{\begin{matrix} a & b & c & \\ \alpha & \beta & \gamma & x \\ \alpha' & \beta' & \gamma' & \end{matrix}\right\}$$

can be interchanged among themselves, also α can be interchanged with α', β with β', and γ with γ, γ'. Further, let

$$x' = \frac{ex+f}{gx+h};$$

where the constants are such that for $x = a, b, c$ we have $x' = a'$, b', c'; then

$$\mathbf{P}\begin{Bmatrix} a & b & c & \\ \alpha & \beta & \gamma & x \\ \alpha' & \beta' & \gamma' & \end{Bmatrix} = \mathbf{P}\begin{Bmatrix} a' & b' & c' & \\ \alpha & \beta & \gamma & x' \\ \alpha' & \beta' & \gamma' & \end{Bmatrix}.$$

It is shown by Riemann that all the **P**-functions having the same values of α, α', β, β', γ, γ', can be led back to the function

$$\mathbf{P}\begin{Bmatrix} 0 & \infty & 1 & \\ \alpha & \beta & \gamma & x \\ \alpha' & \beta' & \gamma' & \end{Bmatrix},$$

which is denoted more briefly by the symbol

$$\mathbf{P}\begin{pmatrix} \alpha & \beta & \gamma & \\ & & & x \\ \alpha' & \beta' & \gamma' & \end{pmatrix}.$$

In such a function the quantities in the exponent-pairs (α, α'), (β, β'), (γ, γ') can be interchanged, and the pairs themselves can be interchanged if we replace x by

$$\frac{ex + f}{gx + h},$$

which for the first, second, and third exponent-pairs gives for the variable the respective values 0, ∞, 1. In this way the functions

$$\mathbf{P}\begin{pmatrix} \alpha & \beta & \gamma & \\ & & & x \\ \alpha' & \beta' & \gamma' & \end{pmatrix}$$

can be expressed by **P**-functions of the arguments

$$x, \quad 1 - x, \quad \frac{1}{x}, \quad \frac{x-1}{x}, \quad \frac{x}{x-1}, \quad \frac{1}{1-x},$$

and having the same exponents, but having them arranged in different orders. From the definition, it is easy to see that we have

$$\mathbf{P}\begin{Bmatrix} a & b & c & \\ \alpha & \beta & \gamma & x \\ \alpha' & \beta' & \gamma' & \end{Bmatrix}\left(\frac{x-a}{x-b}\right)^{\delta} = \mathbf{P}\begin{Bmatrix} a & b & c & \\ \alpha+\delta & \beta+\delta & \gamma & x \\ \alpha'+\delta & \beta'+\delta & \gamma' & \end{Bmatrix},$$

and, consequently, also

$$x^{\delta}(1-x)^{\epsilon}\,\mathbf{P}\begin{pmatrix} \alpha & \beta & \gamma & \\ & & & x \\ \alpha' & \beta' & \gamma' & \end{pmatrix} = \mathbf{P}\begin{pmatrix} \alpha+\delta, & \beta-\delta-\epsilon, & \gamma+\epsilon & \\ & & & x \\ \alpha'+\delta, & \beta'-\delta-\epsilon, & \gamma'+\epsilon & \end{pmatrix}.$$

Through these transformations one sees that the two exponents in the different pairs can take arbitrary values, the sum

$$\alpha+\alpha'+\ldots+\gamma'$$

remaining, however, unity. For any set of exponents we can then replace any other set provided the differences

$$\alpha-\alpha',\quad \beta-\beta',\quad \gamma-\gamma'$$

remain the same. All functions of the form

$$x^{\delta}(1-x)^{\epsilon}\,\mathbf{P}\begin{pmatrix} \alpha & \beta & \gamma & \\ & & & x \\ \alpha' & \beta' & \gamma' & \end{pmatrix}$$

are represented by Riemann by the symbol

$$\mathbf{P}\,(\alpha-\alpha',\quad \beta-\beta',\quad \gamma-\gamma',\quad x).$$

From what precedes the differential equation satisfied by $\mathbf{P}$ is at once seen to be a regular linear differential equation of the second order, having a, b, c as critical points, and whose integrals contain no logarithms. The equation is then of the form

$$(79)\qquad \frac{d^2y}{dx^2}+\frac{Ax^2+Bx+C}{(x-a)(x-b)(x-c)}\frac{dy}{dx}+\frac{Fx^2+Gx+H}{(x-a)^2(x-b)^2(x-c)^2}y=0.$$

(As an additional exercise the student may show that the numerator of the coefficient of y cannot be of a degree higher than the second.) The indicial equations corresponding to the points a, b, c are as follows:

$$(\alpha)\quad \text{For } x = a,\quad r(r-1)+r\frac{Aa^2+Ba+C}{(a-b)(a-c)}+\frac{Fa^2+Ga+H}{(a-b)^2(a-c)^2}=0;$$

$$(\beta)\quad \text{For } x = b,\quad r(r-1)+r\frac{Ab^2+Bb+C}{(b-a)(b-c)}+\frac{Fb^2+Gb+H}{(b-a)^2(b-c)^2}=0;$$

$$(\gamma)\quad \text{For } x = c,\quad r(r-1)+r\frac{Ac^2+Bc+C}{(c-a)(c-b)}+\frac{Fc^2+Gc+H}{(c-a)^2(c-b)^2}=0.$$

From (III) it is obvious that the roots of (α) are (α, α'), the roots of (β) are (β, β'), and the roots of (γ) are (γ, γ'). We have then for these roots the following equations:

$$(80)\quad \begin{cases} \alpha+\alpha' = \dfrac{(a-b)(a-c)-Aa^2-Ba-C}{(a-b)(a-c)}, \\ \beta+\beta' = \dfrac{(b-a)(b-c)-Ab^2-Bb-C}{(b-a)(b-c)}, \\ \gamma+\gamma' = \dfrac{(c-a)(c-b)-Ac^2-Bc-C}{(c-a)(c-b)}; \end{cases}$$

$$(81)\quad \begin{cases} \alpha\alpha' = \dfrac{Fa^2+Ga+H}{(a-b)^2(a-c)^2}, \\ \beta\beta' = \dfrac{Fb^2+Gb+H}{(b-a)^2(b-c)^2}, \\ \gamma\gamma' = \dfrac{Fc^2+Gc+H}{(c-a)^2(c-b)^2}; \end{cases}$$

or

$$(82)\quad \begin{cases} \alpha+\alpha' = 1+\dfrac{(Aa^2+Ba+C)(b-c)}{(a-b)(b-c)(c-a)}, \\ \beta+\beta' = 1+\dfrac{(Ab^2+Bb+C)(c-a)}{(a-b)(b-c)(c-a)}, \\ \gamma+\gamma' = 1+\dfrac{(Ac^2+Bc+C)(a-b)}{(a-b)(b-c)(c-a)}; \end{cases}$$

$$(83)\quad \begin{cases} \alpha\alpha' &= \dfrac{(Fa^2 + Ga + H)(b - c)^2}{(a - b)^2(b - c)^2(c - a)^2}, \\ \beta\beta' &= \dfrac{(Fb^2 + Gb + H)(c - a)^2}{(a - b)^2(b - c)^2(c - a)^2}, \\ \gamma\gamma' &= \dfrac{(Fc^2 + Gc + H)(a - b)^2}{(a - b)^2(b - c)^2(c - a)^2}. \end{cases}$$

Write

$$\Sigma = \alpha + \alpha' + \beta + \beta' + \gamma + \gamma';$$

then on adding equations (82) and observing that

$$a^2(b-c) + b^2(c-a) + c^2(a-b) = -(a-b)(b-c)(c-a),$$
$$a(b-c) + b(c-a) + c(a-b) = 0,$$
$$(b-c) + (c-a) + (a-b) = 0,$$

we have

$$(84)\quad A = 3 - \Sigma, \quad = 1 - \alpha - \alpha' + 1 - \beta - \beta' + 1 - \gamma - \gamma'.$$

Multiplying the first of equations (82) by a, the second by b, the third by c, adding, and observing that

$$\frac{a^3(b - c) + b^3(c - a) + c^3(a - b)}{(a - b)(b - c)(c - a)} = -(a + b + c),$$

we have

$$(85)\quad B = -[(1-\alpha-\alpha')(b+c) + (1-\beta-\beta')(c+a) + (1-\gamma-\gamma')(a+b)].$$

In like manner we find for C the form

$$(86)\quad C = (1 - \alpha - \alpha')bc + (1 - \beta - \beta')ca + (1 - \gamma - \gamma')ab.$$

Multiply now equations (83) in order by

$$(c - a)(a - b), \quad (a - b)(b - c), \quad (b - c)(c - a),$$

and adding, we get the value of F; multiplying again by

$$(c-a)(a-b)(b+c), \quad (a-b)(b-c)(c+a), \quad (b-c)(c-a)(a+b),$$

and adding, we get G; finally, multiplying by

$$(c-a)(a-b)bc, \quad (a-b)(b-c)ca, \quad (b-c)(c-a)ab,$$

and adding, we get H. The values of F, G, H thus found are

$$(87)\quad \begin{cases} F = -\{\alpha\alpha'(c-a)(a-b) + \beta\beta'(a-b)(b-c) \\ \qquad\qquad + \gamma\gamma'(b-c)(c-a)\}, \\ G = \{\alpha\alpha'(c-a)(a-b)(b+c) + \beta\beta'(a-b)(b-c)(c+a) \\ \qquad\qquad + \gamma\gamma'(b-c)(c-a)(a+b)\}, \\ H = -\{\alpha\alpha'(c-a)(a-b)bc + \beta\beta'(a-b)(b-c)ca \\ \qquad\qquad + \gamma\gamma'(b-c)(c-a)ab\}. \end{cases}$$

The values of $A, B, \ldots, H$ from (84), (85), (86), and (87), substituted in equation (79), give us for this equation the form

$$(88)\quad \frac{d^2y}{dx^2} + \left\{ \frac{\begin{array}{l}(1-\alpha-\alpha')[x^2-(b+c)x+bc] \\ \quad + (1-\beta-\beta')[x^2-(c+a)x+ca] \\ \quad + (1-\gamma-\gamma')[x^2-(a+b)x+ab]\end{array}}{(x-a)(x-b)(x-c)} \right\} \frac{dy}{dx}$$

$$+ \left\{ \frac{\begin{array}{l}\alpha\alpha'(a-b)(a-c)[x^2-(b+c)x+bc] \\ \quad + \beta\beta'(b-c)(b-a)[x^2-(c+a)x+ca] \\ \quad + \gamma\gamma'(c-a)(c-b)[x^2-(a+b)x+ab]\end{array}}{(x-a)^2(x-b)^2(x-c)^2} \right\} y = 0;$$

or

$$(89)\quad \frac{d^2y}{dx^2} + \left\{ \frac{1-\alpha-\alpha'}{x-a} + \frac{1-\beta-\beta'}{x-b} + \frac{1-\gamma-\gamma'}{x-c} \right\} \frac{dy}{dx}$$

$$+ \frac{1}{(x-a)(x-b)(x-c)} \left\{ \frac{\alpha\alpha'(a-b)(a-c)}{x-a} + \frac{\beta\beta'(b-c)(b-a)}{x-b} + \frac{\gamma\gamma'(c-a)(c-b)}{x-c} \right\} y = 0.$$

This form of the differential equation satisfied by the **P**-function was first given by E. Papperitz;* Riemann has given indications as

* *Ueber verwandte s-Functionen*, von Erwin Papperitz. *Math. Annalen*, vol. xxv. p. 212.

to the formation of the equation, but does not seem to have worked out the form.

In (79) make $x = \frac{1}{t}$; the equation then takes the form

$$(90)\quad \frac{d^2y}{dt^2} + \left\{ \frac{2}{t} - \frac{A + Bt + Ct^2}{t(1 - at)(1 - bt)(1 - ct)} \right\} \frac{dy}{dt}$$
$$+ \frac{F + Gt + Ht^2}{(1 - at)^2(1 - bt)^2(1 - ct)^2} y = 0.$$

The coefficient of y does not contain t in the denominator as a factor; the coefficient of $\frac{dy}{dt}$ is

$$\frac{1}{t} \cdot \frac{2(1 - at)(1 - bt)(1 - ct) - A - Bt - Ct^2}{(1 - at)(1 - bt)(1 - ct)};$$

substituting for A, B, C their above-found values, we have for the numerator of this coefficient the value

$$2(1 - at)(1 - bt)(1 - ct) - (3 - \alpha - \alpha' - \beta - \beta' - \gamma - \gamma')$$
$$+ [(1 - \alpha - \alpha')(b + c) + (1 - \beta - \beta')(c + a) + (1 - \gamma - \gamma')(a + b)]t$$
$$- [(1 - \alpha - \alpha')bc + (1 - \beta - \beta')ca + (1 - \gamma - \gamma')ab]t^2.$$

Now remembering that $\alpha + \alpha' + \beta + \beta' + \gamma + \gamma' = 1$, we see that this last expression is divisible by t, and consequently that the coefficient of $\frac{dy}{dt}$ in (90) does not contain t as a factor in its denominator. It follows then that $t = 0$ is not a critical point of (90), and consequently that $x = \infty$ is not a critical point of (88) (or (89)). The points a, b, c are then the only critical points of our differential equation.

If in (88) we write

$$a = -1, \quad b = \frac{1}{\epsilon}, \quad c = +1,$$

$$\alpha = \alpha' = \gamma = \gamma' = 0, \quad \beta = -n, \quad \beta' = n + 1, \quad \lim \epsilon = 0,$$

we find, after simple reductions, the equation for zonal spherical harmonics, viz.,

$$(91)\qquad \frac{d^2P}{dx^2} - \frac{2x}{1-x^2}\frac{dP}{dx} + \frac{n(n+1)}{1-x^2}P = 0.$$

Again, writing

$$a = \epsilon,\quad b = \frac{1}{\epsilon},\quad c = \omega,$$

$$\alpha = \nu,\quad \alpha' = -\nu,\quad \beta = i\omega,\quad \beta' = -i\omega,$$

$$\gamma = \tfrac{1}{2}(1 + \sqrt{1-4\omega^2}),\quad \gamma' = \tfrac{1}{2}(1 - \sqrt{1-4\omega^2}),$$

$$\lim \epsilon = 0,\quad \lim \omega = \infty,$$

we have the differential equation for Bessel's functions, viz.,

$$(92)\qquad \frac{d^2J}{dx^2} + \frac{1}{x}\frac{dJ}{dx} + \left(1 - \frac{\nu^2}{x^2}\right)J = 0.$$

If, in space of three dimensions, we consider two points, one at a distance unity and the other at a distance ρ from the origin, and if ω denote the angle between the lines drawn from these points to the origin, we have for the reciprocal of the distance between the points

$$(93)\qquad \frac{1}{R} = (1 - 2\rho\cos\omega + \rho^2)^{-\frac{1}{2}}.$$

The ordinary zonal harmonics are the coefficients of the different powers of ρ in the development of (93) in ascending powers of ρ, viz.,

$$(94)\qquad \frac{1}{R} = P_0 + P_1\rho + P_2\rho^2 + \ldots + P_n\rho^n + \ldots.$$

Writing $\cos\omega = x$, these functions P_n satisfy the differential equation

$$(95)\qquad \frac{d^2y}{dx^2} - \frac{2x}{1-x^2}\frac{dy}{dx} + \frac{n(n+1)}{1-x^2}y = 0.$$

The *second* integral of this equation is the spherical harmonic of the *second kind*, P_n being the zonal harmonic of the first kind.

From Laplace's definite integral form for P_n, viz.,

$$P_n = \frac{1}{\pi}\int_0^\pi (x + \sqrt{x^2 - 1}\cos\phi)d\phi,$$

we derive that P_n is the term independent of ϕ in the development of $(x + \sqrt{x^2 - 1}\cos\phi)^n$ in a series going according to the cosines of multiples of ϕ. The coefficient of $\cos m\phi$, viz., $P_{n,m}$, is the *associated function* of the first kind of *degree* n and *order* m. This function, as is well known, satisfies the differential equation

$$(96) \qquad \frac{d^2y}{dx^2} - \frac{2x}{1 - x^2}\frac{dy}{dx} + \frac{n(n+1)(1 - x^2) - m^2}{(1 - x^2)^2} y = 0.$$

The *second* integral of this equation is the associated function of the second kind, and is denoted by $Q_{n,m}$. If instead of n being an integer it is of the form

$$n = \mu - \tfrac{1}{2},$$

then

$$n + 1 = \mu + \tfrac{1}{2},$$

and (97) becomes

$$(98) \qquad \frac{d^2y}{dx^2} - \frac{2x}{1 - x^2}\frac{dy}{dx} + \frac{(\mu^2 - \frac{1}{4})(1 - x^2) - m^2}{(1 - x^2)^2} y = 0;$$

which is, Basset's Hydrodynamics, Vol. II. page 22, the differential equation for Hicks's *Toroidal Functions*. The spherical harmonics for the cases above considered present themselves in the theory of attraction in space of three dimensions; we will speak of them as being of *rank* 2. In considering the theory of attraction in space of $k + 1$ dimensions, we get spherical harmonics of the rank k; these harmonics appear as the coefficients of the powers of ρ in the development of

$$(1 - 2x\rho + \rho^2)^{-\frac{k-1}{2}}.$$

The differential equation of these functions is known to be

$$(99)\quad \frac{d^2y}{dx^2} - \frac{kx}{1-x^2}\frac{dy}{dx} + \frac{n(n+k-1)(1-x^2) - m(m+k-2)}{(1-x^2)^2}y = 0.$$

For $k = 2$, let $x = \cos\frac{\vartheta}{n}$, and then for an indefinitely increasing value of n we get the Bessel's functions,

$$(100)\quad \begin{cases} J_m = \lim\limits_{n=\infty} \dfrac{2^n}{\sqrt{n\pi}} P_{n,m}\left(\cos\dfrac{\vartheta}{n}\right), \\[2ex] Y_m = \lim\limits_{n=\infty} \dfrac{2^{-n-1}\sqrt{\pi}}{\sqrt{n}} Q_{n,m}\left(\cos\dfrac{\vartheta}{n}\right); \end{cases}$$

the differential equation for which is

$$(101)\quad \frac{d^2y}{d\vartheta^2} + \frac{1}{\vartheta}\frac{dy}{d\vartheta} + \left(1 - \frac{m^2}{\vartheta^2}\right)y = 0.$$

The corresponding equation for the Bessel's functions of rank k is

$$(102)\quad \frac{d^2y}{d\vartheta^2} + \frac{k-1}{\vartheta}\frac{dy}{d\vartheta} + \left(1 - \frac{m(m+k-2)}{\vartheta^2}\right)y = 0.$$

Equation (98), the general equation for toroidal functions, is the only one of the preceding equations involving a parameter which is other than an integer. In (98) replace μ by n, and make $m = 0$; we have now the differential equation of zonal toroidal functions, viz.,

$$(103)\quad \frac{d^2y}{dx^2} - \frac{2x}{1-x^2}\frac{dy}{dx} + \frac{n^2 - \frac{1}{4}}{1-x^2}y = 0.$$

Hicks in his memoir* gives this equation in the form

$$(104)\quad \frac{d^2y}{du^2} + \coth u\frac{dy}{du} - (n^2 - \tfrac{1}{4})y = 0;$$

* Phil. Trans. of the Royal Society; Part III., 1881, p. 617, equation (9).

in this writing

$$x = \cosh u, \quad \sqrt{x^2 - 1} = \sinh u, \quad \frac{x}{\sqrt{x^2 - 1}} = \coth u,$$

we arrive at (103). It is evident that the *first* integral, say P_n, of (104) is a zonal spherical harmonic of degree $\frac{2n+1}{2}$ with a pure imaginary for argument. Writing, for brevity,

$$C = \cosh u, \quad S = \sinh u,$$

Hicks finds for the integral P_n the form

(105)
$$P_n = \int_0^\pi \frac{d\theta}{[C - S\cos\theta]^{\frac{2n+1}{2}}},$$

and shows that the three integrals P_{n-1}, P_n, P_{n+1} are connected by the sequence relation

(106)
$$(2n+1)P_{n+1} - 4nCP_n + (2n-1)P_{n-1} = 0.$$

By aid of this equation Hicks finds the value of P_n as follows: In (106) write

$$P_n = \frac{(2n-2)(2n-4)\ldots 2}{(2n-1)(2n-3)\ldots 3 \cdot 1} u_n,$$

with
$$P_0 = u_0, \quad P_1 = u_1;$$

then
$$u_{n+2} - 2Cu_{n+1} + \frac{(2n+1)^2}{2n(2n+2)} u_n = 0;$$

or, writing

$$c_n = \frac{(2n+1)^2}{2n(2n+2)}, = \frac{(2n+1)^2}{(2n+1)^2 - 1}, = 1 + \frac{1}{2}\left(\frac{1}{2n} - \frac{1}{2n+2}\right),$$

and
$$c_0 = \tfrac{1}{2},$$

$$u_{n+2} = 2Cu_{n+1} - c_n u_n.$$

It is clear from this that u_n is of the form

$$\alpha_m u_1 - \beta_n u_0 ;$$

where α_n, β_n are rational integral functions, algebraic functions, of $2C$; α_n of degree $n - 1$, and β_n of degree $n - 2$. Writing $2C = t$, we have

(107)
$$\begin{cases} u_0 = u_0 ; \\ u_1 = u_1 ; \\ u_2 = tu_1 - \frac{1}{2}u_0 ; \\ u_3 = (t^2 - c_1)u_1 - \frac{1}{2}tu_0 . \end{cases}$$

We can now show that α_n, β_n are of the form

(108) $$\alpha_n = a_{n0}t^{n-1} + a_{n1}t^{n-3} + a_{n2}t^{n-5} + \ldots + a_{nr}t^{n-2r-1} + \ldots$$

For, supposing α_n of this form, *i.e.*, wanting every other power of t, it follows at once that α_{n+1} is of the same form; it is seen above that α_3 is of this form, and so the statement is generally true. The same remarks apply to the function β_n.

Now α_n satisfies the equation

(109) $$\alpha_n = t\alpha_{n-1} - c_{n-2}\alpha_{n-2} ;$$

with $\alpha_0 = 0$, $\alpha_1 = 1$. Hence, substituting the above value for α_n, we must have

(110) $$a_{nr} = a_{n-1,r} - c_{n-2}a_{n-2,r-1} ;$$

also,

(111) $$a_{n0} = a_{n-1,0} = \ldots = 1.$$

Hence

(112) $$a_{nr} = -(c_{n-2}a_{n-2,r-1} + c_{n-3}a_{n-3,r-1} + \ldots + c_{2r-1}a_{2r-1,r-1}).$$

From this we have

$$a_{n1} = -(c_{n-2} + c_{n-3} + \ldots + c_1) ;$$

$$a_{n2} = c_{n-2}(c_{n-4} + \ldots + c_1) + c_{n-3}(c_{n-5} + \ldots + c_1) + \ldots + c_3c_1 .$$

From this last we observe that a_{n2} is equal to the sum of the products $c_i c_j$ taken two and two, with the exception of all products where the subscripts are successive.

Assume that $(-)^r a_{nr} =$ sum of products of the c's taken r at a time up to c_{n-2}, excepting such products in which successive subscripts occur. Then

$$\begin{aligned} a_{n,r+1} = (-)^{r+1}\{ & c_{n-2}(\text{prod. up to } c_{n-4},\ r \text{ at time, etc.}, \ \ldots \ ) \\ & + c_{n-3}(\ \text{"} \quad \text{"} \quad c_{n-5} \quad \text{"} \quad \text{"} \quad \ldots \ ) \\ & + \ldots \qquad\qquad\qquad\qquad \ldots \ \} \\ = (-)^{r+1}\{ & \text{prod. } (r+1) \text{ at a time up to } c_{n-2} \text{ without successive subscripts.}\} \end{aligned}$$

Whence by induction the assumption is seen to be universally true. It may be thus stated: a_{nr} is the sum r at a time of the terms

$$\frac{3^2}{2.4}, \ \frac{5^2}{4.6}, \ \frac{7^2}{6.8}, \ \ldots, \ \frac{(2n-3)^2}{(2n-2)(2n-4)},$$

all products being thrown out in which, regarding the numbers in the denominators as undecomposable, a square occurs in the denominator.

We have

$$a_{n0} = 1;$$

$$a_{n1} = -\frac{(4n-3)(n-2)}{4(n-1)}.$$

This result is of very little use for application. If the coefficients α_n are needed for particular values of t, they can be very rapidly calculated by means of equation (109), while if their general values are to be tabulated, equation (110) will serve to calculate them in succession.

Further, $2\beta_n$ is the same kind of function as α_n in every way, except that it does not contain c_1; in fact, $2\beta_n$ is the same function of $c_2, c_3, \ldots, c_{n-2}$ that α_{n-1} is of $c_1, c_2, \ldots, c_{n-3}$. Calling this α'_{n-1}, we can then write

$$u_n = \alpha_n u_1 - \tfrac{1}{2}\alpha'_{n-1} u_0.$$

The functions n_0 and n_1 are expressible as elliptic integrals, viz.: Writing

$$k^2 = \frac{2S}{C+S}, \quad k'^2 = \frac{1}{(C+S)^2}, \quad (\therefore\ k^2 + k'^2 = 1,)$$

or

$$k^2 = 1 - e^{-2n}, \quad k'^2 = e^{-2n},$$

and

$$t = 2C = k' + \frac{1}{k'},$$

we have

$$(113)\quad \left\{ \begin{aligned} n_0 &= \int_0^\pi \frac{d\theta}{\sqrt{C - S\cos\theta}} = 2\sqrt{k'} \int_0^{\frac{\pi}{2}} \frac{d\theta}{\sqrt{1 - k^2\sin^2\theta}} \\ &= 2\sqrt{k'}F; \\ n_1 &= \int_0^\pi \sqrt{C - S\cos\theta}\,.\,d\theta = \frac{2}{\sqrt{k'}} \int_0^{\frac{\pi}{2}} \sqrt{1 - k^2\sin^2\theta}\,.\,d\theta \\ &= \frac{2}{\sqrt{k'}} E. \end{aligned} \right.$$

Hence

$$(114)\qquad P_n = 2\frac{(2n-2)(2n-4)\ldots 2}{(2n-1)(2n-3)\ldots 1}\left[\frac{\alpha_n}{\sqrt{k'}}E - \tfrac{1}{2}\sqrt{k'}\,.\,\alpha'_{n-1}F\right].$$

For $n = 0$, $P_n = \pi$; for $n = \infty$, $P_n = \infty$.

By very simple operations the second integral Q_n can be shown to have the form

$$(115)\qquad Q_n = 2\frac{(2n-2)\ldots 2}{(2n-1)\ldots 3}\left[\frac{\alpha_n}{\sqrt{k'}}(F' - E') - \tfrac{1}{2}\alpha'_{n-1}\sqrt{k'}F'\right];$$

F' and E' being the complete elliptic integrals of the first and second kinds with modulus k'. The functions P_n and Q_n thus found can obviously be put in the forms of the hypergeometric series. For further information on the Toroidal Functions the reader is referred to Hicks's memoir, and also to Basset's Hydrodynamics, Vol. II.

Chap. XII. It is interesting to derive the above results directly from equation (103), but limits of space will not allow us to enter into the investigation.

Returning now to equation (99), we will write it in the form

$$(116)\quad \frac{d^2y}{dx^2} - \frac{fx}{1-x^2}\frac{dy}{dx} + \frac{g(g+f-1)(1-x^2)-h(h+f-2)}{(1-x^2)^2}y = 0;$$

where the parameters f, g, h are no longer restricted to integral values.* We will define now as spherical harmonics of the first and second kind of rank f, degree g, and order h, two definite linearly independent particular integrals of (116).

These generalized functions therefore depend upon the three arbitrary parameters f, g, h, and have the points $+1$, -1, ∞, as critical points.

The **P**-function above defined depends essentially on the differences of the pairs of exponents (α, α'), (β, β'), (γ, γ'), and upon the critical points a, b, c. At first sight it might appear that these new functions, *i.e.*, the integrals of (116), were identical with the *general* **P**-functions. That this is not the case can, however, be easily seen. The differential equation of the **P**-function is, equation (89),

$$(117)\quad \frac{d^2y}{dx^2} + \left\{\frac{1-\alpha-\alpha'}{x-a} + \frac{1-\beta-\beta'}{x-b} + \frac{1-\gamma-\gamma'}{x-c}\right\}\frac{dy}{dx}$$
$$+ \frac{1}{(x-a)(x-b)(x-c)}\left\{\frac{\alpha\alpha'(a-b)(a-c)}{x-a}\right.$$
$$\left. + \frac{\beta\beta'(b-c)(b-a)}{x-b} + \frac{\gamma\gamma'(c-a)(c-b)}{x-c}\right\} y = 0.$$

This is changed into (116) by the following substitutions:

$$(118)\quad \begin{cases} a = -1, & b = \infty, & c = +1; \\ \alpha = \dfrac{h}{2}, & \beta = -g, & \gamma = \dfrac{h}{2}; \\ \alpha' = -\dfrac{h+f-2}{2}, & \beta' = g+f-1, & \gamma' = -\dfrac{h+f-2}{2}. \end{cases}$$

---

* *Studien über die Kugel- und Cylinderfunctionen*, von R. Olbricht. Halle, 1887.

We have now the **P**-function

$$(119)\quad \mathbf{P}\left\{\begin{matrix} -1, & \infty, & +1 & \\ \frac{h}{2}, & -g, & \frac{h}{2} & x \\ -\frac{h+f-2}{2}, & g+f-1, & -\frac{h+f-2}{2} & \end{matrix}\right\}.$$

In this the differences of the pairs of exponents are

$$h+\frac{f-2}{2}, \quad -2g-f+1, \quad h+\frac{f-2}{2},$$

two of which are equal. We arrive thus at the important theorem:

The theory of the generalized spherical harmonic is identical with the theory of the **P**-*function, in which the differences of two pairs of corresponding exponents are equal.*

We will merely state, without proving, two other theorems found by Olbricht, referring the reader to his memoir for the proofs and also for other interesting results concerning the generalized spherical harmonics and Bessel's functions.

(*a*) *The generalized spherical harmonic of rank f, degree g, and order h are derived from the* **P**-*function*

$$\mathbf{P}\left\{\begin{matrix} -1 & \infty & +1 & \\ \frac{h'}{2} & g' & \frac{h'}{2} & x \\ -\frac{h'}{2} & g'+1 & -\frac{h'}{2} & \end{matrix}\right\},$$

which represents the spherical harmonics of rank 2, degree g', and order h' by replacing g', h' by

$$g' = g+\frac{f-2}{2}, \quad h' = h+\frac{f-2}{2},$$

and multiplying the result by

$$(1 - x^2)^{\frac{2-f}{4}}.$$

(*b*) *The generalized Bessel's functions are obtained from those of the second order by multiplying the* **P**-*function*

$$\lim_{g=\infty} \mathbf{P} \left\{ \begin{matrix} g\pi & \infty & 0 & \\ ig & ig & h & x \\ -ig & -ig & -h & \end{matrix} \right\}$$

by $x^{\frac{2-f}{2}}$, *and replacing h by*

$$\tfrac{1}{2}\sqrt{(2-f)^2 + h(h+f-2)}.$$

In the next chapter, M. Goursat's thesis, will be found an investigation of the differential equation of the second order satisfied by the complete elliptic integral K.

At the time of writing the preceding pages the author had not seen an interesting paper by Humbert,* of which the following is an account, and did not know that some of Heun's results (see above) had been previously given. Humbert investigates the most general polynomials satisfying the equation

$$(120) \qquad \psi(x)\frac{d^2y}{dx^2} + Q_1(x)\frac{dy}{dx} + Q_2(x)y = 0;$$

where, as before, ψ, Q_1, Q_2 are polynomials of the degrees ρ, $\rho - 1$, $\rho - 2$ respectively.

If we wish (120) to have as a solution a polynomial in x of degree n, the three functions ψ, Q_1, Q_2 cannot be arbitrarily taken. Heine

* "Sur l'équation differentielle linéaire du second ordre," par M. G. Humbert. *Journal de l'École Polytechnique*, cahier 48, 1880.

has shown* (see above) that if $\psi(x)$ and $Q_1(x)$ are arbitrary polynomials of degrees ρ and $\rho - 1$ respectively, there exist only

$$\frac{(n+1)(n+2)\ldots(n+\rho-2)}{1\cdot 2\ldots(\rho-2)}$$

integral polynomials $Q_2(x)$ of degree $\rho - 2$, such that equation (120) admits as a solution a polynomial in x of degree n.

Denoting by $x_1, x_2, \ldots, x_\rho$ the finite singular points of the differential equation, we write, as before,

$$\psi(x) = (x - x_1)(x - x_2)\ldots(x - x_\rho);$$

also write

$$Q_1(x) = \psi(x)\left[\frac{\mu_1}{x - x_1} + \frac{\mu_2}{x - x_2} + \ldots + \frac{\mu_\rho}{x - x_\rho}\right],$$

a polynomial of degree $\rho - 1$. Defining a new function $K(x)$ by the equation

$$(121) \qquad \frac{1}{K}\frac{dK}{dx} = \frac{Q_1}{\psi},$$

we have

$$K(x) = (x - x_1)^{\mu_1}(x - x_2)^{\mu_2}\ldots(x - x_\rho)^{\mu_\rho}.$$

This function $K(x)$ has an important property; viz., if we make in (120) the substitution

$$(122) \qquad y = u\frac{\psi(x)}{K(x)},$$

this equation retains the same form; it becomes in fact

$$(123) \quad \psi(x)\frac{d^2u}{dx^2} + [2\psi'(x) - Q_1(x)]\frac{du}{dx} + [Q_2(x) - \psi''(x) - Q_1'(x)]u = 0.$$

* *Handbuch der Kugelfunctionen*, zweite Auflage, p. 473.

For simplicity we will first assume $\rho = 3$; we have then

$$\psi(x) = (x - x_1)(x - x_2)(x - x_3);$$

$$\frac{Q_1(x)}{\psi(x)} = \frac{\mu_1}{x - x_1} + \frac{\mu_2}{x - x_2} + \frac{\mu_3}{x - x_3}.$$

We will also first assume that the constants μ_1, μ_2, μ_3 are all positive. Consider a path of integration between the points x_1 and x_2, and another going from x_1 to x_3; the integrals

$$\int_{x_1}^{x_2} \frac{K(z)}{\psi(z)(x - z)}\,dz, \quad \int_{x_1}^{x_3} \frac{K(z)}{\psi(z)(x - z)}\,dz,$$

will then have a sense and be perfectly determinate so long as these paths of integration do not pass through the point x. Suppose now $P_n(x)$ to be a polynomial of degree n satisfying equation (120); then equation (123) will admit as a solution the function

$$\frac{K(x)}{\psi(x)} P_n(x).$$

Substituting in the left-hand member of this last equation the function

(124) $$v_1 = \int_{x_1}^{x_2} \frac{K(z)P_n(z)}{\psi(z)(x-z)}\,dz,$$

we readily find, by aid of a known formula due to Heine,

(125) $$\psi(x)\frac{d^2v_1}{dx^2} + [2\psi'(x) - Q_1(x)]\frac{dv_1}{dx}$$

$$+ [Q_2(x) + \psi''(x) - Q_1'(x)]v_1 = [\Phi(z)]_{x_1}^{x_2} + \int_{x_1}^{x_2} \Psi(z, x)dz;$$

where

$$\Psi = \frac{K(z)}{\psi(z)} P_n(z) \left\{ \frac{d^2}{dz^2}\left[\frac{\psi(x)-\psi(z)}{x-z}\right] - \frac{d}{dz}\left[\frac{S(x)-S(z)}{x-z}\right] \right\} + \frac{T(x)-T(z)}{x-z},$$

$$\Phi = \frac{K(z)P_n(z)}{(x-z)^2} + \frac{K(z)}{\psi(z)} P_n(z) \frac{\psi'(x)-S(z)}{x-z} - \frac{\psi(z)}{x-z}\left[\frac{K(z)}{\psi(z)} P_n(z)\right]',$$

$$S = 2\psi' - Q_1,$$

$$T = Q_2 + \psi'' - Q_1'.$$

Developing Φ, we find

$$\Phi = \frac{K(z)P_n(z)}{(x-z)^2} - \frac{K(z)P_n'(z)}{x-z}.$$

This is to be taken between the limits x_1 and x_2; and since $K(z)$ vanishes at these limits, it follows that Φ disappears from (115). As to

$$\int_{x_1}^{x_2} \Psi(z, x)dz,$$

we see that it is a constant in x. We conclude then that v_1 is a solution of the equation

$$(126) \qquad \psi(x)\frac{d^2v}{dx^2} + [2\psi'(x) - Q_1(x)]\frac{dv}{dx} + [Q_2(x) + \psi''(x) - Q_1'(x)]v = \text{const.},$$

the constant being determinate. So also the function

$$(127) \qquad v_2 = \int_{x_1}^{x_3} \frac{K(z)}{\psi(z)} \frac{P_n(z)}{x-z} dz$$

is a solution of this same equation with a different value of the constant in the right-hand member.

It follows now that v_1 and v_2 are solutions of the equation

$$(128) \qquad \psi(x)\frac{d^3u}{dx^3} + [3\psi'(x) - Q_1(x)]\frac{d^2u}{dx^2}$$

$$+ [3\psi''(x) - 2Q_1'(x) + Q_2(x)]\frac{du}{dx}$$

$$+ [Q_2'(x) + \psi'''(x) - Q_1''(x)]u = 0,$$

obtained by differentiating (123). The function $\frac{K(x)}{\psi(x)} P_n(x)$ being a solution of (123), is therefore also a solution of (128).

Between these three solutions of this last equation we can show now that there exists no linear relation with constant coefficients.

The functions v_1 and v_2 developed according to descending powers of x commence, at the highest, by a term in $\frac{1}{x}$; on the other hand, the function $\frac{K(x)}{\psi(x)} P_n(x)$ is of the degree

$$\mu_1 + \mu_2 + \mu_3 + n - 3,$$

which is greater than -1, since $n > 1$, and μ_1, μ_2, μ_3 are all positive. If then there exists any linear relation, it must be between the functions v_1 and v_2 alone, and so be of the form

$$v_1 = \lambda v_2,$$

where λ is a constant, or

$$\int_{x_1}^{x_2} \frac{K(z)}{\psi(z)} \frac{P_n(z)}{x - z} dz = \lambda \int_{x_1}^{x_3} \frac{K(z)}{\psi(z)} \frac{P_n(z)}{x - z} dz.$$

We can write

$$\int_{x_1}^{x_2} \frac{K(z)}{\psi(z)} \frac{P_n(z)}{x-z}\, dz = P_n(x) \int_{x_1}^{x} \frac{K(z)}{\psi(z)} \frac{dz}{x-z}$$

$$- \int_{x_1}^{x_2} \frac{K(z)}{\psi(z)} \frac{P_n(x) - P_n(z)}{x-z}\, dz.$$

The second term of the second member of this last equation is an integral polynomial in x of degree $n-1$ at most. We have then

$$P_n(x) \left\{ \int_{x_2}^{x_1} \frac{K(z)}{\psi(z)} \frac{dz}{x-z} - \lambda \int_{x_2}^{x_3} \frac{K(z)}{\psi(z)} \frac{dz}{x-z} \right\} = \Pi_{n-1}(x);$$

$\Pi_{n-1}(x)$ being a polynomial in x of degree $n-1$ at most. It is easy to see that the expression in { } satisfies the differential equation

(129) $$\psi(x) \frac{dy}{dx} = [Q_1(x) - \psi'(x)]\, y + \alpha x + \beta,$$

where α and β are constants. This equation admits then as a solution the rational function

$$\frac{\Pi_{n-1}(x)}{P_n(x)},$$

which after suppressing common factors will be written

$$\frac{\Pi(x)}{P(x)}.$$

We shall have then

$$\psi(\Pi' P - P' \Pi) = (P_1 - \psi') P \Pi + (\alpha x + \beta) P^2,$$

and consequently $\psi \cdot P' \cdot \Pi$ is divisible by P. Now $P(x)$ is prime to $\Pi(x)$, and therefore it must admit at least one of the factors of ψ, say the factor $x - x_1$. Again, the factors of $P(x)$ are also factors of $P_n(x)$; $P_n(x)$ therefore vanishes for $x = x_1$. The differential equation

$$\psi P_n''(x) + Q_1 P_n'(x) + Q_2 P_n(x) = 0$$

shows that $P_n'(x_0) = 0$. Taking the successive derivatives of the first member of this equation, we find in like manner that

$$P_n''(x_0) = 0, \quad P_n'''(x_0) = 0, \quad \ldots, \quad P_n^m(x_0) = 0, \quad \ldots,$$

provided only that none of the polynomials

$$Q_1(x) + p\psi'(x)$$

vanish for $x = x_0$. In order that they should vanish we must have

$$\mu_1 + p = 0,$$

which is impossible, since $\mu_1 > 0$. We arrive thus finally at the conclusion that the constant $P_n^n(x_0)$ is zero, and consequently, as is easily seen, that the relation

$$v_1 = \lambda v_2$$

is impossible.

Denote by Q the coefficient of the first term in $Q_2(x)$; then, as is easily seen, the indicial equation of (120) for the region $x = \infty$ is

$$(130) \qquad r(r-1) + (\mu_1 + \mu_2 + \mu_3)r + Q = 0.$$

Since $P_n(x)$ is to be a solution of (120), n must be a root of (130); the two roots are then

$$n,$$
$$-\mu_1 - \mu_2 - \mu_3 - n + 1.$$

For (128) the roots of the corresponding indicial equation are readily found to be

$$-1,$$
$$n + \mu_1 + \mu_2 + \mu_3 - 3,$$
$$-n - 2.$$

The root $n + \mu_1 + \mu_2 + \mu_3 - 3$ is obviously greater than $- n - 2$. Equation (128) has therefore a solution y_{n+2} developable in a series, of the form

$$(131) \qquad y_{n+2} = \frac{1}{x^{n+2}} \left[\alpha + \frac{\beta}{x} + \frac{\gamma}{x^2} + \ldots \right];$$

consequently

$$(132) \qquad y_{n+2} = v_1 + \omega_1 v_2 + \theta \frac{K(x)}{\psi(x)} P_n(x).$$

In this it is clear that $\theta = 0$, since the degree of $\frac{K(x)}{\psi(x)} P_n(x)$ is greater than $- 1$. There remains, then,

$$(133) \qquad v_1 + \omega_1 v_2 = \left(\frac{1}{x^{n+2}} \right),$$

denoting by $\left(\frac{1}{x^p} \right)$ a series going according to descending powers of x, and commencing with a term in $\frac{1}{x^p}$.

Now,

$$(134) \qquad v_1 = \int_{x_2}^{x_1} \frac{K(z)}{\psi(z)} \frac{P_n(z)}{x - z} dz = P_n(x) \int_{x_2}^{x_1} \frac{K(z)}{\psi(z)} \frac{dz}{x - z}$$
$$+ \int_{x_1}^{x_2} \frac{K(z)}{\psi(z)} \frac{P_n(z) - P_n(x)}{x - z} dz.$$

The second term of the third member of this equation is an integral polynomial of degree $n - 1$ at most.

Write

$$I_1 = \int_{x_1}^{x_2} \frac{K(z)}{\psi(z)} \frac{dz}{x - z},$$

$$I_2 = \int_{x_2}^{x_3} \frac{K(z)}{\psi(z)} \frac{dz}{x - z};$$

the relation (133) now becomes

$$(135) \qquad P_n(x)[I_1 + \omega_1 I_2] = \Pi_{n-1}(x) + \left(\frac{1}{x^{n+2}}\right).$$

From this we see that $I_1 + \omega_1 I_2$ is represented, to terms of the order $\frac{1}{x^{n+2}}$ près, by the quotient

$$\frac{\Pi_{n-1}(x)}{P_n(x)}.$$

In other words, having given the functions I_1 and I_2, if we form the product

$$P(x)[I_1 + \omega_1 I_2],$$

where $P(x)$ is a polynomial of degree n, we can by a proper choice of the coefficients of this polynomial, and of the constant ω_1, cause the terms in $\frac{1}{x}, \frac{1}{x^2}, \ldots, \frac{1}{x^{n+1}}$ of the product to disappear. Equation (135) shows that $P_n(x)$ is such a polynomial.

As we have assumed $Q_2(x)$ to have any one whatever of its determinations, such that (120) admits as a solution an integral polynomial $P_n(x)$, the theorem just proved holds true for all polynomials of degree n which satisfy the differential equation

$$\psi(x)\frac{d^2y}{dx^2} + Q_1(x)\frac{dy}{dx} + Q_2(x)y = 0,$$

whatever be the choice of the polynomial $Q_2(x)$. Heine has given an analogous theorem for polynomials satisfying Lamé's equation.

The previous results are readily extended to the case where the differential equation has ρ finite critical points instead of only 3; for these generalized results the reader is, however, referred to

Humbert's paper. In concluding his paper Humbert gives, without going into details, a theorem for the general case in which the constants

$$\mu_1, \quad \mu_2, \quad \ldots, \quad \mu_p$$

are not all positive; for this also the reader is referred to the paper.

CHAPTER VII.

ON THE LINEAR DIFFERENTIAL EQUATION WHICH ADMITS THE HYPERGEOMETRIC SERIES AS AN INTEGRAL. *

BY M. EDOUARD GOURSAT.

1. The hypergeometric series $F(\alpha, \beta, \gamma, x)$, considered as a function of the fourth element x, is only defined for values of this variable which have moduli less than unity. In order to define it for any value of x, it should be regarded as a particular integral of a linear differential equation of the second order; the problem then comes to finding what this integral becomes when the path described by the variable ends at a point situated *outside* the circle of radius unity and having the origin as centre. By a generalization which immediately presents itself, one is then led to propose the same problem for any integral whatever of the equation, the path followed by the variable being simply subjected to the condition of not passing through any critical point of the equation.

This celebrated equation, studied by Gauss* and Kummer,† admits, as the latter has shown, twenty-four integrals of the form

$$x^p(1 - x)^q F(\alpha', \beta', \gamma', z),$$

where z is one of the variables

$$x, \quad 1 - x, \quad \frac{1}{x}, \quad \frac{1}{1 - x}, \quad \frac{x}{x - 1}, \quad \frac{x - 1}{x},$$

provided none of the numbers γ, $\gamma - \alpha - \beta$, $\beta - \alpha$ is an integer. In the same memoir Kummer gives the linear relations with constant coefficients which connect any three of the integrals; but, as he only considers the case of a real variable, the formulæ which he obtains present some difficulties when we wish to pass to their applications, the more so as he does not always sufficiently define

* Collected Works, vol. iii. p. 207. † Crelle, vol. xv. p. 39.

the sense of his integrals. Suppose, in fact, that we wish to pass from a real value of x, positive and less than unity, to a real value of x, positive and greater than unity; we cannot do this by giving x a series of values all of which are real, because we may not pass through the point $x = 1$, which is a critical point of the differential equation. It is necessary then to give to x a series of imaginary values, which can be done in an infinite number of ways, and, the final value of the function depending on the law of succession of the different values of the variable, we see at once that, for the proposed object, it is necessary to introduce the consideration of imaginary values in the differential equation.

The general theory of this equation, when no restriction is imposed on the value of the variable, has not, so far as the author knows, up to this time been treated completely. Tannery,* however, has, by employing Fuchs's method for linear differential equations, shown that we can find all of Kummer's integrals; in another memoir † he has determined the linear relations between the integrals for a numerical example previously studied by Fuchs. The results obtained are in accord with those deduced from the general case.

2. In PART FIRST the author develops first a method, due to Jacobi‡ for finding the twenty-four integrals of the differential equation when no one of the numbers $\gamma, \gamma - \alpha - \beta, \beta - \alpha$ is an integer; then, by applying Cauchy's theorem to the definite integrals which represent the hypergeometric series, the relations between the integrals themselves are obtained. If among the numbers $\gamma, \gamma - \alpha - \beta$, $\beta - \alpha$ there are one or more integers, there will be a change of the analytical form of certain of the integrals; the new integrals are found by a well-known process, one which occurs particularly often in the theory of differential equations. PART FIRST closes by seeking the conditions to be satisfied by the elements α, β, γ in order that the general integral may be algebraic. We are so conducted to a geometrical representation identical with one already found by Schwarz ‡ in starting from totally different principles.

---

* *Annales de l'École normale supérieure*, 2ᵉ série, t. iv. p. 113.

† Ib., t. viii. p. 169.

‡ Crelle, vol. lvi. p. 149.

PART SECOND is devoted to the investigation of the transformations which the hypergeometric series will admit of when of the three elements α, β, γ, two only, or even one only, remain arbitrary. We arrive at all the transformations, of a very general form, and containing all the transformations indicated by Kummer, and a certain number of others believed to be new. Finally, a number of the numerous formulæ which can be deduced are given.

3. Letting

$$V = u^{\beta - 1}(1 - u)^{\gamma - \beta - 1}(1 - xu)^{-\alpha},$$

we know, according to Euler, that if the integral

$$\int_0^1 V du$$

is capable of representing a function (*i.e.*, a function of x), then this function, say y,

$$y = \int_0^1 V du,$$

satisfies the differential equation

$$(1) \qquad x(1 - x)\frac{d^2y}{dx^2} + [\gamma - (\alpha + \beta + 1)x]\frac{dy}{dx} - \alpha\beta y = 0.$$

In order to demonstrate this it is only necessary to replace in the first member of this equation y, $\frac{dy}{dx}$, $\frac{d^2y}{dx^2}$ by their values. The result of this substitution is evidently

$$\int_0^1 \left\{ x(1 - x)\frac{d^2V}{dx^2} + [\gamma - (\alpha + \beta + 1)x]\frac{dV}{dx} - \alpha\beta V \right\} du.$$

Now by a very simple process we find that the function under the sign of integration is

$$= -\alpha \frac{d}{du}\left[\frac{u(1 - u)}{1 - xu}V\right].$$

The result of the substitution is then

$$-\alpha\left[\frac{u(1-u)}{1-xu}\right]_0^1,$$

which, under the supposition that the definite integral has a meaning, is identically zero.

Consider now the more general integral

$$y=\int_g^h V du,$$

and let us assume that this integral has a meaning, that is, defines a function, whatever particular values we assign to the limits g and h, where g and h are constants. If we substitute this value of y in the first member of equation (1), we have as the result of the substitution

$$-\alpha\left[\frac{u(1-u)}{1-xu}V\right]_g^h.$$

In order that this may vanish it is only necessary to give to g and h one of the values 0, 1, $\pm\infty$. It follows then that each of the three integrals,

$$y=\int_0^1 V du,\qquad y=\int_{-\infty}^{0} V du,\qquad y=\int_1^{\infty} V du,$$

if they have a meaning, satisfies the above differential equation.

Let us consider now, following Jacobi, the definite integral

$$y=\int_g^{\frac{\epsilon}{x}} V du,$$

where g and ϵ are constants, and substitute as before in the first member of equation (1) the values of y, y', and y''. The result of these substitutions is readily found to be

$$-(\gamma-\beta-1)\epsilon^{\beta}(1-\epsilon)^{1-\alpha}x^{1-\gamma}(x-\epsilon)^{\gamma-\beta-1}$$
$$+\alpha g^{\beta}(1-g)^{\gamma-\beta}(1-xg)^{-\alpha-1}.$$

If we take $\epsilon = 1$ and give g one of the values 0, 1, $+\infty$, this result will be zero provided $1 - \alpha$ has its real part positive, and also provided that the product $u^\beta(1 - u)^{\gamma - \beta}(1 - xu)^{-\alpha - 1}$ vanishes for the limit $u = g$; *i.e.*, provided that the integral $\int_g^{\frac{1}{x}} V du$ has a meaning.

We have now, if all of the above conditions are satisfied, six functions, defined by definite integrals, which satisfy the given differential equation. The following is a table of these six integrals together with the conditions which must be satisfied by the quantities α, β, γ in order that the integrals may have a meaning.

The inequality $A > 0$ indicates simply that the real part of A is positive.

(1) $y = \int_0^1 V du \quad \ldots \quad \beta > 0, \qquad \gamma - \beta > 0;$

(2) $y = \int_0^{-\infty} V du \quad \ldots \quad \beta > 0, \qquad \alpha + 1 - \gamma > 0;$

(3) $y = \int_1^{+\infty} V du \quad \ldots \quad \gamma - \beta > 0, \qquad \alpha + 1 - \gamma > 0;$

(4) $y = \int_0^{\frac{1}{x}} V du \quad \ldots \quad \beta > 0, \qquad 1 - \alpha > 0;$

(5) $y = \int_1^{\frac{1}{x}} V du \quad \ldots \quad \gamma - \beta > 0, \qquad 1 - \alpha > 0;$

(6) $y = \int_{\frac{1}{x}}^{+\infty} V du \quad \ldots \quad \alpha + 1 - \gamma > 0, \qquad 1 - \alpha > 0.$

4. Before going farther it is desirable to define exactly those values of the function V which we can take in the above integrals, and also to examine more closely the properties of the functions represented by these definite integrals.

5. The integral

$$y = \int_0^1 u^{\beta - 1}(1 - u)^{\gamma - \beta - 1}(1 - xu)^{-\alpha} du$$

has a meaning provided that the real parts of β and $\gamma - \beta$ are positive, and provided in addition to this that x has no real value greater than unity. We will take this integral along the straight line $0 - 1$ (Fig. 1); we will also take 0 for the value of the arguments of u and $1 - u$, and for the argument of $1 - xu$ we will take that value which reduces to zero when $u = 0$.

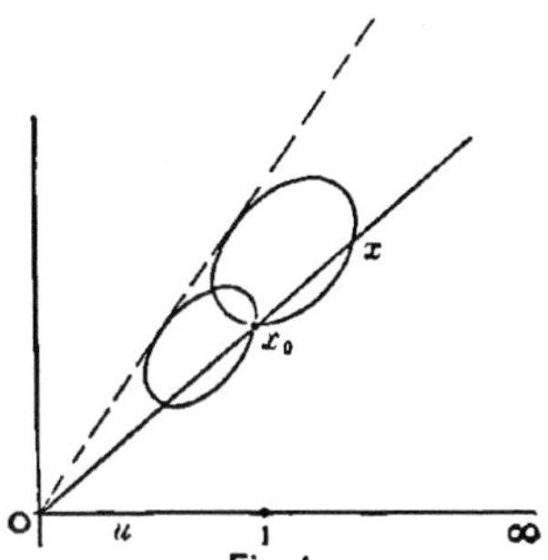

Fig. 1.

The function y thus defined is a uniform function of x in the entire plane if only the path described by the variable does not cut the line $1 \text{———} +\infty$. Suppose now the point x describes a closed curve which does not cut this line; then the point xu, corresponding to a value of u lying between 0 and 1, will describe a closed curve homothetic to the preceding one, and leaving the point 1 outside. The argument of $1 - xu$ will then resume its initial value; the same thing will be true of the element of the integral which corresponds to this value of u, and consequently for the integral itself.

It follows from what precedes that, inside a circle of radius unity and having the origin as centre, the function y can be developed in a series going according to ascending powers of the variable x.

The coefficient of x^m in this development will be

$$\frac{\left(\frac{d^m y}{dx^m}\right)_0}{1 \cdot 2 \cdot 3 \ldots m};$$

that is,

$$\frac{\alpha(\alpha+1)\ldots(\alpha+m-1)}{1 \cdot 2 \cdot 3 \ldots m}\int_0^1 u^{\beta+m-1}(1-u)^{\gamma-\beta-1}du,$$

or

$$\frac{\alpha(\alpha+1)\ldots(\alpha+m-1)}{1 \cdot 2 \cdot 3 \ldots m}\cdot\frac{\Gamma(\beta+m)\Gamma(\gamma-\beta)}{\Gamma(\gamma+m)},$$

or, finally,

$$\frac{\alpha(\alpha+1)\ldots(\alpha+m-1)\beta(\beta+1)\ldots(\beta+m-1)}{1 \cdot 2 \cdot 3 \ldots m \cdot \gamma(\gamma+1)\ldots(\gamma+m-1)}\cdot\frac{\Gamma(\beta)\Gamma(\gamma-\beta)}{\Gamma(\gamma)}.$$

We have then, for every value of x for which $|x| < 1$,

$$\int_0^1 V du = \frac{\Gamma(\beta)\Gamma(\gamma - \beta)}{\Gamma(\gamma)} \cdot F(\alpha, \beta, \gamma, x).$$

The preceding integral can take three other forms, which can be at once obtained by the following transformations due to Jacobi:

$$u = 1 - v;$$

$$u = \frac{v}{1 - x + vx};$$

$$u = \frac{1 - v}{1 - vx}.$$

By the first of these substitutions the integral $\int_0^1 V du$ is changed into

$$(1 - x)^{-\alpha} \int_0^1 v^{\gamma - \beta - 1}(1 - v)^{\beta - 1}\left(1 - \frac{x}{x - 1}v\right)^{-\alpha} dv.$$

This integral is taken along the same path as the one already considered. If we make the same suppositions as above concerning the arguments of v, $1 - v$, $1 - \frac{x}{x - 1}v$, it will be necessary, in order that the two integrals may be identical, to take for argument of $1 - x$ that one which is zero when $x = 0$. We have then

$$\int_0^1 u^{\beta - 1}(1 - u)^{\gamma - \beta - 1}(1 - xu)^{-\alpha} du$$

$$= (1 - x)^{-\alpha} \int_0^1 v^{\gamma - \beta - 1}(1 - v)^{\beta - 1}\left(1 - \frac{x}{x - 1}v\right)^{-\alpha} dv.$$

Make now the second transformation,

$$u = \frac{v}{1 - x + vx};$$

the integral becomes

$$(1 - x)^{-\beta} \int_0^1 v^{\beta - 1}(1 - v)^{\gamma - \beta - 1}\left(1 - \frac{x}{x - 1}v\right)^{-(\gamma - \alpha)} dv.$$

This integral is taken along an arc of circle $0M1$; it is easy to see that the area comprised between this circle and the line $0 - 1$ does not contain the point $\frac{x-1}{x}$, which is a critical point for the new function under the sign of integration. In fact, the value $v = \frac{x-1}{x}$ corresponds to $u = \infty$; the point, say A, which represents $\frac{x-1}{x}$ must thus belong to the circumference of which $0M1$ is an arc. We can consequently take the integral along the straight line $0 - 1$, and, making the same conventions concerning the arguments as before, we can write

$$\int_0^1 Vdu = (1-x)^{-\beta}\int_0^1 v^{\beta-1}(1-v)^{\gamma-\beta-1}\left(1-\frac{x}{x-1}v\right)^{-(\gamma-\alpha)}dv.$$

The transformation

$$u = \frac{1-v}{1-vx}$$

gives in like manner

$$\int_0^1 Vdu = (1-x)^{\gamma-\alpha-\beta}\int_0^1 v^{\gamma-\beta-1}(1-v)^{\beta-1}(1-vx)^{-(\gamma-\alpha)}dv.$$

Each of the new integrals can be developed in a convergent series for values of the variable lying between certain limits. Thus, for values of x such that

$$|x| < 1,$$

we have

$$(1-x)^{\gamma-\alpha-\beta}\int_0^1 v^{\gamma-\beta-1}(1-v)^{\beta-1}(1-vx)^{-(\gamma-\alpha)}dv$$

$$= \frac{\Gamma(\beta)\Gamma(\gamma-\beta)}{\Gamma(\gamma)}(1-x)^{\gamma-\alpha-\beta}F(\gamma-\alpha,\ \gamma-\beta,\ \gamma,\ x);$$

and when

$$\left|\frac{x}{x-1}\right| < 1,$$

we have

$$(1-x)^{-\alpha}\int_0^1 v^{\gamma-\beta-1}(1-v)^{\beta-1}\left(1-\frac{x}{x-1}v\right)^{-\alpha}dv$$

$$=\frac{\Gamma(\beta)\Gamma(\gamma-\beta)}{\Gamma(\gamma)}(1-x)^{-\alpha}F\left(\alpha,\ \gamma-\beta,\ \gamma,\ \frac{x}{x-1}\right);$$

$$(1-x)^{-\beta}\int_0^1 v^{\beta-1}(1-v)^{\gamma-\beta-1}\left(1-\frac{x}{x-1}v\right)^{-(\gamma-\alpha)}dv$$

$$=\frac{\Gamma(\beta)\Gamma(\gamma-\beta)}{\Gamma(\gamma)}(1-x)^{-\beta}F\left(\gamma-\alpha,\ \beta,\ \gamma,\ \frac{x}{x-1}\right).$$

For brevity we will denote by C_0, C_1, the circles of radius unity having the points 0 and 1 respectively as centres. Further, denote by E_0 and E_1 the portions of the plane limited by the common chord (produced indefinitely) of the circles C_0 and C_1, E_0 containing the point 0 and E_1 the point 1. It results from the previous considerations that the proposed differential equation admits an integral which is uniform throughout the plane, provided only the variable is subjected to the condition of not crossing the line $1 \text{———} +\infty$. Denote this particular integral by ϕ_1. In what follows we will suppose additionally that the variable does not cross the line $-\infty \text{———} 0$, though, as will be seen, this restriction is not really necessary.

In the circle C_0 we have

$$\phi_1 = F(\alpha,\ \beta,\ \gamma,\ x) = (1-x)^{\gamma-\alpha-\beta}F(\gamma-\alpha,\ \gamma-\beta, \gamma,\ x),$$

and in the space E_0

$$\phi_1 = (1-x)^{-\alpha}F\left(\alpha,\ \gamma-\beta,\ \gamma,\ \frac{x}{x-1}\right)$$

$$= (1-x)^{-\beta}F\left(\gamma-\alpha,\ \beta,\ \gamma,\ \frac{x}{x-1}\right),$$

the argument of $1-x$ being supposed to lie between $-\pi$ and $+\pi$.

The previous results have only been established on the hypothesis that the real parts of β and $\gamma-\beta$ are positive, but it is clear that these results will still subsist, provided only that γ is not zero or a negative integer. Suppose we wish to verify that one of the preceding functions satisfies equation (1); the *sign* of the quantities β, $\gamma - \beta$ will not enter into the calculation, and consequently the verification will be the same in all four cases. The same is true if we wished to verify that two of these functions are equal for values of x which make the two series convergent. This remark is made once for all in order not to have to refer to it in analogous cases. If the real parts of β and $\gamma - \beta$ are positive, we have

$$\int_0^1 V du = \frac{\Gamma(\beta)\Gamma(\gamma - \beta)}{\Gamma(\gamma)} \phi_1 .$$

6. The integral

$$y = \int_0^{-\infty} u^{\beta - 1}(1 - u)^{\gamma - \beta - 1}(1 - xu)^{-\alpha} du$$

has a sense if only the real parts of β and $\alpha + 1 - \gamma$ are positive, and if x has no real negative value. It is readily demonstrable that the function y is uniform in the entire plane, provided only that the path described by the variable x does not cut the line $-\infty$ —— 0; the same is true if the path described by the variable is further restricted to not cut the line 1 —— $+\infty$. In order to exactly define the sense of this integral it remains to choose the arguments of u, $1 - u$, and $1 - xu$; for the argument of $1 - u$ we will take 0, for the argument of $1 - xu$ that which reduces to zero for $u = 0$; but for u we will take the argument $\pm \pi$. Suppose first we take arg. $u = + \pi$, and make the change of variable defined by

$$u = \frac{v}{v - 1} = (-1)\frac{v}{1 - v},$$

$$du = (-1)\frac{dv}{(1 - v)^2};$$

v varies from 0 to 1, and u from 0 to $-\infty$:

$$1 - u = \frac{1}{1 - v},$$

$$1 - xu = \frac{1 - v(1 - x)}{1 - v}.$$

If we take 0 for the arg. of v and of $1 - v$, we ought to take π for the arg. of (-1), and the arg. of $1 - v(1 - x)$ will be zero for $v = 0$. By this change of the variable the integral becomes

$$e^{\pi\beta i}\int_0^1 v^{\beta-1}(1-v)^{\alpha-\gamma}[1-v(1-x)]^{-\alpha}dv.$$

In this last integral the arguments of v, $1 - v$, $1 - v(1 - x)$ have the same sense as in the integral already studied. We can therefore apply the preceding transformations to this new integral, and so conclude that, subject to the conditions indicated above, while $\alpha + \beta + 1 - \gamma$ is neither zero nor a negative integer, the differential equation (1) admits a new integral, ϕ_2, which is holomorphic in the entire plane.

In the circle C_1 we have

$$\begin{aligned}\phi_2 &= F(\alpha, \beta, \alpha+\beta+1-\gamma, 1-x)\\ &= x^{1-\gamma}F(\alpha+1-\gamma, \beta+1-\gamma, \alpha+\beta+1-\gamma, 1-x),\end{aligned}$$

and in the space E_1

$$\begin{aligned}\phi_2 &= x^{-\alpha}F\left(\alpha, \alpha+1-\gamma, \alpha+\beta+1-\gamma, \frac{x-1}{x}\right)\\ &= x^{-\beta}F\left(\beta, \beta+1-\gamma, \alpha+\beta+1-\gamma, \frac{x-1}{x}\right),\end{aligned}$$

arg. x lying between $-\pi$ and $+\pi$.

When β and $\alpha + 1 - \gamma$ have their real parts positive the function ϕ_2 can be represented by a definite integral, viz.,

$$\int_0^{-\infty} u^{\beta-1}(1-u)^{\gamma-\beta-1}(1-xu)^{-\alpha}du = e^{\pi\beta i}\frac{\Gamma(\beta)\Gamma(\alpha+1-\gamma)}{\Gamma(\alpha+\beta+1-\gamma)}\phi_2,$$

supposing in the integral arg. $u = +\pi$. If we take arg. $u = -\pi$, we will have

$$\int_0^{-\infty} u^{\beta-1}(1-u)^{\gamma-\beta-1}(1-xu)^{-\alpha}du = e^{-\pi\beta i}\frac{\Gamma(\beta)\Gamma(\alpha+1-\gamma)}{\Gamma(\alpha+\beta+1-\gamma)}\phi_2.$$

7. The integral

$$y = \int_1^{+\infty} u^{\beta-1}(1-u)^{\gamma-\beta-1}(1-xu)^{-\alpha}du$$

has a sense provided the real parts of $\gamma - \beta$ and $\alpha + 1 - \gamma$ are positive and x has no real value lying between zero and unity. It is easy to see that if the variable describes a closed path enclosing the line 0 ——— 1, the function y takes its original value multiplied by $e^{\pm 2\pi\alpha i}$; the function can therefore be rendered uniform if we make the convention that the path of the variable must not cut the line 0 ——— $+\infty$. As to the path followed by the variable, then, this new integral presents an essential difference from the two integrals already examined, which arises from the fact that we cannot pass from the upper half of the plane to the lower half, or conversely, by crossing the line 0 ——— 1. The integral ceasing to have a meaning for a point of this line, there is nothing to indicate that the analytical continuation of the function would be represented by the same symbol after crossing the line; in fact, we shall see that the same symbol will not answer after the crossing.

In order to definitely arrive at the sense of the integral we will take 0 for the arg. of u, and for arg. $(1 - xu)$ that which is zero for $u = 0$, and which varies continuously when the variable describes the positive part of the axis of x. As to $1 - u$, we will take

$$\text{arg.}\ (1 - u) = \pm \pi.$$

Suppose arg. $(1 - u) = -\pi$; make

$$u = \frac{1}{v}, \quad du = \frac{-dv}{v^2},$$

$$1 - u = (-1)\frac{1 - v}{v}, \quad 1 - xu = \frac{(-x)}{v}\left(1 - \frac{v}{x}\right).$$

If we take arg. $v = 0$ and arg. $(1 - v) = 0$, and for arg. $\left(1 - \frac{v}{x}\right)$ that which is zero when $v = 0$, we ought to take arg. $(-1) = -\pi$, and for arg. x a value lying between $-\pi$ and $+\pi$.

The integral $\int_1^{+\infty} V du$ now becomes

$$e^{(1+\beta-\gamma)\pi i}(-x)^{-\alpha}\int_0^1 v^{\alpha-\gamma}(1-v)^{\gamma-\beta-1}\left(1-\frac{v}{x}\right)^{-\alpha}dv.$$

This new integral is of the same form as the one first studied, and on applying to it the same transformations we arrive at the following results:

Whenever $\alpha+1-\beta$ is neither zero nor a negative integer, the differential equation (1) admits as an integral a function ϕ_3 which is uniform throughout the plane, provided the path described by the variable does not cut the line $0 \text{———} +\infty$. This function can be developed in a series as follows: Outside the circle C_0 we have

$$\phi_3 = (-x)^{-\alpha}F\left(\alpha,\ \alpha+1-\gamma,\ \alpha+1-\beta,\ \frac{1}{x}\right)$$
$$= (-x)^{\beta-\gamma}(1-x)^{\gamma-\alpha-\beta}F\left(1-\beta,\ \gamma-\beta,\ \alpha+1-\beta,\ \frac{1}{x}\right).$$

Outside the circle C_1 we have

$$\phi_3 = (1-x)^{-\alpha}F\left(\alpha,\ \gamma-\beta,\ \alpha+1-\beta,\ \frac{1}{1-x}\right)$$
$$= (-x)^{1-\gamma}(1-x)^{\gamma-\alpha-1}F\left(\alpha+1-\gamma,\ 1-\beta,\ \alpha+1-\beta,\ \frac{1}{1-x}\right).$$

We suppose arg. $(-x)$, as also arg. $(1-x)$, to lie between $-\pi$ and $+\pi$. If the real parts of $\gamma-\beta$ and of $\alpha+1-\gamma$ are positive, we shall have

$$\int_1^{+\infty} u^{\beta-1}(1-u)^{\gamma-\beta-1}(1-xu)^{-\alpha}du = e^{(1+\beta-\gamma)\pi i}\frac{\Gamma(\alpha+1-\gamma)\Gamma(\gamma-\beta)}{\Gamma(\alpha+1-\beta)}\phi_3$$

on assuming arg. $(1-u) = -\pi$. If arg. $(1-u) = +\pi$, we shall have

$$\int_1^{+\infty} u^{\beta-1}(1-u)^{\gamma-\beta-1}(1-xu)^{-\alpha}du = e^{(\gamma-\beta-1)\pi i}\frac{\Gamma(\alpha+1-\gamma)\Gamma(\gamma-\beta)}{\Gamma(\alpha+1-\beta)}\phi_3.$$

8. We can study in the same way the remaining three integrals:

$$\int_0^{\frac{1}{x}} Vdu, \qquad \int_1^{\frac{1}{x}} Vdu, \qquad \int_{\frac{1}{x}}^{+\infty} Vdu.$$

As this study involves no difficulties it will be sufficient merely to give a summary of the results.

The integral

$$y = \int_0^{\frac{1}{x}} u^{\beta-1}(1-u)^{\gamma-\beta-1}(1-xu)^{-\alpha}\, du$$

has a sense if β and $1-\alpha$ have their real parts positive and if x has no real value between zero and unity. If we subject the path described by the variable to the condition of not cutting the line 0 ——— $+\infty$, the function y will be uniform throughout the plane.

This integral can be thrown into the first form by making $u = \frac{v}{x}$; we have then

$$\int_0^{\frac{1}{x}} u^{\beta-1}(1-u)^{\gamma-\beta-1}(1-xu)^{-\alpha}du$$
$$= e^{\pm\pi\beta i}(-x)^{-\beta}\int_0^1 v^{\beta-1}(1-v)^{-\alpha}\left(1-\frac{v}{x}\right)^{\gamma-\beta-1}dv.$$

In the first integral we take for arguments of $1-u$ and $1-xu$ those which vanish for $u=0$. As to the argument of u, denoting by ω the argument of $(-x)$, we will have for the coefficient of the second integral $e^{\pm\pi\beta i}$, according as we take for this argument the values $(-\omega\pm\pi)$. We have now for our differential equation a new integral, ϕ_4, which, like the preceding ones, can be developed in a series. Outside the circle C_0 we have

$$\phi_4 = (-x)^{-\beta}F\left(\beta+1-\gamma,\ \beta,\ \beta+1-\alpha,\ \frac{1}{x}\right)$$
$$= (-x)^{\alpha-\gamma}(1-x)^{\gamma-\alpha-\beta}F\left(1-\alpha,\ \gamma-\alpha,\ \beta+1-\alpha,\ \frac{1}{x}\right);$$

and outside the circle C_1 we have

$$\phi_4 = (1-x)^{-\beta}F\left(\beta,\ \gamma-\alpha,\ \beta+1-\alpha,\ \frac{1}{1-x}\right)$$

$$= (-x)^{1-\gamma}(1-x)^{\gamma-\beta-1}F\left(\beta+1-\gamma,\ 1-\alpha,\ \beta+1-\alpha,\ \frac{1}{1-x}\right),$$

the arguments of $(-x)$ and $(1-x)$ lying between $-\pi$ and $+\pi$. For properly chosen values of β and $1-\alpha$ we have

$$\int_0^{\frac{1}{x}} u^{\beta-1}(1-u)^{\gamma-\beta-1}(1-xu)^{-\alpha}du = e^{\pm\pi\beta i}\frac{\Gamma(\beta)\Gamma(1-\alpha)}{\Gamma(\beta+1-\alpha)}\phi_4.$$

9. The integral

$$y = \int_{\frac{1}{x}}^{+\infty} u^{\beta-1}(1-u)^{\gamma-\beta-1}(1-xu)^{-\alpha}du$$

has a sense if the real parts of $1-\alpha$ and $\alpha+1-\gamma$ are positive, and if x has no real value greater than unity; further, the function y will be holomorphic if the path described by the variable does not cut either of the infinite lines $-\infty$ ——— 0, 1 ——— $+\infty$. This integral can be put into the first form by the transformation $u = \dfrac{1}{xv}$, and we then have

$$\int_{\frac{1}{x}}^{+\infty} u^{\beta-1}(1-u)^{\gamma-\beta-1}(1-xu)^{-\alpha}du$$

$$= e^{\pm\pi i(\gamma-\beta-1)\pm\pi i\alpha}x^{1-\gamma}\int_0^1 v^{\alpha-\gamma}(1-v)^{-\alpha}(1-xv)^{\gamma-\beta-1}dv.$$

The arguments of u and $1-u$ are fixed by the continuity by supposing that we start from the origin with the argument 0 and describe the infinite radius $0L$ passing through the point $\dfrac{1}{x}$. As to $1-xu$, we can take arg. $(1-xu) = \pm\pi$. In the preceding formula we will take the sign $+$ or the sign $-$ before $\pi i(\gamma-\beta-1)$ according as the point represented by x is in the upper or in the lower half of the plane, and the sign $+$ or the sign $-$ before $\pi\alpha i$ according as we take arg. $(1-xu) = +\pi$ or $-\pi$.

While $2-\gamma$ is neither zero nor a negative integer, the differential equation (1) admits a new integral, ϕ_6, which is uniform under the condition enunciated above as to the path of the variable.

In the circle C_0 we have

$$\phi_6 = x^{1-\gamma}F(\beta+1-\gamma,\ \alpha+1-\gamma,\ 2-\gamma,\ x)$$
$$= x^{1-\gamma}(1-x)^{\gamma-\alpha-\beta}F(1-\alpha,\ 1-\beta,\ 2-\gamma,\ x),$$

and in the space E_0

$$\phi_6 = x^{1-\gamma}(1-x)^{\gamma-\alpha-1}F\left(\alpha+1-\gamma,\ 1-\beta,\ 2-\gamma,\ \frac{x}{x-1}\right)$$
$$= x^{1-\gamma}(1-x)^{\gamma-\beta-1}F\left(\beta+1-\gamma,\ 1-\alpha,\ 2-\gamma,\ \frac{x}{x-1}\right),$$

the arguments of x and $1-x$ lying between $-\pi$ and $+\pi$. For properly chosen values of α, β, γ we will have

$$\int_{\frac{1}{x}}^{+\infty} u^{\beta-1}(1-u)^{\gamma-\beta-1}(1-xu)^{-\alpha}du$$
$$= e^{\pm\pi i(\gamma-\beta-1)\pm\pi\alpha i}\frac{\Gamma(\alpha+1-\gamma)\Gamma(1-\alpha)}{\Gamma(2-\gamma)}\phi_6.$$

10. The integral

$$y = \int_1^{\frac{1}{x}} u^{\beta-1}(1-u)^{\gamma-\beta-1}(1-xu)^{-\alpha}du,$$

which has a sense provided the real parts of $1-\alpha$ and $\gamma-\beta$ are positive, and x has no real negative value, is holomorphic under the same conditions as the preceding one, and can be thrown into the first form by the transformation

$$u = \frac{1-x}{x}v + 1,$$

giving then

$$\int_1^{\frac{1}{x}} u^{\beta-1}(1-u)^{\gamma-\beta-1}(1-xu)^{-\alpha}du$$
$$= e^{\pm\pi i(\gamma-\beta-1)}x^{\beta-\gamma}(1-x)^{\gamma-\alpha-\beta}\int_0^1 v^{\gamma-\beta-1}(1-v)^{-\alpha}\left(1-\frac{x-1}{x}v\right)^{\beta-1}dv.$$

The arguments of u and $1 - xu$ are defined by the continuity; viz., we start from the origin with the initial value 0, and describe the straight line 0——1; then starting from the point $x = 1$, we will describe the straight line joining this point to the point $\frac{1}{x}$ with perfectly determinate values of these arguments. In the case of $1 - u$ there is, however, some ambiguity. In the preceding formula we must take the sign $+$ or the sign $-$ before $\pi i(\gamma - \beta - 1)$ according as we take

$$\arg.(1 - u) = \arg.(1 - x) - \arg. x \pm \pi.$$

While $\gamma + 1 - \alpha - \beta$ is neither zero nor a negative integer, we will have a new integral, ϕ_6, of the differential equation (1), which will be uniform under the same conditions as the preceding one.

In the circle C_1 we have

$$\begin{aligned}\phi_6 &= (1 - x)^{\gamma-\alpha-\beta}F(\gamma - \alpha, \gamma - \beta, \gamma + 1 - \alpha - \beta, 1 - x)\\ &= x^{1-\gamma}(1 - x)^{\gamma-\alpha-\beta}F(1 - \alpha, 1 - \beta, \gamma + 1 - \alpha - \beta, 1 - x),\end{aligned}$$

and in the space E_1

$$\begin{aligned}\phi_6 &= x^{\alpha-\gamma}(1 - x)^{\gamma-\alpha-\beta}F\left(\gamma - \alpha, 1 - \alpha, \gamma + 1 - \alpha - \beta, \frac{x-1}{x}\right)\\ &= x^{\beta-\gamma}(1 - x)^{\gamma-\alpha-\beta}F\left(\gamma - \beta, 1 - \beta, \gamma + 1 - \alpha - \beta, \frac{x-1}{x}\right),\end{aligned}$$

the arguments of x and $1 - x$ lying between $-\pi$ and $+\pi$. For proper values of α, β, γ, we have

$$\int_1^{\frac{1}{x}} u^{\beta-1}(1 - u)^{\gamma-\beta-1}(1 - xu)^{-\alpha}du = e^{\pm\pi i(\gamma-\beta-1)}\frac{\Gamma(\gamma-\beta)\Gamma(1-\alpha)}{\Gamma(\gamma + 1 - \alpha - \beta)}\phi_6.$$

11. In summing up all that precedes we see that, so long as no one of the quantities γ, $\gamma - \alpha - \beta$, $\beta - \alpha$ is an integer, the proposed differential equation admits six particular integrals each of which, in different parts of the plane, can be expressed in four different ways by hypergeometric series: these are Kummer's twenty-four integrals.

We can render these integrals uniform by imposing certain conditions upon the paths described by the variable; thus, for the integrals ϕ_1, ϕ_2, ϕ_5, ϕ_6 these paths must not cut either of the lines $-\infty$ —— 0, 1 —— $+\infty$; for the integrals ϕ_3 and ϕ_4 the path of the variable must not cut the line 0 —— $+\infty$. These six integrals divide into three groups, each group containing the integrals which behave in a simple manner in the region of a critical point. The first group contains the integrals ϕ_1 and ϕ_5; the second group contains ϕ_2 and ϕ_6; and the third group contains ϕ_3 and ϕ_4. Each of these integrals is susceptible of being represented by a definite integral, provided the elements α, β, γ satisfy certain conditions. The following is a table of the twenty-four integrals:

Table of Integrals.

$$\phi_1 \begin{cases} (1) & F(\alpha, \beta, \gamma, x), \\ (2) & (1-x)^{\gamma-\alpha-\beta} F(\gamma-\alpha, \gamma-\beta, \gamma, x), \\ (3) & (1-x)^{-\alpha} F\left(\alpha, \gamma-\beta, \gamma, \dfrac{x}{x-1}\right), \\ (4) & (1-x)^{-\beta} F\left(\beta, \gamma-\alpha, \gamma, \dfrac{x}{x-1}\right). \end{cases}$$

$$\phi_2 \begin{cases} (1) & F(\alpha, \beta, \alpha+\beta+1-\gamma, 1-x), \\ (2) & x^{1-\gamma} F(\alpha+1-\gamma, \beta+1-\gamma, \alpha+\beta+1-\gamma, 1-x), \\ (3) & x^{-\alpha} F\left(\alpha, \alpha+1-\gamma, \alpha+\beta+1-\gamma, \dfrac{x-1}{x}\right), \\ (4) & x^{-\beta} F\left(\beta, \beta+1-\gamma, \alpha+\beta+1-\gamma, \dfrac{x-1}{x}\right). \end{cases}$$

$$\phi_3 \begin{cases} (1) & (-x)^{-\alpha} F\left(\alpha, \alpha+1-\gamma, \alpha+1-\beta, \dfrac{1}{x}\right), \\ (2) & (-x)^{\beta-\gamma}(1-x)^{\gamma-\alpha-\beta} F\left(1-\beta, \gamma-\beta, \alpha+1-\beta, \dfrac{1}{x}\right), \\ (3) & (1-x)^{-\alpha} F\left(\alpha, \gamma-\beta, \alpha+1-\beta, \dfrac{1}{1-x}\right), \\ (4) & (-x)^{1-\gamma}(1-x)^{\gamma-\alpha-1} F\left(\alpha+1-\gamma, 1-\beta, \alpha+1-\beta, \dfrac{1}{1-x}\right). \end{cases}$$

$$\phi_4 \begin{cases} (1) & (-x)^{-\beta}F\left(\beta+1-\gamma,\ \beta,\ \beta+1-\alpha,\ \frac{1}{x}\right), \\ (2) & (-x)^{\alpha-\gamma}(1-x)^{\gamma-\alpha-\beta}F\left(1-\alpha,\ \gamma-\alpha,\ \beta+1-\alpha,\ \frac{1}{x}\right), \\ (3) & (1-x)^{-\beta}F\left(\beta,\ \gamma-\alpha,\ \beta+1-\alpha,\ \frac{1}{1-x}\right), \\ (4) & (-x)^{1-\gamma}(1-x)^{\gamma-\beta-1}F\left(\beta+1-\gamma,\ 1-\alpha,\ \beta+1-\alpha,\ \frac{1}{1-x}\right). \end{cases}$$

$$\phi_5 \begin{cases} (1) & x^{1-\gamma}F(\alpha+1-\gamma,\ \beta+1-\gamma,\ 2-\gamma,\ x), \\ (2) & x^{1-\gamma}(1-x)^{\gamma-\alpha-\beta}F(1-\alpha,\ 1-\beta,\ 2-\gamma,\ x), \\ (3) & x^{1-\gamma}(1-x)^{\gamma-\alpha-1}F\left(\alpha+1-\gamma,\ 1-\beta,\ 2-\gamma,\ \frac{x}{x-1}\right), \\ (4) & x^{1-\gamma}(1-x)^{\gamma-\beta-1}F\left(\beta+1-\gamma,\ 1-\alpha,\ 2-\gamma,\ \frac{x}{x-1}\right). \end{cases}$$

$$\phi_6 \begin{cases} (1) & (1-x)^{\gamma-\alpha-\beta}F(\gamma-\alpha,\ \gamma-\beta,\ \gamma+1-\alpha-\beta,\ 1-x), \\ (2) & x^{1-\gamma}(1-x)^{\gamma-\alpha-\beta}F(1-\alpha,\ 1-\beta,\ \gamma+1-\alpha-\beta,\ 1-x), \\ (3) & x^{\alpha-\gamma}(1-x)^{\gamma-\alpha-\beta}F\left(\gamma-\alpha,\ 1-\alpha,\ \gamma+1-\alpha-\beta,\ \frac{x-1}{x}\right), \\ (4) & x^{\beta-\gamma}(1-x)^{\gamma-\alpha-\beta}F\left(\gamma-\beta,\ 1-\beta,\ \gamma+1-\alpha-\beta,\ \frac{x-1}{x}\right). \end{cases}$$

12. Between three of these integrals there is a linear relation with constant coefficients in all that part of the plane in which the integrals are holomorphic. If we consider three of the integrals denoted by ϕ_1, ϕ_2, ϕ_5, ϕ_6, the relation will be unique throughout the plane. But this will not be true if we take one of the integrals ϕ_3 and ϕ_4 in combination with two of the preceding ones. Take, for example, ϕ_1, ϕ_2, ϕ_3, and let M and M' denote two points, one situated in the upper half of the plane, and the other in the lower half. There exists no path joining M and M' which will not cut at least one of the two lines $-\infty$ ——— 0, 0 ——— $+\infty$; one at least, then, of the three functions represented in the region of M by ϕ_1, ϕ_2, ϕ_3 will

not, after travelling such a path, be represented in the region of M' by the same symbol.

The preceding remark is essential, and it will be useful to develop it. Suppose E, E' (Fig. 2) two areas with simple contours T, T', neither of which encloses in its interior a critical point of a differential equation of the second order; suppose that between these two areas there is a critical point, A, of the differential equation, and suppose further that the areas have in common two *separate* areas, C and C'—that is, such that we cannot pass from a point of the area C to a point of C' without cutting at least one of the contours T, T'. Let P and P' be two linearly independent particular integrals which are uniform in the area E, and let Q and Q' denote two such integrals in the area E'.

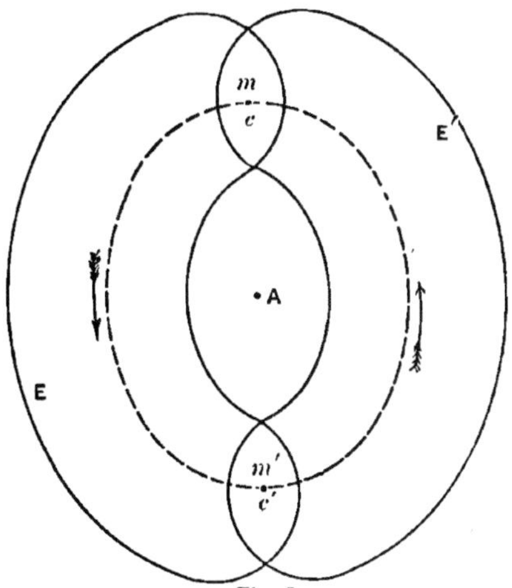

Fig. 2.

In the common part C we have the relations

$$\text{(I)} \qquad \begin{cases} Q = \lambda P + \mu P', \\ Q' = \lambda' P + \mu' P'; \end{cases}$$

and in C' we have

$$\text{(II)} \qquad \begin{cases} Q = \lambda_1 P + \mu_1 P', \\ Q' = \lambda_1' P + \mu_1' P'. \end{cases}$$

It is easy to show that the relations (I) and (II) must be distinct; that is, that we cannot have simultaneously

$$\lambda_1 = \lambda, \quad \mu_1 = \mu, \quad \lambda_1' = \lambda', \quad \mu_1' = \mu'.$$

Suppose, in fact, that these last equations were satisfied; let us start from a point m in C with the particular integral Q and describe a path situated inside the area E, and so arrive at a point m' of C'. All along this path our integral will be given by $\lambda P + \mu P'$, and under our hypothesis will, at the point m', coincide with the particular integral Q. If now we return to the point m by a path lying in E', we will evidently arrive at this point with the original integral Q. We would therefore have described a closed curve containing the point A and have returned the integral to its original value. The same would be true if we had started out with the particular integral Q'. The point A could not, therefore, be a critical point for the differential equation. It is clear that in the case under consideration the areas C and C' coincide respectively with the upper and lower halves of the plane, and the contours T and T' with the lines $-\infty$ ——— 0, 1 ——— $+\infty$, and 0 ——— $+\infty$.

It is thus established that, as they have been defined, there should exist between the integrals ϕ_1, ϕ_2, ϕ_3 two *different* linear relations according as the point x is in the upper or lower half of the plane.

13. Let us assume the point x in the upper half of the plane, and suppose further that the real parts of β, $\gamma - \beta$, $\alpha + 1 - \gamma$ are positive.

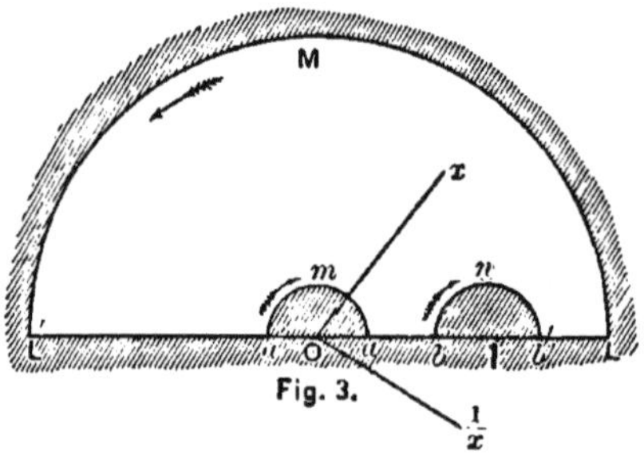

Fig. 3.

Describe around the points $x = 0$ and $x = 1$ two semicircles ama' and bnb' with very small radii r and r' respectively, and around $x = 0$ describe also a semicircle LML' with a very large radius R; the function $V = u^{\beta - 1}(1 - u)^{\gamma - \beta - 1}(1 - xu)^{-\alpha}$ will be holomorphic inside the area bounded by these semicircles and the portions of the straight line $L'a'$, ab, $b'L$.

By Cauchy's theorem we have

$$\int V du = 0,$$

where the integration extends around the entire contour just described.

This can be written in the form

$$\int_{-R}^{-r} Vdu + \int_{+r}^{1-r'} Vdu + \int_{1+r'}^{+R} Vdu = -\int_{LML'} Vdu - \int_{a'ma} Vdu - \int_{bnb'} Vdu.$$

If now the radius R increase indefinitely, and if the radii r and r' tend towards zero at the same time, it is easy to see that each of the three integrals in the second member of this last equation tends towards zero. Take, for example, the integral

$$\int_{LML'} Vdu = \int_{LML'} u^{\beta-1}(1-u)^{\gamma-\beta-1}(1-xu)^{-\alpha}du;$$

for values of u for which mod. u is very great this integral can be written

$$(-x)^{-\alpha}\int_{LML'} u^{\gamma-\alpha-2}(1+\epsilon)du,$$

where ϵ is an infinitely small quantity. Write now

$$u = Re^{i\theta};$$

then $\int_{LML'} Vdu$ becomes

$$i(-x)^{-\alpha}\int_0^{\pi} R^{(\gamma-\alpha-1)i\theta}(1+\epsilon)d\theta.$$

Let now $\gamma - \alpha - 1 = \mu + i\nu$; then

$$\int_{LML'} Vdu = i(-x)^{-\alpha}\int_0^{\pi} e^{[L(R)+\theta i](\mu+\nu i)}(1+\epsilon)d\theta$$

$$= i(-x)^{-\alpha}\int_0^{\pi} e^{\mu L(R)-\nu\theta}e^{i[\mu\theta+\nu L(R)]}(1+\epsilon)d\theta.$$

Since by hypothesis μ is a real negative number, we can take R so great that the maximum modulus of the function under the sign of integration shall be less than an assigned number η; the modulus of the integral will then be less than $\pi\eta[\text{mod. }(-x)^{-\alpha}]$; that is to say, it can be as small as we please. We can show in like manner that each of the two integrals $\int_{ama'} Vdu$, $\int_{bnb'} Vdu$ has zero for its limit. We have therefore the equality

$$\int_{-\infty}^{0} Vdu + \int_0^1 Vdu + \int_1^{\infty} Vdu = 0.$$

If we take 0 for the argument of u and $1-u$ along the path ab, it is clear that we must take $+\pi$ for the argument of u along the path $L'a'$, and $-\pi$ for the argument of $1-u$ along the path $b'L$. By referring then to the definite-integral expressions for the functions ϕ_1, ϕ_2, ϕ_3, we see that the preceding equation gives the following relation connecting these functions:

$$\text{(I)} \qquad e^{\pi\beta i}\frac{\Gamma(\beta)\Gamma(\alpha+1-\gamma)}{\Gamma(\alpha+\beta+1-\gamma)}\phi_3$$
$$=\frac{\Gamma(\beta)\Gamma(\gamma-\beta)}{\Gamma(\gamma)}\phi_1+e^{(1+\beta-\gamma)\pi i}\frac{\Gamma(\alpha+1-\gamma)\Gamma(\gamma-\beta)}{\Gamma(\alpha+1-\beta)}\phi_2.$$

This relation has been established by supposing the real parts of β, $\gamma-\beta$, $\alpha+1-\gamma$ to be positive. In order to demonstrate that the relation is general it is only necessary to employ the well-known method of procedure which consists in showing that if the relation holds for values of β, $\gamma-\beta$, $\alpha+1-\gamma$ comprised between certain limits, it is still true when we diminish one of these values by unity. We will show first that the relation holds whatever be the value of α. If (I) is demonstrated for certain values of α, β, γ, it will hold for $\alpha+1$, β, γ; we can therefore write, replacing ϕ_1, ϕ_2, ϕ_3 by the corresponding series,

$$e^{\pi\beta i}\frac{\Gamma(\beta)\Gamma(\alpha+1-\gamma)}{\Gamma(\alpha+\beta+1-\gamma)}F(\alpha,\ \beta,\ \alpha+\beta+1-\gamma,\ 1-x)$$
$$=\frac{\Gamma(\beta)\Gamma(\gamma-\beta)}{\Gamma(\gamma)}F(\alpha,\ \beta,\ \gamma,\ x)$$
$$+e^{(\beta+1-\gamma)\pi i}\frac{\Gamma(\alpha+1-\gamma)\Gamma(\gamma-\beta)}{\Gamma(\alpha+1-\beta)}\cdot(-x)^{-\alpha}F\left(\alpha,\ \alpha+1-\gamma,\ \alpha+1-\beta,\ \frac{1}{x}\right),$$

$$e^{\pi\beta i}\frac{\Gamma(\beta)\Gamma(\alpha+2-\gamma)}{\Gamma(\alpha+\beta+2-\gamma)}F(\alpha+1,\ \beta,\ \alpha+\beta+2-\gamma,\ 1-x)$$
$$=\frac{\Gamma(\beta)\Gamma(\gamma-\beta)}{\Gamma(\gamma)}F(\alpha+1,\ \beta,\ \gamma,\ x)$$
$$+e^{(\beta+1-\gamma)\pi i}\frac{\Gamma(\alpha+2-\gamma)\Gamma(\gamma-\beta)}{\Gamma(\alpha+2-\beta)}(-x)^{-\alpha-1}$$
$$F\left(\alpha+1,\ \alpha+2-\gamma,\ \alpha+2-\beta,\ \frac{1}{x}\right).$$

Multiply the first of these relations by $(\alpha - \beta)x + \gamma - 2\alpha$, and the second by $\alpha(1 - x)$, and add the results. The coefficient of $\frac{\Gamma(\beta)\Gamma(\gamma-\beta)}{\Gamma(\gamma)}$ will be

$$[(\alpha-\beta)x+\gamma-2\alpha]F(\alpha, \beta, \gamma, x)+\alpha(1-x)F(\alpha+1, \beta, \gamma, x),$$

that is,

$$(\gamma-\alpha)F(\alpha-1, \beta, \gamma, x),$$

by a well-known formula in the theory of hypergeometric series. Further, in the second member of our equation we shall have

$$e^{(\beta+1-\gamma)\pi i}\frac{\Gamma(\alpha+1-\gamma)\Gamma(\gamma-\beta)}{\Gamma(\alpha+1-\beta)}(-x)^{-\alpha}$$

$$\times\left\{[(\alpha-\beta)x+\gamma-2\alpha]F\left(\alpha, \alpha+1-\gamma, \alpha+1-\beta, \frac{1}{x}\right)\right.$$

$$\left.-\frac{\alpha(\alpha+1-\gamma)(1-x)}{(\alpha+1-\beta)x}F\left(\alpha+1, \alpha+2-\gamma, \alpha+2-\beta, \frac{1}{x}\right)\right\}.$$

The quantity in $\{\ \}$ reduces to

$$(\alpha-\beta)xF\left(\alpha-1, \alpha-\gamma, \alpha-\beta, \frac{1}{x}\right),$$

and the coefficient of $F\left(\alpha-1, \alpha-\gamma, \alpha-\beta, \frac{1}{x}\right)$ becomes

$$-e^{(\beta+1-\gamma)\pi i}\frac{\Gamma(\beta+1-\gamma)\Gamma(\gamma-\beta)}{\Gamma(\alpha+1-\beta)}(\alpha-\beta)(-x)^{-(\alpha-1)}$$

$$=e^{(\beta+1-\gamma)\pi i}\frac{\Gamma(\alpha-\gamma)\Gamma(\gamma-\beta)}{\Gamma(\alpha-\beta)}(\gamma-\alpha)(-x)^{-(\alpha-1)}.$$

We find in the same way for the first member of our new equation the value

$$e^{\pi\beta i}\frac{\Gamma(\beta)\Gamma(\alpha-\gamma)}{\Gamma(\alpha+\beta-\gamma)}(\gamma-\alpha)F(\alpha-1, \beta, \alpha+\beta-\gamma, 1-x),$$

and so have finally

$$e^{\pi\beta i}\frac{\Gamma(\beta)\Gamma(\alpha-\gamma)}{\Gamma(\alpha+\beta-\gamma)}F(\alpha-1,\ \beta,\ \alpha+\beta-\gamma,\ 1-x)$$

$$=\frac{\Gamma(\beta)\Gamma(\gamma-\beta)}{\Gamma(\gamma)}F(\alpha-1,\ \beta,\ \gamma,\ x)$$

$$+e^{(\beta+1-\gamma)\pi i}\frac{\Gamma(\alpha-\gamma)\Gamma(\gamma-\beta)}{\Gamma(\alpha-\beta)}(-x)^{-(\alpha-1)}F\left(\alpha-1,\ \alpha-\gamma,\ \alpha-\beta,\ \frac{1}{x}\right)$$

which is simply equation (I) with α changed into $\alpha-1$. This relation is therefore exact whatever be α. In the same way we can show that if (I) holds for two values of γ differing by unity it will also hold for a value of γ differing by unity from the least of the preceding values, and consequently that it holds for all values of α and γ. It remains then only to show that β can also be arbitrary, which is done by a process entirely similar to the above and which need not be reproduced here.

14. Formula (I) is then perfectly general, and we can readily deduce from it the following formulæ:

$$\text{(I)}\quad e^{\pi\beta i}\frac{\Gamma(\beta)\Gamma(\alpha+1-\gamma)}{\Gamma(\alpha+\beta+1-\gamma)}\phi_2=\frac{\Gamma(\beta)\Gamma(\gamma-\beta)}{\Gamma(\gamma)}\phi_1+e^{(\beta+1-\gamma)\pi i}\frac{\Gamma(\alpha+1-\gamma)\Gamma(\gamma-\beta)}{\Gamma(\alpha+1-\beta)}\phi_3,$$

$$\text{(II)}\quad e^{\pi\alpha i}\frac{\Gamma(\alpha)\Gamma(\beta+1-\gamma)}{\Gamma(\alpha+\beta+1-\gamma)}\phi_2=\frac{\Gamma(\alpha)\Gamma(\gamma-\alpha)}{\Gamma(\gamma)}\phi_1+e^{(\alpha+1-\gamma)\pi i}\frac{\Gamma(\beta+1-\gamma)\Gamma(\gamma-\alpha)}{\Gamma(\beta+1-\alpha)}\phi_4,$$

$$\text{(III)}\quad e^{(\gamma-\beta)\pi i}\frac{\Gamma(\gamma-\beta)\Gamma(1-\alpha)}{\Gamma(\gamma+1-\alpha-\beta)}\phi_6=\frac{\Gamma(\beta)\Gamma(\gamma-\beta)}{\Gamma(\gamma)}\phi_1+e^{(1-\beta)\pi i}\frac{\Gamma(1-\alpha)\Gamma(\beta)}{\Gamma(\beta+1-\alpha)}\phi_4,$$

$$\text{(IV)}\quad e^{(\beta+1-\gamma)\pi i}\frac{\Gamma(\beta+1-\gamma)\Gamma(\alpha)}{\Gamma(\alpha+\beta+1-\gamma)}\phi_2=\frac{\Gamma(\beta+1-\gamma)\Gamma(1-\beta)}{\Gamma(2-\gamma)}\phi_5+e^{(\beta+1-\gamma)\pi i}\frac{\Gamma(\alpha)\Gamma(1-\beta)}{\Gamma(\alpha+1-\beta)}\phi_3.$$

Formula (II) is deduced from (I) by permuting α and β; (III) is obtained from (I) by changing α into $\gamma-\alpha$ and β into $\gamma-\beta$ and multiplying by $(1-x)^{\gamma-\alpha-\beta}$, and finally (IV) is obtained from (I) by changing α into $\alpha+1-\gamma$, β into $\beta+1-\gamma$, γ into $2-\gamma$, and multiplying by $x^{1-\gamma}$. Remark that if we take, as we have supposed, the argument of x between $-\pi$ and $+\pi$, and so of course for arg. $(-x)$, we shall have

$$x=(-x)e^{\pi i}.$$

The relations (I), (II), (III), (IV) are distinct, and we can deduce from them all of the linear relations which exist between the six integrals. These relations are twenty in number; among them we may remark particularly those in which there appear two functions of the same group which permit us to pass from the region of one critical point to that of another, and which are useful for the *integration* of the differential equation. There are twelve relations of this sort; the remaining eight relations are between three functions of three different groups.

Upper Half of the Plane.

$$(1)\quad e^{\pi\beta i}\frac{\Gamma(\beta)\Gamma(\alpha+1-\gamma)}{\Gamma(\alpha+\beta+1-\gamma)}\phi_2=\frac{\Gamma(\beta)\Gamma(\gamma-\beta)}{\Gamma(\gamma)}\phi_1$$
$$+e^{(\beta+1-\gamma)\pi i}\frac{\Gamma(\alpha+1-\gamma)\Gamma(\gamma-\beta)}{\Gamma(\alpha+1-\beta)}\phi_3,$$

$$(2)\quad e^{\pi\alpha i}\frac{\Gamma(\alpha)\Gamma(\beta+1-\gamma)}{\Gamma(\alpha+\beta+1-\gamma)}\phi_2=\frac{\Gamma(\alpha)\Gamma(\gamma-\alpha)}{\Gamma(\gamma)}\phi_1$$
$$+e^{(\alpha+1-\gamma)\pi i}\frac{\Gamma(\beta+1-\gamma)\Gamma(\gamma-\alpha)}{\Gamma(\beta+1-\alpha)}\phi_4,$$

$$(3)\quad e^{(\gamma-\beta)\pi i}\frac{\Gamma(\gamma-\beta)\Gamma(1-\alpha)}{\Gamma(\gamma+1-\alpha-\beta)}\phi_6=\frac{\Gamma(\beta)\Gamma(\gamma-\beta)}{\Gamma(\gamma)}\phi_1$$
$$+e^{(1-\beta)\pi i}\frac{\Gamma(1-\alpha)\Gamma(\beta)}{\Gamma(\beta+1-\alpha)}\phi_4,$$

$$(4) \quad e^{(\beta+1-\gamma)\pi i}\frac{\Gamma(\beta+1-\gamma)\Gamma(\alpha)}{\Gamma(\alpha+\beta+1-\gamma)}\phi_2 = \frac{\Gamma(\beta+1-\gamma)\Gamma(1-\beta)}{\Gamma(2-\gamma)}\phi_5 + e^{(\beta+1-\gamma)\pi i}\frac{\Gamma(\alpha)\Gamma(1-\beta)}{\Gamma(\alpha+1-\beta)}\phi_3,$$

$$(5) \quad e^{(\gamma-\alpha)\pi i}\frac{\Gamma(\gamma-\alpha)\Gamma(1-\beta)}{\Gamma(\gamma+1-\alpha-\beta)}\phi_6 = \frac{\Gamma(\alpha)\Gamma(\gamma-\alpha)}{\Gamma(\gamma)}\phi_1 + e^{(1-\alpha)\pi i}\frac{\Gamma(1-\beta)\Gamma(\alpha)}{\Gamma(\alpha+1-\beta)}\phi_3,$$

$$(6) \quad e^{(\alpha+1-\gamma)\pi i}\frac{\Gamma(\alpha+1-\gamma)\Gamma(\beta)}{\Gamma(\alpha+\beta+1-\gamma)}\phi_2 = \frac{\Gamma(\alpha+1-\gamma)\Gamma(1-\alpha)}{\Gamma(2-\gamma)}\phi_5 + e^{(\alpha+1-\gamma)\pi i}\frac{\Gamma(\beta)\Gamma(1-\alpha)}{\Gamma(\beta+1-\alpha)}\phi_4,$$

$$(7) \quad e^{(1-\beta)\pi i}\frac{\Gamma(1-\beta)\Gamma(\gamma-\alpha)}{\Gamma(\gamma+1-\alpha-\beta)}\phi_6 = \frac{\Gamma(1-\beta)\Gamma(\beta+1-\gamma)}{\Gamma(2-\gamma)}\phi_5 + e^{(1-\beta)\pi i}\frac{\Gamma(\gamma-\alpha)\Gamma(\beta+1-\gamma)}{\Gamma(\beta+1-\alpha)}\phi_4,$$

$$(8) \quad e^{(1-\alpha)\pi i}\frac{\Gamma(1-\alpha)\Gamma(\gamma-\beta)}{\Gamma(\gamma+1-\alpha-\beta)}\phi_6 = \frac{\Gamma(1-\alpha)\Gamma(\alpha+1-\gamma)}{\Gamma(2-\gamma)}\phi_5 + e^{(1-\alpha)\pi i}\frac{\Gamma(\gamma-\beta)\Gamma(\alpha+1-\gamma)}{\Gamma(\alpha+1-\beta)}\phi_3,$$

$$(9) \quad \phi_1 = \frac{\Gamma(\gamma)\Gamma(\gamma-\alpha-\beta)}{\Gamma(\gamma-\alpha)\Gamma(\gamma-\beta)}\phi_2 + \frac{\Gamma(\gamma)\Gamma(\alpha+\beta-\gamma)}{\Gamma(\alpha)\Gamma(\beta)}\phi_6,$$

$$(10) \quad \phi_5 = \frac{\Gamma(2-\gamma)\Gamma(\gamma-\alpha-\beta)}{\Gamma(1-\alpha)\Gamma(1-\beta)}\phi_2 + \frac{\Gamma(2-\gamma)\Gamma(\alpha+\beta-\gamma)}{\Gamma(\alpha+1-\gamma)\Gamma(\beta+1-\gamma)}\phi_6,$$

$$(11) \quad \phi_2 = \frac{\Gamma(\alpha+\beta+1-\gamma)\Gamma(1-\gamma)}{\Gamma(\alpha+1-\gamma)\Gamma(\beta+1-\gamma)}\phi_1 + \frac{\Gamma(\alpha+\beta+1-\gamma)\Gamma(\gamma-1)}{\Gamma(\alpha)\Gamma(\beta)}\phi_5,$$

$$(12) \quad \phi_6 = \frac{\Gamma(\gamma+1-\alpha-\beta)\Gamma(1-\gamma)}{\Gamma(1-\alpha)\Gamma(1-\beta)}\phi_1 + \frac{\Gamma(\gamma+1-\alpha-\beta)\Gamma(\gamma-1)}{\Gamma(\gamma-\alpha)\Gamma(\gamma-\beta)}\phi_5,$$

$$(13)\quad \phi_1 = \frac{\Gamma(\gamma)\Gamma(\beta-\alpha)}{\Gamma(\gamma-\alpha)\Gamma(\beta)}\phi_3 + \frac{\Gamma(\gamma)\Gamma(\alpha-\beta)}{\Gamma(\gamma-\beta)\Gamma(\alpha)}\phi_4.$$

$$(14)\quad \phi_6 = \frac{\Gamma(2-\gamma)\Gamma(\beta-\alpha)}{\Gamma(1-\alpha)\Gamma(\beta+1-\gamma)} e^{(1-\gamma)\pi i}\phi_3 + \frac{\Gamma(2-\gamma)\Gamma(\alpha-\beta)}{\Gamma(1-\beta)\Gamma(\alpha+1-\gamma)} e^{(1-\gamma)\pi i}\phi_4.$$

$$(15)\quad \phi_3 = \frac{\Gamma(1-\gamma)\Gamma(\alpha+1-\beta)}{\Gamma(1-\beta)\Gamma(\alpha+1-\gamma)}\phi_1 - \frac{\Gamma(\gamma)\Gamma(1-\gamma)\Gamma(\alpha+1-\beta)}{\Gamma(2-\gamma)\Gamma(\gamma-\beta)\Gamma(\alpha)} e^{(\gamma-1)\pi i}\phi_6.$$

$$(16)\quad \phi_4 = \frac{\Gamma(1-\gamma)\Gamma(\beta+1-\alpha)}{\Gamma(1-\alpha)\Gamma(\beta+1-\gamma)}\phi_1 - \frac{\Gamma(\gamma)\Gamma(1-\gamma)\Gamma(\beta+1-\alpha)}{\Gamma(2-\gamma)\Gamma(\gamma-\alpha)\Gamma(\beta)} e^{(\gamma-1)\pi i}\phi_6.$$

$$(17)\quad \phi_2 = \frac{\Gamma(\alpha+\beta+1-\gamma)\Gamma(\beta-\alpha)}{\Gamma(\beta+1-\gamma)\Gamma(\beta)} e^{-\pi\alpha i}\phi_3 + \frac{\Gamma(\alpha+\beta+1-\gamma)\Gamma(\alpha-\beta)}{\Gamma(\alpha+1-\gamma)\Gamma(\alpha)} e^{-\pi\beta i}\phi_4.$$

$$(18)\quad \phi_6 = \frac{\Gamma(\gamma+1-\alpha-\beta)\Gamma(\beta-\alpha)}{\Gamma(1-\alpha)\Gamma(\gamma-\alpha)} e^{-\pi(\gamma-\beta)i}\phi_3 + \frac{\Gamma(\gamma+1-\alpha-\beta)\Gamma(\alpha-\beta)}{\Gamma(1-\beta)\Gamma(\gamma-\beta)} e^{-\pi(\gamma-\alpha)i}\phi_4.$$

$$(19)\quad \phi_3 = \frac{\Gamma(\gamma+1-\alpha-\beta)\Gamma(\alpha+\beta-\gamma)\Gamma(\alpha+1-\beta)}{\Gamma(1-\beta)\Gamma(\gamma-\beta)\Gamma(\alpha+\beta+1-\gamma)} e^{\pi\alpha i}\phi_2 - \frac{\Gamma(\alpha+\beta-\gamma)\Gamma(\alpha+1-\beta)}{\Gamma(\alpha+1-\gamma)\Gamma(\alpha)} e^{\pi(\gamma-\beta)i}\phi_6.$$

$$(20)\quad \phi_4 = \frac{\Gamma(\gamma+1-\alpha-\beta)\Gamma(\alpha+\beta-\gamma)\Gamma(\beta+1-\alpha)}{\Gamma(1-\alpha)\Gamma(\gamma-\alpha)\Gamma(\alpha+\beta+1-\gamma)} e^{\pi\beta i}\phi_2 - \frac{\Gamma(\alpha+\beta-\gamma)\Gamma(\beta+1-\alpha)}{\Gamma(\beta+1-\gamma)\Gamma(\beta)} e^{\pi(\gamma-\alpha)i}\phi_6.$$

Lower Half of the Plane.

$$(1)' \quad e^{-\pi\beta i}\frac{\Gamma(\beta)\Gamma(\alpha+1-\gamma)}{\Gamma(\alpha+\beta+1-\gamma)}\phi_2 = \frac{\Gamma(\beta)\Gamma(\gamma-\beta)}{\Gamma(\gamma)}\phi_1 + e^{(\gamma-1-\beta)\pi i}\frac{\Gamma(\alpha+1-\gamma)\Gamma(\gamma-\beta)}{\Gamma(\alpha+1-\beta)}\phi_3.$$

$$(2)' \quad e^{-\pi\alpha i}\frac{\Gamma(\alpha)\Gamma(\beta+1-\gamma)}{\Gamma(\alpha+\beta+1-\gamma)}\phi_2 = \frac{\Gamma(\alpha)\Gamma(\gamma-\alpha)}{\Gamma(\gamma)}\phi_1 + e^{(\gamma-1-\alpha)\pi i}\frac{\Gamma(\beta+1-\gamma)\Gamma(\gamma-\alpha)}{\Gamma(\beta+1-\alpha)}\phi_4.$$

$$(3)' \quad e^{(\beta-\gamma)\pi i}\frac{\Gamma(\gamma-\beta)\Gamma(1-\alpha)}{\Gamma(\gamma+1-\alpha-\beta)}\phi_6 = \frac{\Gamma(\beta)\Gamma(\gamma-\beta)}{\Gamma(\gamma)}\phi_1 + e^{(\beta-1)\pi i}\frac{\Gamma(1-\alpha)\Gamma(\beta)}{\Gamma(\beta+1-\alpha)}\phi_4.$$

$$(4)' \quad e^{(\gamma-1-\beta)\pi i}\frac{\Gamma(\beta+1-\gamma)\Gamma(\alpha)}{\Gamma(\alpha+\beta+1-\gamma)}\phi_2 = \frac{\Gamma(\beta+1-\gamma)\Gamma(1-\beta)}{\Gamma(2-\gamma)}\phi_5 + e^{(\gamma-1-\beta)\pi i}\frac{\Gamma(\alpha)\Gamma(1-\beta)}{\Gamma(\alpha+1-\beta)}\phi_3.$$

$$(5)' \quad e^{(\alpha-\gamma)\pi i}\frac{\Gamma(\gamma-\alpha)\Gamma(1-\beta)}{\Gamma(\gamma+1-\alpha-\beta)}\phi_6 = \frac{\Gamma(\alpha)\Gamma(\gamma-\alpha)}{\Gamma(\gamma)}\phi_1 + e^{(\alpha-1)\pi i}\frac{\Gamma(\alpha)\Gamma(1-\beta)}{\Gamma(\alpha+1-\beta)}\phi_3.$$

$$(6)' \quad e^{(\gamma-1-\alpha)\pi i}\frac{\Gamma(\alpha+1-\gamma)\Gamma(\beta)}{\Gamma(\alpha+\beta+1-\gamma)}\phi_2 = \frac{\Gamma(\alpha+1-\gamma)\Gamma(1-\alpha)}{\Gamma(2-\gamma)}\phi_6 + e^{(\gamma-1-\alpha)\pi i}\frac{\Gamma(\beta)\Gamma(1-\alpha)}{\Gamma(\beta+1-\alpha)}\phi_4.$$

$$(7)' \quad e^{(\beta-1)\pi i}\frac{\Gamma(1-\beta)\Gamma(\gamma-\alpha)}{\Gamma(\gamma+1-\alpha-\beta)}\phi_6 = \frac{\Gamma(1-\beta)\Gamma(\beta+1-\gamma)}{\Gamma(2-\gamma)}\phi_5 + e^{(\beta-1)\pi i}\frac{\Gamma(\gamma-\alpha)\Gamma(\beta+1-\gamma)}{\Gamma(\beta+1-\alpha)}\phi_4.$$

$$(8)' \quad e^{(\alpha-1)\pi i}\frac{\Gamma(1-\alpha)\Gamma(\gamma-\beta)}{\Gamma(\gamma+1-\alpha-\beta)}\phi_6 = \frac{\Gamma(1-\alpha)\Gamma(\alpha+1-\gamma)}{\Gamma(2-\gamma)}\phi_5 + e^{(\alpha-1)\pi i}\frac{\Gamma(\gamma-\beta)\Gamma(\alpha+1-\gamma)}{\Gamma(\alpha+1-\beta)}\phi_3 .$$

. . . .

$$(14)' \quad \phi_5 = \frac{\Gamma(2-\gamma)\Gamma(\beta-\alpha)}{\Gamma(1-\alpha)\Gamma(\beta+1-\gamma)}e^{(\gamma-1)\pi i}\phi_3 + \frac{\Gamma(2-\gamma)\Gamma(\alpha-\beta)}{\Gamma(1-\beta)\Gamma(\alpha+1-\gamma)}e^{(\gamma-1)\pi i}\phi_4 .$$

$$(15)' \quad \phi_3 = \frac{\Gamma(1-\gamma)\Gamma(\alpha+1-\beta)}{\Gamma(1-\beta)\Gamma(\alpha+1-\gamma)}\phi_1 - \frac{\Gamma(\gamma)\Gamma(1-\gamma)\Gamma(\alpha+1-\beta)}{\Gamma(2-\gamma)\Gamma(\gamma-\beta)\Gamma(\alpha)}e^{(1-\gamma)\pi i}\phi_5 .$$

$$(16)' \quad \phi_4 = \frac{\Gamma(1-\gamma)\Gamma(\beta+1-\alpha)}{\Gamma(1-\alpha)\Gamma(\beta+1-\gamma)}\phi_1 - \frac{\Gamma(\gamma)\Gamma(1-\gamma)\Gamma(\beta+1-\alpha)}{\Gamma(2-\gamma)\Gamma(\gamma-\alpha)\Gamma(\beta)}e^{(1-\gamma)\pi i}\phi_5 .$$

$$(17)' \quad \phi_2 = \frac{\Gamma(\alpha+\beta+1-\gamma)\Gamma(\beta-\alpha)}{\Gamma(\beta+1-\gamma)\Gamma(\beta)}e^{\pi\alpha i}\phi_3 + \frac{\Gamma(\alpha+\beta+1-\gamma)\Gamma(\alpha-\beta)}{\Gamma(\alpha+1-\gamma)\Gamma(\alpha)}e^{\pi\beta i}\phi_4 .$$

$$(18)' \quad \phi_6 = \frac{\Gamma(\gamma+1-\alpha-\beta)\Gamma(\beta-\alpha)}{\Gamma(1-\alpha)\Gamma(\gamma-\alpha)}e^{\pi(\gamma-\beta)i}\phi_3 + \frac{\Gamma(\gamma+1-\alpha-\beta)\Gamma(\alpha-\beta)}{\Gamma(1-\beta)\Gamma(\gamma-\beta)}e^{\pi(\gamma-\alpha)i}\phi_4 .$$

$$(19)' \quad \phi_3 = \frac{\Gamma(\gamma+1-\alpha-\beta)\Gamma(\alpha+\beta-\gamma)\Gamma(\alpha+1-\beta)}{\Gamma(1-\beta)\Gamma(\gamma-\beta)\Gamma(\alpha+\beta+1-\gamma)}e^{-\pi\alpha i}\phi_2 - \frac{\Gamma(\alpha+\beta-\gamma)\Gamma(\alpha+1-\beta)}{\Gamma(\alpha+1-\gamma)\Gamma(\alpha)}e^{\pi(\beta-\gamma)i}\phi_6 .$$

$$(20)' \quad \phi_4 = \frac{\Gamma(\gamma+1-\alpha-\beta)\Gamma(\alpha+\beta-\gamma)\Gamma(\beta+1-\alpha)}{\Gamma(1-\alpha)\Gamma(\gamma-\alpha)\Gamma(\alpha+\beta+1-\gamma)} e^{-\pi\beta i}\phi_2$$
$$- \frac{\Gamma(\alpha+\beta-\gamma)\Gamma(\beta+1-\alpha)}{\Gamma(\beta+1-\gamma)\Gamma(\beta)} e^{\pi(\alpha-\gamma)i}\phi_6 .$$

15. Formulæ (5) and (6) are deduced from (3) and (4) by permuting α and β. If in (4) we change α into $\gamma-\alpha$, β into $\gamma-\beta$, and multiply by $(1-x)^{\gamma-\alpha-\beta}$, we obtain (7), and from (7) derive (8) by permuting α and β. Formulæ (9) and (10) are obtained by eliminating ϕ_4 between (2) and (3) and between (6) and (7) respectively; solving (9) and (10) for ϕ_2 and ϕ_6 respectively, we find (11) and (12); solving (1) and (2) for ϕ_1 and ϕ_2, we get (13) and (17). If in (13) we change α into $\alpha+1-\gamma$, β into $\beta+1-\gamma$, γ into $2-\gamma$, and multiply by $x^{1-\gamma}$, we obtain (14), and from these last two we then derive (15) and (16). Formula (18) is obtained from (17) by changing α into $\gamma-\alpha$, β into $\gamma-\beta$, and multiplying by $(1-x)^{\gamma-\alpha-\beta}$; then from (17) and (18) we readily get (19) and (20).

As already remarked, (9), (10), (11), and (12) hold throughout the plane. The same is true for (13); for, the function ϕ_1 being uniform in the region of the point $x=0$, we may suppose that the path followed by the variable cuts the line $-\infty$ ——— 0. But the other formulæ for a point in the lower half of the plane will be different; the relation between ϕ_1, ϕ_2, ϕ_3, for example, will then be

$$(1)' \quad e^{-\pi\beta i}\frac{\Gamma(\beta)\Gamma(\alpha+1-\gamma)}{\Gamma(\alpha+\beta+1-\gamma)}\phi_3$$
$$= \frac{\Gamma(\beta)\Gamma(\gamma-\beta)}{\Gamma(\gamma)}\phi_1 + e^{-(\beta+1-\gamma)\pi i}\frac{\Gamma(\alpha+1-\gamma)\Gamma(\gamma-\beta)}{\Gamma(\alpha+1-\beta)}\phi_2 .$$

This differs from (1) in that π is replaced by $-\pi$. All the other formulæ for the lower part of the plane are deduced from this by permutations of the letters just in the same way as in the upper half of the plane.

16. Consider a path of arbitrary form joining any two points M and M' of the plane, but not passing through either of the points 0 or 1. If we start from the point M with a particular solution of the differential equation, this solution will be defined all along the path

described by the variable, and we will arrive at the point M' with a determinate integral. The preceding relations permit us to find this integral when the path from M to M' is given, and when the particular solution with which we start from M is given. Suppose we start from a point A corresponding to a *real* value of x and lying between $x = 0$ and $x = 1$, and go to a point M' of the plane by a direct path which cuts neither of the lines $-\infty$ ——— 0, $+1$ ——— $+\infty$. In the region of A the integral can be represented by $C\phi_1 + C'\phi_2$ by giving the constants C and C' proper values. If the point M' lies in the space E_0, we will replace ϕ_2 by its value in terms of ϕ_1 and ϕ_5, and will employ for the effective calculation of the function the most convenient development of each of these last two functions. So also if M' is in the space E_1, we will replace ϕ_1 by its value in terms of ϕ_2 and ϕ_6. If M' is outside the space common to the circles C_0 and C_1, we can express ϕ_1 and ϕ_2 by means of ϕ_3 and ϕ_4, being careful, however, to use different formulæ according as M' is in the upper or lower half of the plane.

Suppose now we start from A and follow any path up to M', provided the path does not pass through either of the points $x = 0$ or $x = 1$. Such a path can, as we know, be reduced to a series of loops (*lacets*) going round the points 0 and 1, in one sense or the other, followed by a straight path from A to M'.

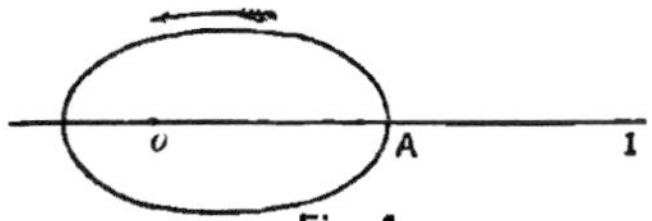

Fig. 4.

Let us start from the point A with a particular integral $C\phi_1 + C'\phi_2$, and describe a loop in the direct sense round the critical point $x = 0$. To see how this integral behaves, we have only to replace ϕ_2 by

$$\frac{\Gamma(\alpha+\beta+1-\gamma)\Gamma(1-\gamma)}{\Gamma(\alpha+1-\gamma)\Gamma(\beta+1-\gamma)}\phi_1 + \frac{\Gamma(\alpha+\beta+1-\gamma)\Gamma(\gamma-1)}{\Gamma(\alpha)\Gamma(\beta)}\phi_5.$$

When the variable describes the loop, ϕ_1 does not change, but ϕ_5 changes into $e^{(1-\gamma)2\pi i}\phi_5$, so that we come back to the point of departure with the integral

$$C\phi_1 + C'\frac{\Gamma(\alpha+\beta+1-\gamma)\Gamma(1-\gamma)}{\Gamma(\alpha+1-\gamma)\Gamma(\beta+1-\gamma)}\phi_1$$
$$+ C'e^{(1-\gamma)2\pi i}\left[\phi_2 - \frac{\Gamma(\alpha+\beta+1-\gamma)\Gamma(1-\gamma)}{\Gamma(\alpha+1-\gamma)\Gamma(\beta+1-\gamma)}\phi_1\right],$$

or
$$C_1\phi_1 + C_1'\phi_2,$$
where

$$C_1 = C + C'\frac{\Gamma(\alpha+\beta+1-\gamma)\Gamma(1-\gamma)}{\Gamma(\alpha+1-\gamma)\Gamma(\beta+1-\gamma)}(1 - e^{(1-\gamma)2\pi i}),$$
$$C_1' = C'e^{-(1-\gamma)2\pi i}.$$

So also for a loop round $x = 1$ the integral $C\phi_1 + C'\phi_2$ changes into $C_1\phi_1 + C_1'\phi_2$, where

$$C_1 = Ce^{\pm 2\pi i(\gamma-\alpha-\beta)},$$
$$C_1' = C' + C\frac{\Gamma(\gamma)\Gamma(\gamma-\alpha-\beta)}{\Gamma(\gamma-\alpha)\Gamma(\gamma-\beta)}(1 - e^{\pm 2\pi i(\gamma-\alpha-\beta)}).$$

The sign $+$ or $-$ goes before $2\pi i(\gamma-\alpha-\beta)$ according as the loop is described in the direct (*i.e.*, positive) sense or in the opposite (*i.e.*, negative) sense.

If the variable describes several loops in succession, the formulæ will have to be applied a corresponding number of times, and we will finally be conducted to results similar to the preceding.

Let us take now the general case where the path of the variable joins any two points of the plane. This path can be replaced by a direct path going from M to A, followed by a perfectly determinate path going from A to M'. In order to be conducted to the preceding case, it will be sufficient to determine with what integral we arrive at the point A in following the direct path from M to A. The above method enables us to do this without difficulty.

Let us consider as an example the differential equation

$$x(1-x)\frac{d^2y}{dx^2} + \left(\frac{1}{2} - \frac{7x}{3}\right)\frac{dy}{dx} - \frac{1}{3}y = 0;$$

this admits as an integral the hypergeometric series

$$F(1, \tfrac{1}{3}, \tfrac{1}{2}, x).$$

Suppose we start from the point A with this solution and describe the closed curve $ABCDA$ (Fig. 5), surrounding the two points 0 and 1.

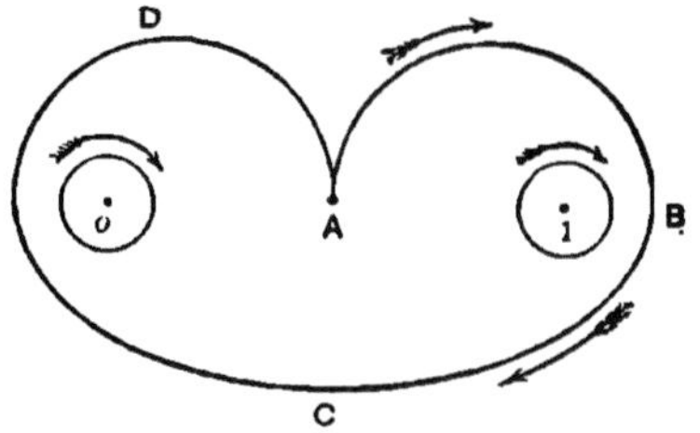

Fig. 5.

This contour reduces to two loops described in the negative sense around the points $x = 1$ and $x = 0$. After describing the first loop around $x = 1$ we return to the starting-point A with the integral

$$e^{-2\pi i(\gamma-\alpha-\beta)} F(1, \tfrac{1}{3}, \tfrac{1}{2}, x)$$
$$+ \frac{\Gamma(\gamma)\Gamma(\gamma-\alpha-\beta)}{\Gamma(\gamma-\alpha)\Gamma(\gamma-\beta)}(1 - e^{-2\pi i(\gamma-\alpha-\beta)})F(1, \tfrac{1}{3}, \tfrac{11}{6}, 1-x);$$

after describing the second loop we will have an integral which, in the region of A, can be represented by

$$CF(1, \tfrac{1}{3}, \tfrac{1}{2}, x) + C_1F(1, \tfrac{1}{3}, \tfrac{11}{6}, 1-x),$$

where

$$C = e^{-2\pi i(\gamma-\alpha-\beta)}$$
$$+ \frac{\sin(\gamma-\alpha)\pi \sin(\gamma-\beta)\pi}{\sin\gamma\pi \sin(\gamma-\alpha-\beta)\pi}(1 - e^{-2\pi i(\gamma-\alpha-\beta)})(1 - e^{-2\pi i(1-\gamma)});$$

$$C_1 = \frac{\Gamma(\gamma)\Gamma(\gamma-\alpha-\beta)}{\Gamma(\gamma-\alpha)\Gamma(\gamma-\beta)} e^{-2\pi i(1-\gamma)}(1 - e^{-2\pi i(\gamma-\alpha-\beta)}).$$

Making now $\alpha = 1$, $\beta = \frac{1}{3}$, $\gamma = \frac{1}{2}$, we have

$$C = e^{\frac{5\pi i}{3}} + 2(1 - e^{\frac{5\pi i}{3}}) = 2 - e^{\frac{5\pi i}{3}};$$

$$C_1 = -\tfrac{3}{5}(1 - e^{\frac{5\pi i}{3}}).$$

17. When γ is an integer, or indeed when $\gamma - \alpha - \beta$, or $\alpha - \beta$, is an integer, we have only one integral in one of the groups; in order to find a new integral we employ the following well-known process. Let

$$y = F(x, r), \qquad y_1 = F_1(x, r)$$

be two distinct integrals of the differential equation which become equal for a particular value, $r = r_1$, of the constant r. We will obtain another integral by seeking the limit for $r = r_1$ of the expression

$$\frac{F(x, r) - F_1(x, r)}{r - r_1},$$

which is also an integral of the differential equation whatever be the value of r.

Let us suppose γ to be an integer; we may also assume γ positive; for, if it were negative, we could make the transformation $y = x^{1-\gamma} y_1$. There are two cases to be distinguished according as γ is equal to unity or greater than unity.

First Case: $\gamma = 1$.—The two integrals

$$F(\alpha, \beta, \gamma, x) \quad \text{and} \quad x^{1-\gamma} F(\alpha + 1 - \gamma, \beta + 1 - \gamma, 2 - \gamma, x)$$

become identical for $\gamma = 1$. From what has been said we must seek now the limit for $\gamma = 1$ of the expression

$$\frac{x^{1-\gamma} F(\alpha + 1 - \gamma, \beta + 1 - \gamma, 2 - \gamma, x) - F(\alpha, \beta, \gamma, x)}{1 - \gamma}.$$

This limit is obviously equal

$$\phi_1 \log x + \frac{d\phi_1}{d\alpha} + \frac{d\phi_1}{d\beta} + 2\frac{d\phi_1}{d\gamma},$$

in which we make $\gamma = 1$. As the function ϕ_1 is susceptible of four different forms, there must equally be four different forms for the new integral. Let

$$\phi_1 = F(\alpha, \beta, \gamma, x),$$

and write

$$\psi_1(x) = \frac{d\phi_1}{d\alpha} + \frac{d\phi_1}{d\beta} + 2\frac{d\phi_1}{d\gamma},$$

and denote by A_m the coefficient of x^m in the development of

$$F(\alpha, \beta, \gamma, x).$$

We find then readily

$$\psi_1(x) = \sum_{m=1}^{m=+\infty} A_m B_m x^m,$$

where

$$B_m = \frac{1}{\alpha} + \frac{1}{\alpha+1} + \ldots + \frac{1}{\alpha+m-1} + \frac{1}{\beta} + \frac{1}{\beta+1} + \ldots + \frac{1}{\beta+m-1} - 2\left(1 + \frac{1}{2} + \frac{1}{3} + \ldots + \frac{1}{m}\right).$$

If we take $\phi_1 = (1-x)^{\gamma-\alpha-\beta}F(\gamma-\alpha, \gamma-\beta, \gamma, x)$, and denote as above by A_m the coefficient of the general term in the series $F(\gamma-\alpha, \gamma-\beta, \gamma, x)$, we shall have for $\psi_1(x)$ the new form

$$\psi_1(x) = \sum_{m=1}^{m=+\infty} A_m B_m x^m (1-x)^{1-\alpha-\beta},$$

where

$$B_m = \frac{1}{1-\alpha} + \frac{1}{2-\alpha} + \ldots + \frac{1}{m-\alpha} + \frac{1}{1-\beta} + \frac{1}{2-\beta} + \ldots + \frac{1}{m-\beta} - 2\left(1 + \frac{1}{2} + \frac{1}{3} + \ldots + \frac{1}{m}\right).$$

Each of the two expressions for ϕ_1 will give a different expression for ψ_1. Thus we find

$$\psi_1(x) = -\log(1-x)(1-x)^{-\alpha}F\left(\alpha, 1-\beta, 1, \frac{x}{x-1}\right) + (1-x)^{-\alpha}\sum_{m=1}^{m=\infty} A_m B_m \left(\frac{x}{x-1}\right)^m,$$

$$A_m = \frac{\alpha(\alpha+1)\ldots(\alpha+m-1)(1-\beta)(2-\beta)\ldots(m-\beta)}{(1 \cdot 2 \ldots m)^2},$$

$$B_m = \frac{1}{\alpha} + \frac{1}{\alpha+1} + \ldots + \frac{1}{\alpha+m-1} + \frac{1}{1-\beta} + \frac{1}{2-\beta} + \ldots$$
$$+ \frac{1}{m-\beta} - 2\left(1 + \frac{1}{2} + \frac{1}{3} + \ldots + \frac{1}{m}\right).$$

Also,

$$\psi_1(x) = -\log(1-x)(1-x)^{-\beta}F\left(\beta,\ 1-\alpha,\ 1,\ \frac{x}{x-1}\right)$$
$$+ (1-x)^{-\beta}\sum_{m=1}^{m=\infty} A_m B_m \left(\frac{x}{x-1}\right)^m,$$

$$A_m = \frac{\beta(\beta+1)\ldots(\beta+m-1)(1-\alpha)(2-\alpha)\ldots(m-\alpha)}{(1\,.\,2\ldots m)^2},$$

$$B_m = \frac{1}{\beta} + \frac{1}{\beta+1} + \ldots + \frac{1}{\beta+m-1} + \frac{1}{1-\alpha} + \frac{1}{2-\alpha} + \ldots$$
$$+ \frac{1}{m-\alpha} - 2\left(1 + \frac{1}{2} + \frac{1}{3} + \ldots + \frac{1}{m}\right).$$

Whichever be the expression adopted for $\psi_1(x)$, we will denote by Q the new integral, viz.,

$$Q = \phi_1 \log x + \psi_1(x).$$

This new integral, like the preceding ones, will be uniform throughout the plane provided the path described by the variable does not cut either of the lines $-\infty$ ——— 0, 1 ——— $+\infty$. The method employed for determining the new integral enables us also to find the linear relations connecting it and the integrals already known. To fix the ideas, suppose that the sum $\alpha+\beta$ is neither zero nor a negative integer, and give to γ a value differing but little from unity. We know the three integrals

$$F(\alpha,\beta,\gamma,x), \qquad x^{1-\gamma}F(\alpha+1-\gamma,\ \beta+1-\gamma,\ 2-\gamma,\ x),$$
$$F(\alpha,\ \beta,\ \alpha+\beta+1-\gamma,\ 1-x),$$

between which exists the relation

$$\phi_3 = \frac{\Gamma(\alpha+\beta+1-\gamma)\Gamma(1-\gamma)}{\Gamma(\alpha+1-\gamma)\Gamma(\beta+1-\gamma)}\phi_1 + \frac{\Gamma(\alpha+\beta+1-\gamma)\Gamma(\gamma-1)}{\Gamma(\alpha)\Gamma(\beta)}\phi_6.$$

Replace in this relation ϕ_6 by $\phi_1 + (1 - \gamma)Q_1$, and let γ tend to the value unity; ϕ_1 and ϕ_2 will reduce respectively to $F(\alpha, \beta, 1, x)$ and $F(\alpha, \beta, \alpha + \beta, 1 - x)$, and Q_1 will become equal to Q. We must show now what the values of the coefficients of ϕ_1 and Q_1 become under this hypothesis. We have

$$\phi_2 = \frac{\Gamma(\alpha + \beta + 1 - \gamma)}{\Gamma(\alpha)\Gamma(\beta)\Gamma(\alpha + 1 - \gamma)\Gamma(\beta + 1 - \gamma)}$$

$$[\Gamma(1 - \gamma)\Gamma(\alpha)\Gamma(\beta) + \Gamma(\gamma - 1)\Gamma(\alpha + 1 - \gamma)\Gamma(\beta + 1 - \gamma)]\phi_1$$

$$- \frac{\Gamma(\alpha + \beta + 1 - \gamma)}{\Gamma(\alpha)\Gamma(\beta)} \Gamma(\gamma)Q_1.$$

Let γ tend to unity; the coefficient of Q reduces to $-\dfrac{\Gamma(\alpha + \beta)}{\Gamma(\alpha)\Gamma(\beta)}$; the first factor of the coefficient of ϕ_1 has $\dfrac{\Gamma(\alpha + \beta)}{[\Gamma(\alpha)\Gamma(\beta)]^2}$ for limit; the second factor can be written in the form

$$\frac{\Gamma(2 - \gamma)\Gamma(\alpha)\Gamma(\beta) - \Gamma(\gamma)\Gamma(\alpha + 1 - \gamma)\Gamma(\beta + 1 - \gamma)}{1 - \gamma}.$$

The limit of this will be found by taking the derivative of the numerator for $\gamma = 1$, and is

$$2\Gamma'(1)\Gamma(\alpha)\Gamma(\beta) - \Gamma(\alpha)\Gamma'(\beta) - \Gamma(\beta)\Gamma'(\alpha).$$

We have, therefore,

(21) $F(\alpha, \beta, \alpha + \beta, 1 - x)$

$$= \frac{\Gamma(\alpha + \beta)}{\Gamma(\alpha)\Gamma(\beta)}\left[2\Gamma'(1) - \frac{\Gamma'(\beta)}{\Gamma(\beta)} - \frac{\Gamma'(\alpha)}{\Gamma(\alpha)}\right]F(\alpha, \beta, 1, x) - \frac{\Gamma(\alpha + \beta)}{\Gamma(\alpha)\Gamma(\beta)} Q.$$

Second Case.—Let us suppose γ greater than unity, and write $\gamma = 2m$ where m is an integer and may be zero. Let us examine the series $F(\alpha + 1 - \gamma, \beta + 1 - \gamma, 2 - \gamma, x)$ when γ, supposed at first to differ a little from $2 + m$, tends towards this value.

The first terms of the series are

$$1+\frac{(\alpha+1-\gamma)(\beta+1-\gamma)}{1\cdot(2-\gamma)}x$$

$$+\frac{(\alpha+1-\gamma)(\alpha+2-\gamma)(\beta+1-\gamma)(\beta+2-\gamma)}{1\cdot 2(2-\gamma)(3-\gamma)}x^2+\dots$$

$$+\left\{\frac{(\alpha+1-\gamma)(\alpha+2-\gamma)\dots(\alpha+m+1-\gamma)\times(\beta+1-\gamma)(\beta+2-\gamma)(\beta+m+1-\gamma)}{1\cdot 2\dots(m+1)(2-\gamma)(3-\gamma)\dots(m+2-\gamma)}\right\}x^{m+1}+\dots$$

As γ tends to the value $2+m$, the terms in x, x^2, . . . , x^m preserve finite values, while the coefficient of x^{m+1} becomes infinitely great unless some factor in the numerator vanishes. In order that this may happen it is necessary and sufficient that α or β take one of the values 1, 2, . . . , $m+1$. In this case all of the other terms will retain finite values for $\gamma=2+m$, and we shall still have an integral. The integral is uniform in the region of the origin, and admits this point as a pole; the integral is thus seen to be different from $F(\alpha, \beta, \gamma, x)$. Starting from the term in x^m, the aggregate of the remaining terms may manifestly be written

$$CF(\alpha, \beta, m+2, x)x^{m+1};$$

as to the preceding terms, they may be replaced by

$$C_1F\left(\alpha, \alpha+1-\gamma, \alpha+1-\beta, \frac{1}{x}\right)x^{m+1-\alpha},$$

so that after multiplying by $x^{-(m+1)}$ the integral becomes

$$C_1x^{-\alpha}F\left(\alpha, \alpha+1-\gamma, \alpha+1-\beta, \frac{1}{x}\right)+CF(\alpha, \beta, \gamma, x).$$

Take for example the differential equation

$$x(1-x)\frac{d^2y}{dx^2}+2\frac{dy}{dx}+2y=0,$$

which corresponds to

$$\alpha=1, \quad \beta=-2, \quad \gamma=2.$$

This equation admits

$$y_1 = \frac{1}{x}, \quad y_2 = 1 - x + \frac{x^2}{3}$$

as particular integrals; the general integral is then

$$y = \frac{C}{x} + C_1\left(1 - x + \frac{x^2}{3}\right).$$

It is to be remarked that when the circumstance above signalized presents itself, the origin is not a critical point for the general integral, but may be a pole. Removing this special case, we see that the series

$$F(\alpha + 1 - \gamma, \beta + 1 - \gamma, 2 - \gamma, x)$$

presents terms which increase indefinitely when γ tends to the value $2 + m$.

We will now seek the limit of this integral when γ tends to $2 + m$:

$$\begin{aligned}(m + 2 - \gamma)&F(\alpha + 1 - \gamma, \quad \beta + 1 - \gamma, \quad 2 - \gamma, \quad x) \\ &= (m + 2 - \gamma)\left[1 + \frac{(\alpha + 1 - \gamma)(\beta + 1 - \gamma)}{1 \, . \, (2 - \gamma)} x + \ldots\right].\end{aligned}$$

The first term in the [ ] to become infinite is

$$\frac{(\alpha + 1 - \gamma) \mid (\alpha + m + 1 - \gamma)(\beta + 1 - \gamma) \mid (\beta + m + 1 - \gamma)}{\underline{(m + 1)|}(2 - \gamma) \mid (m + 2 - \gamma)} x^{m+1}.$$

Letting now γ tend to the value $2 + m$, the terms in $1, x, x^2, \ldots, x^m$ vanish, and the remaining terms are finite; for example, the term in x^{m+1} becomes

$$\frac{(\alpha - m - 1) \mid (\alpha - 1)(\beta - m - 1) \mid (\beta - 1)}{\underline{m + 1|} \, . \, \underline{-m|}} x^{m+1},$$

where $\underline{-m|} = (-m)(-m + 1) \ldots (-1)$. The series tends therefore towards the value

$$x^{m+1} \frac{(\alpha - m - 1) \mid (\alpha - 1)(\beta - m + 1) \mid (\beta - 1)}{(-1)^m \, \underline{m|} \, . \, \underline{m + 1|}} F(\alpha, \beta, \gamma, x).$$

If, therefore, we consider the integral

$$F_1 = \frac{(-1)^m \underline{m|}\,.\,\underline{m+1|}\,(m+2-\gamma)}{(\alpha-m-1)\cdots(\alpha-1)(\beta-m-1)\cdots(\beta-1)} x^{1-\gamma} F(\alpha+1-\gamma, \beta+1-\gamma, 2-\gamma, x),$$

we see that for $\gamma = 2 + m$ this becomes $F(\alpha, \beta, \gamma, x)$. As before, a new integral can be found by seeking the limit for $\gamma = 2 + m$ of the expression

$$\frac{F_1 - F(\alpha, \beta, \gamma, x)}{2 + m - \gamma};$$

this is

$$\log x F(\alpha, \beta, \gamma, x) + \frac{dF}{d\alpha} + \frac{dF}{d\beta} + 2\frac{dF}{d\gamma},$$

in which $\gamma = 2 + m$. We can then express this new integral in terms of the already known integrals ϕ_1 and ϕ_2 by the same process as that above employed. We operate in exactly the same manner if $\gamma - \alpha - \beta$, or $\alpha - \beta$, is an integer.

18. We will now apply the general theory to the case of the equation

$$(1)' \qquad x(1-x)\frac{d^2y}{dx^2} + (1-2x)\frac{dy}{dx} - \frac{1}{4}y = 0,$$

which is obtained by making $\alpha = \beta = \frac{1}{2}$, $\gamma = 1$. This equation presents itself in the theory of the elliptic functions when we wish to define the complete integral of the first kind,

$$K = \int_0^1 \frac{dx}{\sqrt{(1-x^2)(1-k^2x^2)}},$$

as a function of the modulus for imaginary values of the latter. The equation has been studied by Fuchs* from this point of view, and studied directly by Tannery.†

The results which these writers have obtained are derived without difficulty from the general case. For this example we will employ Tannery's notation.

* Crelle, vol. 71, p. 91.

† *Annales de l'École normale supérieure*, 2e série, t. viii.

Equation (1)′ admits an integral, P say, which is uniform throughout the plane, provided the path of the variable never crosses the line 1 ——— $+\infty$. Inside the circle C_0 we have

$$P = F(\tfrac{1}{2}, \tfrac{1}{2}, 1, x) = 1 + \sum_{m=1}^{m=\infty} \left[\frac{1 \cdot 3 \cdot 5 \ldots (2m-1)}{2 \cdot 4 \cdot 6 \ldots 2m}\right]^2 x^m.$$

Let

$$\phi(x) = F(\tfrac{1}{2}, \tfrac{1}{2}, 1, x).$$

In the space E_0 we shall have, in like manner,

$$P = \frac{1}{\sqrt{1-x}} \phi\left(\frac{x}{x-1}\right).$$

Since $\gamma = 1$, we shall have a new integral containing a logarithm. Denote this integral by Q. This integral will be uniform throughout the plane if the path of the variable does not cut either of the lines $-\infty$ ——— 0, 1 ——— $+\infty$. Write

$$\psi(x) = \sum_{m=1}^{m=\infty} \left[\frac{1 \cdot 3 \ldots (2m-1)}{2 \cdot 4 \ldots 2m}\right]^2 B_m x^m,$$

$$B_m = 1 + \frac{1}{3} + \frac{1}{5} + \ldots + \frac{1}{2m-1} - \frac{1}{2} - \frac{1}{4} - \ldots - \frac{1}{2m};$$

then inside the circle C_0 we have

$$Q = 4\psi(x) + \phi(x) \log x,$$

and in the space E_0,

$$Q = \frac{1}{\sqrt{1-x}} \left\{ \phi\left(\frac{x}{x-1}\right) [\log x - \log(1-x)] + 4\psi\left(\frac{x}{x-1}\right) \right\}.$$

Equation (1)′ does not change form when we replace x by $1-x$, and it therefore admits two other integrals, P' and Q', which, subject to the same conditions as the preceding ones, are uniform throughout the plane. In the circle C_1 we shall have

$$P' = \phi(1-x),$$

$$Q' = \phi(1-x) \log(1-x) + 4\psi(1-x);$$

in the space E_1,

$$P' = \frac{1}{\sqrt{x}}\phi\left(\frac{x-1}{x}\right),$$

$$Q' = \frac{1}{\sqrt{x}}\left\{\phi\left(\frac{x-1}{x}\right)[\log(1-x) - \log x] + 4\psi\left(\frac{x-1}{x}\right)\right\}.$$

These four integrals suffice to express any integral whatever in the entire extent of the plane. In order to find the linear relations existing between these integrals, write in formula (21) $\alpha = \frac{1}{2}$, $\beta = \frac{1}{2}$, $\gamma = 1$; we find then

$$P' = \frac{2}{\pi}\left[\Gamma(1) - \frac{\Gamma'(\frac{1}{2})}{\Gamma(\frac{1}{2})}\right]P - \frac{1}{\pi}Q,$$

or

$$P' = \frac{4\log 2}{\pi}P - \frac{1}{\pi}Q;$$

also,

$$P = \frac{4\log 2}{\pi}P' - \frac{1}{\pi}Q'.$$

We have now

$$(22)\quad \begin{cases} P' = \dfrac{4\log 2}{\pi}P - \dfrac{1}{\pi}Q, \\[2ex] Q' = \dfrac{16\log^2 2 - \pi^2}{\pi}P - \dfrac{4\log 2}{\pi}Q, \\[2ex] P = \dfrac{4\log 2}{\pi}P' - \dfrac{1}{\pi}Q', \\[2ex] Q = \dfrac{16\log^2 2 - \pi^2}{\pi}P' - \dfrac{4\log 2}{\pi}Q'. \end{cases}$$

These are the relations found directly by Tannery. These formulæ suffice to *integrate* the equation (1)′, a fact which has been already observed in the general case. As an application, we will take the example treated in Tannery's memoir. Suppose we start from a point α (Fig. 6) very near the point $x = 2$, and in the upper half of the plane, and describe a closed path including in its interior the points 0 and 1.

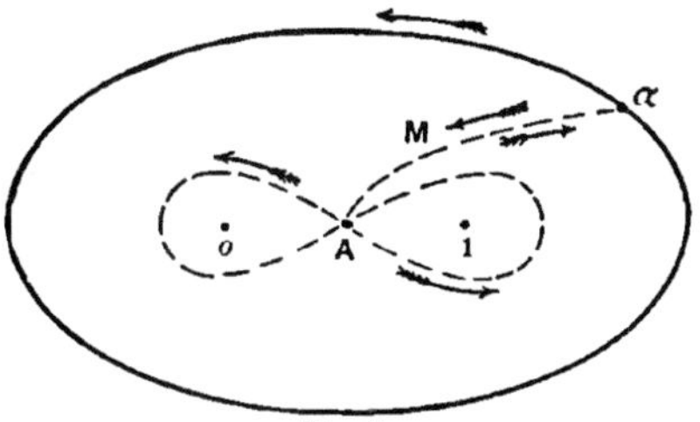

Fig. 6.

This path can be reduced to a path αMA, where A is a point of the line 0 —— 1, say the point $x = \frac{1}{2}$, followed by two loops described successively in the direct sense round the points $x = 0$ and $x = 1$, and then the path $AM\alpha$. Let us start from the point α with the integral P'; we will of course arrive at the point A with the same integral. Now to find the change in P' when we go round the loop $x = 0$, we replace P' by its value,

$$P' = \frac{4 \log 2}{\pi} P - \frac{1}{\pi} Q.$$

After describing the loop 0, P returns to its original value, but Q changes into $Q + 2\pi i P$, and therefore P' becomes

$$P' - 2iP.$$

Now to see how this integral behaves on going round the loop 1, we replace P by

$$P = \frac{4 \log 2}{\pi} P' - \frac{1}{\pi} Q';$$

P' will not change, but Q' becomes $Q' + 2\pi i P'$, so that $P' - 2iP$ changes into

$$-3P' - 2iP.$$

We return to α, therefore, with the integral

$$-\frac{3\pi + 8i \log 2}{\pi} P' + \frac{2i}{\pi} Q'.$$

If we start from α with Q', we arrive at A with

$$\frac{16\log^2 2-\pi^2}{\pi}P-\frac{4\log 2}{\pi}Q.$$

After the loop 0 the integral will be represented by

$$\frac{16\log^2 2-\pi^2}{\pi}P-\frac{4\log 2}{\pi}Q-8i\log 2\,.\,P;$$

or, what comes to the same thing, by

$$Q'-8i\log 2\,.\,P;$$

or, again, by

$$Q'\frac{\pi+8i\log 2}{\pi}-\frac{32i\log^2 2}{\pi}P'.$$

After the loop 1 we shall have

$$Q'\frac{\pi+8i\log 2}{\pi}+\frac{2i\pi(\pi+8i\log 2)-32i\log^2 2}{\pi}P';$$

that is,

$$\frac{2i(\pi+4i\log 2)^2}{\pi}P'+\frac{\pi+8i\log 2}{\pi}Q'.$$

We will of course return to α with this same integral.

The equation (1)′ admits also two other integrals, susceptible of development in series, analogous to the preceding ones. First we have the integral

$$\frac{1}{\sqrt{-x}}\phi\left(\frac{1}{x}\right)=\frac{1}{\sqrt{1-x}}\phi\left(\frac{1}{1-x}\right),$$

which is uniform throughout the plane provided the path of the variable does not cut the line 0 —— $+\infty$. Denote this integral by P''. To find another integral, suppose first $\gamma=1$, $\alpha=\frac{1}{2}$, and let β differ a little from the value $\frac{1}{2}$. The differential equation will admit the two integrals

$$\frac{1}{\sqrt{-x}}F\left(\frac{1}{2},\frac{1}{2},\frac{3}{2}-\beta,\frac{1}{x}\right),\quad(-x)^{-\beta}F\left(\beta,\beta,\beta+\frac{1}{2},\frac{1}{x}\right).$$

The expression

$$\frac{\dfrac{1}{\sqrt{-x}}F\left(\frac{1}{2}, \frac{1}{2}, \frac{3}{2}-\beta, \frac{1}{x}\right) - (-x)^{-\beta}F\left(\beta, \beta, \beta+\frac{1}{2}, \frac{1}{x}\right)}{\frac{1}{2}-\beta}$$

will also be an integral. The limit of this expression when $\beta = \frac{1}{2}$ will be a new integral Q'', viz.,

$$Q'' = -\log(-x)\phi\left(\frac{1}{x}\right)\frac{1}{\sqrt{-x}} + \frac{4}{\sqrt{-x}}\psi\left(\frac{1}{x}\right);$$

also,

$$Q'' = \frac{1}{\sqrt{1-x}}\left[-\phi\left(\frac{1}{1-x}\right)\log(1-x) + 4\psi\left(\frac{1}{1-x}\right)\right].$$

Q'' is uniform throughout the plane under the same conditions as P''.

For all values of x in the upper half of the plane we have the relations

$$(23)\qquad \begin{cases} P = \dfrac{4\log 2}{\pi}P'' - \dfrac{1}{\pi}Q'', \\ P' = \dfrac{\pi - 4i\log 2}{\pi}P'' + \dfrac{i}{\pi}Q''; \end{cases}$$

for values of x in the lower half of the plane we have

$$(24)\qquad \begin{cases} P = \dfrac{4\log 2}{\pi}P'' - \dfrac{1}{\pi}Q'', \\ P' = \dfrac{\pi + 4i\log 2}{\pi}P'' - \dfrac{i}{\pi}Q''. \end{cases}$$

These formulæ are established, like (21), by starting with the relations (13) and (17) which exist in the general case.

Remark.—Formulæ (23) and (24) do not appear to be in accord with the formulæ given by Tannery (*loc. cit.* p. 188). This arises from the fact that the functions P'' and Q'' here used are not precisely the same as the corresponding ones employed by Tannery.

Instead of the system of integrals P'' and Q'' consider the following system:

$$P_1'' = \frac{1}{\sqrt{x}}\,\phi\left(\frac{1}{x}\right),$$

$$Q_1'' = \frac{1}{\sqrt{x}}\left[-\phi\left(\frac{1}{x}\right)\log x + 4\,\psi\left(\frac{1}{x}\right)\right],$$

where the argument of x is supposed to lie between $-\pi$ and $+\pi$. We find easily for the upper half of the plane

$$P'' = iP_1'', \qquad Q'' = iQ_1'' - \pi P_1'';$$

and substituting these values in the second of (23), we have

$$P' = \frac{4\log 2}{\pi}P_1'' - \frac{1}{\pi}Q_1'',$$

which is identical with Tannery's relation.

19. The formulæ established in what precedes enable us to determine whether or not the differential equation admits an algebraic integral and whether its general integral is or is not algebraic. This question has been treated by Schwarz (Crelle, t. 73, p. 292), who, by employing the differential equation of the third order satisfied by the ratio of two particular integrals of equation (1) and by the aid of Riemann's surfaces, was led to a question in spherical geometry where regular polyhedra presented themselves. Later, Klein considered the more general case of a linear differential equation of the second order with rational coefficients. In the following, Schwarz's results are arrived at by quite elementary considertitions.

If equation (1) possesses a single algebraic integral, this integral must reproduce itself to a constant factor près, when we turn round a critical point. In the domain of the point $x = 0$ it will be represented by one of the integrals ϕ_1, ϕ_5, and in the domain of $x = 1$ it will coincide, to a factor près, with either ϕ_2 or ϕ_6. Referring now to the relations existing among these four integrals, we see that we cannot have an integral of the required kind unless one of the numbers α, β, $\gamma - \alpha$, $\gamma - \beta$ is an integer. In each of these cases one of the hypergeometric series which express the general integral will have a limited number of terms. Suppose for example $\gamma - \alpha = -m$; equation (1) will then have the integral

$$(1 - x)^{\gamma-\alpha-\beta}P,$$

P being an integral function of x. In order that this integral shall be algebraic, it is further necessary that β be a real and rational number. If equation (1) admits more than one algebraic integral, the general integral will be algebraic. We propose now to find all the cases where this is true. We can exclude from our research the cases where one of the numbers γ, $\gamma - \alpha - \beta$ is an integer; in fact, we have seen that in these cases there exists a logarithm in the complete integral in the region of one of the critical points; as an exceptional case this logarithm may disappear, but then the point considered is no longer a branch-point for the general integral, and it will suffice to examine how an integral behaves in the region of the other critical point in order to be certain of its nature. We see further that the three numbers α, β, γ must be real and rational; if this were not so, then in the domain of the critical points there would exist an integral which could take an infinite number of values. This being granted, suppose now that y_1 and y_2 are two linearly independent particular integrals, uniform throughout the plane provided the path of the variable does not cross either of the lines $-\infty$ ——— 0, 1 ——— $+\infty$. Start from an arbitrary point A of the plane with an integral $cy_1 + c'y_2$ and describe an arbitrary closed path which does not pass through either of the points $x = 0$ or $x = 1$. We return to the point of departure with an integral $Cy_1 + C'y_2$. If this integral is an algebraic one the coefficients C and C' admit only a limited number of values whatever be the path described. If this is true, then, whatever be the original coefficients c and c', the general integral is algebraic. Consider in fact the logarithmic derivative of the preceding integral, viz.,

$$\frac{Cy_1' + C'y_2'}{Cy_1 + C'y_2} = \frac{y_1' + \dfrac{C'}{C}y_2'}{y_1 + \dfrac{C'}{C}y_2}.$$

The ratio $\frac{C'}{C}$ admitting only a limited number of values, the same is true of this logarithmic derivative, and as besides it admits of no other singular points than critical points and poles, it follows that it is an algebraic function of x.

This being true for any integral, let z_1 and z_2 denote two integrals of (1); we have then the relation

$$z_2 z_1' - z_1 z_2' = Hx^{-\gamma}(1 - x)^{\gamma-\alpha-\beta-1},$$

which can also be written

$$z_1 z_2\left(\frac{z_1'}{z_1} - \frac{z_2'}{z_2}\right) = Hx^{-\gamma}(1 - x)^{\gamma-\alpha-\beta-1},$$

$$z_1 z_2 = \frac{Hx^{-\gamma}(1 - x)^{\gamma-\alpha-\beta-1}}{\dfrac{z_1'}{z_1} - \dfrac{z_2'}{z_2}}.$$

The product of any two integrals, z_1, z_2, is then an algebraic function of x. If z_1 and z_2 are two integrals, then z_1 and $z_1 + z_2$ will also be two integrals. The product $z_1^2 + z_1 z_2$, and consequently z_1^2, will be an algebraic function of the variable.

Every closed path starting from and returning to a point A can be reduced to a series of loops described round the points $x = 0$ and $x = 1$; we are therefore led to examine the behavior of the preceding ratio when we describe one of these loops. We will take for y_1 and y_2 the integrals ϕ_1 and ϕ_2; then starting from a point A with the integral $C\phi_1 + C'\phi_2$ and describing the loop $x = 0$ in the direct sense, we shall, as we have seen above, return to A with the integral $C_1\phi_1 + C_1'\phi_2$, where

$$C_1 = C + C'\frac{\Gamma(\alpha+\beta+1-\gamma)\Gamma(1-\gamma)}{\Gamma(\alpha+1-\gamma)\Gamma(\beta+1-\gamma)}(1 - e^{2\pi i(1-\gamma)}),$$

$$C_1' = C'e^{2\pi i(1-\gamma)}.$$

Let ρ be the initial value of $\frac{C}{C'}$, and ρ' its final value; then

$$\rho' = \rho e^{-2\pi i(1-\gamma)} + \frac{\Gamma(\alpha+\beta+1-\gamma)\Gamma(1-\gamma)}{\Gamma(\alpha+1-\gamma)\Gamma(\beta+1-\gamma)}\left(\frac{1-e^{2\pi i(1-\gamma)}}{e^{2\pi i(1-\gamma)}}\right).$$

Making

$$\frac{\Gamma(\alpha+\beta+1-\gamma)\Gamma(1-\gamma)}{\Gamma(\alpha+1-\gamma)\Gamma(\beta+1-\gamma)} = -a,$$

$$e^{-2\pi i(1-\gamma)} = K,$$

this last formula becomes

$$(A) \qquad \rho' - a = K(\rho - a).$$

If the loop were described in the negative sense we should have

$$(A)' \qquad \rho' - a = \frac{1}{K}(\rho - a).$$

Now let the variable describe the positive loop around $x = 1$; the integral $C\phi_1 + C'\phi_2$ changes into $C_1\phi_1 + C_1'\phi_2$, where

$$C_1 = Ce^{2\pi i(\gamma-\alpha-\beta)},$$

$$C_1' = C' + C\frac{\Gamma(\gamma)\Gamma(\gamma-\alpha-\beta)}{\Gamma(\gamma-\alpha)\Gamma(\gamma-\beta)}(1 - e^{2\pi i(\gamma-\alpha-\beta)}).$$

Denoting again by ρ and ρ' the initial and final values of the ratio $\frac{C}{C'}$, we have

$$\frac{1}{\rho'} = \frac{1}{\rho}e^{-2\pi i(\gamma-\alpha-\beta)} + \frac{\Gamma(\gamma)\Gamma(\gamma-\alpha-\beta)}{\Gamma(\gamma-\alpha)\Gamma(\gamma-\beta)}\left(\frac{1-e^{2\pi i(\gamma-\alpha-\beta)}}{e^{2\pi i(\gamma-\alpha-\beta)}}\right).$$

Making

$$\frac{\Gamma(\gamma)\Gamma(\gamma-\alpha-\beta)}{\Gamma(\gamma-\alpha)\Gamma(\gamma-\beta)} = -b,$$

$$e^{-2\pi i(\gamma-\alpha-\beta)} = K',$$

we have, from the last equation,

$$(B) \qquad \left(\frac{1}{\rho'} - b\right) = K'\left(\frac{1}{\rho} - b\right);$$

if the loop is described in the negative sense, we have

$$(B)' \qquad \left(\frac{1}{\rho'} - b\right) = \frac{1}{K'}\left(\frac{1}{\rho} - b\right).$$

The problem which we seek to solve can be stated as follows:

Having given a series of quantities such that each is deduced from the preceding by one of the formulæ (A), $(A)'$, (B), $(B)'$, *in what cases can we arrive at a limited number of different quantities, the order in which these formulæ are successively applied being perfectly arbitrary?*

The geometrical method seems to conduct most readily to the sought result.

(20) Draw in the plane two rectangular axes $O\xi$ and $O\eta$. Represent as usual the quantity $\rho = \xi + i\eta$ by the point whose co-ordinates are ξ and η; let A and B denote the points (on the axis $O\xi$) which represent the quantities a and b respectively. Let M be the point representing the initial value of ρ, and M' be the point representing the value of ρ' obtained by letting the variable travel round the loop $x = 0$ in the direct sense:

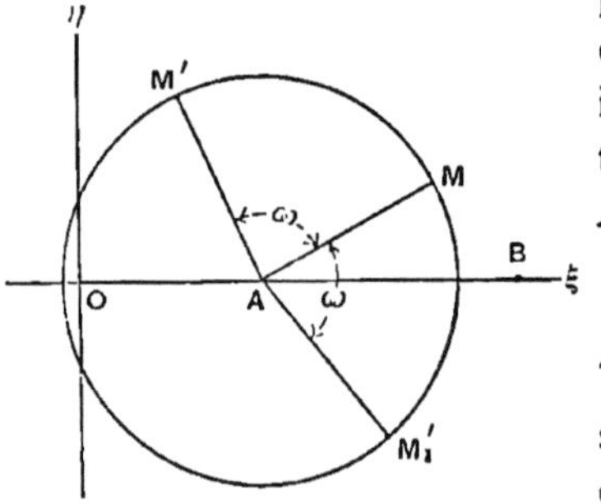

Fig. 7.

$$\rho' - a = K(\rho - a).$$

The quantity $\rho - a$ will be represented by the extremity of a segment, drawn through the origin, equal and parallel to the segment AM; if this segment is turned round the point A through an angle ω ($\omega = 2\pi(\gamma - 1)$), we obtain the segment AM', of which the extremity M' represents the quantity ρ'. If the loop were described in the negative sense, we should come to the point M_1' obtained, as before, by making the segment AM turn round the point A through an angle ω, the turning, however, being in the negative sense.

When the variable describes a series of loops round O, the different values of ρ will be represented by the vertices of a regular polygon inscribed in the circle of radius AM, one of these vertices being the point M.

Suppose now (Fig. 8) that the variable describes a loop round the critical point $x = 1$. Let M be the point denoting the initial value of ρ; $\frac{1}{\rho}$ will be represented by the point M_1, $\frac{1}{\rho} - b$ by the extremity of a segment drawn through the origin equal and parallel to BM_1, and $K'\left(\frac{1}{\rho} - b\right)$ by the extremity of a segment equal and parallel to the segment BM_1', which is simply the segment BM_1 turned round B through an angle $\omega' = 2\pi(\alpha + \beta - \gamma)$.

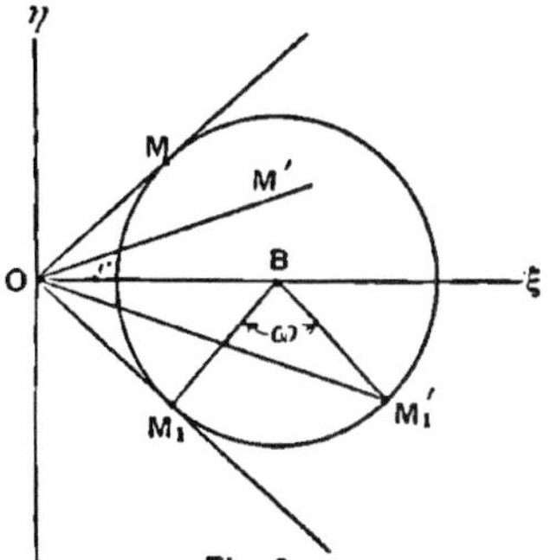

Fig. 8.

Consequently the point M_1' represents the quantity $\frac{1}{\rho'}$, and the point M' represents the quantity ρ'. Let the variable describe several successive loops round the point $x = 1$; the point $\frac{1}{\rho}$ will then coincide successively with vertices of a regular polygon inscribed in the circle whose centre is at B; the point ρ itself will be on the circumference of a circle transformed from the preceding by reciprocal radii vectores having the origin for the pole of the transformation, and unity for the modulus. If we apply this transformation to all of the circles having B as the centre, we evidently obtain a system of circles passing through two fixed imaginary points. It is easy to see what are the point-circles of this system: first is the origin, which corresponds to the circle of infinite radius with centre at B, and then the point C, $= \frac{1}{b}$, corresponding to the circle of radius zero. All the circles of this system are conjugate with respect to these two points, which are two double points of the homographic transformation

$$\frac{1}{\rho'} - b = K'\left(\frac{1}{\rho} - b\right).$$

Before going farther we shall demonstrate the following geometrical theorem :

Having given a circumference and n points, A, B, C, . . . , L, upon this circumference which are so disposed that on making an inversion, taking for the pole a point O of the plane, the corresponding points A′, B′, . . . , L′ are the vertices of a regular polygon, if O′ is the conjugate point to O with respect to the circumference, all the points of the circumference described upon OO′ as a diameter and perpendicular to the plane of the figure will possess the same property as the point O.

Fig. 9.

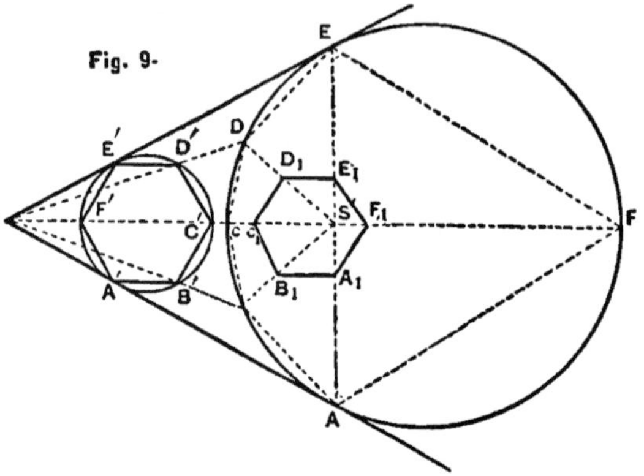

Let S be a point of this circumference; then, from an elementary property of the inversion,

$$A'B' = \frac{AB}{OA \cdot OB} P,$$

$$A_1B_1 = \frac{AB}{SA \cdot SB} P_1;$$

consequently

$$\frac{A_1B_1}{A'B'} = \left(\frac{P_1}{P}\right) \times \frac{OA}{SA} \times \frac{OB}{SB}.$$

Each of the ratios $\frac{OA}{SA}$, $\frac{OB}{SB}$ is constant, and so the same is true of the ratio $\frac{A_1B_1}{A'B'}$. If the polygon $A'B'C'D'E'F'$ has its sides equal, the same will be true of the polygon $A_1B_1C_1D_1E_1F_1$. Q. E. D.

Returning now to the proposed question, consider the circumference described upon OC (Fig. 10) as diameter in a plane perpendicular to the plane of the figure. Let S and S' be the points of intersection of this circumference with the perpendicular to the plane of the figure through the point A. Conceive a sphere described upon SS' as diameter, and make a projection upon this sphere, taking S as the point of sight.

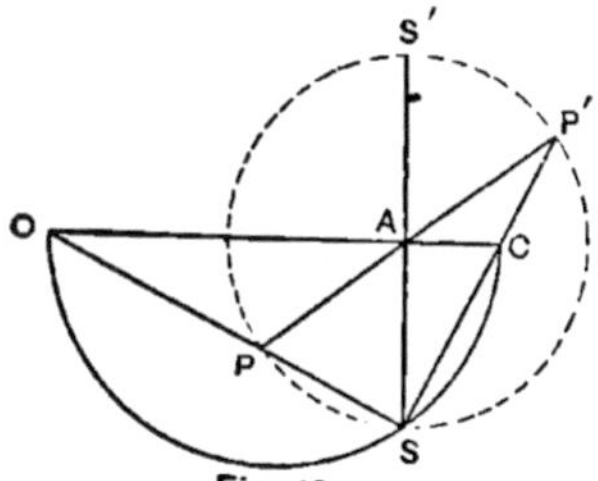

Fig. 10.

All circles having their centres at A are projected upon circles having for poles the points S and S'. As to circles conjugate with respect to two points O and C, they are projected upon circles having PP' for axis. It follows from what precedes that having given upon the surface of the sphere a point m representing the value of ρ, we shall find the point m' representing a new value of ρ by turning the point m through an angle ω round SS' or through an angle ω' round PP'. This is evident for the axis SS'; as to the axis PP' it suffices to remark that, having given two points M, M' in the plane representing two values of ρ, of which one is deduced from the other by one of the formulæ (B), $(B)'$, if we make an inversion with the point O for pole, the angle M_1BM_1' is equal to ω', and, from the preceding theorem, the angle mpm' must have the same value, p denoting the foot of the perpendicular let fall from m upon the axis PP'.

We can now replace the above enunciation of our problem by the following:

Having given two diameters SS' and PP' in a sphere and a series of points upon the surface which succeed one another by a law such that we pass from any one to the following one by turning through an angle ω round SS' or an angle ω' round PP', in what cases can we arrive at a limited number of points, the order in which the constructions are applied being perfectly arbitrary?

All depends, evidently, upon the order of symmetry of the axes SS' and PP' and on the angle, V, between them. Let

$$\gamma = \frac{p}{q}, \quad \gamma - \alpha - \beta = \frac{p'}{q'}.$$

The fractions $\frac{p}{q}$ and $\frac{p'}{q'}$ being irreducible, SS' will be an axis of symmetry of order q, and PP' one of order q'. As to the angle V, we have from the preceding figure

$$V = 2AOS,$$

$$\cos AOS = \frac{OA}{OS} = \frac{OA}{\sqrt{OA \times OC}} = \sqrt{\frac{OA}{OC}},$$

$$\cos V = 2\frac{OA}{OC} - 1 = 2ab - 1.$$

Replacing a and b by their values, we get

$$\cos V = 2\frac{\Gamma(\alpha+\beta+1-\gamma)\Gamma(1-\gamma)}{\Gamma(\alpha+1-\gamma)\Gamma(\beta+1-\gamma)} \times \frac{\Gamma(\gamma-\alpha-\beta)\Gamma(\gamma)}{\Gamma(\gamma-\alpha)\Gamma(\gamma-\beta)} - 1,$$

or

$$\cos V = 2\frac{\sin(\gamma-\alpha)\pi \sin(\gamma-\beta)\pi}{\sin\gamma\pi \sin(\gamma-\alpha-\beta)\pi} - 1.$$

Remark.—When we diminish or increase the value of one of the quantities α, β, γ by any number of units, q and q' do not change, neither does cos V. We may then suppose $1-\gamma$ and $\gamma-\alpha-\beta$ to lie between 0 and 1, and can thus conduct $\alpha-\beta$ to lie between -1 and $+1$; but as there is symmetry between the elements α and β, we may suppose $\alpha-\beta$ to lie between 0 and 1, a supposition which will be adopted in what follows.

(21) The geometrical representation upon the sphere is not possible unless the point A lies between the points O and C, or, what comes to the same thing, unless the product

$$ab = \frac{\sin(\gamma-\alpha)\pi \sin(\gamma-\beta)\pi}{\sin\gamma\pi \sin(\gamma-\alpha-\beta)\pi}$$

is comprised between 0 and 1 In the contrary case we can demonstrate directly that the ratio ρ can take an infinite number of values in each point of the plane. The establishment of this last statement is based upon the following remarks:

I. There exists a circle having its centre at A, and conjugate with respect to the segment OC. If the point M which represents the initial value of ρ is on this circle, all the other points derived from it will likewise be on the circle. If the point M is outside or inside the circle, the derived points will themselves be all outside or all inside.

II. Supposing that we have taken the point A as origin, let us denote by z and z' the quantities which are represented by the points M and M' in the new system.

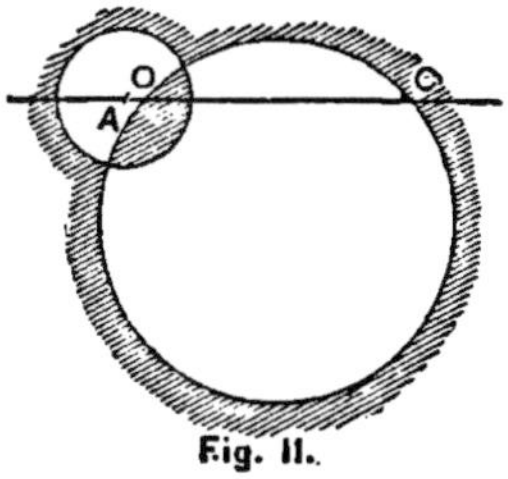

Fig. II.

The first transformation is

$$z' = kz;$$

in the second transformation z' is connected with z by a relation

$$z' = \frac{az + b}{cz + d},$$

where

$$a = \alpha + i\beta, \qquad c = \alpha' + i\beta',$$
$$b = \gamma + i\delta, \qquad d = \gamma' + i\delta'.$$

Let us seek the locus of the points z such that mod z = mod z'; we have

$$\text{mod } z' = \frac{\text{mod }(az + b)}{\text{mod }(cz + d)},$$

$$az + b = (\alpha x - \beta y + \gamma) + i(\beta x + \alpha y + \delta),$$

$$\text{mod }(az + b) = \sqrt{(\alpha^2 + \beta^2)(x^2 + y^2) + mx + ny + p},$$

$$\text{mod }(cz + d) = \sqrt{(\alpha'^2 + \beta'^2)(x^2 + y^2) + m'x + n'y + p'}.$$

The equation of the sought locus is

$$(\alpha^2 + \beta^2)(x^2 + y^2) + mx + ny + p$$
$$= (\alpha'^2 + \beta'^2)(x^2 + y^2)^2 + (m'x + n'y + p')(x^2 + y^2).$$

This is a quartic curve having the circular points at infinity for double points. The circle with centre A (Fig. 11), conjugate to the segment OC, is evidently part of the locus, and as the points O and C also belong to it the quartic breaks up into two circles, viz., the circle (A) and another circle through the two points O and C.

If the point M, which represents the value of z, is on the exterior or the interior of the two circles, we have

$$\text{mod } z' < \text{mod } z;$$

but if z' is outside one circle and inside the other, we have

$$\text{mod } z' > \text{mod } z.$$

This granted, let M be the point representing the inital value of ρ, and suppose M outside the two circles (A), (A)'.

Applying to the point M the second transformation as often as it gives us different points, we shall arrive at a certain number of points upon a circle conjugate to the segment OC. Suppose M_1 that one of these last points which is nearest A; M_1 must therefore lie inside the circle A', and we have $AM_1 < AM$. Apply now the first transformation to the point M_1 so as to obtain a point M_2 outside the circle A'; we shall have $AM_2 = AM_1$. Apply the first transformation to the point M_2; we shall find as before a point M_3 nearer A than the point M_2, etc. We shall find by continuing this process a series of points $M, M_1, M_2, M_3, \ldots$ following one another according to such a law that we shall never have a new point coinciding with one of the preceding ones, and this will go on indefinitely. The ratio ρ takes therefore an infinite number of values.

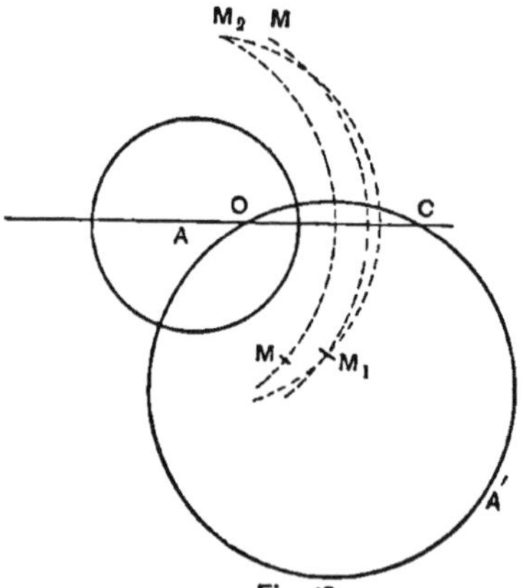

Fig. 12.

(22) The question in spherical geometry to which we have been led, which, as we shall presently see, is identical with the question to

which Schwarz was led, has been solved by Steiner.* We can also deduce the solution from a memoir by Jordan† upon Eulerian polyhedra.

There are two particular cases where the solution is immediately perceived.

1. If we have two axes of binary symmetry, the angle between them must be commensurable with 2π.

2. If we have one axis of binary symmetry, and one axis of order of symmetry $= K$ perpendicular to the first, we shall always end with a limited number of points.

Discarding these particular cases, let us suppose that we have two axes of symmetry PP', SS', one of which, PP', is of order $K(K \geqq 3)$. The repetitions by symmetry of the axis PP' will be axes of symmetry of order K of the figure formed by the symmetric repetitions of a point M of the sphere. These repetitions ought to be limited in number. This comes to studying the figure formed by taking P itself as the point of departure. We can demonstrate without difficulty that, if we arrive at a limited number of points, these points are the vertices of a regular polyhedron. The axes PP', SS' ought then to be the axes of symmetry of a regular polyhedron having one of its vertices at P. It is evident that the plane of the two axes will be a plane of symmetry for the polyhedron. This necessary condition is also sufficient. Conceive in fact this regular polyhedron, and consider the spherical polygons on the circumscribed sphere which correspond to its different faces. The repetitions by symmetry of one of these polygons will always lead to an analogous polygon; as, further, one of these polygons can only coincide with itself in a limited number of ways, it follows that the symmetrical representations of a point will be limited in number.

Example I.—Let $\gamma = \frac{1}{2}$, $\alpha + \beta = 0$. We have two axes of binary symmetry,

$$\cos V = 2 \cos \pi\alpha \cos \pi\beta - 1 = \cos 2\alpha\pi.$$

It is sufficient that α be commensurable. This can be verified directly. The differential equation is

$$x(1-x)\frac{d^2y}{dx^2} + \left(\frac{1}{2} - x\right)\frac{dy}{dx} + \alpha^2 y = 0,$$

* Crelle, vol. xviii. p. 295.

† Ibid. vol. lxvi. p. 22.

or

$$2x(1-x)\frac{d^2y}{dx^2}\frac{dy}{dx}+(1-2x)\left(\frac{dy}{dx}\right)^2+2\alpha^2y\frac{dy}{dx}=0.$$

From this we have

$$d\left[x(1-x)\left(\frac{dy}{dx}\right)^2\right]=\alpha^2d(c^2-y^2),$$

$$\frac{dy}{\sqrt{c^2-y^2}}=\frac{\alpha dx}{\sqrt{x(1-x)}},$$

$$\sin^{-1}\frac{y}{c}=\alpha\sin^{-1}\sqrt{x}+h.$$

If α is commensurable, we deduce from this an algebraic relation between x and y.

Example II.—Schwarz's example:

$$\gamma=\tfrac{2}{3},\quad \alpha=-\tfrac{1}{12},\quad \beta=\tfrac{1}{4},$$

$$\gamma-\alpha=\tfrac{3}{4},\quad \gamma-\beta=\tfrac{5}{12},\quad \gamma-\alpha-\beta=\tfrac{1}{2}.$$

There is one axis of binary and one of ternary symmetry; their angle is given by

$$\cos V=\frac{2\sin\frac{3\pi}{4}\sin\frac{5\pi}{12}}{\sin\frac{2\pi}{3}}-1=\frac{1}{\sqrt{3}}.$$

This is exactly the angle which in a regular tetrahedron is included between the altitude and the line joining the middle points of two opposite edges: the integral is then algebraic.

(23) The preceding results can be placed in a little different form and so bring better into evidence their identity with Schwarz's results.

Suppose, as explained above, that we have conducted the three numbers $1-\gamma$, $\gamma-\alpha-\beta$, $\alpha-\beta$ to values lying between 0 and 1, and denote by λ, μ, ν the three positive numbers, less than unity, so obtained:

$$\lambda \equiv 1 - \gamma,$$
$$\mu \equiv \gamma - \alpha - \beta,$$
$$\nu \equiv \alpha - \beta.$$

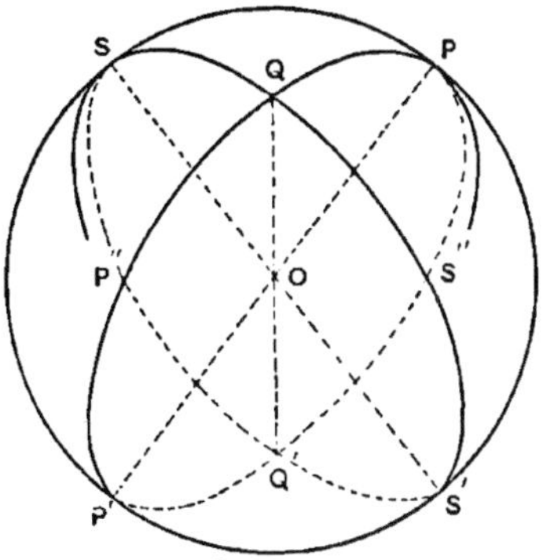

Fig. 13.

We will now find out for what systems of values of these numbers the integrals are algebraic. As before, let SS' and PP' (Fig. 13) be the two axes of symmetry.

Construct a triangle having for base SP, and for angles $PSQ = (1-\gamma)\pi$, $SPQ = (\gamma - \alpha - \beta)\pi$.

For the angle Q we have

$$\cos Q = \sin P \sin S \cos (SP) - \cos P \cos S$$
$$= \sin \gamma\pi \sin (\gamma - \alpha - \beta)\pi \left[\frac{2 \sin (\gamma - \alpha)\pi \sin (\gamma - \beta)\pi}{\sin \gamma\pi \sin (\gamma - \alpha - \beta)\pi} - 1\right]$$
$$+ \cos \gamma\pi \cos (\gamma - \alpha - \beta)\pi,$$

or, finally,

$$\cos Q = \cos (\alpha - \beta)\pi.$$

The angle Q is comprised between 0 and π, and further we have supposed $\alpha - \beta$ comprised between 0 and 1;

$$\therefore \quad Q = (\alpha - \beta)\pi.$$

The three angles of SPQ are then

$$S = \lambda\pi, \quad P = \mu\pi, \quad Q = \nu\pi.$$

What conditions should this triangle satisfy? In the particular case where $\gamma = \frac{1}{2}$, $\alpha + \beta = 0$, S and P are right angles; the point Q is then one of the poles of the great circle $SPS'P'$. If we construct a double pyramid having for vertices the poles Q and Q', and for base the regular polygon of which S and P are two vertices, the three planes of the trihedron SPQ will be three planes of symmetry for this double pyramid. If the axis SS' is a binary axis, and if the arc SP is a quadrant, the three planes of the trihedron will still be three planes of symmetry for a double pyramid whose vertices are at the points P and P', and whose base is in the plane $SQS'Q'$.

These singular cases being examined, let us see what takes place in the general case, supposing there exists a regular polyhedron having SS' and PP' as axes of symmetry and having one vertex at P. The three planes OSP, OPQ, OQS will be three planes of symmetry for this polyhedron; it is evident from what precedes that OSP is a plane of symmetry. The point P'', symmetrical to P with respect to the plane OSQ, is one of the vertices of the polyhedron; as we can start from P'' instead of P to find the remaining vertices, it is clear that the plane OSQ is a plane of symmetry. So for OQP; viz., the point S'' is an extremity of an axis of symmetry of the same order as the axis SS'; if we replace the axis OS by the axis OS'', we shall obviously form the same figure, and consequently OQP is a plane of symmetry.

Thus the three faces of the trihedron $OPQS$ are three planes of symmetry of a regular polyhedron. The converse is easily demonstrated; in fact, every body admitting these three planes of symmetry OPS, OSQ, OQP will also admit OSP'' as a plane of symmetry. If we take the body which with respect to the plane OSQ is symmetric to the first body, then take the body symmetric with respect to the plane OSP'' of this new body, this last will coincide with the first; but these two operations are equivalent to turning the body through an angle $PSP'' = 2(1-\gamma)\pi$ round OS; $\therefore$ OS is an axis of symmetry for the body. Finally, then, in order that the general integral shall be algebraic, it is necessary and sufficient that the three planes of the trihedron $OSPQ$ shall be the three planes of symmetry of a double pyramid or of a regular polyhedron. It is necessary first that the three angles $\lambda\pi$, $\mu\pi$, $\nu\pi$ shall be the angles of a spherical triangle; this requires

$$\lambda + \mu + \nu > 1,$$
$$\lambda + 1 > \mu + \nu,$$
$$\mu + 1 > \nu + \lambda,$$
$$\nu + 1 > \lambda + \mu.$$

These conditions are equivalent to the above-found condition that

$$\frac{\sin(\gamma-\alpha)\pi \sin(\gamma-\beta)\pi}{\sin\gamma\pi \sin(\gamma-\alpha-\beta)\pi}$$

shall lie between 0 and 1.

It evidently comes to the same thing whether we consider the triangle PQS or one of the triangles PQS', $QP'S'$, QSP'; the angles of these triangles have the following values:

PQS	$\lambda\pi$	$\mu\pi$	$\nu\pi$,
PQS'	$\lambda\pi$	$(1-\mu)\pi$	$(1-\nu)\pi$,
$P'QS$	$(1-\lambda)\pi$	$\mu\pi$	$(1-\nu)\pi$,
$P'QS'$	$(1-\lambda)\pi$	$(1-\mu)\pi$	$\nu\pi$.

We will choose the triangle for which the sum of the angles is the least. Let $\lambda'\pi$, $\mu'\pi$, $\nu'\pi$ be the angles of this triangle, and let λ'', μ'', ν'' denote the numbers λ', μ', ν' arranged in descending order of magnitude. The following table gives the systems of values of λ'', μ'', ν'' in order that the general integral shall be algebraic:

	λ''	μ''	ν''	
I, . . .	$\frac{1}{2}$	$\frac{1}{2}$	“	Double pyramids.
II, . . .	$\frac{1}{2}$	$\frac{1}{3}$	$\frac{1}{3}$	Tetrahedron.
III, . . .	$\frac{2}{3}$	$\frac{1}{3}$	$\frac{1}{3}$	
IV, . . .	$\frac{1}{2}$	$\frac{1}{3}$	$\frac{1}{4}$	Cube and octahedron.
V, . . .	$\frac{2}{3}$	$\frac{1}{4}$	$\frac{1}{4}$	
VI, . . .	$\frac{1}{2}$	$\frac{1}{3}$	$\frac{1}{5}$	Dodecahedron and icosahedron.
VII, . . .	$\frac{2}{5}$	$\frac{1}{3}$	$\frac{1}{3}$	
VIII, . . .	$\frac{2}{3}$	$\frac{1}{5}$	$\frac{1}{5}$	
IX, . . .	$\frac{1}{2}$	$\frac{2}{5}$	$\frac{1}{5}$	
X, . . .	$\frac{3}{5}$	$\frac{1}{3}$	$\frac{1}{5}$	
XI, . . .	$\frac{2}{5}$	$\frac{2}{5}$	$\frac{2}{5}$	
XII, . . .	$\frac{2}{3}$	$\frac{1}{3}$	$\frac{1}{5}$	
XIII, . . .	$\frac{4}{5}$	$\frac{1}{5}$	$\frac{1}{5}$	
XIV, . . .	$\frac{1}{2}$	$\frac{2}{5}$	$\frac{1}{3}$	
XV, . . .	$\frac{3}{5}$	$\frac{2}{5}$	$\frac{1}{3}$	

(24) We have neglected the intermediate case where the circle

described upon OC as diameter is tangent to the perpendicular SS'; that is to say, the case where the two homographic transformations,

$$(A)\qquad \rho' - a = K(\rho - a),$$

$$(B)\qquad \frac{1}{\rho'} - b = K'\left(\frac{1}{\rho} - b\right),$$

have a common double point. The double points of the first are $\xi = a$ and $\xi = \infty$; those of the second are $\xi = 0$, $\xi = \frac{1}{b}$. If we refer to the values of a and b, which may become zero, but not infinite, we can see that there are three different ways in which the two homographic transformations may have a common double point.

(1) $a = 0$; this will happen when one of the numbers $\alpha + 1 - \gamma$, $\beta + 1 - \gamma$ is zero or a negative integer.

(2) $b = 0$; this will happen if $\gamma - \alpha$ or $\gamma - \beta$ is zero or a negative integer.

(3) $ab = 1$; this condition gives

$$\sin(\gamma - \alpha)\pi \sin(\gamma - \beta)\pi = \sin\gamma\pi \sin(\gamma - \alpha - \beta)\pi,$$

or

$$\cos(\alpha - \beta)\pi = \cos(\alpha + \beta)\pi,$$

from which

$$(\alpha - \beta) \pm (\alpha + \beta) = 2m;$$

one of the numbers α, β must then be an integer. In order to see what takes place in each of these cases, suppose the common double point at infinity; the homographic transformations will be defined by

$$z' - z_0 = K(z - z_0),$$
$$z' - z_1 = K'(z - z_1),$$

the points z_0 and z_1 being the other two double points. These give rise to a simple geometrical construction. Having given in the plane a point M representing z, we shall find the point M' representing z' by turning the radius z_0M through an angle ω round z_0, or the radius z_1M through an angle ω' round z_1. Starting from any point of the plane, it is clear that we can apply the constructions successively in such an order that we shall never arrive at a point already found; for example, we can arrange so that in following the process the radius z_0M never decreases. The common double point is a case of exception, as it will always coincide with itself. There will then be a particular integral whose logarithmic derivative has only a single value in each point of the plane; this derivative is therefore a rational fraction. Under the adopted hypothesis, *i.e.*, where the numbers α, β, γ are real and rational, the corresponding particular integral will be an algebraic function. We see, further, that there is no other such integral. We must, however, remark that this will cease to be true if the other two double points are the same; in this case *all* the integrals will be algebraic. This case arises when we have simultaneously $a = 0$, $b = 0$.

For example, consider the differential equation

$$x(1-x)\frac{d^2y}{dx^2} + (\tfrac{3}{2} - 2x)\frac{dy}{dx} - \tfrac{1}{4}y = 0,$$

where

$$\alpha = \beta = \tfrac{1}{2}, \quad \gamma = \tfrac{3}{2}.$$

The equation admits a particular algebraic integral $y = \frac{1}{\sqrt{x}}$, but the general integral,

$$y = \frac{C}{\sqrt{x}} + \frac{C' \sin^{-1}\sqrt{x}}{\sqrt{x}},$$

is transcendental. On the contrary, take the equation

$$x(1-x)\frac{d^2y}{dx^2} + \left(\frac{1}{3} - \frac{2x}{3}\right)\frac{dy}{dx} + \frac{20}{9}y = 0,$$

where

$$\alpha = -\tfrac{5}{3}, \quad \beta = \tfrac{4}{3}, \quad \gamma = \tfrac{1}{3}.$$

We have simultaneously $a = 0$, $b = 0$; the equation admits the two particular integrals,

$$x^{\frac{2}{3}}\left(1 - \frac{6x}{5}\right), \quad (1-x)^{\frac{2}{3}}\left[1 - \frac{6(1-x)}{5}\right],$$

and so the general integral is algebraic.

(25) In what precedes we have supposed the two integrals ϕ_1 and ϕ_2 to be distinct; *i.e.*, neither of the numbers α, β is zero or a negative integer. If this circumstance does, however, arise, we can replace ϕ_2 by another integral, for example ϕ_6, and operating as above, we shall be led to consider two homographic transformations. These two transformations will have a common double point, since equation (1) admits as an integral an entire function whose logarithmic derivative is a rational fraction. If the other two double points are different (which will generally be the case), there will not be any other algebraic integral. But if these two double points coincide, the general integral will be algebraic.

In order that this may be so, one of the elements α, β must be zero or a negative integer, and the other a positive integer.

Thus the differential equation

$$x(1-x)\frac{d^2y}{dx^2} + \frac{1}{2}\frac{dy}{dx} + 2y = 0,$$

where

$$\alpha = -2, \quad \beta = 1, \quad \gamma = \tfrac{1}{2},$$

admits two particular algebraic integrals, and therefore the general integral will be algebraic, viz.:

$$y = C\left(1 - 4x + \frac{8x^2}{3}\right) + C' x^{\frac{1}{2}}(1-x)^{\frac{3}{2}}.$$

PART SECOND.

(I) The memoirs of Gauss and Kummer on the hypergeometric series contain a great many formulæ not found among those given above, and which only exist when the constants α, β, γ satisfy certain conditions. The general type of these formulæ is

$$x^{-p}(1-x)^{-q}F(\alpha, \beta, \gamma, x) = t^{p'}(1-t)^{q'}F(\alpha', \beta', \gamma', t),$$

where t is an algebraic function of x. The function

$$x^{p}(1-x)^{q}t^{p'}(1-t)^{q'}F(\alpha', \beta', \gamma', t)$$

is then an integral of the differential equation

$$(1) \qquad x(1-x)\frac{d^2y}{dx^2}+[\gamma-(\alpha+\beta+1)x]\frac{dy}{dx}-\alpha\beta y=0.$$

We are thus led to seek the cases in which equation (1) admits integrals of the above form. Such is, very nearly at least, the path followed by Kummer. Kummer, however, indicates no means of finding all the cases where such integrals exist. This question we propose to treat in what follows.

(2) We will adopt a slightly different point of view from that of Kummer. Let us denote, as Riemann does, by $P(x)$ a non-uniform function of x possessing the following properties:

(1) It admits in the entire extent of the plane, or of the sphere, only three critical points, viz., $x=0$, $x=1$, $x=\infty$; it is holomorphic in any region of the plane having a simple contour which does not contain either of the points $x=0$, $x=1$.

(2) Between any three branches, P', P'', P''', of the function there exists a linear homogeneous relation

$$C'P'+C''P''+C'''P'''=0$$

with constant coefficients.

(3) Each branch of the function is finite for $x=0$, $x=1$, and also for $x=\infty$ when we multiply it by a proper power of x or $1-x$.

Riemann has shown that certain branches of the function P can be expressed by products such as $x^{-p}(1-x)^{-q}F(\alpha, \beta, \gamma, x)$. In the light of the more recent analysis, this can be demonstrated by a simpler method than that adopted by Riemann. It results, in fact, from a theorem given by Tannery,* viz.:

The different branches of the function P are integrals of a linear differential equation of the second order having uniform coefficients, and having no other critical points than the points 0, 1, ∞; *further, all these integrals are regular in the region of a critical point.*

This differential equation, as Fuchs has shown, is of the form

$$(2) \quad x^2(1-x)^2\frac{d^2P}{dx^2}+[l-(l+m)x]x(1-x)\frac{dP}{dx} \\ +(Ax^2+Bx+C)P=0.$$

* *Annales de l'École normale supérieure*, 2e série, t. iv. p. 130.

We pass from (1) to (2) by writing

$$y = x^p (1 - x)^q P;$$

A, B, C, l, m are given by the formulæ

$$(3)\quad \begin{cases} l = 2p + \gamma,\ m = 2q + \alpha + \beta + 1 - \gamma,\ A = (p+q+\alpha)(p+q+\beta), \\ C = p(p - 1 + \gamma),\ A + B + C = q(q + \alpha + \beta - \gamma). \end{cases}$$

Conversely, we pass from (2) to (1) by making

$$P = x^{-p}(1 - x)^{-q} y;$$

$\alpha, \beta, \gamma, p, q$ will now be determined by aid of equations (3) in terms of A, B, C, l, m. Riemann's theorem can be established in the same way.

[In order to completely define the function P it is necessary to add the following condition, viz.: if P', P'' are two linearly distinct branches, the determinant

$$D = \begin{vmatrix} P', & P'' \\ \dfrac{dP'}{dx}, & \dfrac{dP''}{dx} \end{vmatrix}$$

must be different from zero for every point of the plane other than the points $x = 0$, $x = 1$. If, in fact, this determinant vanished for $x = a$, the point a would be an *apparent singular point* for the differential equation. Consider, for example, the integrals of equation (1); it is clear that they satisfy the conditions which serve to define the function P. At first sight, the same is true of the products obtained by multiplying each of these integrals by the factor $(x - 2)$; nevertheless, these new functions satisfy the differential equation

$$\begin{aligned} & x(1 - x)(x - 2)^2 \frac{d^2z}{dx^2} \\ & + \{[\gamma - (\alpha + \beta + 1)x](x - 2) - 2x(1 - x)\}(x - 2)\frac{dz}{dx} \\ & + \{2x(1 - x) + [(\alpha + \beta + 1)x - \gamma](x - 2) - \alpha\beta(x - 2)^2\} z = 0, \end{aligned}$$

which is not comprised in the form (2); but the determinant D is evidently zero for the point $x = 2$.]

Remark that having given a system of values for A, B, C, l, m, there result four systems of values for $\alpha, \beta, \gamma, p, q$; equation (2), and consequently equation (1), admits then four integrals of the form $x^{-p}(1-x)^{-q}F(\alpha, \beta, \gamma, x)$—a well-known result.

Let t be a new variable given by the equation $x = \phi(t)$. If, when in equation (2) we change the variable by the relation $x = \phi(t)$, the function P satisfies the same relations as relatively to x, we ought to find a new differential equation (4) analogous to (2), viz.:

$$(4)\quad t^2(1-t)^2\frac{d^2P}{dt^2} + [l' - (l'+m')t]t(1-t)\frac{dP}{dt} + (A't^2 + B't + C')P = 0.$$

It is easy to see that the problem treated by Kummer is herein contained; if, in fact, equation (1) admits the integral

$$x^p(1-x)^q t^{p'}(1-t)^{q'}F(\alpha', \beta', \gamma', t),$$

equation (2) will admit the integral

$$t^{p'}(1-t)^{q'}F(\alpha', \beta', \gamma', t).$$

If then in equation (2) we change the variable by the relation $x = \phi(t)$, we must obtain an equation of the form (4).

The problem to be studied may now be stated as follows:

Required to determine for what values of the constants A, B, C, l, m, there exist transformations such as $x = \phi(t)$, for which equation (2) does not change its form, A, B, C, l, m being simply replaced by the new constants A', B', C', l', m'.

When we have found the conditions which the constants A, B, C, l, m must satisfy, equations (3) will give us the conditions to be satisfied by the elements α, β, γ themselves.

(3) In equation (2), making the change of variable $x = \phi(t)$, and writing $x' = \frac{dx}{dt}$, $x'' = \frac{d^2x}{dt^2}$, we have

$$\frac{dP}{dx} = \frac{1}{x'}\frac{dP}{dt},$$

$$\frac{d^2P}{dx^2} = \frac{1}{x'^3}\left[x'\frac{d^2P}{dt^2} - x''\frac{dP}{dt}\right],$$

and the new equation is

$$\frac{x^2(1-x)^2}{x'^2}\frac{d^2P}{dt^2}+\left[\frac{l-(l+m)x}{x'}x(1-x)-\frac{x''x^2(1-x)^2}{x'^3}\right]\frac{dP}{dt}$$
$$+(Ax^2+Bx+C)P=0,$$

where x, x', x'' are to be replaced by their values as functions of t.

In order that the new equation shall have the desired form, we must have

$$(5)\qquad \frac{x^2(1-x)^2}{x'^2(Ax^2+Bx+C)}=\frac{t^2(1-t)^2}{A't^2+B't+C'},$$

$$(6)\qquad \frac{l-(l+m)x}{x(1-x)}x'-\frac{x''}{x'}=\frac{l'-(l'+m')t}{t(1-t)}.$$

This last equation can be integrated once, since it can be written in the form

$$\frac{d}{dt}\log\left[x^l(1-x)^m\right]-\frac{d}{dt}\log x'=\frac{d}{dt}\log\left[t^{l'}(1-t)^{m'}\right].$$

We can then replace equations (5) and (6) by a system of two differential equations of the first order, viz.:

$$(5)\qquad \frac{\sqrt{Ax^2+Bx+C}}{x(1-x)}dx=\frac{\sqrt{A't^2+B't+C'}}{t(1-t)}dt,$$

$$(7)\qquad \frac{dx}{x^l(1-x)^m}=\frac{Kdt}{t^{l'}(1-t)^{m'}}.$$

The constants A', B', C', l', m', K are arbitrary, and we have now to see in what cases we can determine them so that equations (5) and (7) shall have a common integral.

(4) The following five transformations are readily seen to exist in all cases, whatever be the values of A, B, C, l, m; viz.:

$$x=1-t,\quad x=\frac{1}{t},\quad x=\frac{1}{1-t},\quad x=\frac{t}{t-1},\quad x=\frac{t-1}{t}.$$

This result is also evident if we refer to the definition of the function P. If a multiple-valued function satisfies the required conditions when we take x as the variable, it is clear that it will also satisfy these conditions when we take as variable one of the quantities $1 - x$, $\frac{1}{x}$, $\frac{1}{1-x}$, $\frac{x}{x-1}$, $\frac{x-1}{x}$; for, the values of any one of these quantities for $x = 0, 1, \infty$ will also be $0, 1, \infty$ taken in a certain order. From these considerations arises a very simple method for finding Kummer's twenty-four integrals, but we will not take up that point here.*

We can show now that there can exist no transformation of the first order between the two variables other than the above-mentioned ones. Consider, in fact, the transformation

$$x = \frac{at + b}{ct + d}.$$

If the values of t corresponding to the values $x = 0$, $x = 1$, $x = \infty$ are also $0, 1, \infty$, taken in a certain order, the transformation will obviously be one of the preceding ones. Suppose on the contrary that for $x = 0$, for example, t takes a finite value t_1 different from 0 and from 1. Let x (Fig. 14) describe a small loop round the origin; the

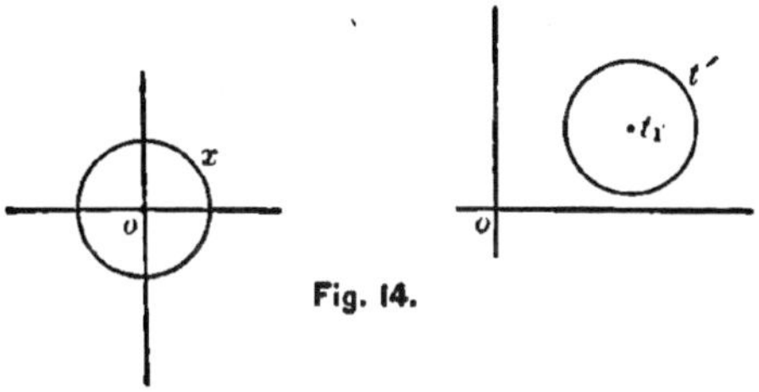

Fig. 14.

point t will describe a small loop surrounding t_1. After describing such a path, any integral whatever of (4) will return to its initial value: the point $x = 0$ itself can then not be a critical point for equation (2), which is contrary to hypothesis.

We see further that, if for proper values of A, B, C, l, m, we can make the transformation $x = \phi(t)$, we can also, for the same values of the constants, make the five other transformations

$$x=\phi(1-t),\quad x=\phi\left(\frac{1}{t}\right),\quad x=\phi\left(\frac{1}{1-t}\right),\quad x=\phi\left(\frac{t}{t-1}\right),\quad x=\phi\left(\frac{t-1}{t}\right).$$

* This method has been given in Chapter VI.—AUTHOR.

From the form of equations (5) and (7) we can deduce still further consequences. If for the values of A, B, C, l, m we can make two different transformations $x = \phi(t)$, $x = \psi(t)$, then for properly chosen values of the constants we can make the transformation $\phi(x) = \psi(t)$. Thus, from a transformation $x = \phi(t)$ we would be able to deduce all those which would be obtained either in replacing x by $1 - x$, $\frac{1}{x}$, $\frac{1}{1-x}$, $\frac{x}{x-1}$, $\frac{x-1}{x}$, or in replacing t by $1 - t$, $\frac{1}{t}$, $\frac{1}{1-t}$, $\frac{t}{t-1}$, $\frac{t-1}{t}$, or in making the two transformations simultaneously, making thus in all thirty-six transformations, not all of them, however, being different. We will determine, at the same time, all of these transformations, together with the inverse transformations.

(5) If equations (5) and (7) admit a common integral, this integral will also satisfy the equation obtained by dividing the equations (5) and (7) member by member, viz.,

$$\sqrt{Ax^2+Bx+C}\,.\,x^{l-1}(1-x)^{m-1} = K\sqrt{A't^2+B't+C'}\,.\,t^{l'-1}(1-t)^{m'-1}.$$

Taking the logarithmic derivative, we have

$$dx\left[\frac{2Ax+B}{2(Ax^2+Bx+C)} + \frac{l-1}{x} + \frac{m-1}{x-1}\right]$$
$$= dt\left[\frac{2A't+B'}{2(A't^2+B't+C')} + \frac{l'-1}{t} + \frac{m'-1}{t-1}\right].$$

Replacing, from equation (5), dx and dt by the quantities to which they are proportional, and squaring, we have, finally,

$$(8)\qquad \frac{\{(2Ax+B)x(x-1) + 2(Ax^2+Bx+C)[(l-1)(x-1)+(m-1)x]\}^2}{(Ax^2+Bx+C)^3}$$
$$= \frac{\{(2A't+B')t(t-1) + 2(A't^2+B't+C')[(l'-1)(t-1)+(m'-1)t]\}^2}{(A't^2+B't+C')^3}.$$

If this relation is not an identity, we see that x and t are connected by an equation of at most the sixth degree in each of the variables.

Equation (8) will be an identity if the two members reduce to zero or to equal constants.

1. In order that each member of (8) shall be identically zero it is necessary and sufficient that

$$\sqrt{Ax^2 + Bx + C} \,.\, x^{l-1}(1 - x)^{m-1}$$

and

$$\sqrt{A't^2 + B't + C'} \,.\, t^{l'-1}(1 - t)^{m'-1}$$

reduce to constants; $Ax^2 + Bx + C = 0$ should then admit of no other roots than 0 and 1, and the same should be true of

$$A't^2 + B't + C' = 0.$$

All of the possible combinations are given in the following table:

1st case.	2d case.	3d case.	4th case.	5th case.	6th case.
$A+B=0$,	$A=0$,	$A=0$,	$A=C$,	$A=0$,	$B=0$,
$C=0$,	$C=0$,	$B+C=0$,	$B-2A=0$,	$B=0$,	$C=0$,
$l=m=\frac{1}{2}$,	$l=\frac{1}{2}$, $m=1$,	$l=1$, $m=\frac{1}{2}$,	$l=1$, $m=0$,	$l=m=1$,	$l=0$, $m=1$,
$\lambda=\pm\frac{1}{2}$,	$\lambda=\pm\frac{1}{2}$,	$\mu=\pm\frac{1}{2}$,	$\mu=\pm 1$,	$\nu=\pm 1$,	$\lambda=\pm 1$,
$\mu=\pm\frac{1}{2}$,	$\nu=\pm\frac{1}{2}$,	$\nu=\pm\frac{1}{2}$,	$\lambda=\pm\nu$,	$\lambda=\pm\mu$,	$\mu=\pm\nu$.

In the last two horizontal lines we have given the corresponding conditions to be satisfied by the elements α, β, γ, where

$$\lambda = 1 - \gamma, \quad \mu = \gamma - \alpha - \beta, \quad \nu = \beta - \alpha.$$

The conditions for A', B', C', l', m' will be identical in form with the above. In each of these cases equations (5) and (7) reduce to the same, and so an infinite number of changes of the variable can be made which will leave (2) unaltered in form. For example, take $A + B = 0$, $C = 0$, $l = m = \frac{1}{2}$; it is sufficient now to take for x

an integral of one of the following differential equations, in which K is arbitrary:

$$\frac{dx}{\sqrt{x(1-x)}} = \frac{Kdt}{\sqrt{t(1-t)}}, \qquad \frac{dx}{\sqrt{x(1-x)}} = \frac{Kdt}{(1-t)\sqrt{t}},$$

$$\frac{dx}{\sqrt{x(1-x)}} = \frac{Kdt}{t\sqrt{1-t}}, \qquad \frac{dx}{\sqrt{x(1-x)}} = \frac{Kdt}{t},$$

$$\frac{dx}{\sqrt{x(1-x)}} = \frac{Kdt}{1-t}, \qquad \frac{dx}{\sqrt{x(1-x)}} = \frac{Kdt}{t(1-t)}.$$

It is easy to assure one's self that, in all these cases, the general integral of (4), and consequently of (2), is expressed by means of elementary functions. Remark first that the second and third cases can be conducted to the first by changing respectively x into $\frac{x}{x-1}$ or into $\frac{1}{x}$; so also the fifth and sixth cases can be led to the fourth by changing x into $\frac{1}{1-x}$ or into $1-x$. It remains then to consider the two equations

$$x(1-x)\frac{d^2P}{dx^2} + (\tfrac{1}{2} - x)\frac{dP}{dx} + AP = 0,$$

$$x^2\frac{d^2P}{dx^2} + x\frac{dP}{dx} + AP = 0.$$

The general integral of the first of these is

$$\sin^{-1}\frac{P}{C_1} = \sqrt{A}\sin^{-1}\sqrt{x} + C_2,$$

and of the second is

$$P = C_1x^r + C_2x^{-r},$$

where $r = \sqrt{A}$. In particular, if $A = 0$, the last integral is

$$P = C_1 + C_2\log x.$$

2. It may happen that the two members of (8) reduce to constants; for this it is necessary that Ax^2+Bx+C admits x or $x-1$ as double factors or that it reduces to a constant. The following are all the possible combinations:

1st case.	2d case.	3d case.
$B=0$,	$A=C$,	$A=0$,
$C=0$,	$B-2A=0$	$B=0$,
$l=0$,	$m=0$,	$l+m=2$.

We find for α, β, γ the same conditions as in the last three examined cases.

Suppose, for example, $A=C$, $B-2A=0$, $m=0$; we can make any one of the transformations

$$x=K_1t^n,\quad x=K_1(1-t)^n,\quad x=K_1\left(\frac{t}{t-1}\right)^n,$$

where K_1 and n are arbitrary. It is easy to see now that the general integral of (2) is of the form

$$P=C_1x^r+C_2x^{r'}.$$

Summing up: When (8) reduces to an identity we can effect an infinite number of changes of the variable in (2) so as to leave the form of this equation unaltered, and can so deduce an infinity of relations included in the type of Kummer's formulæ; but it is to be noticed that all of these relations exist between functions which are expressed by means of the elementary functions, exponential, circular, or logarithmic.

(6) Discarding these particular cases, suppose that (8) does not reduce to an identity. If now there exists an integral common to (5) and (7), x will be an algebraic function of t defined by an equation of, at most, the sixth degree in x and in t. It is therefore only among algebraic functions that we must seek the functions which will permit us to transform (2) in the desired manner. We shall determine first the rational transformations, and after will show that all the other transformations can be conducted to these.

Let $x=\dfrac{P}{S}$ be a rational transformation; R and S are two polynomials of degrees at most $= 6$, and at least one of them is of a degree higher than unity. Among the values of t which correspond to the values 0, 1, ∞ of x there will be at least one which is different from 0, 1, ∞. Suppose, for example, that for $x = 0$ we have $t = 0$, and for $x = \infty$, $t = 1$; then R will be equal to Kt^r, and S to $K'(1 - t)^s$. The values of t which correspond to $x = 1$ will then be the roots of the equation $Kt^r - K'(1 - t)^s = 0$. One, at least, of the numbers r, s being greater than unity, the left-hand member of this equation cannot reduce to a constant; further, the equation does not admit either 0 or 1 as a root. There are therefore finite values of t which are neither 0 nor 1 and which correspond to $x = 1$. Suppose then that for $x = 0$, for example, t takes a value t_1 which is neither zero nor unity. The point $x = 0$ will be a critical point for that value of t which becomes t_1. In fact, let the variable x describe a small loop round the origin (Fig. 14): if t were a holomorphic function of x in the region of the point $x = 0$, it would return to its original value after having described a small loop round t_1, and we could then conclude, as above, that the origin would not be a critical point for equation (2), which is not the case. It is therefore necessary that several values of t become equal at t_1 when $x = 0$. Suppose n to be the number of these values; then

$$x = (t - t_1)^n f(t),$$

$f(t)$ being a rational function of t which is neither zero nor infinite for $t = t_1$. In the same way we shall have

$$\frac{dx}{dt} = (t - t_1)^{n-1} f_1(t),$$

$f_1(t)$ possessing, relatively to the point t_1, the same properties as $f(t)$. From equation (5) we have, further,

$$\frac{dx}{dt} = \frac{x(1-x)}{t(1-t)} \frac{\sqrt{A't^2 + B't + C'}}{\sqrt{Ax^2 + Bx + C}}.$$

The quotient $\dfrac{1-x}{t(1-t)}$ is different from zero for $t = t_1$; if neither

$Ax^2 + Bx + C$ nor $A't^2 + B't + C'$ were zero for $x = 0$ and $t = t_1$, we should have

$$\frac{dx}{dt} = (t - t_1)^n \psi(t),$$

$\psi(t)$ being a uniform function of t in the region of the point t_1 and different from zero for $t = t_1$, which is impossible. If $A't^2 + B't + C'$ were zero for $t = t_1$ and C *not* zero, we should have

$$\frac{dx}{dt} = (t - t_1)^{n+\frac{1}{2}} \psi(t), \quad \text{or} \quad \frac{dx}{dt} = (t - t_1)^{n+1} \psi(t),$$

$\psi(t)$ having the same meaning as above. This is also impossible. We therefore conclude that: *If for $x = 0$, t takes a finite value t_1 which is neither zero nor unity, it is necessary that the constant C be zero.*

(7) It remains to find the values which the integer n may take. Several cases are to be examined, as follows:

First hypothesis:

$$A't_1^2 + B't_1 + C' \gtrless 0, \quad B \gtrless 0;$$

then

$$\frac{dx}{dt} = \sqrt{x}\, \psi(t) = (t - t_1)^{\frac{n}{2}} \psi_1(t);$$

also,

$$\frac{dx}{dt} = (t - t_1)^{n-1} f_1(t).$$

We must have, therefore,

$$\frac{n}{2} = n - 1, \quad \text{whence } n = 2.$$

Second hypothesis:

$$A't_1^2 + B't_1 + C' \gtrless 0, \quad B = 0;$$

consequently

$$\frac{dx}{dt} = \psi(t), \quad \text{whence } n = 1.$$

This hypothesis is therefore to be rejected.

Third hypothesis:

$$A't_1^2 + B't_1 + C' = 0, \quad 2A't_1 + B' \lessgtr 0, \quad B \lessgtr 0;$$

it follows that

$$\frac{dx}{dt} = (t - t_1)^{\frac{n+1}{2}} \psi(t), \quad \text{whence} \frac{n+1}{2} = n - 1, \; n = 3.$$

Fourth hypothesis:

$$A't_1^2 + B't_1 + C' = 0, \quad 2A't_1 + B' \lessgtr 0, \quad B = 0;$$

which gives

$$\frac{dx}{dt} = (t - t_1)^{\frac{1}{2}} \psi(t), \quad \text{whence } n = \tfrac{3}{2}.$$

The transformation being supposed rational, this hypothesis is to be rejected.

Fifth hypothesis:

$$A't_1^2 + B't_1 + C' = 0, \quad 2A't_1 + B' = 0, \quad B \lessgtr 0;$$

therefore

$$\frac{dx}{dt} = (t - t_1)^{\frac{n}{2}+1} \psi(t),$$

from which we derive

$$\frac{n}{2} + 1 = n - 1, \quad n = 4.$$

Sixth hypothesis:

$$A't_1^2 + B't_1 + C' = 0, \quad 2A't_1 + B' = 0, \quad B = 0;$$

then

$$\frac{dx}{dt} = (t - t_1)\psi(t), \quad \text{whence } n = 2.$$

The only admissible values for the integer n are consequently, 2, 3, 4.

We can find the corresponding values of l by remarking that the ratio $\frac{1}{x^l}\frac{dx}{dt}$ must be finite for $x = 0$, $t = t_1$; this requires that we have $l = \frac{n-1}{n}$.

(8) We can prove in like manner that if t takes a value different from 0, 1, ∞ for $x = 1$ or for $x = \infty$, we must have $A + B + C = 0$ in the first case, and $A = 0$ in the second. To verify this it is only necessary to change x into $1 - x$ for the first case, and into $\frac{1}{x}$ for the second. Discarding the special case where we have simultaneously $A = B = C = 0$, we find that, if we give x the values 0, 1, ∞, there is at least one of these values for which none of the values of t is different from 0, 1, ∞. Suppose the case $x = \infty$, and suppose further that for $x = 0$, t takes a finite value t_1 which is neither 0 nor 1; we shall always have $C = 0$. There are two cases to distinguish according as among the values of t corresponding to $x = 1$ there are or are not values which are neither 0, 1, ∞.

The transformation will have one of the following forms:

(a) $$x = R^n t^r (1 - t)^s,$$

(b) $$x = R^n t^r,$$

(c) $$x = R^n (1 - t)^s,$$

(d) $$x = R^n,$$

(e) $$x = \frac{R^n t^r}{(1 - t)^s},$$

(f) $$x = \frac{R^n (1 - t)^s}{t^r},$$

(g) $$x = \frac{R^n}{t^r},$$

(h) $$x = \frac{R^n}{(1 - t)^s},$$

(i) $$x = \frac{R^n}{t^r (1 - t)^s},$$

R denoting an entire function which has no double factor, and which is not zero for either $t = 0$ or $t = 1$; n is one of the numbers 2, 3, 4, and r and s are positive integers. We can reduce the number of these transformations. Thus, we can suppress the forms (c), (f),

(h), which by changing t into $1 - t$ are conducted to the forms (b), (c), (g). Take now the transformation (g) and let r' be the degree of the numerator. If $r' \leqq r$, then, by changing t into $\frac{1}{t}$, $\frac{R^n}{t^r}$ changes into $\frac{R_1{}^n}{t^{r'-r}}$ or into $R_1{}^n t^{r-r'}$, and we are thus conducted to the form (b) or the form (d). If $r' > r$, then on changing t into $\frac{t-1}{t}$, $\frac{R^n}{t^r}$ changes into $\frac{R_1{}^n}{t^{r'-r}(1-t)^r}$, and so we are conducted to the form (i). Take now the form (c) and let s' be the degree of the numerator. If $s' \leqq s$, changing t into $\frac{t}{t-1}$ changes $\frac{R^n t^r}{(1-t)^s}$ into

$$\frac{R_1{}^n t^r}{(1-t)^{s'-s}} = R_1{}^n t^r (1-t)^{s-s'},$$

and we are conducted to the form (a) or to the form (b). If $s' > s$, changing t into $\frac{1}{t}$ changes $\frac{R^n t^r}{(1-t)^s}$ into $\frac{R_1{}^n}{t^{s'-s}(1-t)^s}$, and so the transformation is conducted to the form (i). We are therefore confined to considering the four following forms:

(a) $$x = R^n t^r (1-t)^s,$$

(b) $$x = R^n t^r,$$

(d) $$x = R^n,$$

(i) $$x = \frac{R^n}{t^r(1-t)^s}.$$

(9) We now proceed to calculate the unknown coefficients which enter into these transformations. Suppose first that for $x = 1$ there is no value of t different from $0, 1, \infty$. If the transformation has the form (a), the values of t, for $x = 1$, are given by the equation

$$R^n t^r (1-t)^s - 1 = 0.$$

It is clear that this equation does not admit either $t = 0$ or $t = 1$ as roots, and that the first member does not reduce to a constant. Take in the same way the form (b); the equation $R^n t^r - 1 = 0$ must only admit of the root $t = 1$; consequently

$$R^n t^r = 1 + H(1 - t)^s.$$

Now the first member of this admits multiple factors, while the second does not; the equality is therefore impossible. If we take the form (d), the equation $R^n - 1 = 0$ ought to admit no other root than 0 and 1; this requires that n be equal to 2, and that R be of the first degree. It is easy to see that we must take $R = 2t - 1$; from this results the transformation

$$x = (2t - 1)^2.$$

For the form (i) we must have that

$$R^n - t^r(1 - t)^s$$

reduces to a constant H. The equation $R^n - H = 0$ cannot admit $t = 0$ or $t = 1$ as roots. We are thus led back to the preceding case, which gives now the transformation

$$x = \frac{(2t - 1)^2}{4t(t - 1)}.$$

In order that these transformations may be effected we must have $C = 0$ and $l = \frac{1}{2}$. Following is the table of transformations, deduced by the above process, together with the inverse transformations. On the left-hand side of the table are given the conditions to be satisfied by the constants A, B, C, l, m and the elements α, β, γ themselves. The quantities λ, μ, ν have the same meaning as before.

$C = 0$, $l = \frac{1}{2}$, $\lambda = \pm \frac{1}{2}$.	I... $x = (2t - 1)^2$, $x = \left(\frac{2 - t}{t}\right)^2$, $x = \left(\frac{1 + t}{1 - t}\right)^2$,
	II... $x = \frac{(2t - 1)^2}{4t(t - 1)}$, $x = \frac{(2 - t)^2}{4(1 - t)}$, $x = \frac{(1 + t)^2}{4t}$,

$A+B+C=0$, $m=\frac{1}{2}$, $\mu=\pm\frac{1}{2}$,

$$\text{III}\ldots\ x=4t(1-t),\quad x=\frac{4(t-1)}{t^2},\quad x=\frac{-4t}{(1-t)^2},$$

$$\text{IV}\ldots\ x=\frac{1}{4t(1-t)},\quad x=\frac{t^2}{4(t-1)},\quad x=\frac{(1-t)^2}{-4t},$$

$A=0$, $l+m=\frac{3}{2}$, $\nu=\pm\frac{1}{2}$.

$$\text{V}\ldots\ x=\frac{1}{(2t-1)^2},\quad x=\left(\frac{t}{2-t}\right)^2,\quad x=\left(\frac{1-t}{1+t}\right)^2,$$

$$\text{VI}\ldots\ x=\frac{4t(t-1)}{(2t-1)^2},\quad x=\frac{4(1-t)}{(2-t)^2},\quad x=\frac{4t}{(1+t)^2},$$

$A+B=0$, $l=m$, $\lambda=\pm\mu$.

$$\text{VII}\ldots\ \begin{cases} x=\dfrac{1+\sqrt{t}}{2},\quad x=\dfrac{1+\sqrt{1-t}}{2},\quad x=\dfrac{1+\sqrt{t}}{2\sqrt{t}},\\[2ex] x=\dfrac{1+\sqrt{1-t}}{2\sqrt{1-t}},\quad x=\dfrac{\sqrt{t-1}+\sqrt{t}}{2\sqrt{t-1}},\\[2ex] \qquad\qquad x=\dfrac{\sqrt{t}+\sqrt{t-1}}{2\sqrt{t}},\end{cases}$$

$$\text{VIII}\ldots\ \begin{cases} x=\dfrac{(\sqrt{t}+\sqrt{t-1})^2}{4\sqrt{t(t-1)}},\quad x=\dfrac{(1+\sqrt{1-t})^2}{4\sqrt{1-t}},\\[2ex] \qquad\qquad x=\dfrac{(1+\sqrt{t})^2}{4\sqrt{t}},\end{cases}$$

$B+C=0$, $l+2m=2$, $\mu=\pm\nu$.

$$\text{IX}\ldots\ \begin{cases} x=\dfrac{2}{1+\sqrt{t}},\quad x=\dfrac{2}{1+\sqrt{1-t}},\quad x=\dfrac{2\sqrt{t}}{1+\sqrt{t}},\\[2ex] x=\dfrac{2\sqrt{1-t}}{1+\sqrt{1-t}},\quad x=\dfrac{2\sqrt{t-1}}{\sqrt{t-1}+\sqrt{t}},\\[2ex] \qquad\qquad x=\dfrac{2\sqrt{t}}{\sqrt{t}+\sqrt{t-1}}.\end{cases}$$

$$\text{X}\ldots\ \begin{cases} x=\dfrac{4\sqrt{t(t-1)}}{(\sqrt{t}+\sqrt{t-1})^2},\quad x=\dfrac{4\sqrt{1-t}}{(1+\sqrt{1-t})^2},\\[2ex] \qquad\qquad x=\dfrac{4\sqrt{t}}{(1+\sqrt{t})^2},\end{cases}$$

$$A - C = 0,\quad 2l + m = 2,\quad \overline{\lambda = \pm \nu}.\quad \left\{ \begin{array}{l} \text{XI}\ldots \left\{ \begin{array}{l} x = \dfrac{\sqrt{t}-1}{\sqrt{t}+1},\ x = \dfrac{\sqrt{1-t}-1}{\sqrt{1-t}+1},\ x = \dfrac{1-\sqrt{t}}{1+\sqrt{t}}, \\ x = \dfrac{1-\sqrt{1-t}}{1+\sqrt{1-t}},\ x = \dfrac{\sqrt{t}-\sqrt{t-1}}{\sqrt{t}+\sqrt{t-1}}, \\ x = \dfrac{\sqrt{t-1}-\sqrt{t}}{\sqrt{t-1}+\sqrt{t}}, \end{array} \right. \\ \text{XII}\ldots \left\{ \begin{array}{l} x = \left(\dfrac{\sqrt{t-1}+\sqrt{t}}{\sqrt{t-1}-\sqrt{t}}\right)^2,\ x = \left(\dfrac{1+\sqrt{1-t}}{1-\sqrt{1-t}}\right)^2, \\ x = \left(\dfrac{1+\sqrt{t}}{1-\sqrt{t}}\right)^2. \end{array} \right. \end{array} \right.$$

The transformations VII, IX, XI are the inverses of the rational transformations. Transformations VIII, X, XII are obtained by combining transformations I and II in the manner already explained. They can also be obtained by combining III and IV or V and VI.

The preceding table contains all the transformations given by Kummer in the case where two of the three elements α, β, γ are arbitrary.

(10) Suppose now that, for $x = 1$, several values of t are different from 0, 1, ∞; we shall have simultaneously $C = 0$, $A + B = 0$. Examine now the corresponding forms of transformation. Let

$$x = R^n t^r (1 - t)^s;$$

the values of t for $x = 1$ are given by the equation

$$R^n t^r (1 - t)^s - 1 = 0,$$

admitting neither 0 nor 1 as roots. We must then have

$$R^n t^r (1 - t)^s = 1 + S^{n'},$$

where S is an integral function of the same nature as R; n and n' are each one of the numbers 2, 3, 4, and r and s are positive integers. Remark further that each of the members of this equality must be of a degree less than or at most equal to the sixth, and that we cannot suppose $n = n' = 2$; for we would have at the same time $l = m = \frac{1}{2}$, and should thus be led back to a particular case already examined. These remarks made, we shall now demonstrate the impossibility of the above equality. We see at first that S cannot be of the first degree, for then the second member could have no multiple factor, whereas the first member has such factors. Let us assume

$$S = at^2 + bt + c, \quad n' = 2;$$

n will be equal to 3 or 4, and the first member will be of a degree higher than the fourth. Let

$$S = at^2 + bt + c, \quad n' = 3;$$

both members will be of the sixth degree; the second member admits a multiple factor of order of multiplicity at most $= 2$. The same cannot be true of the first member, for, if R is of the first degree and $n = 2$, one at least of the integers r, s will be greater than unity. Suppose, finally,

$$S = at^3 + bt^2 + ct + d, \quad n' = 2;$$

n will be equal to 3 or to 4, and the first member will admit either a quadruple factor or a triple and a double factor. The second member cannot admit a quadruple factor, neither can it admit a double and a triple factor; in fact, the double factor would have to be either t or $1 - t$. Suppose it to be t; we must then have $c = 0$, $d = 1$, but in such a case the equation

$$at^3 + bt^2 + 2 = 0$$

could have no triple root.

Examine in the same way form (d). In this case we must have

$$R^n - 1 = S^{n'}t^r(1 - t)^s.$$

This equality is identical with the preceding one, save that here r and s may be zero. If R is of the second degree and $n = 2$, we should have $n' = 3$ or $n' = 4$, and then the second member would admit a triple or a quadruple factor, while the first member has no such factor. The remainder of the discussion is the same as above, and we can show in the same way that the preceding equality is impossible.

In order that the form (b) may hold we must have

$$R^n t^r - 1 = S^{n'}(1 - e)^s,$$

where s may be zero. If s is zero, we are led to the preceding case. Suppose then r and s not zero. The functions R and S will be at most of the second degree; and as one of the numbers n, n' must be greater than 2, it is necessary that at least one of the functions be of the first degree. Let

$$R = at + b;$$

$R^n t^r - 1$ can only have one multiple factor, and that must be a double factor. In fact, every multiple factor must be a divisor of the derivative, that is, of

$$R^{n-1} t^{r-1}[nat + r(at + b)];$$

the only suitable root of the derivative is given by the equation of the first degree

$$nat + r(at + b) = 0;$$

further, this root does not annul the second derivative

$$n(n - 1)a^2t^2 + 2nrat(at + b) + r(r - 1)(at + b)^2 = 0.$$

From the first we get

$$\frac{at}{r} = \frac{at + b}{-n},$$

and substituting in the second we find

$$-nr(r + n),$$

which is always different from zero, since r and n are positive integers. $S^{n'}(1 - t)^s$ can then only admit one double factor; now, if $n' = 2$, n is greater than 2, and the first member is at least of the fourth degree. Then S should be of the second degree or $s > 1$; in both cases the second member admits more than one multiple factor.

The only form which can hold, therefore, is the form (i),

$$x = \frac{R^n}{t^r(1-t)^s}.$$

The values of t for $x = 1$ can be neither 0 nor 1, and they must all be roots of

$$R^n - t^r(1-t)^s = 0$$

of the same degree of multiplicity. We are thus conducted to the following problem:

Required to find two entire functions R and S, and two integers n and n', so that the equation

$$R^n - S^{n'} = 0$$

admits only the two roots 0 *and* 1; R, S, n, n' being subjected to the restrictions already indicated.

As shown above, we cannot suppose $n = n' = 2$; neither can we suppose $n = n' = 4$, because the equation $R^4 - S^4 = 0$ always admits more than two distinct roots. The only admissible hypotheses are, supposing $n \leqq n'$,

$$\begin{array}{cccc} n = 3, & n = 2, & n = 2, & n = 3, \\ n' = 3, & n' = 4, & n' = 3, & n' = 4. \end{array}$$

(11) Let $n = n' = 3$; the equation $R^3 - S^3 = 0$ is equivalent to the three equations

$$R = S, \qquad R = jS, \qquad R = j^2S.$$

The first member of one of these must reduce to a constant, and the other two ought each to admit a distinct root: this requires that they shall be of the first degree. Let

$$R = t + u, \qquad S = t + v;$$

we must have

$$u = j^2v, \qquad 1 + u = j + jv,$$

from which

$$v = j^2, \qquad u = j.$$

In fact, we have the identity

$$(t+j)^3 - (t+j^2)^3 = 3j(j-1)t(1-t),$$

and we deduce the transformation

$$x = \frac{(t+j)^3}{3j(j-1)t(1-t)}.$$

The following are the transformations which conduct to this form j is of course one of the imaginary cube roots of unity:

$$\text{XIII}\ldots\left\{\begin{array}{c} A+B=0,\ C=0, \\ \underline{l=m=\tfrac{2}{3},} \\ \lambda=\pm\tfrac{1}{3},\ \mu=\pm\tfrac{1}{3}. \end{array}\right\} \qquad x = \frac{(t+j)^3}{3j(j-1)t(1-t)},$$

$$\text{XIV}\ldots\left\{\begin{array}{c} A=0,\ C=0, \\ \underline{l=m=\tfrac{2}{3},} \\ \lambda=\pm\tfrac{1}{3},\ \nu=\pm\tfrac{1}{3}. \end{array}\right\} \qquad x = \frac{(t+j)^3}{(t+j^2)^3},$$

$$\text{XV}\ldots\left\{\begin{array}{c} A=0,\ B+C=0, \\ \underline{l=m=\tfrac{2}{3},} \\ \mu=\pm\tfrac{1}{3},\ \nu=\pm\tfrac{1}{3}. \end{array}\right\} \qquad x = \frac{3j(j-1)t(1-t)}{(t+j)^3}.$$

Inverse Transformations.

$$\text{XVI}\ldots\left\{\begin{array}{c} A=C=-B, \\ \underline{l=m=\tfrac{2}{3},} \\ \lambda=\pm\mu, \\ \lambda=\pm\nu. \end{array}\right. \left\{\begin{array}{l} x=\dfrac{j^2\sqrt[3]{t}-j\sqrt[3]{t-1}}{\sqrt[3]{t-1}-\sqrt[3]{t}},\ x=\dfrac{\sqrt[3]{t-1}-\sqrt[3]{t}}{j^2\sqrt[3]{t}-j\sqrt[3]{t-1}}, \\ x=\dfrac{j^2\sqrt[3]{t}-j}{1-\sqrt[3]{t}},\ x=\dfrac{1-\sqrt[3]{t}}{j^2\sqrt[3]{t}-j}, \\ x=\dfrac{j^2\sqrt[3]{1-t}-j}{1-\sqrt[3]{1-t}},\ x=\dfrac{1-\sqrt[3]{1-t}}{j^2\sqrt[3]{1-t}-j}. \end{array}\right.$$

(12) Let $n = 2$, $n' = 4$. It is required to determine two polynomials R and S such that

$$R^2 - S^4 = H t^r (1 - t)^s,$$

where H is a constant introduced for convenience. From this we deduce

$$R - S^2 = H' t^r, \quad R + S^2 = H''(1 - t)^s.$$

In fact, the equations $R - S^2 = 0$, $R + S^2 = 0$ can have no other roots than 0 and 1; besides, these equations must admit no common roots, because a common root would annul both R and S. One of the equations admits then only the root $t = 0$, and the other only the root $t = 1$, at least unless the first member of one of them reduces to a constant, a case which will be examined later. S is compelled to be of the first degree, but R may be of the first, second, or third degree. Suppose R to be of the first degree,

$$R = at + b, \qquad S = mt + n,$$

$$R - S^2 = -m^2t^2 + (a - 2mn)t + b - n^2,$$

$$R + S^2 = m^2t^2 + (a + 2mn)t + b + n^2.$$

As we can always suppose $n = 1$, we ought to have simultaneously

$$b = 1, \quad a = 2m, \quad m^2 + a + 2m + 2 = 0, \quad 2m^2 + a + 2m = 0,$$

which is impossible.

Take now R of the second degree,

$$R = at^2 + bt + c.$$

If neither $R - S^2$ nor $R + S^2$ reduces to the first degree, we shall have

$$c = 1, \quad b = 2m, \quad a + m^2 + 4m + 2 = 0, \quad 2a + 2m^2 + 4m = 0,$$

giving $m = -1$, and consequently $S = 1 - t$.

Suppose $R - S^2$ reduces to the first degree; we must then have

$$c = 1, \quad a = m^2, \quad 2m^2 + 2m + 2 + b = 0, \quad 4m^2 + b + 2m = 0,$$

giving $m = 1$, $b = -6$, $a = 1$, and consequently the identity

$$(t^2 - 6t + 1)^2 + (t + 1)^2 = -16t(1 - t)^2.$$

If R is of the third degree, we must have

$$R - (mt + n)^2 = t^3, \quad R + (mt + n)^2 = (1 - t)^3,$$

from which we derive

$$[(1 - t)^3 - t^3] = 2(mt + n)^2,$$

which is impossible since the two roots of the equation

$$(1 - t)^3 - t^3 = 0$$

are distinct. If one of the functions $R - S^2$, $R + S^2$ reduces to a constant, R will necessarily be of the second degree. Let

$$R = m^2t^2 + 2mt + c, \quad S = mt + 1,$$

$$R + S^2 = 2m^2t^2 + 4mt + c + 1.$$

We must have

$$c = -1, \quad m = -2,$$

and consequently the identity

$$(4t^2 - 4t - 1)^2 - (2t - 1)^4 = 16t(1 - t);$$

this can be deduced from the preceding identity by changing t into $\dfrac{t}{t - 1}$.

The following is a table of the corresponding transformations:

$$\text{XVII}\ldots\left\{\begin{array}{c} A+B=0,\ C=0, \\ l=\frac{1}{2},\ m=\frac{3}{4}, \\ \hline \lambda=\pm\frac{1}{2},\ \mu=\pm\frac{1}{4}. \end{array}\right\} \begin{array}{l} x=\dfrac{(t^2-6t+1)^2}{-16t(1-t)^2},\ x=\dfrac{(t^2+4t-4)^2}{-16t^2(1-t)}, \\ x=\dfrac{(1+4t-4t^2)^2}{16t(1-t)}, \end{array}$$

$$\text{XVIII}\ldots\left\{\begin{array}{c} A+B=0,\ C=0, \\ l=\frac{3}{4},\ m=\frac{1}{2}, \\ \hline \lambda=\pm\frac{1}{4},\ \mu=\pm\frac{1}{2}. \end{array}\right\} \begin{array}{l} x=\dfrac{(1+t)^4}{16t(1-t)^2},\ x=\dfrac{(2-t)^4}{16t^2(1-t)}, \\ x=\dfrac{2t-1)^4}{-16t(1-t)}, \end{array}$$

$$\text{XIX}\ldots\left\{\begin{array}{c} A=0,\ C=0, \\ l=\tfrac{1}{2},\ m=\tfrac{3}{4}, \\ \hline \lambda=\pm\tfrac{1}{2},\ \nu=\pm\tfrac{1}{4}, \end{array}\right\} \begin{array}{l} x=\dfrac{(t^2-6t+1)^2}{(1+t)^4},\ x=\dfrac{(t^2+4t-4)^2}{(2-t)^4}, \\ x=\dfrac{(1+4t-4t^2)^2}{2t-1)^4}, \end{array}$$

$$\text{XX}\ldots\left\{\begin{array}{c} A=0,\ C=0, \\ l=m=\tfrac{3}{4}, \\ \hline \lambda=\pm\tfrac{1}{4},\ \nu=\pm\tfrac{1}{2}. \end{array}\right\} \begin{array}{l} x=\dfrac{(1+t)^4}{(t^2-6t+1)^2},\ x=\dfrac{(2-t)^4}{(t^2+4t-4)^2}, \\ x=\dfrac{(2t-1)^4}{(1+4t-4t^2)^2}, \end{array}$$

$$\text{XXI}\ldots\left\{\begin{array}{c} A=0,\ B+C=0, \\ l=\tfrac{3}{4},\ m=\tfrac{1}{2}, \\ \hline \mu=\pm\tfrac{1}{2},\ \nu=\pm\tfrac{1}{4}. \end{array}\right\} \begin{array}{l} x=\dfrac{16t(1-t)^2}{(1+t)^4},\ x=\dfrac{16t^2(1-t)}{(2-t)^4}, \\ x=\dfrac{-16t(1-t)}{(2t-1)^4}, \end{array}$$

$$\text{XXII}\ldots\left\{\begin{array}{c} A=0,\ B+C=0, \\ l=m=\tfrac{3}{4}, \\ \hline \mu=\pm\tfrac{1}{4},\ \nu=\pm\tfrac{1}{2}. \end{array}\right\} \begin{array}{l} \dfrac{-16(1-t)^2}{(t^2-6t+1)^2},\ x=\dfrac{-16t^2(1-t)}{(t^2+4t-4)^2}, \\ x=\dfrac{16t(1-t)}{(1+4t-4t^2)^2}, \end{array}$$

Inverse Transformations.

$$\text{XXIII}\ldots\left\{\begin{array}{c} A=C,\ B=2A, \\ l=\tfrac{3}{4},\ m=\tfrac{1}{2}, \\ \hline \lambda=\pm\dfrac{\mu}{2}=\pm\nu. \end{array}\right. \left\{\begin{array}{l} \dfrac{(x^2-6x+1)^2}{-16x(1-x)^2}=t,\ \dfrac{(1+x)^4}{16x(1-x)^2}=t, \\ \dfrac{(x^2-6x+1)^2}{(1+x)^4}=t,\ \dfrac{(1+x)^4}{(x^2-6x+1)^2}=t, \\ \dfrac{16x(1-x)^2}{(1+x)^4}=t,\ \dfrac{-16x(1-x)^2}{(x^2-6x+1)^2}=t, \end{array}\right.$$

$$\text{XXIV}\ldots\left\{\begin{array}{c} C=4A,\ B=-4A, \\ l=\tfrac{1}{2},\ m=\tfrac{3}{4}, \\ \hline \dfrac{\lambda}{2}=\pm\mu=\pm\nu. \end{array}\right. \left\{\begin{array}{l} \dfrac{(x^2+4x-4)^2}{-16x^2(1-x)}=t,\ \dfrac{(2-x)^4}{16x^2(1-x)}=t, \\ \dfrac{(x^2+4x-4)^2}{(2-x)^4}=t,\ \dfrac{(2-x)^4}{(x^2+4x-4)^2}=t, \\ \dfrac{16x^2(1-x)}{(2-x)^4}=t,\ \dfrac{-16x^2(1-x)}{(x^2+4x-4)^2}=t, \end{array}\right.$$

$$\text{XXV}\ldots\begin{cases} A=4C,\ B=-4C, \\ l=\frac{3}{4},\ m=\frac{3}{4}, \\ \overline{} \\ \lambda=\pm\mu=\pm\frac{\nu}{2}. \end{cases}\quad \begin{cases} \dfrac{(1+4x-4x^2)^2}{16x(1-x)}=t, & \dfrac{(2x-1)^4}{-16x(1-x)}=t, \\ \dfrac{(1+4x-4x^2)^2}{(2x-1)^4}=t, & \dfrac{(2x-1)^4}{(1+4x-4x^2)^2}=t, \\ \dfrac{-16x(1-x)}{(2x-1)^4}=t, & \dfrac{16x(1-x)}{(1+4x-4x^2)^2}=t. \end{cases}$$

(13) Let $n=2$, $n'=3$. R and S are to be determined in such a way that we shall have

$$R^2-S^3=Ht^r(1-t)^s.$$

R can be of the first, second, or third degree, and S of the first or second degree, giving in all six cases to examine.

First Case.—Suppose R and S each of the first degree,

$$R=at+b, \quad S=mt+n.$$

The polynomial R^2-S^3 will be of the third degree, and will admit one double factor and one simple factor. Let t be the double factor; we must then have

$$b^2=n^3, \quad 2ab=3mn^2, \quad (a+b)^2=(m+n)^3.$$

We can always take $b=1$, $n=1$. On doing this we have $a=\dfrac{3m}{2}$ and consequently

$$\left(\frac{3m}{2}+1\right)^2=(m+1)^3;$$

this gives $m=-\frac{3}{4}$, $a=-\frac{9}{8}$, and therefore the identity

$$(9t-8)^2-(4-3t)^3=-27t^2(1-t).$$

Second Case.—Suppose R of the second and S of the first degree, viz.,

$$R=at^2+bt+c, \quad S=mt+n.$$

If we had

$$R^2-S^3=Ht^2(1-t)^2,$$

we should have

$$S^3 = R^2 - Ht^2(1-t)^2.$$

Now the equation $R^2 - Ht^2(1-t)^2 = 0$ breaks up into two equations each of which is at most of the second degree, viz.,

$$R = \pm\sqrt{H}\,.\,t(1-t),$$

and consequently cannot admit of a triple root; $S^3 - R^2$ will then have a triple factor and a simple factor. If t be the triple factor, we shall have the conditions

$$c^2 = n^3, \quad 2bc = 3mn^2, \quad b^2 + 2ac = 3m^2n, \quad (a+b+c)^2 = (m+n)^3.$$

Taking $n = 1$, $c = 1$, we get

$$b = \frac{3m}{2}, \quad a = \frac{3m^2}{8}, \quad \frac{9m^4}{64} + \frac{9m^3}{8} = m^3,$$

and so

$$m = -\tfrac{8}{9}, \quad b = -\tfrac{4}{3}, \quad a = \tfrac{8}{27},$$

from which results the identity

$$(8t^2 - 36t + 27)^2 - (9 - 8t)^3 = -64t^3(1-t).$$

Third Case.—Let R be of the third and S of the first degree,

$$R = at^3 + bt^2 + ct + d, \quad S = mt + n.$$

$R^2 - S^3$ will admit of a quadruple factor and a double factor, or two triple factors, or a quintuple factor and a simple factor. We cannot have $R^2 - S^3 = t^4(1-t)^2$, for then

$$S^3 = R^2 - t^4(1-t)^2.$$

In order that

$$R^2 - t^4(1-t)^2 = 0$$

should admit only one triple root, we ought to have

$$R = t^2(1-t) + H;$$

but then the equation

$$R + t^2(1 - t) = 0$$

becomes $2t^2(1 - t) + H = 0$ and has no triple root. Neither can we have $R^2 - S^3 = Ht^3(1 - t)^3$, for then

$$R^2 = S^3 + Ht^3(1 - t)^3,$$

and in no case can the second member of this admit three double factors. It remains to be seen whether we can have

$$R^2 - S^3 = Ht^6(1 - t).$$

Taking $n = 1$, $d = 1$, we must have

$$2c = 3m, \quad 2c^2 + 4b = 6m^2, \quad 2bc + 2a = m^3, \quad b^2 + ac = 0,$$

$$(a + b + c + 1)^2 = (m + 1)^3,$$

giving

$$c = \frac{3m}{2}, \quad b = \frac{3m^2}{8}, \quad a = -\frac{m^3}{16};$$

and substituting in the fourth relation this becomes

$$\frac{9m^4}{64} - \frac{3m^4}{16} = 0, \text{ giving } m = 0.$$

This hypothesis must then be rejected.

Fourth Case.—Suppose R of the first and S of the second degree,

$$R = at + b, \quad S = mt^2 + nt + p.$$

$R^2 - S^3$ cannot admit two triple factors; if we had

$$R^2 - S^3 = t^3(1 - t)^3,$$

then

$$R^2 = S^3 + t^3(1 - t)^3,$$

and the second member evidently admits more than one distinct factor. If we had $S^3 = R^2 - t^4(1-t)^2$, each of the two equations $R = \pm t^2(1-t)$ should admit a triple root:

$$R - t^2(1-t) = (ut+v)^3,$$

$$R + t^2(1-t) = (u_1t+v_1)^3.$$

Then

$$2R = (ut+v)^3 + (u_1t+v_1)^3;$$

R being of the first degree, this is impossible. If we had

$$S^2 - R^3 = Ht^6(1-t),$$

we should find as condition $n = 0$, $ab = 0$. This hypothesis therefore gives nothing.

Fifth Case.—R and S of the second degree. We can demonstrate as in the preceding cases that this hypothesis is to be rejected.

Sixth Case.—R is of the third and S of the second degree,

$$R = at^3 + bt^2 + ct + d, \quad S = mt^2 + nt + p.$$

$R^2 - S^3$ may be of a degree lower than the sixth. By simple transformations the second member of the identity

$$R^2 - S^3 = Ht^r(1-t)^s$$

is conducted to one of the following forms:

$$Ht^3, \quad Ht^2, \quad Ht, \quad Ht^2(1-t)^2, \quad Ht^2(1-t), \quad Ht(1-t).$$

If, for example, we had

$$R^2 - S^3 = Ht^4(1-t),$$

changing t into $\frac{1}{t}$ and multiplying by t^6, we deduce

$$R_1^2 - S_1^3 = -Ht(1-t).$$

If we had

$$R^2 - S^3 = Ht^3,$$

we should conclude that the equation

$$R^2 - Ht^3 = S^3$$

had three double roots, which is impossible. Supposing we have

$$R^2 - S^3 = Ht^2,$$

then

$$R^2 - Ht^2 = S^3,$$

and we should have

$$R - \sqrt{H}\,.\,t = (ut + v)^3,$$

$$R + \sqrt{H}\,.\,t = (u_1t + v_1)^3,$$

and consequently

$$2\sqrt{H}\,.\,t = (ut + v)^3 - (u_1t + v_1)^3,$$

which is inadmissible. If we had

$$R^2 - S^3 = Ht^2(1 - t)^2,$$

then we should have

$$R + \sqrt{H}\,.\,t(1 - t) = (ut + v)^3,$$

$$R - \sqrt{H}\,.\,t(1 - t) = (u_1t + v_1)^3,$$

and consequently

$$(ut + v)^3 - (u_1t + v_1)^3 = 2\sqrt{H}\,.\,t(1 - t).$$

That this may be so we must have

$$ut + v = t + j, \quad u_1t + v_1 = t + j^2;$$

from which

$$R = \tfrac{1}{2}(t + j)^3 + \tfrac{1}{2}(t + j^2)^3 = t^3 - \tfrac{3}{2}t^2 - \tfrac{3}{2}t + 1,$$

$$S = (t + j)(t + j^2) = t^2 - t + 1,$$

and consequently the identity

$$(2t^3 - 3t^2 - 3t + 2)^2 - 4(t^2 - t + 1)^3 = -27t^2(1 - t)^2.$$

If we wished to have

$$R^2 - S^3 = Ht^2(1 - t),$$

we should, in supposing $d = p = 1$, get the conditions

$$m^3 = a^2, \quad 3m^2n = 2ab, \quad 3m^2 + 3mn^2 = b^2 + 2ac, \quad 3n = 2c,$$
$$(m + n + 1)^3 = (a + b + c + 1)^2;$$

these give

$$a = m\sqrt{m}, \quad b = \frac{3\sqrt{mn}}{2}, \quad c = \frac{3n}{2};$$

substituting these in the third relation, we have $m(2\sqrt{m} - 2)^2 = 0$. If we take $n = 2\sqrt{m}$, we get

$$a = m\sqrt{m}, \quad b = 3m, \quad c = 3\sqrt{m},$$

and are so led to the identity

$$(mt + 2\sqrt{m}\,t + 1)^3 - (m\sqrt{m}\,t^3 + 3mt^2 + 3\sqrt{m}\,t + 1)^2 = 0$$

this hypothesis gives therefore no transformation. Suppose

$$R^2 - S^3 = Ht(1 - t);$$

the conditions now are

$$m^0 = a^2, \quad 3m^2n = 2ab, \quad 3m^2 + 3mn^2 = b^2 + 2ac,$$
$$6mn + n^3 = 2a + 2bc, \quad 3m + 3n^2 + 3n = c^2 + 2b + 2c;$$

these give

$$a = m\sqrt{m}, \quad b = \frac{3n\sqrt{m}}{2}, \quad c = \frac{6mn + n^3 - 2m\sqrt{m}}{3n\sqrt{m}};$$

substituting these in the third equation gives

$$n^3 - 12mn + 16m\sqrt{m} = 0,$$

or, making $n = u\sqrt{m}$,

$$u^3 - 12u + 16 = (u + 4)(u - 2)^2 = 0.$$

We have therefore either $n = 2\sqrt{m}$, or $n = -4\sqrt{m}$. If we take $n = 2\sqrt{m}$, we come to the same identity as above. Take $n = -4\sqrt{m}$; now

$$a = m\sqrt{m}, \quad b = -6m, \quad c = \tfrac{15}{2}\sqrt{m}.$$

Substituting these in the last relation, we find $\sqrt{m} = 4$; therefore

$$m = 16, \quad n = -16, \quad a = 64, \quad b = -96, \quad c = 30,$$

giving the identity

$$(64t^3 - 96t^2 + 30t + 1)^2 - (16t^2 - 16t + 1)^3 = -108t(1-t).$$

If $R^2 - S^3 = Ht$, the conditions are

$$a^2 = m^3, \quad 3m^2n = 2ab, \quad 3m^2 + 3mn^2 = b^2 + 2ac,$$

$$6mn + n^3 = 2a + 2bc, \quad c^2 + 2b = 3m + 3n^2,$$

giving

$$n = -4\sqrt{m}, \quad a = m\sqrt{m}, \quad b = -6m, \quad c = \frac{15\sqrt{m}}{2};$$

these substituted in the last relation give $m = 0$.

The following is a table of the transformations deduced from the preceding identities:

$$A + B = 0, \quad C = 0, \quad l = \tfrac{1}{2}, \quad m = \tfrac{2}{3}, \quad \lambda = \pm\tfrac{1}{2}, \quad \mu = \pm\tfrac{1}{3}.$$

$$\text{XXVI}\ldots\begin{cases} x = \dfrac{(9t-8)^2}{-27t^2(1-t)}, \; x = \dfrac{(1-9t)^2}{-27t(1-t)^2}, \; x = \dfrac{(9-8t)^2 t}{27(1-t)}, \\ x = \dfrac{(1+8t)^2(1-t)}{27t}, \; x = \dfrac{(t-9)^2 t}{-27(1-t)^2}, \\ x = \dfrac{(t+8)^2(1-t)}{-27t^2}, \end{cases}$$

$$\text{XXVII}\ldots\begin{cases} x = \dfrac{(8t^2-36t+27)^2}{-64t^3(1-t)}, & x = \dfrac{(8t^2+20t-1)^2}{-64t(1-t)^3}, \\ x = \dfrac{(8-36t+27t^2)^2}{64(1-t)}, & x = \dfrac{(27t^2-18t-1)^2}{64t}, \\ x = \dfrac{(t^2+18t-27)^2}{64t^3}, & x = \dfrac{(t^2-20t-8)^2}{64(1-t)^3}, \end{cases}$$

$$\text{XXVIII}\ldots\ldots x = \frac{(2t^3-3t^2-3t+2)^2}{-27t^2(1-t)^2},$$

$$\text{XXIX}\ldots\begin{cases} x = \dfrac{(64t^3-96t^2+30t+1)^2}{-108t(1-t)}, & x = \dfrac{(t^3+30t^2-96t+64)^2}{108t^4(1-t)}, \\ x = \dfrac{(t^3-33t^2-33t+1)^2}{108t(1-t)^4}. \end{cases}$$

$A+B=0$, $C=0$, $l=\frac{2}{3}$, $m=\frac{1}{2}$, $\lambda=\pm\frac{1}{3}$, $\mu=\pm\frac{1}{2}$.

$$\text{XXX}\ldots\begin{cases} x = \dfrac{(3t-4)^3}{-27t^2(1-t)}, & x = \dfrac{(3t+1)^3}{27t(1-t)^2}, & x = \dfrac{(3-4t)^3}{27(1-t)}, \\ x = \dfrac{(4t-1)^3}{27t}, & x = \dfrac{(4-t)^3}{27t^2}, & x = \dfrac{(t+3)^3}{27(1-t)^2}, \end{cases}$$

$$\text{XXXI}\ldots\begin{cases} x = \dfrac{(8t-9)^3}{-64t^3(1-t)}, & x = \dfrac{(8t+1)^3}{64t(1-t)^3}, & x = \dfrac{(8-9t)^3t}{64(1-t)}, \\ x = \dfrac{(9t-1)^3(1-t)}{64t}, & x = \dfrac{(t-9)^3(1-t)}{64t^3}, \\ x = \dfrac{(8+t)^3t}{-64(1-t)^3}, \end{cases}$$

$$\text{XXXII}\ldots\ldots x = \frac{4(t^2-t+1)^3}{27t^2(1-t)^2},$$

$$\text{XXXIII}\ldots\begin{cases} x = \dfrac{(16t^2-16t+1)^3}{108t(1-t)}, & x = \dfrac{(t^2-16t+16)^3}{-108t^4(1-t)}, \\ x = \dfrac{(1+14t+t^2)^3}{-108t(1-t)^4}. \end{cases}$$

$A=0$, $C=0$, $l=\frac{1}{2}$, $m=\frac{5}{6}$, $\lambda=\pm\frac{1}{2}$, $\nu=\pm\frac{1}{3}$.

$$\text{XXXIV}\ldots\begin{cases} x=\dfrac{(9t-8)^2}{-(3t-4)^3},\ x=\dfrac{(1-9t)^2}{(3t+1)^3},\ x=\dfrac{(9-8t)^2t}{(4t-3)^3}, \\ x=\dfrac{(t-9)^2t}{(t+3)^3},\ x=\dfrac{(t+8)^2(1-t)}{(4-t)^3},\ x=\dfrac{(1+8t)^2(1-t)}{(1-4t)^3}, \end{cases}$$

$$\text{XXXV}\ldots\begin{cases} x=\dfrac{(8t^2-36t+27)^2}{(9-8t)^3},\ x=\dfrac{(8t^2+20t-1)^2}{(8t+1)^3}, \\ x=\dfrac{(8-36t+27t^2)^2}{(9t-8)^3t},\ x=\dfrac{(27t^2-18t-1)^2}{(1-9t)^3(1-t)}, \\ x=\dfrac{(t^2+18t-27)^2}{(9-t)^3(1-t)},\ x=\dfrac{(t^2-20t-8)^2}{t(8+t)^3}, \end{cases}$$

$$\text{XXXVI}\ldots\ldots x=\frac{(2t^3-3t^2-3t+2)^2}{4(t^2-t+1)^3},$$

$$\text{XXXVII}\ldots\begin{cases} x=\dfrac{(64t^3-96t^2+30t+1)^2}{(16t^2-16t+1)^3},\ x=\dfrac{(t^3+30t^2-96t+64)^2}{(t^2-16t+16)^3}, \\ x=\dfrac{(t^3-33t^2-33t+1)^2}{(1+14t+t^2)^3}. \end{cases}$$

$$A=0,\quad C=0,\quad l=\tfrac{2}{3},\quad m=\tfrac{5}{6},\quad \lambda=\pm\tfrac{1}{3},\quad \nu=\pm\tfrac{1}{2}.$$

$$\text{XXXVIII}\ldots\begin{cases} x=\dfrac{-(3t-4)^3}{(9t-8)^2},\ x=\dfrac{(3t+1)^3}{(1-9t)^2},\ x=\dfrac{(4t-3)^3}{(9-8t)^2t}, \\ x=\dfrac{(t+3)^3}{(t-9)^2t},\ x=\dfrac{(4-t)^3}{(t+8)^2(1-t)},\ x=\dfrac{(1-4t)^3}{(1+8t)^2(1-t)}, \end{cases}$$

$$\text{XXXIX}\ldots\begin{cases} x=\dfrac{(9-8t)^3}{(8t^2-36t+27)^2},\ x=\dfrac{(8t+1)^3}{(8t^2+20t-1)^2}, \\ x=\dfrac{(9t-8)^3t}{(8-36t+27t^2)^2},\ x=\dfrac{(1-9t)^3(1-t)}{(27t^2-18t-1)^2}, \\ x=\dfrac{(9-t)^3(1-t)}{(t^2+18t-27)^2},\ x=\dfrac{(8+t)^3t}{(t^2-20t-8)^2}, \end{cases}$$

$$\text{XL}\ldots\ldots x=\frac{4(t^2-t+1)^3}{(2t^3-3t^2-3t+2)^2},$$

$$\text{XLI}\ldots\begin{cases} x = \dfrac{(16t^2 - 16t + 1)^3}{(64t^3 - 96t^2 + 30t + 1)^2}, \quad x = \dfrac{(t^2 - 16t + 16)^3}{(t^3 + 30t^2 - 96t + 64)^2}, \\ x = \dfrac{(1 + 14t + t^2)^3}{(t^3 - 33t^2 - 33t + 1)^2}. \end{cases}$$

$A = 0$, $B + C = 0$, $l = \frac{5}{6}$, $m = \frac{1}{2}$, $\mu = \pm \frac{1}{2}$, $\nu = \pm \frac{1}{3}$.

$$\text{XLII}\ldots\begin{cases} x = \dfrac{-27t^2(1 - t)}{(3t - 4)^3}, \quad x = \dfrac{27t(1 - t)}{(3t + 1)^3}, \quad x = \dfrac{27(1 - t)}{(3 - 4t)^3}, \\ x = \dfrac{27t}{(4t - 1)^3}, \quad x = \dfrac{27t^2}{(4 - t)^3}, \quad x = \dfrac{27(1 - t)^2}{(t + 3)^3}. \end{cases}$$

$$\text{XLIII}\ldots\begin{cases} x = \dfrac{64t^3(1 - t)}{(9 - 8t)^3}, \quad x = \dfrac{64t(1 - t)^3}{(8t + 1)^3}, \quad x = \dfrac{64(1 - t)}{(8 - 9t)^3 t}, \\ x = \dfrac{64t}{(9t - 1)^3(1 - t)}, \quad x = \dfrac{64t^3}{(t - 9)^3(1 - t)}, \\ x = \dfrac{-64(1 - t)^3}{(8 + t)^3 t}, \end{cases}$$

$$\text{XLIV}\ldots\ldots \; x = \frac{27t^2(1 - t)^2}{4(t^2 - t + 1)^3},$$

$$\text{XLV}\ldots\begin{cases} x = \dfrac{108t(1 - t)}{(16t^2 - 16t + 1)^3}, \quad x = \dfrac{-108t^4(1 - t)}{(t^2 - 16t + 16)^3}, \\ x = \dfrac{-108t(1 - t)^4}{(1 + 14t + t^2)^3}. \end{cases}$$

$A = 0$, $B + C = 0$, $l = \frac{5}{6}$, $m = \frac{2}{3}$, $\mu = \pm \frac{1}{3}$, $\nu = \pm \frac{1}{2}$.

$$\text{XLVI}\ldots\begin{cases} x = \dfrac{-27t^2(1 - t)}{(9t - 8)^2}, \quad x = \dfrac{-27t(1 - t)^2}{(1 - 9t)^2}, \quad x = \dfrac{27(1 - t)}{(9 - 8t)^2 t}, \\ x = \dfrac{27t}{(1 + 8t)^2(1 - t)}, \quad x = \dfrac{27(1 - t)^2}{(t - 9)^2 t}, \\ x = \dfrac{-27t^2}{(t + 8)^2(1 - t)}, \end{cases}$$

$$\text{XLVII}\ldots\begin{cases} x=\dfrac{-64t^3(1-t)}{(8t^2-36t+27)^2},\quad x=\dfrac{-64t(1-t)^3}{(8t^2+20t-1)^2},\\[2ex] x=\dfrac{64(1-t)}{(8-36t+27t^2)^2},\quad x=\dfrac{64t}{(27t^2-18t-1)^2},\\[2ex] x=\dfrac{64t^3}{(t^2+18t-27)^2},\quad x=\dfrac{64(1-t)^3}{(t^2-20t-8)^2},\end{cases}$$

$$\text{XLVIII}\ldots\ldots\ x=\frac{-27t^2(1-t)^2}{(2t^3-3t^2-3t+2)^2},$$

$$\text{XLIX}\ldots\begin{cases} x=\dfrac{-108t(1-t)}{(64t^3-96t^2+30t+1)^2},\quad x=\dfrac{108t^4(1-t)}{(t^3+30t^2-96t+64)^2},\\[2ex] x=\dfrac{108t(1-t)^4}{(t^3-33t^2-33t+1)^2}.\end{cases}$$

Inverse Transformations.

$$\text{L}\ldots\left\{\begin{matrix} A=0,\ 4B+3C=0,\\ l=\frac{2}{3},\ m=\frac{5}{6},\\ \overline{\nu=\pm\frac{1}{2},}\\ \dfrac{\lambda}{2}=\pm\mu.\end{matrix}\right\}\begin{matrix} \dfrac{(9x-8)^2}{-27x^2(1-x)}=t,\ \dfrac{(3x-4)^3}{-27x^2(1-x)}=t,\\[2ex] \dfrac{(9x-8)^2}{(4-3x)^3}=t,\ \dfrac{(4-3x)^3}{(9x-8)^2}=t,\\[2ex] \dfrac{-27x^2(1-x)}{(3x-4)^3}=t,\ \dfrac{-27x^2(1-x)}{(9x-8)^2}=t.\end{matrix}$$

$$\text{LI}\ldots\left\{\begin{matrix} A=0,\ B=3C,\\ l=\frac{5}{6},\ m=\frac{2}{3},\\ \overline{\nu=\pm\frac{1}{2},}\\ \lambda=\pm\dfrac{\mu}{2}.\end{matrix}\right\}\begin{matrix} \dfrac{(1-9x)^2}{-27x(1-x)^2}=t,\ \dfrac{(3x+1)^3}{27x(1-x)^2}=t,\\[2ex] \dfrac{(1-9x)^2}{(3x+1)^3}=t,\ \dfrac{(3x+1)^3}{(1-9x)^2}=t,\\[2ex] \dfrac{27x(1-x)^2}{(3x+1)^3}=t,\ \dfrac{-27x(1-x)^2}{(1-9x)^2}=t.\end{matrix}$$

$$\text{LII}\ldots\left\{\begin{matrix} C=0,\ 4B+3A=0,\\ l=\frac{1}{2},\ m=\frac{5}{6},\\ \overline{\lambda=\pm\frac{1}{2},}\\ \dfrac{\nu}{2}=\pm\mu.\end{matrix}\right\}\begin{matrix} \dfrac{(9-8x)^2x}{27(1-x)}=t,\ \dfrac{(3-4x)^3}{27(1-x)}=t,\\[2ex] \dfrac{(9-8x)^2x}{(4x-3)^3}=t,\ \dfrac{(4x-3)^3}{(9-8x)^2x}=t,\\[2ex] \dfrac{27(1-x)}{(3-4x)^3}=t,\ \dfrac{27(1-x)}{(9-8x)^2x}=t.\end{matrix}$$

$$\text{LIII}\ldots\left\{\begin{array}{c} C=0,\ B=3A, \\ l=\frac{1}{2},\ m=\frac{2}{3}, \\ \hline \lambda=\pm\frac{1}{2}, \\ \nu=\pm\frac{\mu}{2}. \end{array}\right\}\begin{array}{l} \dfrac{(x-9)^2x}{-27(1-x)^2}=t,\quad \dfrac{(x+3)^3}{27(1-x)^2}=t, \\ \dfrac{(x-9)^2x}{(x+3)^3}=t,\quad \dfrac{(x+3)^3}{(x-9)^2x}=t, \\ \dfrac{27(1-x)^2}{(x+3)^3}=t,\quad \dfrac{-27(1-x)^2}{(x-9)^2x}=t. \end{array}$$

$$\text{LIV}\ldots\left\{\begin{array}{c} C=4A,\ B+5A=0, \\ l=\frac{2}{3},\ m=\frac{1}{2}, \\ \hline \mu=\pm\frac{1}{2}, \\ \frac{\lambda}{2}=\pm\nu. \end{array}\right\}\begin{array}{l} \dfrac{(x+8)^2(1-x)}{-27x^2}=t,\quad \dfrac{(4-x)^3}{27x^2}=t, \\ \dfrac{(x+8)^2(1-x)}{(4-x)^3}=t,\quad \dfrac{(4-x)^3}{(x+8)^2(1-x)}=t, \\ \dfrac{27x^2}{(4-x)^3}=t,\quad \dfrac{-27x^2}{(x+8)^2(1-x)}=t. \end{array}$$

$$\text{LV}\ldots\left\{\begin{array}{c} A=4C,\ B+5C=0, \\ l=\frac{5}{6},\ m=\frac{1}{2}, \\ \hline \mu=\pm\frac{1}{2}, \\ \lambda=\pm\frac{\nu}{2}. \end{array}\right\}\begin{array}{l} \dfrac{(1+8x)^2(1-x)}{27x}=t,\quad \dfrac{(4x-1)^3}{27x}=t, \\ \dfrac{(1+8x)^2(1-x)}{(1-4x)^3}=t,\quad \dfrac{(1-4x)^3}{(1+8x)^2(1-x)}=t, \\ \dfrac{27x}{(4x-1)^3}=t,\quad \dfrac{27x}{(1+8x)^2(1-x)}=t. \end{array}$$

$$\text{LVI}\ldots\left\{\begin{array}{c} A=0,\ 9B+8C=0, \\ l=\frac{1}{2},\ m=\frac{5}{6}, \\ \hline \nu=\pm\frac{1}{3}, \\ \frac{\lambda}{3}=\pm\mu. \end{array}\right\}\left\{\begin{array}{l} \dfrac{(8x^2-36x+27)^2}{-64x^3(1-x)}=t, \\ \dfrac{(8x-9)^3}{-64x^3(1-x)}=t, \\ \dfrac{(8x^2-36x+27)^2}{(9-8x)^3}=t, \\ \dfrac{(9-8x)^3}{(8x^2-36x+27)^2}=t, \\ \dfrac{-64x^3(1-x)}{(8x-9)^3}=t, \\ \dfrac{-64x^3(1-x)}{(8x^2-36x+27)^2}=t. \end{array}\right.$$

$$\text{LVII}\ldots\left\{\begin{array}{c} A = 0,\ B = 8C, \\ l = \frac{5}{6},\ m = \frac{1}{2}, \\ \hline \nu = \pm \frac{1}{3}, \\ \lambda = \pm \dfrac{\mu}{3}. \end{array}\right. \qquad \left\{\begin{array}{l} \dfrac{(8x^2 + 20x - 1)^2}{-64x(1-x)^3} = t, \\ \dfrac{(8x+1)^3}{64x(1-x)^3} = t, \\ \dfrac{(8x^2 + 20x - 1)^2}{(8x+1)^3} = t, \\ \dfrac{(8x+1)^3}{(8x^2 + 20x - 1)^2} = t, \\ \dfrac{64x(1-x)^3}{(8x+1)^3} = t, \\ \dfrac{-64x(1-x)^3}{(8x^2 + 20x - 1)^2} = t. \end{array}\right.$$

$$\text{LVIII}\ldots\left\{\begin{array}{c} C = 0,\ 9B + 8A = 0, \\ l = \frac{2}{3},\ m = \frac{5}{6}, \\ \hline \lambda = \pm \frac{1}{3}, \\ \dfrac{\nu}{3} = \pm \mu. \end{array}\right. \qquad \left\{\begin{array}{l} \dfrac{(1 - 36x + 27x^2)^2}{64(1-x)} = t, \\ \dfrac{(8-9x)^3 x}{64(1-x)} = t, \\ \dfrac{(8 - 36x + 27x^2)^2}{(9x-8)^3 x} = t, \\ \dfrac{(9x-8)^3 x}{(8 - 36x + 27x^2)^2} = t, \\ \dfrac{64(1-x)}{(8-9x)^3 x} = t, \\ \dfrac{64(1-x)}{(8 - 36x + 27x^2)^2} = t. \end{array}\right.$$

$$\text{LIX}\ldots\left\{\begin{array}{c} C = 0,\ B = 8A, \\ l = \frac{2}{3},\ m = \frac{1}{2}, \\ \hline \lambda = \pm \frac{1}{3}, \\ \nu = \pm \dfrac{\mu}{3}. \end{array}\right. \qquad \left\{\begin{array}{l} \dfrac{(8 + 20x - x^2)^2}{64(1-x)^3} = t, \\ \dfrac{-(8+x)^3 x}{64(1-x)^3} = t, \\ \dfrac{(8 + 20x - x^2)^2}{(8+x)^3 x} = t, \\ \dfrac{(8+x)^3 x}{(8 + 20x - x^2)^2} = t, \\ \dfrac{64(1-x)^3}{-(8+x)^3 x} = t, \\ \dfrac{64(1-x)^3}{(8 + 20x - x^2)^2} = t. \end{array}\right.$$

$$\text{LX}\ldots\left\{\begin{array}{c} A=9C,\ B+10C=0, \\ l=\frac{5}{6},\ m=\frac{2}{3}, \\ \hline \mu=\pm\frac{1}{3}, \\ \lambda=\pm\frac{\nu}{3}. \end{array}\right. \quad \left\{\begin{array}{l} \dfrac{(27x^2-18x-1)^2}{64x}=t, \\ \dfrac{(9x-1)^3(1-x)}{64x}=t, \\ \dfrac{(27x^2-18x-1)^2}{(1-9x)^3(1-x)}=t, \\ \dfrac{(1-9x)^3(1-x)}{(27x^2-18x-1)^2}=t, \\ \dfrac{64x}{(9x-1)^3(1-x)}=t, \\ \dfrac{64x}{(27x^2-18x-1)^2}=t. \end{array}\right.$$

$$\text{LXI}\ldots\left\{\begin{array}{c} C=9A,\ B+10A=0, \\ l=\frac{1}{2},\ m=\frac{2}{3}, \\ \hline \mu=\pm\frac{1}{3}, \\ \frac{\lambda}{3}=\pm\nu. \end{array}\right. \quad \left\{\begin{array}{l} \dfrac{(x^2+18x-27)^2}{64x^3}=t, \\ \dfrac{(x-9)^3(1-x)}{64x^3}=t, \\ \dfrac{(x^2+18x-27)^2}{(9-x)^3(1-x)}=t, \\ \dfrac{(9-x)^3(1-x)}{(x^2+18x-27)^2}=t, \\ \dfrac{64x^3}{(x-9)^3(1-x)}=t, \\ \dfrac{64x^3}{(x^2+18x-27)^2}=t. \end{array}\right.$$

$$\text{LXII}\ldots\left\{\begin{array}{c} A=C=-B, \\ l=m=\frac{2}{3}, \\ \hline \lambda=\pm\mu=\pm\nu. \end{array}\right. \quad \left\{\begin{array}{l} \dfrac{(2x^3-3x^2-3x+2)^2}{-27x^2(1-x)^2}=t, \\ \dfrac{4(x^2-x+1)^3}{27x^2(1-x)^2}=t, \\ \dfrac{(2x^3-3x^2-3x+2)^2}{4(x^2-x+1)^3}=t \\ \dfrac{4(x^2-x+1)^3}{(2x^3-3x^2-3x+2)^2}=t, \\ \dfrac{27x^2(1-x)^2}{4(x^2-x+1)^3}=t, \\ \dfrac{-27x^2(1-x)^2}{(2x^3-3x^2-3x+2)^2}=t. \end{array}\right.$$

LXIII...
$$
\begin{cases}
1+B=0,\ A=16C,\\[4pt]
l=m=\tfrac{5}{6},\\[4pt]
\text{———}\\[4pt]
\lambda=\pm\mu=\pm\dfrac{\nu}{4}.
\end{cases}
\qquad
\begin{cases}
\dfrac{(64x^3 - 96x^2 + 30x + 1)^2}{-108x(1-x)} = t,\\[10pt]
\dfrac{(16x^2 - 16x + 1)^3}{108x(1-x)} = t,\\[10pt]
\dfrac{(64x^3 - 96x^2 + 30x + 1)^2}{16x^2 - 16x + 1)^3} = t,\\[10pt]
\dfrac{(16x^2 - 16x + 1)^3}{(64x^3 - 96x^2 + 30x + 1)^2} =\\[10pt]
\dfrac{108x(1-x)}{(16x^2 - 16x + 1)^3} = t,\\[10pt]
\dfrac{-108x(1-x)}{(64x^3 - 96x^2 + 30x + 1)^2} = t.
\end{cases}
$$

LXIV...
$$
\begin{cases}
B+C=0,\ C=16A,\\[4pt]
l=\tfrac{1}{3},\ m=\tfrac{5}{6},\\[4pt]
\text{———}\\[4pt]
\dfrac{\lambda}{4}=\pm\mu=\pm\nu.
\end{cases}
\qquad
\begin{cases}
\dfrac{(x^3 + 30x^2 - 96x + 64)^2}{108x^4(1-x)} = t,\\[10pt]
\dfrac{(x^2 - 16x + 16)^3}{-108x^4(1-x)} = t,\\[10pt]
\dfrac{(x^3 + 30x^2 - 96x + 64)^2}{(x^2 - 16x + 16)^3} = t,\\[10pt]
\dfrac{(x^2 - 16x + 16)^3}{(x^3 + 30x - 96x + 64)^2} = t,\\[10pt]
\dfrac{-108x^4(1-x)}{(x^2 - 16x + 16)^3} = t,\\[10pt]
\dfrac{108x^4(1-x)}{(x^3 + 30x^2 - 96x + 64)^2} = t.
\end{cases}
$$

LXV...
$$
\begin{cases}
A=C,\ B=14C,\\[4pt]
l=\tfrac{5}{6},\ m=\tfrac{1}{3},\\[4pt]
\text{———}\\[4pt]
\lambda=\pm\dfrac{\mu}{4}=\pm\nu.
\end{cases}
\qquad
\begin{cases}
\dfrac{(x^3 - 33x^2 - 33x + 1)^2}{108x(1-x)^4} = t,\\[10pt]
\dfrac{(1 + 14x + x^2)^3}{-108x(1-x)^4} = t,\\[10pt]
\dfrac{(x^3 - 33x^2 - 33x + 1)^2}{(1 + 14x + x^2)^3} = t,\\[10pt]
\dfrac{1 + 14x + x^2)^3}{(x^3 - 33x^2 - 33x + 1)^2} = t,\\[10pt]
\dfrac{-108x(1-x)^4}{(1 + 14x + x^2)^3} = t,\\[10pt]
\dfrac{108x(1-x)^4}{(x^3 - 33x^2 - 33x + 1)^2} = t.
\end{cases}
$$

(14) We still have the hypothesis $n = 3$, $n' = 4$ to examine. It is easily shown that, under the imposed restrictions, it is impossible to have an identity

$$R^3 - S^4 = Ht^r(1 - t)^s;$$

there exist therefore no other rational transformations than those already determined.

(15) In combining among themselves the transformations which we can effect for the same values of A, B, C, l, m, we obtain new algebraic, but not rational, transformations. It remains to demonstrate that all such transformations are obtained in this way.

Resume the two differential equations (5) and (7) which are to be satisfied by the function x of t.

$$(5) \qquad \frac{\sqrt{Ax^2 + Bx + C}}{x(1 - x)} dx = \frac{\sqrt{A't^2 + B't + C'}}{t(1 - t)} dt.$$

$$(7) \qquad \frac{dx}{x^l(1 - x)^m} = \frac{Kdt}{t^{l'}(1 - t)^{m'}}.$$

The relation deduced from these is

$$(8) \qquad \frac{[\phi(x)]^2}{(Ax^2 + Bx + C)^3} = \frac{[\phi_1(t)]^2}{(A't^2 + B't + C')^3},$$

where

$$\phi(x) = x(x-1)(2Ax + B) + 2(Ax^2 + Bx + C)[(l-1)(x-1) + (m-1)x],$$

$$\phi_1(t) = t(t-1)(2A't + B') + 2(A't^2 + B't + C')[(l'-1)(t-1) + (m'-1)t].$$

Differentiating (8) and taking account of (5), we get the new relation

$$(9) \qquad \frac{x(x-1)[2\phi'(x)(Ax^2 + Bx + C) - 3\phi(x)(2Ax + B)]}{(Ax^2 + Bx + C)^3}$$

$$= \frac{t(t-1)[2\phi_1'(t)(A't^2 + B't + C') - 3\phi_1(t)(2A't + B')]}{(A't^2 + B't + C')^3}.$$

Every integral common to (5) and (7) satisfies (8) and (9). Conversely, every algebraic function which satisfies simultaneously (8) and (9) is a common integral to (5) and (7). In fact, every function which satisfies (8) satisfies also

$$\frac{2\phi'(x)(Ax^2+Bx+C)-3\phi(x)(2Ax+B)}{(Ax^2+Bx+C)^3}\sqrt{Ax^2+Bx+C}\,.\,dx$$

$$=\frac{2\phi_1'(t)(A't^2+B't+C')-3\phi_1(t)(2A't+B')}{(A't^2+B't+C')^3}\sqrt{A't^2+B't+C'}\,.\,dt,$$

or, taking (9) into account, equation (5). Taking account of (5), (8) can be written

$$d\,.\log\left[\sqrt{Ax^2+Bx+C}\,.\,x^{l-1}(1-x)^{m-1}\right]$$

$$=d\,.\log\left[\sqrt{A't^2+B't+C'}\,.\,t^{l'-1}(1-t)^{m'-1}\right].$$

We have then

$$\sqrt{Ax^2+Bx+C}\,.\,x^{l-1}(1-x)^{m-1}=\sqrt{A't^2+B't+C'}\,.\,t^{l'-1}(1-t)^{m'-1},$$

and consequently, still taking account of (5),

$$\frac{dx}{x^l(1-x)^m}=\frac{Kdt}{t^{l'}(1-t)^{m'}}.$$

We are thus reduced to the problem of finding the conditions to be satisfied in order that equations (8) and (9) shall have one or more common roots.

(16) To this end we shall first demonstrate the following theorem:

Whenever (8) *and* (9) *have a common factor of degree higher than the second with respect to one of the variables, they are identical.*

Suppose, for example, that they have a common factor of degree n in x ($n \geqq 3$). Let x and x_1 be two values of x corresponding to the same value of t. Between x and x_1 there will be a symmetric relation of degree n, at least, which should be contained in each of the following:

$$(10)\qquad \frac{[\phi(x)]^2}{(Ax^2+Bx+C)^3}=\frac{[\phi(x)]^2}{(Ax_1^2+Bx_1+C)^3}$$

$$(11)\quad \frac{x(x-1)[2\phi'(x)(Ax^2+Bx+C)-3\phi(x)(2Ax+B)]}{(Ax^2+Bx+C)^3}$$
$$=\frac{x_1(x_1-1)[2\phi(x_1)(Ax_1^2+Bx_1+C)-3\phi(x_1)(2Ax_1+B)]}{(Ax_1^2+Bx_1+C)^3}.$$

Multiply the first of these by 3 and add:

$$(12)\quad \frac{6\phi(x)[(l-1)(x-1)+(m-1)x]+2x(x-1)\phi'(x)}{(Ax^2+Bx+C)^3}$$
$$=\frac{6\phi(x_1)[(l-1)(x_1-1)+(m-1)x_1]+2x_1(x_1-1)\phi'(x_1)}{(Ax_1^2+Bx_1+C)^3}.$$

We can replace the system (10) and (11) by (10) and (12). For brevity write

$$\psi(x)=6\phi(x)[(l-1)(x-1)+(m-1)x]+2x(x-1)\phi'(x).$$

If $n=5$ or 6, the relation (12), which is at most of the fourth degree, must be an identity, and consequently equations (10) and (11) become the same. Let $n=4$. If (12) is not an identity, it must be an equation of the fourth degree; this requires

$$A\gtrless 0,\quad A+B+C\gtrless 0,\quad C\gtrless 0.$$

All the roots of (12) should belong to (10); this requires that Ax^2+Bx+C be a perfect square. In fact, let $x=\alpha$ be a simple root of $Ax^2+Bx+C=0$. Consider the two equations (10) and (12). For $x_1=\alpha$ two values of x become equal to α in (12), and, from the form of this equation, we have for these values

$$\lim\left(\frac{x-\alpha}{x_1-\alpha}\right)^2=1.$$

To prove this it suffices to write $x=\alpha+z$, $x_1=\alpha+z_1$; equation (12) can then be written

$$\frac{1+\epsilon}{z^2}=\frac{1+\epsilon_1}{z_1^2},$$

where ϵ and ϵ_1 are infinitely small at the same time as z and z_1; from this we evidently deduce

$$\lim \left(\frac{z}{z_1}\right) = 1.$$

In like manner, in equation (10), for $x_1 = \alpha$ three values of x become equal to α, and we have for these values

$$\lim \left(\frac{x-\alpha}{x_1-\alpha}\right)^3 = 1.$$

The two values of x which satisfy the first condition can evidently not satisfy the second. It is necessary, then, that we have

$$Ax^2 + Bx + C = A(x-\alpha)^2;$$

consequently

$$\phi(x) = 2A(x-\alpha)\{x(x-1) + (x-\alpha)[(l-1)(x-1) + (m-1)x]\}.$$

Relation (10) reduces to the fourth degree:

$$(10) \quad \frac{4A\{x(x-1) + (x-\alpha)[(l-1)(x-1) + (m-1)x]\}^2}{A^2(x-\alpha)^4} = \frac{4A\{x_1(x_1-1) + (x_1-\alpha)[(l-1)(x_1-1) + (m-1)x_1]\}^2}{A^2(x_1-\alpha)^4}.$$

In like manner (12) can be written

$$(12) \quad \frac{4Ax^2(x-1)^2 + (x-\alpha)f_1(x)}{A^2(x-\alpha)^4} = \frac{4Ax_1^2(x_1-1)^2 + (x_1-\alpha)f_1(x_1)}{A^2(x_1-\alpha)^4}.$$

Relations (10) and (12) should be the same. Subtracting them, member from member, we get a new relation,

$$(13) \quad \frac{f_2(x)}{A^2(x-\alpha)^3} = \frac{f_2(x_1)}{A^2(x_1-\alpha)^3},$$

which should be an identity, as it is at most of the third degree.

Suppose $n = 3$; if (12) is not an identity it should be, at least, of the third degree, since two of the quantities A, $A+B+C$, C cannot vanish at the same time. Let

$$A \gtrless 0, \quad A+B+C \gtrless 0.$$

There are several cases to be examined.

First Case.—$Ax^2 + Bx + C = 0$ admits two simple roots $x = \alpha$, $x = \beta$ different from zero. $\phi(\alpha)\phi(\beta) \gtrless 0$. We cannot have simultaneously $\psi(\alpha) = 0$ and $\psi(\beta) = 0$, for then (12) would be of a degree lower than the third. Suppose $\psi(\alpha)\psi(\beta) \gtrless 0$; for $x_1 = \alpha$, two values of x become equal to α and two to β, by (12); let us assume that the two values of x which become equal to α belong to (10). From (12) we should have

$$\lim \left(\frac{x-\alpha}{x_1-\alpha}\right)^2 = 1,$$

and from (10)

$$\lim \left(\frac{x-\alpha}{x_1-\alpha}\right)^3 = 1,$$

two incompatible conditions. We reason in the same way if $\psi(\beta) = 0$. This hypothesis must therefore be rejected.

Second Case.—$Ax^2 + Bx + C = 0$ admits the simple root $x = 0$ and another root $x = \alpha$ different from unity. Equations (10) and (12) become

$$(10) \qquad \frac{[\phi(x)]^2}{x(x-\alpha)^3} = \frac{[\phi(x_1)]^2}{x_1(x_1-\alpha)^3},$$

$$(12) \qquad \frac{\psi(x)}{x(x-\alpha)^2} = \frac{\psi(x_1)}{x_1(x_1-\alpha)^2};$$

$\psi(x)$ is of the third degree, and we should have $\psi(\alpha) \gtrless 0$, $\phi(\alpha) \gtrless 0$. Reasoning as before, we find that this hypothesis must also be rejected.

Third Case.—$Ax^2 + Bx + C = A(x-\alpha)^2$. Replace the system (10) and (12) by (10) and (13). This last equation must be an identity, otherwise it would be of the third degree, and we should derive from it

$$\lim \left(\frac{x-\alpha}{x_1-\alpha}\right)^3 = 1,$$

while (12) gives

$$\lim \left(\frac{x-\alpha}{x_1-\alpha}\right)^4 = 1.$$

Summing up: When n is equal to or greater than 3, one of the equations (12) or (13) reduces to an identity which requires that one of the rational fractions

$$\frac{\psi(x)}{(Ax^2+Bx+C)^2}, \quad \frac{f_2(x)}{A^2(x-\alpha)^3}$$

reduces to zero or a constant. It is easy to see directly that the numerators $\psi(x)$, $f_2(x)$ cannot be zero while (8) is not an identity. One of the two expressions ought then to have a constant value different from zero.

(17) It follows now that, if $\dfrac{\psi(x)}{(Ax^2+Bx+C)^2}$ reduces to a constant, we can make all the transformations

$$\frac{[\phi(x)]^2}{(Ax^2+Bx+C)^3} = \frac{[\phi_1(t)]^2}{(A't^2+B't+C')^3},$$

where A', B', C', l', m' have values such that the quotient

$$\frac{\psi_1(t)}{(A't^2+B't+C')^2}$$

reduces to the same constant. Among these transformations we can find one which is of the first degree in t; it is sufficient to take

$$A'+B'=0, \quad C'=0, \quad l'=\tfrac{1}{2}, \quad m'=\tfrac{2}{3};$$

then

$$\phi_1(t) = \frac{A'}{3} t^2 (t - 1),$$

$$\psi_1(t) = - \frac{A' t^2 (1 - t)^2}{3},$$

$$\frac{\psi_1(t)}{(A't^2 + B't + C')^2} = - \frac{1}{3A'}.$$

If we take a proper value for A', we shall be able to make the transformation

$$\frac{[\phi(x)]^2}{[Ax^2 + Bx + C]^3} = \frac{t}{9A'(t-1)},$$

which is of the first degree with respect to t. It is clear that every transformation which we can make in the same case will conduct to the preceding followed by a new transformation, equally of the first degree with respect to t.

If the ratio $\dfrac{f_2(x)}{A^2(x-\alpha)^3}$ has a constant value, we will take in like manner

$$A' + B' = 0, \quad C' = 0, \quad l' = \tfrac{1}{2}, \quad m' = \tfrac{3}{4},$$

$$\phi(t) = - \frac{A'}{2} t^2 (t - 1), \quad \phi_1'(t) = - \frac{A'}{2} t(3t - 2).$$

Now

$$\frac{[\phi_1(t)]^2}{(A't^2 + B't + C')^3} = \frac{t}{4A'(t-1)}$$

$$\frac{\psi_1(t)}{(A't^2 + B't + C')^2} = \frac{3t - 2}{4A'(t-1)}.$$

The difference of these is $\dfrac{-1}{2A'}$; if we take a proper value for A', we can effect the transformation

$$\frac{[\phi(x)]^2}{(Ax^2 + Bx + C)^3} = \frac{t}{4A'(t-1)},$$

from which we derive the same conclusion as above.

In order then to obtain the new transformations, it suffices to combine the rational transformations which can be effected for the two systems of values of the constants A, B, C, l, m,

$$C = 0, \quad A + B = 0, \quad l = \tfrac{1}{2}, \quad m = \tfrac{2}{3};$$

$$C = 0, \quad A + B = 0, \quad l = \tfrac{1}{2}, \quad m = \tfrac{3}{4}.$$

(18) It remains to examine the case where there exists a transformation of the second degree with respect to the two variables. Let x_0 be a value of x, and t_0, t_1 the corresponding values of t. To the value t_0 correspond for x the values x_0 and x_1; to the value t_1 correspond for x the values x_0 and x_2. If $x_1 \gtrless x_2$, the relation $F(x_0, x_1) = 0$ between two values of x which answer to the same value of t will be of the third degree, and we thus come back to the preceding hypothesis. If $x_1 = x_2$, the relation $F(x_0, x_1) = 0$ is of the second degree and breaks up into two relations of the first degree, viz.,

$$x_1 = x_0, \quad x_1 = \frac{ax_0 + b}{cx_0 + d}.$$

From what we have seen above this last relation should have one of the forms

$$x_1 = 1 - x_0, \quad x_1 = \frac{1}{x_0}, \quad x_1 = \frac{x_0}{x_0 - 1}.$$

Suppose, for example, $x_1 = 1 - x_0$. The relation between x and t will now be

$$(2x - 1)^2 = f_2(t),$$

$f_2(t)$ being a rational function of t of the second degree. The two differential equations

$$\frac{\sqrt{Ax^2 + Bx + C}}{x(1 - x)}\,dx = \frac{\sqrt{Ax_1^2 + Bx_1 + C}}{x_1(1 - x_1)}\,dx_1,$$

$$\frac{dx}{x^l(1-x)^m} = \frac{Kdx_1}{x_1^{\,l}(1-x_1)^m}$$

should then have for integral $x = 1 - x_1$; for this we must have $l = m$, $A + B = 0$; but then we could make the transformation

$$(2x-1)^2 = u,$$

and the proposed transformation would be a combination of the two following:

$$(2x-1)^2 = u, \quad u = f_2(t),$$

which is one of the transformations indicated above (VIII, X, XII).

(19) For brevity we will denote by U, V, W, respectively, any one of the transformations contained in the following three tables:

$$U\ldots\ldots \ (2t-1)^2, \ \left(\frac{2-t}{t}\right)^2, \ \left(\frac{1+t}{1-t}\right)^2, \ \frac{(2t-1)^2}{4t(t-1)}, \ \frac{(2-t)^2}{4(1-t)}, \ \frac{(1+t)^2}{4t}.$$

$$V\ldots\ldots \ \frac{(t^2-6t+1)^2}{-16t(1-t)^2}, \ \frac{(t^2+4t-4)^2}{-16t^2(1-t)}, \ \frac{(1+4t-4t^2)^2}{16t(1-t)}.$$

$$W\ldots \left\{\begin{array}{l}
\dfrac{(9t-8)^2}{-27t^2(1-t)}, \ \dfrac{(1-9t)^2}{-27t(1-t)^2}, \ \dfrac{t(9-8t)^2}{27(1-t)}, \ \dfrac{(1-t)(1+8t)^2}{27t}, \\[2ex]
\dfrac{t(t-9)^2}{-27(1-t)^2}, \ \dfrac{(t+8)^2(1-t)}{-27t^2}, \ \dfrac{(8t^2-36t+27)^2}{-64t^3(1-t)}, \\[2ex]
\dfrac{(8t^2+20t-1)^2}{-64t(1-t)^3}, \ \dfrac{(8-36t+27t^2)^2}{64(1-t)}, \ \dfrac{(27t^2-18t-1)^2}{64t}, \\[2ex]
\dfrac{(t^2+18t-27)^2}{64t^3}, \ \dfrac{(t^2-20t-8)^2}{64(1-t)^3}, \ \dfrac{(2t^3-3t^2-3t+2)^2}{-27t^2(1-t)^2}, \\[2ex]
\dfrac{(64t^3-96t^2+30t+1)^2}{-108t(1-t)}, \ \dfrac{(t^3+30t^2-96t+64)^2}{108t^4(1-t)}, \\[2ex]
\dfrac{(t^3-33t^2-33t+1)^2}{108t(1-t)^4}.
\end{array}\right.$$

Employing these conventions, all the new transformations will be found in the following table:

$$\text{LXVI}\ldots\left\{\begin{array}{c} A+B=0,\ C=0, \\ l=m=\tfrac{3}{4}, \\ \hline \lambda=\pm\mu=\pm\tfrac{1}{4}. \end{array}\right\}(2x-1)^2=V,$$

$$\text{LXVII}\ldots\left\{\begin{array}{c} A=0,\ C=0, \\ l=\tfrac{3}{4},\ m=\tfrac{1}{2}, \\ \hline \lambda=\pm\nu=\pm\tfrac{1}{4}. \end{array}\right\}\frac{(1+x)^2}{(1-x)^2}=V,$$

$$\text{LXVIII}\ldots\left\{\begin{array}{c} A=0,\ B+C=0, \\ l=\tfrac{1}{2},\ m=\tfrac{3}{4}, \\ \hline \mu=\pm\nu=\pm\tfrac{1}{4}. \end{array}\right\}\left(\frac{2-x}{x}\right)^2=V,$$

$$\text{LXIX}\ldots\left\{\begin{array}{c} A+B=0,\ C=0, \\ l=m=\tfrac{2}{3}, \\ \hline \lambda=\pm\mu=\pm\tfrac{1}{3}. \end{array}\right\}(2x-1)^2=W,$$

$$\text{LXX}\ldots\left\{\begin{array}{c} A=0,\ C=0, \\ l=m=\tfrac{2}{3}, \\ \hline \lambda=\pm\nu=\pm\tfrac{1}{3}. \end{array}\right\}\left(\frac{1+x}{1-x}\right)^2=W,$$

$$\text{LXXI}\ldots\left\{\begin{array}{c} A=0,\ B+C=0, \\ l=m=\tfrac{2}{3}, \\ \hline \mu=\pm\nu=\pm\tfrac{1}{3}. \end{array}\right\}\left(\frac{2-x}{x}\right)^2=W,$$

$$\text{LXXII}\ldots\left\{\begin{array}{c} A=0,\ B=0, \\ l=m=\tfrac{5}{6}, \\ \hline \nu=\pm\tfrac{2}{3},\ \lambda=\pm\mu. \end{array}\right\}\frac{(2x-1)^2}{4x(x-1)}=W,$$

$$\text{LXXIII}\ldots\left\{\begin{array}{c} B=0,\ C=0, \\ l=\tfrac{1}{3},\ m=\tfrac{5}{6}, \\ \hline \lambda=\pm\tfrac{2}{3},\ \mu=\pm\nu. \end{array}\right\}\frac{(2-x)^2}{4(1-x)}=W,$$

$$\text{LXXIV}\ldots\left\{\begin{array}{c} A = C,\ B+2C = 0, \\ l = \frac{5}{6},\ m = \frac{1}{3}, \\ \hline \mu = \pm\frac{2}{3},\ \lambda = \pm\nu. \end{array}\right\} \frac{(1+x)^2}{4x} = W,$$

$$\text{LXXV}\ldots\left\{\begin{array}{c} A = C,\ B = 2A, \\ l = \frac{3}{4},\ m = \frac{1}{2}, \\ \hline \lambda = \frac{\mu}{2} = \pm\nu. \end{array}\right\} \begin{array}{l} \frac{(x^2-6x+1)^2}{-16x(1-x)^2} = U, \\ \frac{(x^2-6x+1)^2}{-16x(1-x)^2} = V, \end{array}$$

$$\text{LXXVI}\ldots\left\{\begin{array}{c} C = 4A,\ B = -4A, \\ l = \frac{1}{2},\ m = \frac{3}{4}, \\ \hline \frac{\lambda}{2} = \pm\mu = \pm\nu. \end{array}\right\} \begin{array}{l} \frac{(x^2+4x-4)^2}{-16x^2(1-x)} = U, \\ \frac{(x^2+4x-4)^2}{-16x^2(1-x)} = V, \end{array}$$

$$\text{LXXVII}\ldots\left\{\begin{array}{c} A = 4C,\ B = -4C, \\ l = m = \frac{3}{4}, \\ \hline \lambda = \pm\mu = \pm\frac{\nu}{2}. \end{array}\right\} \begin{array}{l} \frac{(1+4x-4x^2)^2}{76x(1-x)} = U, \\ \frac{(1+4x-4x^2)^2}{16x(1-x)} = V, \end{array}$$

$$\text{LXXVIII}\ldots\left\{\begin{array}{c} A = 0,\ 4B+3C = 0, \\ l = \frac{2}{3},\ m = \frac{5}{6}, \\ \hline \nu = \pm\frac{1}{2},\ \frac{\lambda}{2} = \pm\mu. \end{array}\right\} \begin{array}{l} \frac{(9x-8)^2}{-27x^2(1-x)} = U, \\ \frac{(9x-8)^2}{-27x^2(1-x)} = W, \end{array}$$

$$\text{LXXIX}\ldots\left\{\begin{array}{c} A = 0,\ B = 3C, \\ l = \frac{5}{6},\ m = \frac{2}{3}, \\ \hline \nu = \pm\frac{1}{2},\ \lambda = \pm\frac{\mu}{2}. \end{array}\right\} \begin{array}{l} \frac{(1-9x)^2}{-27x(1-x)^2} = U, \\ \frac{(1-9x)^2}{-27x(1-x)^2} = W, \end{array}$$

$$\text{LXXX}\ldots\left\{\begin{array}{c} C=0,\ 4B+3A=0, \\ l=\tfrac{1}{2},\ m=\tfrac{5}{6}, \\ \hline \lambda=\pm\tfrac{1}{2},\ \mu=\pm\dfrac{\nu}{2}. \end{array}\right\}\begin{array}{l} \dfrac{(9-8x)^2x}{27(1-x)}=U, \\ \dfrac{(9-8x)^2x}{27(1-x)}=W, \end{array}$$

$$\text{LXXXI}\ldots\left\{\begin{array}{c} C=0,\ B=3A, \\ l=\tfrac{1}{2},\ m=\tfrac{2}{3}, \\ \hline \lambda=\pm\tfrac{1}{2},\ \dfrac{\mu}{2}=\pm\nu. \end{array}\right\}\begin{array}{l} \dfrac{(x-9)^2x}{-27(1-x)^2}=U, \\ \dfrac{(x-9)^2x}{-27(1-x)^2}=W, \end{array}$$

$$\text{LXXXII}\ldots\left\{\begin{array}{c} C=4A,\ B+5A=0, \\ l=\tfrac{2}{3},\ m=\tfrac{1}{2}, \\ \hline \mu=\pm\tfrac{1}{2},\ \dfrac{\lambda}{2}=\pm\nu. \end{array}\right\}\begin{array}{l} \dfrac{(x+8)^2(1-x)}{-27x^2}=U, \\ \dfrac{(x+8)^2(1-x)}{-27x^2}=W, \end{array}$$

$$\text{LXXXIII}\ldots\left\{\begin{array}{c} A=4C,\ B+5C=0, \\ l=\tfrac{5}{6},\ m=\tfrac{1}{2}, \\ \hline \mu=\pm\tfrac{1}{2},\ \lambda=\pm\dfrac{\nu}{2}. \end{array}\right\}\begin{array}{l} \dfrac{(1+8x)^2(1-x)}{27x}=U, \\ \dfrac{(1+8x)^2(1-x)}{27x}=W, \end{array}$$

$$\text{LXXXIV}\ldots\left\{\begin{array}{c} A=0,\ 9B+8C=0, \\ l=\tfrac{1}{2},\ m=\tfrac{5}{6}, \\ \hline \nu=\pm\tfrac{1}{3},\ \dfrac{\lambda}{3}=\pm\mu. \end{array}\right\}\begin{array}{l} \dfrac{(8x^2-36x+27)^2}{-64x^3(1-x)}=U, \\ \dfrac{(8x^2-36x+27)^2}{-64x^3(1-x)}=W, \end{array}$$

$$\text{LXXXV}\ldots\left\{\begin{array}{c} A=0,\ B=8C, \\ l=\tfrac{5}{6},\ m=\tfrac{1}{2}, \\ \hline \nu=\pm\tfrac{1}{3},\ \lambda=\pm\dfrac{\mu}{3}. \end{array}\right\}\begin{array}{l} \dfrac{(8x^2+20x-1)^2}{-64x(1-x)^3}=U, \\ \dfrac{(8x^2+20x-1)^2}{-64x(1-x)^3}=W, \end{array}$$

$$\text{LXXXVI}\ldots\left\{\begin{array}{c} C = 0,\ 9B+8A = 0, \\ l = \tfrac{2}{3},\ m = \tfrac{5}{6}, \\ \overline{} \\ \lambda = \pm \tfrac{1}{3},\ \mu = \pm \dfrac{\nu}{3}. \end{array}\right\} \begin{array}{l} \dfrac{(8 - 36x + 27x^2)^2}{64(1 - x)} = U, \\ \dfrac{(8 - 36x + 27x^2)^2}{64(1 - x)} = W, \end{array}$$

$$\text{LXXXVII}\ldots\left\{\begin{array}{c} C = 0,\ B = 8A, \\ l = \tfrac{2}{3},\ m = \tfrac{1}{2}, \\ \overline{} \\ \lambda = \pm \tfrac{1}{3},\ \dfrac{\mu}{3} = \pm \nu. \end{array}\right\} \begin{array}{l} \dfrac{(8 + 20x - x^2)^2}{64(1 - x)^3} = U, \\ \dfrac{(8 + 20x - x^2)^2}{64(1 - x)^3} = W, \end{array}$$

$$\text{LXXXVIII}\ldots\left\{\begin{array}{c} A = 9C,\ B + 10C = 0, \\ l = \tfrac{5}{6},\ m = \tfrac{2}{3}, \\ \overline{} \\ \mu = \pm \tfrac{1}{3},\ \lambda = \pm \dfrac{\nu}{3}. \end{array}\right\} \begin{array}{l} \dfrac{(27x^2 - 18x - 1)^2}{64x} = U, \\ \dfrac{(27x^2 - 18x - 1)^2}{64x} = W, \end{array}$$

$$\text{LXXXIX}\ldots\left\{\begin{array}{c} C = 9A,\ B + 10A = 0, \\ l = \tfrac{1}{2},\ m = \tfrac{2}{3}, \\ \overline{} \\ \mu = \pm \tfrac{1}{3},\ \dfrac{\lambda}{3} = \pm \nu. \end{array}\right\} \begin{array}{l} \dfrac{(x^2 + 18x - 27)^2}{64x^3} = U, \\ \dfrac{(x^2 + 18x - 27)^2}{64x^3} = W, \end{array}$$

$$\text{XC}\ldots\left\{\begin{array}{c} A = C,\ B + C = 0, \\ l = m = \tfrac{2}{3}, \\ \overline{} \\ \lambda = \pm \mu = \pm \nu. \end{array}\right\} \begin{array}{l} \dfrac{(2x^3 - 3x^2 - 3x + 2)^2}{-27x^2(1 - x)^2} = U, \\ \dfrac{(2x^3 - 3x^2 - 3x + 2)^2}{-27x^2(1 - x)^2} = W, \end{array}$$

$$\text{XCI}\ldots\left\{\begin{array}{c} A + B = 0,\ A = 16C, \\ l = m = \tfrac{5}{6}, \\ \overline{} \\ \lambda = \pm \mu = \pm \dfrac{\nu}{4}. \end{array}\right\} \begin{array}{l} \dfrac{(64x^3 - 96x^2 + 30x + 1)^2}{-108x(1 - x)} = U, \\ \dfrac{(64x^3 - 96x^2 + 30x + 1)^2}{-108x(1 - x)} = W, \end{array}$$

$$\text{XCII}\ldots\left\{\begin{array}{c} B+C=0,\ C=16A, \\ l=\frac{1}{3},\ m=\frac{5}{6}, \\ \overline{\qquad} \\ \frac{\lambda}{4}=\pm\,\mu=\pm\,\nu. \end{array}\right\}\begin{array}{l} \dfrac{(x^3+30x^2-96x+64)^2}{108x^4(1-x)}=U, \\ \dfrac{(x^3+30x^2-96x+1)^2}{108x^4(1-x)}=W, \end{array}$$

$$\text{XCIII}\ldots\left\{\begin{array}{c} A=C,\ B=14C, \\ l=\frac{5}{6},\ m=\frac{1}{3}, \\ \overline{\qquad} \\ \lambda=\pm\,\frac{\mu}{4}=\pm\,\nu. \end{array}\right\}\begin{array}{l} \dfrac{(x^3-33x^2-33x+1)^2}{108x(1-x)^4}=U, \\ \dfrac{(x^3-33x^2-33x+1)^2}{108x(1-x)^4}=W. \end{array}$$

The transformations thus obtained are not all irreducible. Thus, it is clear that the relation

$$\frac{(9x-8)^2}{-27x^2(1-x)}=\frac{(9t-8)^2}{-27t^2(1-t)}$$

is equivalent to two distinct relations, of which one is $x = t$, and the other is of the second order with respect to each of the variables. So, also, the transformation

$$(2x-1)^2=\frac{(2t^3-3t^2-3t+2)^2}{-27t^2(1-t)^2}$$

breaks up into two rational transformations which have already been found directly:

$$2x-1=\pm\,\frac{2t^3-3t^2-3t+2}{3i\sqrt{3}\,.\,t(1-t)}.$$

These simplifications are readily perceived in each particular case, so no more need be said of them.

(20) Suppose that for proper values of A, B, C, l, m we can effect the transformation $x=\phi(t)$, and thus pass from equation (2) to equation (4). If for $x=0$ several values of t become equal to t_1, different from zero and unity, these values, as we have seen, will be two,

three, or four in number. Suppose first that there are two such values, say t' and t'', and suppose further mod. $t_1 < 1$. Equation (4), and consequently equation (2), will then admit the integral

$$t'^{p'}(1-t')^{q'}F(\alpha', \beta', \gamma', t') + t''^{p'}(1-t'')^{q'}F(\alpha', \beta', \gamma', t'').$$

This function is evidently holomorphic in the region of the point $x=0$; for, if we make the variable x describe a loop about this point, the roots t' and t'' will simply be interchanged. Equation (2) will then admit a uniform integral in the region of the origin. This integral will be of the form

$$(1-x)^{-q}F(\alpha, \beta, \gamma, x);$$

and, denoting by A a suitable constant, we shall have

$$\begin{aligned} &A(1-x)^{-q}F(\alpha, \beta, \gamma, x) \\ &\qquad = t'^{p'}(1-t')^{q'}F(\alpha', \beta', \gamma', t') + t''^{p'}(1-t'')^{q'}F(\alpha', \beta', \gamma', t''). \end{aligned}$$

Equation (2) also admits the integral

$$t'^{p'}(1-t')^{q'}F(\alpha', \beta', \gamma', t') - t''^{p'}(1-t'')^{q'}F(\alpha', \beta', \gamma', t''),$$

which is reproduced, to sign près, when x describes a loop round the origin. This integral will be of the form

$$x^{\frac{1}{2}}(1-x)^{-q_1}F(\alpha_1, \beta_1, \gamma_1, x);$$

and so, denoting by B a constant, we shall have

$$\begin{aligned} &Bx^{\frac{1}{2}}(1-x)^{-q_1}F(\alpha_1, \beta_1, \gamma_1, x) \\ &\qquad = t'^{p'}(1-t')^{q'}F(\alpha', \beta', \gamma', t') - t''^{p'}(1-t'')^{q'}F(\alpha', \beta', \gamma', t''). \end{aligned}$$

If the values of t which become $=t_1$ for $x=0$ are in number greater than 2, it is easy to form the integrals of (2) which, when the variable describes a loop round $x=0$, reproduce themselves multiplied by a constant. Suppose, for example, that three values of t become equal to t_1 when $x=0$, and let t', t'', t''' be these three roots arranged in the order in which they present themselves when the variable describes, in the direct sense, successive loops round the origin. The integral

$$t'^{p'}(1-t')^{q'}F(\alpha', \beta', \gamma', t')+t''^{p'}(1-t'')^{q'}F(\alpha', \beta', \gamma', t'')$$
$$+t'''^{p'}(1-t''')^{q'}F(\alpha', \beta', \gamma', t''')$$

is evidently uniform in the region of the origin while the integral

$$t'^{p'}(1-t')^{q'}F(\alpha', \beta', \gamma', t')+j^2t''^{p'}(1-t'')^{q'}F(\alpha', \beta', \gamma', t'')$$
$$+jt'''^{p'}(1-t''')^{q'}F(\alpha', \beta', \gamma', t''')$$

reproduces itself multiplied by j when x describes a loop in the direct sense round the origin. We deduce now relations analogous to the preceding. One or several values of t may be zero at the same time as x. If the value of t which vanishes for $x=0$ is unique, the integral $t^{p'}(1-t)^{q'}F(\alpha', \beta', \gamma', t)$ reproduces itself, to a constant factor près, when x describes a closed path surrounding the origin. It ought then to be equal to an integral of the form

$$Cx^{-p}(1-x)^{-q}F(\alpha, \beta, \gamma, x).$$

The case when several values of t are zero at the same time as x requires a fuller explanation. Suppose first that $x=\phi(t)$, where ϕ denotes a rational fraction. Equation (2) admits an integral of the form

$$x^{-p}(1-x)^{-q}F(\alpha, \beta, \gamma, x),$$

and the same is true of equation (4). If we consider t as the independent variable, then when t describes a loop round $t=0$, x will return to its initial value after having described several loops round $x=0$, and the preceding integral will reproduce itself multiplied by a constant factor. We have then

$$x^{-p}(1-x)^{-q}F(\alpha, \beta, \gamma, x)=C't'^{p'}(1-t')^{q'}F(\alpha', \beta', \gamma', t'),$$

t' denoting one of the roots of the equation $x=\phi(t)$ which are zero when $x=0$. Since every transformation can be led back to rational transformations, the preceding conclusion holds when x is no longer a rational function of t.

The calculation of the constants which enter into our formulæ is effected without difficulty, when for $x = 0$ we have at the same time $t = 0$; it is sufficient to seek the limit of the ratio $\frac{t'^{p'}}{x^{-p}}$ for $x = 0$. Now take the case when for $x = 0$, t takes a finite value, t_1, different from zero and unity. In the formulæ written above make $x = 0$, $t = t_1$; we get

$$A = 2t_1^{p'}(1 - t_1)^{q'} F(\alpha', \beta', \gamma', t_1),$$

$$B = \lim. \frac{t'^{p'}(1-t')^{q'}F(\alpha', \beta', \gamma', t') - t''^{p'}(1-t'')^{q'}F(\alpha', \beta', \gamma', t'')}{x^{\frac{1}{2}}},$$

for $x = 0$, $t = t_1$,

$$B = 2\left(\frac{dt'}{dx^{\frac{1}{2}}}\right)_{x=0} \frac{d}{dt}[t'^{p'}(1 - t')^{q'} F(\alpha', \beta', \gamma', t']_{t=t_1}.$$

The following examples show the method to be followed in each case.

(21) Consider the differential equation

$$x(1 - x)\frac{d^2y}{dx^2} + [\tfrac{1}{2} - (\alpha + \beta + 1)x]\frac{dy}{dx} - \alpha\beta y = 0.$$

Writing $x = (2t - 1)^2$, this becomes

$$t(t - 1)\frac{d^2y}{dt^2} + [\alpha + \beta + \tfrac{1}{2} - (2\alpha + 2\beta + 1)t]\frac{dy}{dt} - 4\alpha\beta y = 0.$$

For $x = 0$ the two values of t are equal to $\frac{1}{2}$. We shall then have

$$aF(\alpha, \beta, \tfrac{1}{2}, x)$$
$$= F\left(2\alpha, 2\beta, \alpha + \beta + \tfrac{1}{2}, \frac{1 + \sqrt{x}}{2}\right) + F\left(2\alpha, 2\beta, \alpha + \beta + \tfrac{1}{2}, \frac{1 - \sqrt{x}}{2}\right);$$

$$b\sqrt{x}F(\alpha + \tfrac{1}{2}, \beta + \tfrac{1}{2}, \tfrac{3}{2}, x)$$
$$= F\left(2\alpha, 2\beta, \alpha + \beta + \tfrac{1}{2}, \frac{1 + \sqrt{x}}{2}\right) - F\left(2\alpha, 2\beta, \alpha + \beta + \tfrac{1}{2}, \frac{1 - \sqrt{x}}{2}\right).$$

For $x = 1$ one of the values of t is $= 0$; then

$$F(\alpha, \beta, \alpha+\beta+\tfrac{1}{2}, 1-x) = F\left(2\alpha, 2\beta, \alpha+\beta+\tfrac{1}{2}, \frac{1-\sqrt{x}}{2}\right).$$

This last formula enables us to calculate a and b. We have in fact

$$a = 2F(2\alpha, 2\beta, \alpha+\beta+\tfrac{1}{2}, \tfrac{1}{2}),$$

or, from the preceding formula,

$$a = 2F(\alpha, \beta, \alpha+\beta+\tfrac{1}{2}, 1), = \frac{2\sqrt{\pi}\,\Gamma(\alpha+\beta+\frac{1}{2})}{\Gamma(\alpha+\frac{1}{2})\Gamma(\beta+\frac{1}{2})}.$$

In like manner,

$$b = \frac{4\alpha\beta}{\alpha+\beta+\frac{1}{2}} F(2\alpha+1, 2\beta+1, \alpha+\beta+\tfrac{3}{2}, \tfrac{1}{2}).$$

In the value of a change α into $\alpha+\frac{1}{2}$ and β into $\beta+\frac{1}{2}$; we find then

$$b = \frac{4\alpha\beta}{\alpha+\beta+\frac{1}{2}} \frac{\sqrt{\pi}\,\Gamma(\alpha+\beta+\frac{3}{2})}{\Gamma(\alpha+1)\Gamma(\beta+1)} = \frac{4\sqrt{\pi}\,\Gamma(\alpha+\beta+\frac{1}{2})}{\Gamma(\alpha)\Gamma(\beta)}.$$

(22) Consider again the differential equation

$$x(1-x)\frac{d^2y}{dx^2} + \left(\frac{1}{2} - \frac{5x}{4}\right)\frac{dy}{dx} - \alpha(\tfrac{1}{4}-\alpha)y = 0.$$

Writing
$$x = \frac{(t^2-6t+1)^2}{-16t(1-t)^2},$$

and at the same time making

$$y = t^{\alpha}(1-t)^{2\alpha}z,$$

we have the new differential equation

$$t(1-t)\frac{d^2z}{dt^2} + \left[2\alpha+\tfrac{3}{4} - (6\alpha+\tfrac{5}{4})t\right]\frac{dz}{dt} - 4\alpha(2\alpha+\tfrac{1}{4})z = 0.$$

For $x = 0$ two values of t become equal to $3 - 2\sqrt{2}$.

Let t' and t'' denote these values; in the region of $x = 0$ they are developable in a convergent series going according to ascending powers of $x^{\frac{1}{2}}$. Let

$$t' = 3 - 2\sqrt{2} + \sqrt{-1}\,(3\sqrt{2} - 4)x^{\frac{1}{2}} + \ldots,$$
$$t'' = 3 - 2\sqrt{2} - \sqrt{-1}\,(3\sqrt{2} - 4)x^{\frac{1}{2}} + \ldots.$$

The integral

$$t'^{\alpha}(1 - t')^{2\alpha}F(4\alpha, 2\alpha + \tfrac{1}{4}, 2\alpha + \tfrac{3}{4}, t') \\ + t''^{\alpha}(1 - t'')^{2\alpha}F(4\alpha, 2\alpha + \tfrac{1}{4}, 2\alpha + \tfrac{3}{4}, t'')$$

is uniform in the region of the origin. We have then, A being a constant,

$$AF(\alpha, \tfrac{1}{4} - \alpha, \tfrac{1}{2}, x) = t'^{\alpha}(1 - t')^{2\alpha}F(4\alpha, 2\alpha + \tfrac{1}{4}, 2\alpha + \tfrac{3}{4}, t') \\ + t''^{\alpha}(1 - t'')^{2\alpha}F(4\alpha, 2\alpha + \tfrac{1}{4}, 2\alpha + \tfrac{3}{4}, t'').$$

In like manner, denoting by B a new constant,

$$Bx^{\frac{1}{2}}F(\alpha + \tfrac{1}{2}, \tfrac{3}{4} - \alpha, \tfrac{3}{2}, x) = t'^{\alpha}(1 - t')^{2\alpha}F(4\alpha, 2\alpha + \tfrac{1}{4}, 2\alpha + \tfrac{3}{4}, t') \\ - t''^{\alpha}(1 - t'')^{2\alpha}F(4\alpha, 2\alpha + \tfrac{1}{4}, 2\alpha + \tfrac{3}{4}, t'').$$

Making $x = 0$, we get

$$A = 2(3 - 2\sqrt{2})^{\alpha}(2\sqrt{2} - 2)^{2\alpha}F(4\alpha, 2\alpha + \tfrac{1}{4}, 2\alpha + \tfrac{3}{4}, 3 - 2\sqrt{2}),$$
$$B = 2\sqrt{-1}\,(3\sqrt{2} - 4)\frac{d}{dt}\left[t^{\alpha}(1 - t)^{2\alpha}F(4\alpha, 2\alpha + \tfrac{1}{4}, 2\alpha + \tfrac{3}{4}, t)\right]$$
$$\text{for } t = 3 - 2\sqrt{2}.$$

Observe now that, for $x = \infty$, one of the values of t becomes zero. The proposed differential equation admits the two integrals

$$t^{\alpha}(1 - t)^{2\alpha}F(4\alpha, 2\alpha + \tfrac{1}{4}, 2\alpha + \tfrac{3}{4}, t),$$
$$x^{-\alpha}F\left(\alpha, \alpha + \tfrac{1}{2}, 2\alpha + \tfrac{3}{4}, \frac{1}{x}\right)$$

which belong to the same exponent in the region of the point $x = \infty$.

They ought then to be identical, to a constant près. We deduce from them the relation

$$F(4\alpha, 2\alpha+\tfrac{1}{4}, 2\alpha+\tfrac{3}{4}, t)$$
$$= (t^2-6t+1)^{-2\alpha} F\left(\alpha, \alpha+\tfrac{1}{2}, 2\alpha+\tfrac{3}{4}, \frac{-16t(1-t)^2}{(t^2-6t+1)^2}\right),$$

which can also be written

$$F(4\alpha, 2\alpha+\tfrac{1}{4}, 2\alpha+\tfrac{3}{4}, t)$$
$$= (1+t)^{-4\alpha} F\left(\alpha, \alpha+\tfrac{1}{4}, 2\alpha+\tfrac{3}{4}, \frac{16t(1-t)^2}{(1+t)^4}\right).$$

In this last formula make $t = 3 - 2\sqrt{2}$; then

$$F(4\alpha, 2\alpha+\tfrac{1}{4}, 2\alpha+\tfrac{3}{4}, 3-2\sqrt{2}) = (4-2\sqrt{2})^{-4\alpha} F(\alpha, \alpha+\tfrac{1}{4}, 2\alpha+\tfrac{3}{4}, 1),$$

and so

$$A = \left[\frac{(3-2\sqrt{2})(2-2\sqrt{2})^2}{(4-2\sqrt{2})^4}\right]^{\alpha} \frac{2\sqrt{\pi}\,\Gamma(2\alpha+\tfrac{3}{4})}{\Gamma(\alpha+\tfrac{1}{2})\Gamma(\alpha+\tfrac{3}{4})}$$
$$= \left(\frac{1}{16}\right)^{\alpha} \frac{2\sqrt{\pi}\,\Gamma(2\alpha+\tfrac{3}{4})}{\Gamma(\alpha+\tfrac{1}{2})\Gamma(\alpha+\tfrac{3}{4})}.$$

In the same way we have

$$t^{\alpha}(1-t)^{2\alpha} F(4\alpha, 2\alpha+\tfrac{1}{4}, 2\alpha+\tfrac{3}{4}, t)$$
$$= \left[\frac{t(1-t)^2}{(1+t)^4}\right]^{\alpha} F\left[\alpha, \alpha+\tfrac{1}{4}, 2\alpha+\tfrac{3}{4}, \frac{16t(1-t)^2}{(1+t)^4}\right]$$
$$= \left[\frac{t(1-t)^2}{(1+t)^4}\right]^{\alpha} \left\{ \frac{\sqrt{\pi}\,\Gamma(2\alpha+\tfrac{3}{4})}{\Gamma(\alpha+\tfrac{1}{2})\Gamma(\alpha+\tfrac{3}{4})} F\left[\alpha, \alpha+\tfrac{1}{4}, \tfrac{1}{2}, \frac{(t^2-6t+1)^2}{(1+t)^4}\right]\right.$$
$$\left. + \frac{\Gamma(-\tfrac{1}{2})\Gamma(2\alpha+\tfrac{3}{4})}{\Gamma(\alpha)\Gamma(\alpha+\tfrac{1}{4})} \frac{t^2-6t+1}{(1+t)^2} F\left[\alpha+\tfrac{1}{2}, \alpha+\tfrac{3}{4}, \tfrac{3}{2}, \frac{(t^2-6t+1)^2}{(1+t)^4}\right]\right\}.$$

Taking the derivative of the second member and neglecting all terms which are zero for $t = 3 - 2\sqrt{2}$, we find

$$B = 2\sqrt{-1}\left[\frac{(3-2\sqrt{2})(2\sqrt{2}-2)^2}{(4-2\sqrt{2})^{\alpha}}\right]^{\alpha}$$
$$\frac{(3\sqrt{2}-4)(-4\sqrt{2})}{(4-2\sqrt{2})^2} \cdot \frac{\Gamma(-\tfrac{1}{2})\Gamma(2\alpha+\tfrac{3}{4})}{\Gamma(\alpha)\Gamma(\alpha+\tfrac{1}{4})},$$

or, on reducing,

$$B = \sqrt{-1}\left(\frac{1}{16}\right)^{a} \frac{4\sqrt{\pi}\Gamma(2\alpha + \frac{3}{4})}{\Gamma(\alpha)\Gamma(\alpha + \frac{1}{4})}.$$

23. The following formulæ have been obtained in an analogous way:

(25) $aF(\alpha, \beta, \frac{1}{2}, x)$

$$= F\left(2\alpha, 2\beta, \alpha + \beta + \tfrac{1}{2}, \frac{1+\sqrt{x}}{2}\right) + F\left(2\alpha, 2\beta, \alpha + \beta + \tfrac{1}{2}, \frac{1-\sqrt{x}}{2}\right),$$

$$a = \frac{2\sqrt{\pi}\Gamma(\alpha + \beta + \frac{1}{2})}{\Gamma(\alpha + \frac{1}{2})\Gamma(\beta + \frac{1}{2})},$$

(26) $cF(\alpha, \beta, \frac{1}{2}, x)$

$$= (1-x)^{-a} F\left(2\alpha, 1 - 2\beta, \alpha + 1 - \beta, \frac{\sqrt{x-1} + \sqrt{x}}{2\sqrt{x-1}}\right)$$

$$+ (1-x)^{-a} F\left(2\alpha, 1 - 2\beta, \alpha + 1 - \beta, \frac{\sqrt{x-1} - \sqrt{x}}{2\sqrt{x-1}}\right),$$

$$c = \frac{2\sqrt{\pi}\Gamma(\alpha + 1 - \beta)}{\Gamma(\alpha + \frac{1}{2})\Gamma(1 - \beta)},$$

(27) $b\sqrt{x}F(\alpha, \beta, \frac{3}{2}, x) = F\left(2\alpha - 1, 2\beta - 1, \alpha + \beta - \frac{1}{2}, \frac{1+\sqrt{x}}{2}\right)$

$$- F\left(2\alpha - 1, 2\beta - 1, \alpha + \beta - \tfrac{1}{2}, \frac{1-\sqrt{x}}{2}\right).$$

(28) $d\sqrt{x}F(\alpha, \beta, \frac{3}{2}, x)$

$$= (1-x)^{-a} F\left(2\alpha - 1, 2 - 2\beta, \alpha + 1 - \beta, \frac{\sqrt{x-1} + \sqrt{x}}{2\sqrt{x-1}}\right)$$

$$- (1-x)^{-a} F\left(2\alpha - 1, 2 - 2\beta, \alpha + 1 - \beta, \frac{\sqrt{x-1} - \sqrt{x}}{2\sqrt{x-1}}\right),$$

$$b = \frac{4\sqrt{\pi}\Gamma(\alpha + \beta - \frac{1}{2})}{\Gamma(\alpha - \frac{1}{2})\Gamma(\beta - \frac{1}{2})}, \quad d = \frac{4\sqrt{\pi}\Gamma(\alpha + 1 - \beta)}{\Gamma(\alpha - \frac{1}{2})\Gamma(1 - \beta)},$$

(29) $F(\alpha, \beta, \alpha + \beta + \tfrac{1}{2}, x) = F\left(2\alpha, 2\beta, \alpha + \beta + \tfrac{1}{2}, \dfrac{1 - \sqrt{1 - x}}{2}\right),$

(30) $F(\alpha, \beta, \alpha + \beta + \tfrac{1}{2}, x)$

$$= \left(\frac{1 + \sqrt{1 - x}}{2}\right)^{-2\alpha} F\left(2\alpha, \alpha - \beta + \tfrac{1}{2}, \alpha + \beta + \tfrac{1}{2}, \frac{\sqrt{1 - x} - 1}{\sqrt{1 + x} + 1}\right),$$

(31) $F(\alpha, \beta, \alpha + \beta + \tfrac{1}{2}, x)$

$$= (\sqrt{1 - x} + \sqrt{-x})^{-2\alpha} F\left(2\alpha, \alpha + \beta, 2\alpha + 2\beta, \frac{2\sqrt{-x}}{\sqrt{1 - x} + \sqrt{-x}}\right),$$

(32) $F(\alpha, \beta, \alpha + \beta - \tfrac{1}{2}, x)$

$$= (1 - x)^{-\frac{1}{2}} F\left(2\alpha - 1, 2\beta - 1, \alpha + \beta - \tfrac{1}{2}, \frac{1 - \sqrt{1 - x}}{2}\right),$$

(33) $F(\alpha, \beta, \alpha + \beta - \tfrac{1}{2}, x) = (1 - x)^{-\frac{1}{2}} \left(\dfrac{1 + \sqrt{1 - x}}{2}\right)^{1 - 2\alpha}$

$$F\left(2\alpha - 1, \alpha - \beta + \tfrac{1}{2}, \alpha + \beta - \tfrac{1}{2}, \frac{\sqrt{1 - x} - 1}{\sqrt{1 - x} + 1}\right),$$

(34) $F(\alpha, \beta, \alpha + \beta - \tfrac{1}{2}, x) = (1 - x)^{-\frac{1}{2}} (\sqrt{1 - x} + \sqrt{-x})^{1 - 2\alpha}$

$$F\left(2\alpha - 1, \alpha + \beta - 1, 2\alpha + 2\beta - 2, \frac{2\sqrt{-x}}{\sqrt{1 - x} + \sqrt{-x}}\right),$$

(35) $F(\alpha, \alpha + \tfrac{1}{2}, \gamma, x)$

$$= \left(\frac{1 + \sqrt{1 - x}}{2}\right)^{-2\alpha} F\left(2\alpha, 2\alpha + 1 - \gamma, \gamma, \frac{1 - \sqrt{1 - x}}{1 + \sqrt{1 - x}}\right),$$

(36) $F(\alpha, \alpha + \tfrac{1}{2}, \gamma, x) = (1 - x)^{-\alpha} F\left(2\alpha, 2\gamma - 2\alpha - 1, \gamma, \dfrac{\sqrt{1 - x} - 1}{2\sqrt{1 - x}}\right),$

(37) $F(\alpha, \alpha + \tfrac{1}{2}, \gamma, x) = (1 + \sqrt{x})^{-2\alpha} F\left(2\alpha, \gamma - \tfrac{1}{2}, 2\gamma - 1, \dfrac{2\sqrt{x}}{1 + \sqrt{x}}\right),$

$$(38)\quad F\left(\alpha,\ \beta,\ \frac{\alpha+\beta+1}{2},\ x\right)=F\left[\frac{\alpha}{2},\frac{\beta}{2},\ \frac{\alpha+\beta+1}{2},\ 4x(1-x)\right]$$

$$=(1-2x)F\left[\frac{\alpha+1}{2},\ \frac{\beta+1}{2},\ \frac{\alpha+\beta+1}{2},\ 4x(1-x)\right],$$

$$(39)\quad F\left(\alpha,\ \beta,\ \frac{\alpha+\beta+1}{2},\ x\right)=(1-2x)^{-\alpha}$$

$$F\left[\frac{\alpha}{2},\ \frac{\alpha+1}{2},\ \frac{\alpha+\beta+1}{2},\ \frac{4x(x-1)}{(2x-1)^2}\right],$$

$$(40)\quad F\left(\alpha,\ \beta,\ \frac{\alpha+\beta+1}{2},\ x\right)$$

$$=(\sqrt{1-x}+\sqrt{-x})^{-2\alpha}F\left[\alpha,\ \frac{\alpha+\beta}{2},\ \alpha+\beta,\ \frac{4\sqrt{x(x-1)}}{(\sqrt{1-x}+\sqrt{-x})^2}\right],$$

$$(41)\quad F(\alpha,\ 1-\alpha,\ \gamma,\ x)=(1-x)^{\gamma-1}F\left[\frac{\gamma-\alpha}{2},\ \frac{\gamma+\alpha-1}{2},\ \gamma,\ 4x(1-x)\right]$$

$$=(1-x)^{\gamma-1}(1-2x)F\left[\frac{\gamma+\alpha}{2},\ \frac{\gamma+1-\alpha}{2},\ \gamma,\ 4x(1-x)\right],$$

$$(42)\quad F(\alpha,\ 1-\alpha,\ \gamma,\ x)$$

$$=(1-x)^{\gamma-1}(1-2x)^{\alpha-\gamma}F\left[\frac{\gamma-\alpha}{2},\ \frac{\gamma+1-\alpha}{2},\ \gamma,\ \frac{4x(x-1)}{(2x-1)^2}\right],$$

$$(43)\quad F(\alpha,\ 1-\alpha,\ \gamma,\ x)=(1-x)^{\gamma-1}(\sqrt{1-x}+\sqrt{-x})^{2-2\alpha-2\gamma}$$

$$F\left[\gamma+\alpha-1,\ \gamma-\tfrac{1}{2},\ 2\gamma-1,\ \frac{4\sqrt{x(x-1)}}{(\sqrt{1-x}+\sqrt{-x})^2}\right],$$

$$(44)\quad F(\alpha,\ \beta,\ 2\beta,\ x)=(1-x)^{\frac{\alpha}{2}}F\left[\frac{\alpha}{2},\ \beta-\frac{\alpha}{2},\ \beta+\frac{1}{2},\ \frac{x^2}{4(x-1)}\right]$$

$$=\left(1-\frac{x}{2}\right)(1-x)^{-\frac{\alpha+1}{2}}F\left[\beta+\frac{1-\alpha}{2},\ \frac{1+x}{2},\ \beta+\frac{1}{2},\ \frac{x^2}{4(x-1)}\right],$$

$$(45)\quad F(\alpha,\ \beta,\ 2\beta,\ x)=\left(1-\frac{x}{2}\right)^{-\alpha}F\left[\frac{\alpha}{2},\ \frac{\alpha+1}{2},\ \beta+\frac{1}{2},\ \left(\frac{x}{2-x}\right)^2\right]$$

$$=(1-x)^{\beta-\alpha}\left(1-\frac{x}{2}\right)^{\alpha-2\beta}F\left[\beta-\frac{\alpha}{2},\ \beta+\frac{1-\alpha}{2},\ \beta+\frac{1}{2},\ \left(\frac{x}{2-x}\right)^2\right],$$

(46) $F(\alpha,\ \beta,\ 2\beta,\ x)$

$$= (1 - x)^{-\frac{\alpha}{2}} F\left[\alpha,\ 2\beta - \alpha,\ \beta + \tfrac{1}{2},\ \frac{(1 - \sqrt{1 - x})^2}{-4\sqrt{1 - x}}\right],$$

(47) $F(\alpha,\ \beta,\ 2\beta,\ x)$

$$= \left(\frac{1 + \sqrt{1 - x}}{2}\right)^{-2\alpha} F\left[\alpha,\ \alpha - \beta + \tfrac{1}{2},\ \beta + \tfrac{1}{2},\ \left(\frac{1 - \sqrt{1 - x}}{1 + \sqrt{1 - x}}\right)^2\right],$$

(48) $F(\alpha,\ \beta,\ \alpha - \beta + 1,\ x)$

$$= (1 - x)^{-\alpha} F\left[\frac{\alpha}{2},\ \frac{\alpha + 1 - 2\beta}{2},\ \alpha - \beta + 1,\ \frac{-4x}{(1 - x)^2}\right],$$

(49) $F(\alpha,\ \beta,\ \alpha - \beta + 1,\ x)$

$$= (1 +)(1 - x)^{-\alpha - 1} F\left[\frac{\alpha + 1}{2},\ \frac{\alpha}{2} + 1 - \beta,\ \alpha - \beta + 1,\ \frac{-4x}{(1 - x)^2}\right],$$

(50) $F(\alpha.\ \beta,\ \alpha - \beta + 1,\ x)$

$$= (1 + x)^{-\alpha} F\left[\frac{\alpha}{2},\ \frac{\alpha + 1}{2} \alpha - \beta - 1,\ \frac{4x}{(1 + x)^2}\right],$$

(51) $F(\alpha,\ \beta,\ \alpha - \beta + 1,\ x) = (1 - x)^{1 - 2\beta}(1 + x)^{2\beta - \alpha - 1}$

$$F\left[\frac{\alpha + 1 - 2\beta}{2},\ \frac{\alpha - 2\beta + 2}{2},\ \alpha + 1 - \beta,\ \frac{4x}{(1 + x)^2}\right],$$

(52) $F(\alpha,\ \beta,\ \alpha - \beta + 1,\ x)$

$$= (1 + \sqrt{x})^{-2\alpha} F\left[\alpha,\ \alpha - \beta + \tfrac{1}{2},\ 2\alpha - 2\beta + 1,\ \frac{4\sqrt{x}}{(1 + \sqrt{x})^2}\right].$$

Formulæ furnished by the Rational Transformations of the Fourth Degree.

(53) $AF(\alpha,\ \tfrac{1}{4} - \alpha,\ \tfrac{1}{2},\ x)$

$$= A(1 - x)^{\frac{1}{4}} F(\tfrac{1}{2} - \alpha,\ \alpha + \tfrac{1}{4},\ \tfrac{1}{2},\ x)$$

$$= t'^{\alpha}(1 - t')^{2\alpha} F(4\alpha,\ 2\alpha + \tfrac{1}{4},\ 2\alpha + \tfrac{3}{4},\ t')$$

$$+ t''^{\alpha}(1 - t'')^{2\alpha} F(4\alpha,\ 2\alpha + \tfrac{1}{4},\ 2\alpha + \tfrac{3}{4},\ t''),$$

$$\begin{aligned}(54)\quad & Bx^{\frac{1}{2}}F(\alpha+\tfrac{1}{2},\ \tfrac{3}{4}-\alpha,\ \tfrac{3}{2},\ x)\\ &= Bx^{\frac{1}{2}}(1-x)^{\frac{1}{4}}F(1-\alpha,\ \alpha+\tfrac{3}{4},\ \tfrac{3}{2},\ x)\\ &= t'^{\alpha}(1-t')^{2\alpha}F(4\alpha,\ 2\alpha+\tfrac{1}{4},\ 2\alpha+\tfrac{3}{4},\ t')\\ &\qquad - t''^{\alpha}(1-t'')^{2\alpha}F(4\alpha,\ 2\alpha+\tfrac{1}{4},\ 2\alpha+\tfrac{3}{4},\ t'');\end{aligned}$$

t', t'' denote those two roots of the equation

$$(t^2-6t+1)^2+16t(1-t)^2x=0$$

which are equal to $3-2\sqrt{2}$ for $x=0$:

$$t' = 3-2\sqrt{2}+\sqrt{-1}(3\sqrt{2}-4)x^{\frac{1}{4}}+\ldots,$$

$$t'' = 3-2\sqrt{2}-\sqrt{-1}(3\sqrt{2}-4)x^{\frac{1}{4}}+\ldots,$$

$$A=(\tfrac{1}{16})^{\alpha}\frac{2\sqrt{\pi}\,\Gamma(2\alpha+\frac{3}{4})}{\Gamma(\alpha+\frac{1}{2})\,\Gamma(\alpha+\frac{3}{4})},\quad B=\sqrt{-1}(\tfrac{1}{16})^{\alpha}\frac{4\sqrt{\pi}\,\Gamma(2\alpha+\frac{3}{4})}{\Gamma(\alpha)\Gamma(\alpha+\frac{1}{4})}$$

$$\begin{aligned}(55)\quad & AF(\alpha,\ \tfrac{1}{4}-\alpha,\ \tfrac{1}{2},\ x)\\ &= A(1-x)^{\frac{1}{4}}F(\tfrac{1}{2}-\alpha,\ \alpha+\tfrac{1}{4},\ \tfrac{1}{2},\ x)\\ &= (-t')^{\alpha}(1-t')^{\alpha}F(4\alpha,\ \tfrac{1}{2},\ 2\alpha+\tfrac{3}{4},\ t')\\ &\qquad + (-t'')^{\alpha}(1-t'')^{\alpha}F(4\alpha,\ \tfrac{1}{2},\ 2\alpha+\tfrac{3}{4},\ t''),\end{aligned}$$

$$\begin{aligned}(56)\quad & Bx^{\frac{1}{2}}F(\alpha+\tfrac{1}{2},\ \tfrac{3}{4}-\alpha,\ \tfrac{3}{2},\ x)\\ &= Bx^{\frac{1}{2}}(1-x)^{\frac{1}{4}}F(1-\alpha,\ \alpha+\tfrac{3}{4},\ \tfrac{3}{2},\ x)\\ &= (-t')^{\alpha}(1-t')^{\alpha}F(4\alpha,\ \tfrac{1}{2},\ 2\alpha+\tfrac{3}{4},\ t')\\ &\qquad - (-t'')^{\alpha}(1-t'')^{\alpha}F(4\alpha,\ \tfrac{1}{2},\ 2\alpha+\tfrac{3}{4},\ t'');\end{aligned}$$

t', t'' are those two roots of the equation

$$(1+4t-4t^2)^2-16t(1-t)x=0$$

which are equal to $\frac{1-\sqrt{2}}{2}$ for $x=0$.

$$t' = \frac{1 - \sqrt{2}}{2} - \sqrt{-1}\,\frac{\sqrt{2}}{4}\,x^{\frac{1}{2}} + \ldots,$$

$$t'' = \frac{1 - \sqrt{2}}{2} + \sqrt{-1}\,\frac{\sqrt{2}}{4}\,x^{\frac{1}{2}} + \ldots .$$

(57) $CF(\alpha,\ \tfrac{1}{4} - \alpha,\ \tfrac{1}{2},\ x)$

$$= C(1 - x)^{\frac{1}{4}}F(\tfrac{1}{2} - \alpha,\ \alpha + \tfrac{1}{4},\ \tfrac{1}{2},\ x)$$
$$= t'^{2\alpha}(1 - t')^{\alpha}F(4\alpha,\ 2\alpha + \tfrac{1}{4},\ 4\alpha + \tfrac{1}{2},\ t')$$
$$+ t''^{2\alpha}(1 - t'')^{\alpha}F(4\alpha,\ 2\alpha + \tfrac{1}{4},\ 4\alpha + \tfrac{1}{2},\ t''),$$

(58) $Dx^{\frac{1}{2}}F(\alpha + \tfrac{1}{2},\ \tfrac{3}{4} - \alpha,\ \tfrac{3}{2},\ x)$

$$= Dx^{\frac{1}{2}}(1 - x)^{\frac{1}{4}}F(1 - \alpha,\ \alpha + \tfrac{3}{4},\ \tfrac{3}{2},\ x)$$
$$= t'^{2\alpha}(1 - t')^{\alpha}F(4\alpha,\ 2\alpha + \tfrac{1}{4},\ 4\alpha + \tfrac{1}{2},\ t')$$
$$- t''^{2\alpha}(1 - t'')^{\alpha}F(4\alpha,\ 2\alpha + \tfrac{1}{4},\ 4\alpha + \tfrac{1}{2},\ t'');$$

t', t'' are those two roots of the equation

$$(t^2 + 4t - 4)^2 + 16t^2(1 - t)x = 0$$

which are equal to $2\sqrt{2} - 2$ for $x = 0$.

$$t' = 2\sqrt{2} - 2 + \sqrt{-1}(3\sqrt{2} - 4)x^{\frac{1}{2}} + \ldots,$$
$$t'' = 2\sqrt{2} - 2 - \sqrt{-1}(3\sqrt{2} - 4)x^{\frac{1}{2}} + \ldots,$$

$$C = \frac{2\sqrt{\pi}\,\Gamma(2\alpha + \frac{3}{4})}{\Gamma(\alpha + \frac{1}{2})\Gamma(\alpha + \frac{3}{4})}, \quad D = \sqrt{-1}\,\frac{4\sqrt{\pi}\,\Gamma(2\alpha + \frac{3}{4})}{\Gamma(\alpha)\Gamma(\alpha + \frac{1}{4})}.$$

(59) $EF(\alpha,\ \alpha + \tfrac{1}{4},\ \tfrac{1}{2},\ x)$

$$= (1 + t')^{4\alpha}F(4\alpha,\ 2\alpha + \tfrac{1}{4},\ 2\alpha + \tfrac{3}{4},\ t')$$
$$+ (1 + t'')^{4\alpha}F(4\alpha,\ 2\alpha + \tfrac{1}{4},\ 2\alpha + \tfrac{3}{4},\ t''),$$

(60) $Gx^{\frac{1}{2}}F(\alpha + \frac{1}{2}, \alpha + \frac{3}{4}, \frac{3}{2}, x)$

$$= (1 + t')^{4\alpha}F(4\alpha, 2\alpha + \tfrac{1}{4}, 2\alpha + \tfrac{3}{4}, t')$$
$$- (1 + t'')^{4\alpha}F(4\alpha, 2\alpha + \tfrac{1}{4}, 2\alpha + \tfrac{3}{4}, t'');$$

t', t'' are those two roots of the equation

$$(t^2 - 6t + 1)^2 - (1 + t)^4 x = 0$$

which are equal to $3 - 2\sqrt{2}$ for $x = 0$.

$$t' = 3 - 2\sqrt{2} + (3\sqrt{2} - 4)x^{\frac{1}{2}} + \ldots,$$

$$t'' = 3 - 2\sqrt{2} - (3\sqrt{2} - 4)x^{\frac{1}{2}} + \ldots,$$

$$E = 2\frac{\sqrt{\pi}\,\Gamma(2\alpha + \frac{3}{4})}{\Gamma(\alpha + \frac{1}{2})\Gamma(\alpha + \frac{3}{4})}, \quad G = \frac{4\sqrt{\pi}\,\Gamma(2\alpha + \frac{3}{4})}{\Gamma(\alpha)\Gamma(\alpha + \frac{1}{4})}.$$

(61) $EF(\alpha, \alpha + \frac{1}{4}, \frac{1}{2}, x)$

$$= (1 - 2t')^{4\alpha}F(4\alpha, \tfrac{1}{2}, 2\alpha + \tfrac{3}{4}, t')$$
$$+ (1 - 2t'')^{4\alpha}F(4\alpha, \tfrac{1}{2}, 2\alpha + \tfrac{3}{4}, t''),$$

(62) $Gx^{\frac{1}{2}}F(\alpha + \frac{1}{2}, \alpha + \frac{3}{4}, \frac{2}{3}, x)$

$$= (1 - 2t')^{4\alpha}F(4\alpha, \tfrac{1}{2}, 2\alpha + \tfrac{3}{4}, t')$$
$$- (1 - 2t'')^{4\alpha}F(4\alpha, \tfrac{1}{2}, 2\alpha + \tfrac{3}{4}, t'');$$

t', t'' are those two roots of the equation

$$(1 + 4t - 4t^2)^2 - (2t - 1)^4 x = 0$$

which are equal to $\frac{1 - \sqrt{2}}{2}$ for $x = 0$.

$$t' = \frac{1 - \sqrt{2}}{2} - \frac{\sqrt{2}}{4}x^{\frac{1}{2}} + \ldots,$$

$$t'' = \frac{1 - \sqrt{2}}{2} + \frac{\sqrt{2}}{4}x^{\frac{1}{2}} + \ldots.$$

(63) $EF(\alpha,\ \alpha+\frac{1}{4},\ \frac{1}{2},\ x)$

$$= \left(\frac{2-t'}{2}\right)^{4\alpha} F(4\alpha,\ 2\alpha+\tfrac{1}{4},\ 4\alpha+\tfrac{1}{2},\ t')$$

$$+ \left(\frac{2-t''}{2}\right)^{4\alpha} F(4\alpha,\ 2\alpha+\tfrac{1}{4},\ 4\alpha+\tfrac{1}{2},\ t''),$$

(64) $Gx^{\frac{1}{2}}F(\alpha+\frac{1}{2},\ \alpha+\frac{3}{4},\ \frac{3}{2},\ x)$

$$= \left(\frac{2-t'}{2}\right)^{4\alpha} F(4\alpha,\ 2\alpha+\tfrac{1}{4},\ 4\alpha+\tfrac{1}{2},\ t')$$

$$- \left(\frac{2-t''}{2}\right)^{4\alpha} F(4\alpha,\ 2\alpha+\tfrac{1}{4},\ 4\alpha+\tfrac{1}{2},\ t'');$$

$t',\ t''$ are those two roots of the equation

$$(t^2+4t-4)^2-(2-t)^4x=0$$

which are equal to $2\sqrt{2}-2$ for $x=0$.

$$t' = 2\sqrt{2}-2+(3\sqrt{2}-4)x^{\frac{1}{2}}+\ldots,$$

$$t'' = 2\sqrt{2}-2-(3\sqrt{2}-4)x^{\frac{1}{2}}+\ldots.$$

(65) $HF(\alpha,\ \frac{1}{4}-\alpha,\ \frac{3}{4},\ x)$

$$= H(1-x)^{\frac{1}{2}}F(\tfrac{3}{4}-\alpha,\ \alpha+\tfrac{1}{2},\ \tfrac{3}{4},\ x)$$

$$= t'^{\alpha}(1-t')^{\alpha}F(4\alpha,\ \tfrac{1}{2},\ 2\alpha+\tfrac{3}{4},\ t')$$

$$+ t''^{\alpha}(1-t'')^{\alpha}F(4\alpha,\ \tfrac{1}{2},\ 2\alpha+\tfrac{3}{4},\ t'')$$

$$+ t'''^{\alpha}(1-t''')^{\alpha}F(4\alpha,\ \tfrac{1}{2},\ 2\alpha+\tfrac{3}{4},\ t''')$$

$$+ t^{\text{iv}\,\alpha}(1-t^{\text{iv}})^{\alpha}F(4\alpha,\ \tfrac{1}{2},\ 2\alpha+\tfrac{3}{4},\ t^{\text{iv}}),$$

(66) $Kx^{\frac{1}{4}}F(\alpha+\frac{1}{4},\ \frac{1}{2}-\alpha,\ \frac{5}{4},\ x)$

$$= Kx^{\frac{1}{4}}(1-x)^{\frac{1}{2}}F(1-\alpha,\ \alpha+\tfrac{3}{4},\ \tfrac{5}{4},\ x)$$

$$= t'^{\alpha}(1-t')^{\alpha}F(4\alpha,\ \tfrac{1}{2},\ 2\alpha+\tfrac{3}{4},\ t')$$

$$- \sqrt{-1}\ t''^{\alpha}(1-t'')^{\alpha}F(4\alpha,\ \tfrac{1}{2},\ 2\alpha+\tfrac{3}{4},\ t'')$$

$$- t'''^{\alpha}(1-t''')^{\alpha}F(4\alpha,\ \tfrac{1}{2},\ 2\alpha+\tfrac{3}{4},\ t''')$$

$$+ \sqrt{-1}\ t^{\text{iv}\,\alpha}(1-t^{\text{iv}})^{\alpha}F(4\alpha,\ \tfrac{1}{2},\ 2\alpha+\tfrac{3}{4},\ t^{\text{iv}});$$

$t', t'', t''', t^{\text{iv}}$ are the four roots of the equation

$$(2t - 1)^4 + 16t(1 - t)x = 0,$$

taken in the order in which they present themselves when the variable x describes successive loops round the origin in the positive sense.

$$t' = \tfrac{1}{2} + \frac{1}{\sqrt{2}}\left(\cos\frac{\pi}{4} + \sqrt{-1}\sin\frac{\pi}{4}\right)x^{\frac{1}{4}} + \ldots,$$

$$t'' = \tfrac{1}{2} + \frac{1}{\sqrt{2}}\left(\cos\frac{3\pi}{4} + \sqrt{-1}\sin\frac{3\pi}{4}\right)x^{\frac{1}{4}} + \ldots,$$

$$t''' = \tfrac{1}{2} + \frac{1}{\sqrt{2}}\left(\cos\frac{5\pi}{4} + \sqrt{-1}\sin\frac{5\pi}{4}\right)x^{\frac{1}{4}} + \ldots,$$

$$t^{\text{iv}} = \tfrac{1}{2} + \frac{1}{\sqrt{2}}\left(\cos\frac{7\pi}{4} + \sqrt{-1}\sin\frac{7\pi}{4}\right)x^{\frac{1}{4}} + \ldots,$$

$$H = (\tfrac{1}{16})^{\alpha}\frac{4\Gamma(\frac{1}{4})\Gamma(2\alpha + \frac{3}{4})}{\Gamma(\alpha + \frac{1}{4})\Gamma(\alpha + \frac{3}{4})},$$

$$K = (\tfrac{1}{16})^{\alpha}\left(\cos\frac{\pi}{4} + \sqrt{-1}\sin\frac{\pi}{4}\right)\frac{16\Gamma(\frac{3}{4})\Gamma(2\alpha + \frac{3}{4})}{\Gamma(\alpha)\Gamma(\alpha + \frac{1}{2})}.$$

$$(67)\quad HF(\alpha, \alpha + \tfrac{1}{2}, \tfrac{3}{4}, x) = \left(\frac{1 + 4t' - 4t'^2}{4}\right)^{2\alpha} F(4\alpha, \tfrac{1}{2}, 2\alpha + \tfrac{3}{4}, t')$$
$$+ \left(\frac{1 + 4t'' - 4t''^2}{4}\right)^{2\alpha} F(4\alpha, \tfrac{1}{2}, 2\alpha + \tfrac{3}{4}, t'')$$
$$+ \left(\frac{1 + 4t''' - 4t'''^2}{4}\right)^{2\alpha} F(4\alpha, \tfrac{1}{2}, 2\alpha + \tfrac{3}{4}, t''')$$
$$+ (1 + 4t^{\text{iv}} - 4t^{\text{iv}2})^{2\alpha} F(4\alpha, \tfrac{1}{2}, 2\alpha + \tfrac{3}{4}, t^{\text{iv}}),$$

$$(68)\quad Lx^{\frac{1}{4}}F(\alpha + \tfrac{1}{4}, \alpha + \tfrac{3}{4}, \tfrac{5}{4}, x) = \left(\frac{1 + 4t' - 4t'^2}{4}\right)^{2\alpha} F(4\alpha, \tfrac{1}{2}, 2\alpha + \tfrac{3}{4}, t')$$
$$- \sqrt{-1}\left(\frac{1 + 4t'' - 4t''^2}{4}\right)^{2\alpha} F(4\alpha, \tfrac{1}{2}, 2\alpha + \tfrac{3}{4}, t'')$$
$$- \left(\frac{1 + 4t''' - 4t'''^2}{4}\right)^{2\alpha} F(4\alpha, \tfrac{1}{2}, 2\alpha + \tfrac{3}{4}, t''')$$
$$+ \sqrt{-1}\left(\frac{1 + 4t^{\text{iv}} - 4t^{\text{iv}2}}{4}\right)^{2\alpha} F(4\alpha, \tfrac{1}{2}, 2\alpha + \tfrac{3}{4}, t^{\text{iv}});$$

$t', t'', t''', t^{\text{iv}}$ are the four roots of the equation

$$(2t - 1)^4 - x(1 + 4t - 4t^2)^2 = 0.$$

Thus

$$t' = \tfrac{1}{2} + \frac{1}{\sqrt{2}} x^{\frac{1}{4}} + \ldots,$$

$$t'' = \tfrac{1}{2} + \frac{\sqrt{-1}}{\sqrt{2}} x^{\frac{1}{4}} + \ldots,$$

$$t''' = \tfrac{1}{2} - \frac{1}{\sqrt{2}} x^{\frac{1}{4}} + \ldots,$$

$$t^{\text{iv}} = \tfrac{1}{2} - \frac{\sqrt{-1}}{\sqrt{2}} x^{\frac{1}{4}} + \ldots,$$

$$L = (\tfrac{1}{16})\alpha \frac{16\Gamma(\frac{3}{4})\Gamma(2\alpha + \frac{3}{4})}{\Gamma(\alpha)\Gamma(\alpha + \frac{1}{2})}.$$

$$\text{(69)} \quad F(\alpha, \alpha + \tfrac{1}{2}, 2\alpha + \tfrac{3}{4}, x) = (1 - x)^{\frac{1}{4}} F(\alpha + \tfrac{1}{4}, \alpha + \tfrac{3}{4}, 2\alpha + \tfrac{3}{4}, x)$$
$$= \left(\frac{t'^2 + 4t' - 4}{4}\right)^{2\alpha} F(4\alpha, 2\alpha + \tfrac{1}{4}, 4\alpha + \tfrac{1}{2}, t');$$

t' being one of those two roots of the equation

$$16t^2(1 - t) + x(t^2 + 4t - 4)^2 = 0$$

which are zero for $x = 0$.

$$\text{(70)} \quad F(\alpha, \alpha + \tfrac{1}{4}, 2\alpha + \tfrac{3}{4}, x) = (1 - x)^{\frac{1}{4}} F(\alpha + \tfrac{1}{2}, \alpha + \tfrac{3}{4}\ 2\alpha + \tfrac{3}{4}, x)$$
$$= \left(\frac{2 - t'}{2}\right)^{2\alpha} F(4\alpha, 2\alpha + \tfrac{1}{4}, 4\alpha + \tfrac{1}{2}. t');$$

t' is one of those two roots of the equation

$$16t^2(1 - t) - x(2 - t)^4 = 0$$

which are zero for $x = 0$.

(71) $$\begin{aligned} F(\alpha, \alpha+\tfrac{1}{2}, 2\alpha+\tfrac{3}{4}, x) &= (1-x)^{\frac{1}{4}} F(\alpha+\tfrac{1}{4}, \alpha+\tfrac{3}{4}, 2\alpha+\tfrac{3}{4}, x) \\ &= (t^2 - 6t + 1)^{2\alpha} F(4\alpha, 2\alpha+\tfrac{1}{4}, 2\alpha+\tfrac{3}{4}, t); \end{aligned}$$

t denotes that root of the equation

$$16t(1-t^2) + x(t^2 - 6t + 1)^2 = 0$$

which is zero for $x = 0$.

(72) $$\begin{aligned} F(\alpha, \alpha+\tfrac{1}{2}, 2\alpha+\tfrac{3}{4}, x) &= (1-x)^{\frac{1}{4}} F(\alpha+\tfrac{1}{4}, \alpha+\tfrac{3}{4}, 2\alpha+\tfrac{3}{4}, x) \\ &= (1 + 4t - 4t^2)^{2\alpha} F(4\alpha, \tfrac{1}{2}, 2\alpha+\tfrac{3}{4}, t); \end{aligned}$$

t denotes that root of the equation

$$16t(1-t) - (1 + 4t - 4t^2)^2 x = 0$$

which is zero for $x = 0$.

(73) $$\begin{aligned} F(\alpha, \alpha+\tfrac{1}{4}, 2\alpha+\tfrac{3}{4}, x) &= (1-x)^{\frac{1}{2}} F(\alpha+\tfrac{1}{2}, \alpha+\tfrac{3}{4}, 2\alpha+\tfrac{3}{4}, x) \\ &= (1 + t)^{4\alpha} F(4\alpha, 2\alpha+\tfrac{1}{4}, 2\alpha+\tfrac{3}{4}, t); \end{aligned}$$

t denotes that root of the equation

$$16t(1-t)^2 - x(1+t)^4 = 0$$

which is zero for $x = 0$.

(74) $$\begin{aligned} F(\alpha, \alpha+\tfrac{1}{4}, 2\alpha+\tfrac{3}{4}, x) &= (1-x)^{\frac{1}{2}} F(\alpha+\tfrac{1}{2}, \alpha+\tfrac{3}{4}, 2\alpha+\tfrac{3}{4}, x) \\ &= (1 - 2t)^{4\alpha} F(4\alpha, \tfrac{1}{2}, 2\alpha+\tfrac{3}{4}, t); \end{aligned}$$

t denotes that root of the equation

$$16t(1-t) + x(2t-1)^4 = 0$$

which is zero for $x = 0$.

Formulæ furnished by the Inverse Transformations.

(75) $$\begin{aligned} F(4\alpha, 2\alpha+\tfrac{1}{4}, 2\alpha+\tfrac{3}{4}, x) &= (1-x)^{\frac{1}{2}-4\alpha} F(\tfrac{3}{4}-2\alpha, \tfrac{1}{2}, 2\alpha+\tfrac{3}{4}, x) \\ &= \begin{cases} (x^2 - 6x + 1)^{-2\alpha} F\left[\alpha, \alpha+\tfrac{1}{2}, 2\alpha+\tfrac{3}{4}, \dfrac{-16x(1-x)^2}{(1-6x+x^2)^2}\right], \\ (1+x)^{-4\alpha} F\left[\alpha, \alpha+\tfrac{1}{4}, 2\alpha+\tfrac{3}{4}, \dfrac{16x(1-x)^2}{(1+x)^4}\right]; \end{cases} \end{aligned}$$

$$(76)\quad F(4\alpha,\ 2\alpha+\tfrac{1}{4},\ 4\alpha+\tfrac{1}{2},\ x)=(1-x)^{\frac{1}{4}-2\alpha}F(2\alpha+\tfrac{1}{4},\ \tfrac{1}{2},\ 4\alpha+\tfrac{1}{2},\ x)$$

$$=\begin{cases}\left(\dfrac{4-4x-x^2}{4}\right)^{-2\alpha}F\left[\alpha,\ \alpha+\tfrac{1}{2},\ 2\alpha+\tfrac{3}{4},\ \dfrac{-16x^2(1-x)}{(x^2+4x-4)^2}\right],\\[2ex] \left(1-\dfrac{x}{2}\right)^{-4\alpha}F\left[\alpha,\ \alpha+\tfrac{1}{4},\ 2\alpha+\tfrac{8}{4},\ \dfrac{16x^2(1-x)}{(2-x^4)}\right];\end{cases}$$

$$(77)\quad F(4\alpha,\ \tfrac{1}{2},\ 2\alpha+\tfrac{3}{4},\ x)=(1-x)^{\frac{1}{4}-2\alpha}F(2\alpha+\tfrac{1}{4},\ \tfrac{3}{4}-2\alpha,\ 2\alpha+\tfrac{3}{4},\ x)$$

$$=\begin{cases}(1+4x-4x^2)^{-2\alpha}F\left[\alpha,\ \alpha+\tfrac{1}{2},\ 2\alpha+\tfrac{3}{4},\ \dfrac{16x(1-x)}{(1+x-4x^2)^2}\right],\\[2ex] (1-2x)^{-4\alpha}F\left[\alpha,\ \alpha+\tfrac{1}{4},\ 2\alpha+\tfrac{8}{4},\ \dfrac{-16x(1-x)}{(1-2x)^4}\right].\end{cases}$$

Formulæ furnished by the Rational Transformations of the Third Degree.

$$(78)\quad A_1F(\alpha,\ \tfrac{1}{6}-\alpha,\ \tfrac{1}{2},\ x)=A_1(1-x)^{\frac{1}{3}}F(\tfrac{1}{2}-\alpha,\ \alpha+\tfrac{1}{3},\ \tfrac{1}{2},\ x)$$
$$=t'^{2\alpha}(1-t')^{\alpha}F(3\alpha,\ 3\alpha+\tfrac{1}{2},\ 4\alpha+\tfrac{2}{3},\ t')$$
$$+\,t''^{2\alpha}(1-t'')^{\alpha}F(3\alpha,\ 3\alpha+\tfrac{1}{2},\ 4\alpha+\tfrac{2}{3},\ t''),$$

$$(79)\quad B_1x^{\frac{1}{2}}F(\alpha+\tfrac{1}{2},\ \tfrac{2}{3}-\alpha,\ \tfrac{3}{2},\ x)=B_1(1-x)^{\frac{1}{3}}F(1-\alpha,\ \alpha+\tfrac{5}{6},\ \tfrac{3}{2},\ x)$$
$$=t'^{2\alpha}(1-t')^{\alpha}F(3\alpha,\ 3\alpha+\tfrac{1}{2},\ 4\alpha+\tfrac{2}{3},\ t')$$
$$-\,t''^{2\alpha}(1-t'')^{\alpha}F(3\alpha,\ 3\alpha+\tfrac{1}{2},\ 4\alpha+\tfrac{2}{3},\ t'');$$

t', t'' denote those two roots of the equation

$$(9t-8)^2+27t^2(1-t)x=0$$

which are equal to $\frac{8}{9}$ for $x=0$:

$$t'=\tfrac{8}{9}+\sqrt{-1}\,\frac{8\sqrt{3}}{81}\,x^{\frac{1}{2}}+\ldots,$$
$$t''=\tfrac{8}{9}-\sqrt{-1}\,\frac{8\sqrt{3}}{81}\,x^{\frac{1}{2}}+\ldots,$$

$$A_1=(\tfrac{1}{27})\alpha\,\frac{2\sqrt{\pi}\,\Gamma(2\alpha+\frac{5}{6})}{\Gamma(\alpha+\frac{1}{2})\Gamma(\alpha+\frac{5}{6})},\quad B_1=\sqrt{-1}(\tfrac{1}{27})\alpha\,\frac{4\sqrt{\pi}\,\Gamma(2\alpha+\frac{5}{6})}{\Gamma(\alpha)\,\Gamma(\alpha+\frac{1}{3})}.$$

(80) $A_1 F(\alpha, \frac{1}{6} - \alpha, \frac{1}{2}, x) = A_1(1 - x)^{\frac{1}{3}} F(\frac{1}{2} - \alpha, \alpha + \frac{1}{3}, \frac{1}{2}, x)$

$$= t'^{\alpha}(1 - t')^{2\alpha} F(3\alpha, 3\alpha + \tfrac{1}{2}, 2\alpha + \tfrac{5}{6}, t')$$
$$+ t''^{\alpha}(1 - t'')^{2\alpha} F(3\alpha, 3\alpha + \tfrac{1}{2}, 2\alpha + \tfrac{5}{6}, t''),$$

(81) $B_1 x^{\frac{1}{2}} F(\alpha + \frac{1}{2}, \frac{2}{3} - \alpha, \frac{3}{2}, x) = B_1 x^{\frac{1}{2}}(1 - x)^{\frac{1}{3}} F(1 - \alpha, \alpha + \frac{5}{6}, \frac{3}{2}, x)$

$$= t'^{\alpha}(1 - t')^{2\alpha} F(3\alpha, 3\alpha + \tfrac{1}{2}, 2\alpha + \tfrac{5}{6}, t')$$
$$- t''^{\alpha}(1 - t'')^{2\alpha} F(3\alpha, 3\alpha + \tfrac{1}{2}, 2\alpha + \tfrac{5}{6}, t'');$$

t', t'' are those two roots of the equation

$$(1 - 9t)^2 + 27t(1 - t)^2 x = 0$$

which are equal to $\frac{1}{9}$ for $x = 0$:

$$t' = \tfrac{1}{9} + \sqrt{-1}\,\frac{8\sqrt{3}}{81} x^{\frac{1}{2}} + \ldots,$$

$$t'' = \tfrac{1}{9} - \sqrt{-1}\,\frac{8\sqrt{3}}{81} x^{\frac{1}{2}} + \ldots.$$

(82) $A_1 F(\alpha, \frac{1}{6} - \alpha, \frac{1}{2}, x) = A_1(1 - x)^{\frac{1}{3}} F(\frac{1}{2} - \alpha, \alpha + \frac{1}{3}, \frac{1}{2}, x)$

$$= (- t')^{\alpha} F(3\alpha, \tfrac{1}{3} - \alpha, 2\alpha + \tfrac{5}{6}, t')$$
$$+ (- t'')^{\alpha} F(3\alpha, \tfrac{1}{3} - \alpha, 2\alpha + \tfrac{5}{6}, t'');$$

(83) $B_1 x^{\frac{1}{2}} F(\alpha + \frac{1}{2}, \frac{2}{3} - \alpha, \frac{3}{2}, x) = B_1 x^{\frac{1}{2}}(1 - x)^{\frac{1}{3}} F(1 - \alpha, \alpha + \frac{5}{6}, \frac{3}{2}, x)$

$$= (- t')^{\alpha} F(3\alpha, \tfrac{1}{3} - \alpha, 2\alpha + \tfrac{5}{6}, t')$$
$$- (- t'')^{\alpha} F(3\alpha, \tfrac{1}{3} - \alpha, 2\alpha + \tfrac{5}{6}, t'');$$

t', t'' are those two roots of the equation

$$(1 + 8t)^2(1 - t) - 27tx = 0$$

which are equal to $- \frac{1}{8}$ for $x = 0$:

$$t' = - \tfrac{1}{8} - \sqrt{-1}\,\frac{\sqrt{3}}{8} x^{\frac{1}{2}} + \ldots,$$

$$t'' = - \tfrac{1}{8} + \sqrt{-1}\,\frac{\sqrt{3}}{8} x^{\frac{1}{2}} + \ldots.$$

$$(84)\quad F(\alpha, \tfrac{1}{6} - \alpha, \tfrac{1}{2}, x) = (1 - x)^{\frac{1}{3}} F(\tfrac{1}{2} - \alpha, \alpha + \tfrac{1}{3}, \tfrac{1}{2}, x)$$
$$= (1 - t)^{\alpha} F(3\alpha, \tfrac{1}{3} - \alpha, \tfrac{1}{2}, t),$$

$$(85)\quad F(\alpha + \tfrac{1}{2}, \tfrac{2}{3} - \alpha, \tfrac{3}{2}, x) = (1 - x)^{\frac{1}{3}} F(1 - \alpha, \alpha + \tfrac{5}{6}, \tfrac{3}{2}, x)$$
$$= (1 - t)^{\alpha + \frac{1}{3}} \frac{9}{9 - 8t} F(3\alpha + \tfrac{1}{2}, 56 - \alpha, \tfrac{3}{2}, t);$$

t being that root of the equation

$$(9 - 8t)^2 t - 27(1 - t)x = 0$$

which is zero for $x = 0$.

$$(86)\quad F(\alpha, \tfrac{1}{6} - \alpha, \tfrac{1}{2}, x) = (1 - x)^{\frac{1}{3}} F(\tfrac{1}{2} - \alpha, \alpha + \tfrac{1}{3}, \tfrac{1}{2}, x)$$
$$= (1 - t)^{2\alpha} F(3\alpha, \alpha + \tfrac{1}{6}, \tfrac{1}{2}, t),$$

$$87)\quad F(\alpha + \tfrac{1}{2}, \tfrac{2}{3} - \alpha, \tfrac{3}{2}, x) = (1 - x)^{\frac{1}{3}} F(1 - \alpha, \alpha + \tfrac{5}{6}, \tfrac{3}{2}, x)$$
$$= (1 - t)^{2\alpha + 1} \frac{9}{9 - t} F(3\alpha + \tfrac{1}{2}, \alpha + \tfrac{2}{3}, \tfrac{3}{2}, t);$$

t being that root of the equation

$$(t - 9)^2 t + 27(1 - t)^2 x = 0$$

which is zero for $x = 0$.

$$(88)\quad C_1 F(\alpha, \alpha + \tfrac{1}{8}, \tfrac{1}{2}, x) = \left(\frac{4 - 3t'}{4}\right)^{3\alpha} F(3\alpha, 3\alpha + \tfrac{1}{2}, 4\alpha + \tfrac{2}{3}, t')$$
$$+ \left(\frac{4 - 3t''}{4}\right)^{3\alpha} F(3\alpha, 3\alpha + \tfrac{1}{2}, 4\alpha + \tfrac{2}{8}, t''),$$

$$(89)\quad D_1 x^{\frac{1}{2}} F(\alpha + \tfrac{1}{2}, \alpha + \tfrac{5}{6}, \tfrac{3}{2}, x) = \left(\frac{4 - 3t'}{4}\right)^{3\alpha} F(3\alpha, 3\alpha + \tfrac{1}{2}, 4\alpha + \tfrac{2}{8}, t')$$
$$- \left(\frac{4 - 3t''}{4}\right)^{3\alpha} F(3\alpha, 3\alpha + \tfrac{1}{2}, 4\alpha + \tfrac{2}{8}, t'');$$

t', t'' are those two roots of the equation

$$(9t - 8)^2 + (3t - 4)^3 x = 0$$

which are equal to $\frac{8}{9}$ for $x = 0$.

$$t' = \frac{8}{9} + \frac{8\sqrt{3}}{81} x^{\frac{1}{2}} + \ldots,$$

$$t'' = \frac{8}{9} - \frac{8\sqrt{3}}{81} x^{\frac{1}{2}} + \ldots,$$

$$C_1 = \frac{2\sqrt{\pi}\,\Gamma(2\alpha + \frac{5}{6})}{\Gamma(\alpha + \frac{1}{2})\Gamma(\alpha + \frac{5}{6})}, \quad D_1 = \frac{4\sqrt{\pi}\,\Gamma(2\alpha + \frac{5}{6})}{\Gamma(\alpha)\Gamma(\alpha + \frac{1}{8})}.$$

$$(90) \quad C_1 F(\alpha, \alpha + \tfrac{1}{3}, \tfrac{1}{2}, x) = (1 + 3t')^{3\alpha} F(3\alpha, 3\alpha + \tfrac{1}{2}, 2\alpha + \tfrac{5}{6}, t') + (1 + 3t'')^{3\alpha} F(3\alpha, 3\alpha + \tfrac{1}{2}, 2\alpha + \tfrac{5}{6}, t''),$$

$$(91) \quad D_1 x^{\frac{1}{2}} F(\alpha + \tfrac{1}{2}, \alpha + \tfrac{5}{6}, \tfrac{3}{2}, x) = (1 + 3t')^{3\alpha} F(3\alpha, 3\alpha + \tfrac{1}{2}, 2\alpha + \tfrac{5}{6}, t') - (1 + 3t'')^{3\alpha} F(3\alpha, 3\alpha + \tfrac{1}{2}, 2\alpha + \tfrac{5}{6}, t'');$$

t', t'' are those two roots of the equation

$$(1 - 9t)^2 - x(3t + 1)^3 = 0$$

which are equal to $\frac{1}{9}$ for $x = 0$:

$$t' = \frac{1}{9} + \frac{8\sqrt{3}}{81} x^{\frac{1}{2}} + \ldots,$$

$$t'' = \frac{1}{9} - \frac{8\sqrt{3}}{81} x^{\frac{1}{2}} + \ldots.$$

$$(92) \quad C_1 F(\alpha, \alpha + \tfrac{1}{3}, \tfrac{1}{2}, x) = (1 - 4t')^{3\alpha} F(3\alpha, \tfrac{1}{3} - \alpha, 2\alpha + \tfrac{5}{6}, t') + (1 - 4t'')^{3\alpha} F(3\alpha, \tfrac{1}{3} - \alpha, 2\alpha + \tfrac{5}{6}, t''),$$

$$(93) \quad D_1 x^{\frac{1}{2}} F(\alpha + \tfrac{1}{2}, \alpha + \tfrac{5}{6}, \tfrac{3}{2}, x) = (1 - 4t')^{3\alpha} F(3\alpha, \tfrac{1}{3} - \alpha, 2\alpha + \tfrac{5}{6}, t') - (1 - 4t'')^{3\alpha} F(3\alpha, \tfrac{1}{3} - \alpha, 2\alpha + \tfrac{5}{6}, t'');$$

t', t'' are those two roots of the equation

$$(1 + 8t)^2(1 - t) - (1 - 4t)^3 x = 0$$

which are equal to $(-\frac{1}{8})$ for $x = 0$:

$$t' = -\tfrac{1}{8} - \frac{\sqrt{3}}{8}x^{\frac{1}{2}} + \ldots,$$

$$t' = -\tfrac{1}{8} + \frac{\sqrt{3}}{8}x^{\frac{1}{2}} + \ldots.$$

(94) $$F(\alpha, \alpha + \tfrac{1}{3}, \tfrac{1}{2}, x) = \left(1 - \frac{4t}{3}\right)^{3\alpha} F(3\alpha, \tfrac{1}{3} - \alpha, \tfrac{1}{2}, t).$$

(95) $$F(\alpha+\tfrac{1}{2}, \alpha+\tfrac{5}{6}, \tfrac{3}{2}, x) = \left(1-\frac{4t}{3}\right)^{3\alpha+\frac{3}{2}} \frac{9}{9-8t} F(3\alpha+\tfrac{1}{2}, \tfrac{5}{6} - \alpha, \tfrac{3}{2}, t);$$

t denoting that root of the equation

$$(9 - 8t)^2 t + (3 - 4t)^3 x = 0$$

which is zero for $x = 0$.

(96) $$F(\alpha, \alpha + \tfrac{1}{3}, \tfrac{1}{2}, x) = \left(1 + \frac{t}{3}\right)^{3\alpha} F(3\alpha, \alpha + \tfrac{1}{6}, \tfrac{1}{2}, t),$$

(97) $$F(\alpha+\tfrac{1}{2}, \alpha+\tfrac{5}{6}, \tfrac{3}{2}, x) = \left(1+\frac{t}{3}\right)^{3\alpha+\frac{3}{2}} \frac{9}{9-t} F(3\alpha + \tfrac{1}{2}, \alpha + \tfrac{2}{3}, \tfrac{3}{2}, t);$$

t being that root of the equation

$$(t - 9)^2 t - (t + 3)^3 x = 0$$

which is zero for $x = 0$.

(98) $$\begin{aligned} E_1 F(\alpha, \tfrac{1}{6} - \alpha, \tfrac{2}{3}, x) &= E_1 (1 - x)^{\frac{1}{2}} F(\tfrac{2}{3} - \alpha, \alpha + \tfrac{1}{2}, \tfrac{2}{3}, x) \\ &= (1 - t')^{\alpha} F(3\alpha, \tfrac{1}{3} - \alpha, \tfrac{1}{2}, t') \\ &\quad + (1 - t'')^{\alpha} F(3\alpha, \tfrac{1}{3} - \alpha, \tfrac{1}{2}, t'') \\ &\quad + (1 - t''')^{\alpha} F(3\alpha, \tfrac{1}{3} - \alpha, \tfrac{1}{2}, t'''). \end{aligned}$$

(99) $$\begin{aligned} G_1 x^{\frac{1}{3}} F(\alpha + \tfrac{1}{3}, \tfrac{1}{2} - \alpha, \tfrac{4}{3}, x) &= G_1 (1 - x)^{\frac{1}{2}} x^{\frac{1}{3}} F(1 - \alpha, \alpha + \tfrac{5}{6}, \tfrac{4}{3}, x) \\ &= (1 - t')^{\alpha} F(3\alpha, \tfrac{1}{3} - \alpha, \tfrac{1}{2}, t') \\ &\quad + j^2 (1 - t'')^{\alpha} F(3\alpha, \tfrac{1}{3} - \alpha, \tfrac{1}{2}, t'') \\ &\quad + j (1 - t''')^{\alpha} F(3\alpha, \tfrac{1}{3} - \alpha, \tfrac{1}{2}, t'''); \end{aligned}$$

t', t'', t''' are the three roots of the equation

$$(3 - 4t)^3 - 27(1 - t)x = 0,$$

taken in the order in which they present themselves when the variable x describes successive loops round the origin in the direct sense:

$$t' = \tfrac{3}{4} - \frac{3\sqrt[3]{2}}{8}x^{\frac{1}{3}} + \ldots, \quad j = -\tfrac{1}{2} + \sqrt{-1}\frac{\sqrt{3}}{2},$$

$$t'' = \tfrac{3}{4} - \frac{3\sqrt[3]{2}}{8}jx^{\frac{1}{3}} + \ldots, \quad j^2 = -\tfrac{1}{2} - \sqrt{-1}\frac{\sqrt{3}}{2}.$$

$$t''' = \tfrac{3}{4} - \frac{3\sqrt[3]{2}}{8}j^2x^{\frac{1}{3}} + \ldots,$$

$$E_1 = \frac{3\sqrt{\pi}\,\Gamma(\frac{1}{3})}{\Gamma(\frac{1}{2} - \alpha)\Gamma(\alpha + \frac{1}{3})}, \quad G_1 = \frac{-9\sqrt{\pi}\,\Gamma(\frac{2}{3})}{\Gamma(\alpha)\Gamma(\frac{1}{6} - \alpha)}.$$

$$\begin{aligned}(100)\quad H_1F(\alpha, \tfrac{1}{6} - \alpha, \tfrac{2}{3}, x) &= H_1(1 - x)^{\frac{1}{2}}F(\tfrac{2}{3} - \alpha, \alpha + \tfrac{1}{2}, \tfrac{2}{3}, x)\\ &= t'^{\alpha}F(3\alpha, \tfrac{1}{3} - \alpha, 2\alpha + \tfrac{5}{6}, t')\\ &+ t''^{\alpha}F(3\alpha, \tfrac{1}{3} - \alpha, 2\alpha + \tfrac{5}{6}, t'')\\ &+ t'''^{\alpha}F(3\alpha, \tfrac{1}{3} - \alpha, 2\alpha + \tfrac{5}{6}, t'''),\end{aligned}$$

$$\begin{aligned}(101)\quad K_1x^{\frac{1}{3}}F(\alpha+\tfrac{1}{3}, \tfrac{1}{2}-\alpha, \tfrac{4}{3}, x) &= K_1x^{\frac{1}{3}}(1-x)^{\frac{1}{2}}F(1-\alpha, \alpha+\tfrac{5}{6}, \tfrac{4}{3}, x)\\ &= t'^{\alpha}F(3\alpha, \tfrac{1}{3} - \alpha, 2\alpha + \tfrac{5}{6}, t')\\ &+ j^2t''^{\alpha}F(3\alpha, \tfrac{1}{3} - \alpha, 2\alpha + \tfrac{5}{6}, t'')\\ &+ jt'''^{\alpha}F(3\alpha, \tfrac{1}{3} - \alpha, 2\alpha + \tfrac{5}{6}, t''');\end{aligned}$$

t', t'', t''' are the three roots of the equation

$$(1 - 4t)^3 + 27tx = 0:$$

$$t' = \tfrac{1}{4} + \frac{3\sqrt[3]{2}}{8}x^{\frac{1}{3}} + \ldots,$$

$$t'' = \tfrac{1}{4} + \frac{3\sqrt[3]{2}}{8}jx^{\frac{1}{3}} + \ldots,$$

$$t''' = \tfrac{1}{4} + \frac{3\sqrt[3]{2}}{8}j^2x^{\frac{1}{3}} + \ldots,$$

$$H_1 = (\tfrac{1}{27})^{\alpha}\frac{3\Gamma(\frac{1}{3})\Gamma(2\alpha + \frac{5}{6})}{\Gamma(\alpha + \frac{1}{3})\Gamma(\alpha + \frac{5}{6})}, \quad K_1 = -(\tfrac{1}{27})^{\alpha}\frac{9\Gamma(\frac{2}{3})\Gamma(2\alpha + \frac{5}{6})}{\Gamma(\alpha)\Gamma(\alpha + \frac{1}{2})}.$$

$$(102)\quad H_1F(\alpha,\ \tfrac{1}{6}-\alpha,\ \tfrac{2}{3},\ x) = H_1(1-x)^{\frac{1}{2}}F(\tfrac{2}{3}-\alpha,\ \alpha+\tfrac{1}{2},\ \tfrac{2}{3},\ x)$$
$$= (-t')^{\alpha}(1-t')^{2\alpha}F(3\alpha,\ 3\alpha+\tfrac{1}{2},\ 2\alpha+\tfrac{5}{6},\ t')$$
$$+(-t'')^{\alpha}(1-t'')^{2\alpha}F(3\alpha,\ 3\alpha+\tfrac{1}{2},\ 2\alpha+\tfrac{5}{6},\ t'')$$
$$+(1-t''')^{\alpha}(1-t''')^{2\alpha}F(3\alpha,\ 3\alpha+\tfrac{1}{2},\ 2\alpha+\tfrac{5}{6},\ t'''),$$

$$(103)\quad K_1x^{\frac{1}{3}}F(\alpha+\tfrac{1}{3},\ \tfrac{1}{2}-\alpha,\ \tfrac{4}{3},\ x)$$
$$= K_1x^{\frac{1}{3}}(1-x)^{\frac{1}{2}}F(1-\alpha,\ \alpha+\tfrac{5}{6},\ \tfrac{4}{3},\ x)$$
$$= (-t')^{\alpha}(1-t')^{2\alpha}F(3\alpha,\ 3\alpha+\tfrac{1}{2},\ 2\alpha+\tfrac{5}{6},\ t')$$
$$+j^2(-t'')^{\alpha}(1-t'')^{2\alpha}F(3\alpha,\ 3\alpha+\tfrac{1}{2},\ 2\alpha+\tfrac{5}{6},\ t'')$$
$$+j(-t''')^{\alpha}(1-t''')^{2\alpha}F(3\alpha,\ 3\alpha+\tfrac{1}{2},\ 2\alpha+\tfrac{5}{6},\ t''');$$

t', t'', t''' are the three roots of the equation

$$(3t+1)^3 - 27t(1-t)^2x = 0:$$

$$t' = -\tfrac{1}{3} - \frac{2\sqrt[3]{2}}{3}x^{\frac{1}{3}} + \ldots,$$

$$t'' = -\tfrac{1}{3} - \frac{2\sqrt[3]{2}}{3}jx^{\frac{1}{3}} + \ldots,$$

$$t''' = -\tfrac{1}{3} - \frac{2\sqrt[3]{2}}{3}j^2x^{\frac{1}{3}} + \ldots.$$

$$(104)\quad L_1F(\alpha,\ \alpha+\tfrac{1}{2},\ \tfrac{2}{3},\ x) = (9-8t')^{2\alpha}t'^{\alpha}F(3\alpha,\ \tfrac{1}{3}-\alpha,\ \tfrac{1}{2},\ t')$$
$$+(9-8t'')^{2\alpha}t''^{\alpha}F(3\alpha,\ \tfrac{1}{3}-\alpha,\ \tfrac{1}{2},\ t'')$$
$$+(9-8t''')^{2\alpha}t'''^{\alpha}F(3\alpha,\ \tfrac{1}{3}-\alpha,\ \tfrac{1}{2},\ t'''),$$

$$(105)\quad M_1x^{\frac{1}{3}}F(\alpha+\tfrac{1}{3},\ \alpha+\tfrac{5}{6},\ \tfrac{4}{3},\ x) = (9-8t')^{2\alpha}t'^{\alpha}F(3\alpha,\ \tfrac{1}{3}-\alpha,\ \tfrac{1}{2},\ t')$$
$$+j^2(9-8t)^{2\alpha}t''^{\alpha}F(3\alpha,\ \tfrac{1}{3}-\alpha,\ \tfrac{1}{2},\ t'')$$
$$+j(9-8t''')^{2\alpha}t'''^{\alpha}F(3\alpha,\ \tfrac{1}{3}-\alpha,\ \tfrac{1}{2},\ t''');$$

t', t'', t''' are the three roots of the equation

$$(4t-3)^3 - (9-8t)^2tx = 0:$$

$$t' = \tfrac{3}{4} + \frac{3\sqrt[3]{2}}{8}x^{\frac{1}{3}} + \ldots,$$

$$t'' = \tfrac{3}{4} + \frac{3\sqrt[3]{2}}{8} jx^{\frac{1}{3}} + \ldots,$$

$$t''' = \tfrac{3}{4} + \frac{3\sqrt[3]{2}}{8} j^2x^{\frac{1}{3}} + \ldots,$$

$$L_1 = (27)^\alpha \frac{3\sqrt{\pi}\,\Gamma(\frac{1}{3})}{\Gamma(\frac{1}{2}-\alpha)\Gamma(\alpha+\frac{1}{3})}, \quad M_1 = (27)^\alpha \frac{9\sqrt{\pi}\,\Gamma(\frac{2}{3})}{\Gamma(\alpha)\Gamma(\frac{1}{6}-\alpha)}.$$

(106) $PF(\alpha, \alpha+\frac{1}{2}, \frac{2}{3}, x)$

$$\begin{aligned} &= (1+8t')^{2\alpha}(1-t')^\alpha F(3\alpha, \tfrac{1}{3}-\alpha, 2\alpha+\tfrac{5}{6}, t') \\ &+ (1+8t'')^{2\alpha}(1-t'')^\alpha F(3\alpha, \tfrac{1}{3}-\alpha, 2\alpha+\tfrac{5}{6}, t'') \\ &+ (1+8t''')^{2\alpha}(1-t''')^\alpha F(3\alpha, \tfrac{1}{3}-\alpha, 2\alpha+\tfrac{5}{6}, t'''), \end{aligned}$$

(107) $Qx^{\frac{1}{3}}F(\alpha+\frac{1}{3}, \alpha+\frac{5}{6}, \frac{4}{3}, x)$

$$\begin{aligned} &= (1+8t')^{2\alpha}(1-t')^\alpha F(3\alpha, \tfrac{1}{3}-\alpha, 2\alpha+\tfrac{5}{6}, t') \\ &+ j^2(1+8t'')^{2\alpha}(1-t'')^\alpha F(3\alpha, \tfrac{1}{3}-\alpha, 2\alpha+\tfrac{5}{6}, t'') \\ &+ j(1+8t''')^{2\alpha}(1-t''')^\alpha F(3\alpha, \tfrac{1}{3}-\alpha, 2\alpha+\tfrac{5}{6}, t'''); \end{aligned}$$

t', t'', t''' are the three roots of the equation

$$(1-4t)^3 - (1+8t)^2(1-t)x = 0:$$

$$t' = \tfrac{1}{4} - \frac{3\sqrt[3]{2}}{8} x^{\frac{1}{3}} + \ldots,$$

$$t'' = \tfrac{1}{4} - \frac{3\sqrt[3]{2}}{8} jx^{\frac{1}{3}} + \ldots,$$

$$t''' = \tfrac{1}{4} - \frac{3\sqrt[3]{2}}{8} j^2x^{\frac{1}{3}} + \ldots,$$

$$P = \frac{3\Gamma(\frac{1}{3})\Gamma(2\alpha+\frac{5}{6})}{\Gamma(\alpha+\frac{1}{3})\Gamma(\alpha+\frac{5}{6})}, \quad Q = -\frac{9\,\Gamma(\frac{2}{3})\Gamma(2\alpha+\frac{5}{6})}{\Gamma(\alpha)\,\Gamma(\alpha+\frac{1}{2})}.$$

(108) $PF(\alpha, \alpha+\frac{1}{2}, \frac{2}{3}, x) = (1-9t')^{2\alpha}F(3\alpha, 3\alpha+\frac{1}{2}, 2\alpha+\frac{5}{6}, t')$

$$\begin{aligned} &+ (1-9t'')^{2\alpha}F(3\alpha, 3\alpha+\tfrac{1}{2}, 2\alpha+\tfrac{5}{6}, t'') \\ &+ (1-9t''')^{2\alpha}F(3\alpha, 3\alpha+\tfrac{1}{2}, 2\alpha+\tfrac{5}{6}, t'''), \end{aligned}$$

(109) $Qx^{\frac{1}{3}}F(\alpha+\frac{1}{3}, \alpha+\frac{5}{6}, \frac{4}{3}, x)$

$$= (1-9t')^{2\alpha}F(3\alpha, 3\alpha+\tfrac{1}{2}, 2\alpha+\tfrac{5}{6}, t')$$
$$+ j^2(1-9t'')^{2\alpha}F(3\alpha, 3\alpha+\tfrac{1}{2}, 2\alpha+\tfrac{5}{6}, t'')$$
$$+ j(1-9t''')^{2\alpha}F(3\alpha, 3\alpha+\tfrac{1}{2}, 2\alpha+\tfrac{5}{6}, t''');$$

t', t'', t''' are the three roots of the equation

$$(3t+1)^3-(1-9t)^2x=0:$$

$$t' = -\tfrac{1}{3}+\frac{2\sqrt[3]{2}}{3}x^{\frac{1}{3}}+\ldots,$$

$$t'' = -\tfrac{1}{3}+\frac{2\sqrt[3]{2}}{3}jx^{\frac{1}{3}}+\ldots,$$

$$t''' = -\tfrac{1}{3}+\frac{2\sqrt[3]{2}}{3}j^2x^{\frac{1}{3}}+\ldots.$$

(110) $F(\alpha, \alpha+\frac{1}{2}, 2\alpha+\frac{5}{6}, x) = (1-x)^{\frac{1}{3}}F(\alpha+\frac{1}{3}, \alpha+\frac{5}{6}, 2\alpha+\frac{5}{6}, x)$

$$= \left(1-\frac{9t'}{8}\right)^{2\alpha}F(3\alpha, 3\alpha+\tfrac{1}{2}, 4\alpha+\tfrac{2}{3}, t');$$

t' is one of those two roots of the equation

$$27t^2(1-t)+(9t-8)^2x=0$$

which are zero for $x=0$.

(111) $F(\alpha, \alpha+\frac{1}{2}, 2\alpha+\frac{5}{6}, x) = (1-x)^{\frac{1}{3}}F(\alpha+\frac{1}{3}, \alpha+\frac{5}{6}, 2\alpha+\frac{5}{6}, x)$

$$= \left(1+\frac{t'}{8}\right)^{2\alpha}(1-t')^{\alpha}F(3\alpha, \alpha+\tfrac{1}{6}, 4\alpha+\tfrac{2}{3}, t');$$

t' is one of those two roots of the equation

$$27t^2+(t+8)^2(1-t)x=0$$

which are zero for $x=0$.

(112) $F(\alpha, \alpha+\frac{1}{2}, 2\alpha+\frac{5}{6}, x) = (1-x)^{\frac{1}{3}}F(\alpha+\frac{1}{3}, \alpha+\frac{5}{6}, 2\alpha+\frac{5}{6}, x)$

$$= (1-9t)^{2\alpha}F(3\alpha, 3\alpha+\tfrac{1}{2}, 2\alpha+\tfrac{5}{6}, t);$$

t being that root of the equation

$$27t(1-t)^2+(1-9t)^2x=0$$

which is zero for $x=0$.

(113) $F(\alpha, \alpha+\frac{1}{2}, 2\alpha+\frac{5}{6}, x)=(1-x)^{\frac{1}{3}}F(\alpha+\frac{1}{3}, \alpha+\frac{5}{6}, 2\alpha+\frac{5}{6}, x)$

$$=(1+8t)^{2\alpha}(1-t)^{\alpha}F(3\alpha, \tfrac{1}{3}-\alpha, 2\alpha+\tfrac{5}{6}, t);$$

t being that root of the equation

$$27t-(1+8t)^2(1-t)x=0$$

which is zero for $x=0$.

(114) $F(\alpha, \alpha+\frac{1}{3}, 2\alpha+\frac{5}{6}, x)=(1-x)^{\frac{1}{4}}F(\alpha+\frac{1}{2}, \alpha+\frac{5}{6}, 2\alpha+\frac{5}{6}, x)$

$$=\left(1-\frac{3t'}{4}\right)^{3\alpha}F(3\alpha, 3\alpha+\tfrac{1}{2}, 4\alpha+\tfrac{2}{3}, t');$$

t' is one of those two roots of the equation

$$27t^2(1-t)-(4-3t)^3x=0$$

which are zero for $x=0$.

(115) $F(\alpha, \alpha+\frac{1}{3}, 2\alpha+\frac{5}{6}, x)=(1-x)^{\frac{1}{4}}F(\alpha+\frac{1}{2}, \alpha+\frac{5}{6}, 2\alpha+\frac{5}{6}, x)$

$$=\left(1-\frac{t'}{4}\right)^{3\alpha}F(3\alpha, \alpha+\tfrac{1}{6}, 4\alpha+\tfrac{2}{3}, t');$$

t' is one of those two roots of the equation

$$27t^2-(4-t)^3x=0$$

which are zero for $x=0$.

(116) $F(\alpha, \alpha+\frac{1}{3}, 2\alpha+\frac{5}{6}, x)=(1-x)^{\frac{1}{4}}F(\alpha+\frac{1}{2}, \alpha+\frac{5}{6}, 2\alpha+\frac{5}{6}, x)$

$$=(1+3t)^{3\alpha}F(3\alpha, 3\alpha+\tfrac{1}{2}, 2\alpha+\tfrac{5}{6}, t);$$

t being that root of the equation

$$27t(1-t)^2-(3t+1)^3x=0.$$

which is zero for $x = 0$.

(117) $$F(\alpha, \alpha + \tfrac{1}{3}, 2\alpha + \tfrac{5}{6}, x) = (1 - x)^{\frac{1}{2}} F(\alpha + \tfrac{1}{2}, \alpha + \tfrac{5}{6}, 2\alpha + \tfrac{5}{6}, x)$$
$$= (1 - 4t)^{3\alpha} F(3\alpha, \tfrac{1}{3} - \alpha, 2\alpha + \tfrac{5}{6}, t);$$

t being that root of the equation

$$27t + (1 - 4t)^3 x = 0$$

which is zero for $x = 0$.

Formulæ furnished by the Inverse Transformations.

(118) $$F(3\alpha, 3\alpha + \tfrac{1}{2}, 4\alpha + \tfrac{2}{3}, x)$$
$$= (1 - x)^{\frac{1}{6} - 2\alpha} F(\alpha + \tfrac{1}{6}, \alpha + \tfrac{2}{3}, 4\alpha + \tfrac{2}{3}, x)$$
$$= \begin{cases} \left(1 - \dfrac{9x}{8}\right)^{-2\alpha} F\left[\alpha, \alpha + \tfrac{1}{2}, 2\alpha + \tfrac{5}{6}, \dfrac{-27x^2(1 - x)}{(9x - 8)^2}\right], \\ \left(1 - \dfrac{3x}{4}\right)^{-3\alpha} F\left[\alpha, \alpha + \tfrac{1}{3}, 2\alpha + \tfrac{5}{6}, \dfrac{-27x^2(1 - x)}{(3x - 4)^3}\right]; \end{cases}$$

(119) $$F(3\alpha, 3\alpha + \tfrac{1}{2}, 2\alpha + \tfrac{5}{6}, x)$$
$$= (1 - x)^{\frac{1}{3} - 4\alpha} F(\tfrac{1}{3} - \alpha, \tfrac{5}{6} - \alpha, 2\alpha + \tfrac{5}{6}, x)$$
$$= \begin{cases} (1 - 9x)^{-2\alpha} F\left[\alpha, \alpha + \tfrac{1}{2}, 2\alpha + \tfrac{5}{6}, \dfrac{-27x(1 - x)^2}{(1 - 9x)^2}\right], \\ (1 + 3x)^{-3\alpha} F\left[\alpha, \alpha + \tfrac{1}{3}, 2\alpha + \tfrac{5}{6}, \dfrac{27x(1 - x)^2}{(1 + 3x)^3}\right]; \end{cases}$$

(120) $$F(3\alpha, \alpha + \tfrac{1}{6}, 4\alpha + \tfrac{2}{3}, x)$$
$$= (1 - x)^{\frac{1}{2}} F(\alpha + \tfrac{2}{3}, 3\alpha + \tfrac{1}{2}, 4\alpha + \tfrac{2}{3}, x)$$
$$= \begin{cases} \left(1 + \dfrac{x}{8}\right)^{-2\alpha} (1 - x)^{-\alpha} F\left[\alpha, \alpha + \tfrac{1}{2}, 2\alpha + \tfrac{5}{6}, \dfrac{-27x^2}{(x + 8)^2(1 - x)}\right], \\ \left(1 - \dfrac{x}{4}\right)^{-3\alpha} F\left[\alpha, \alpha + \tfrac{1}{3}, 2\alpha + \tfrac{5}{6}, \dfrac{-27x^2}{(x - 4)^3}\right]; \end{cases}$$

(121) $F(3\alpha,\ \tfrac{1}{3}-\alpha,\ 2\alpha+\tfrac{5}{6},\ x)$

$$= (1-x)^{\frac{1}{2}} F(\tfrac{5}{6}-\alpha,\ 3\alpha+\tfrac{1}{2},\ 2\alpha+\tfrac{5}{6},\ x)$$

$$= \begin{cases} (1+8x)^{-2\alpha}(1-x)^{-\alpha} F\left[\alpha,\ \alpha+\tfrac{1}{2},\ 2\alpha+\tfrac{5}{6},\ \dfrac{27x}{(1+8x)^2(1-x)}\right], \\ (1-4x)^{-3\alpha} F\left[\alpha,\ \alpha+\tfrac{1}{3},\ 2\alpha+\tfrac{5}{6},\ \dfrac{27x}{(4x-1)^3}\right]; \end{cases}$$

(122) $F(3\alpha,\ \tfrac{1}{3}-\alpha,\ \tfrac{1}{2},\ x)$

$$= (1-x)^{\frac{1}{6}-2\alpha} F(\tfrac{1}{2}-3\alpha,\ \tfrac{1}{6}+\alpha,\ \tfrac{1}{2},\ x)$$

$$= \begin{cases} (1-x)^{-\alpha} F\left[\alpha,\ \tfrac{1}{6}-\alpha,\ \tfrac{1}{2},\ \dfrac{(9-8x)^2 x}{27(1-x)}\right], \\ \left(1+\dfrac{4x}{3}\right)^{-3\alpha} F\left[\alpha,\ \alpha+\tfrac{1}{3},\ \tfrac{1}{2},\ \dfrac{(9-8x)^2 x}{(4x-3)^3}\right]; \end{cases}$$

(123) $F(3\alpha,\ \alpha+\tfrac{1}{6},\ \tfrac{1}{2},\ x)$

$$= (1-x)^{\frac{1}{3}-4\alpha} F(\tfrac{1}{2}-3\alpha,\ \tfrac{1}{3}-\alpha,\ \tfrac{1}{2},\ x)$$

$$= \begin{cases} (1-x)^{-2\alpha} F\left[\alpha,\ \tfrac{1}{6}-\alpha,\ \tfrac{1}{2},\ \dfrac{(x-9)^2 x}{-27(1-x)^2}\right], \\ \left(1+\dfrac{x}{3}\right)^{-3\alpha} F\left[\alpha,\ \alpha+\tfrac{1}{3},\ \tfrac{1}{2},\ \dfrac{(x-9)^2 x}{(x+3)^3}\right]; \end{cases}$$

(124) $F(3\alpha+\tfrac{1}{2},\ \tfrac{5}{6}-\alpha,\ \tfrac{3}{2},\ x)$

$$= (1-x)^{\frac{1}{6}-2\alpha} F(\tfrac{1}{2}-3\alpha,\ \alpha+\tfrac{2}{3},\ \tfrac{3}{2},\ x)$$

$$= \begin{cases} \left(1-\dfrac{8x}{9}\right)(1-x)^{-\alpha-\frac{1}{3}} F\left[\alpha+\tfrac{1}{2},\ \tfrac{2}{3}-\alpha,\ \tfrac{3}{2},\ \dfrac{(9-8x)^2 x}{27(1-x)}\right], \\ \left(1-\dfrac{8x}{9}\right)\left(1-\dfrac{4x}{3}\right)^{-3\alpha-\frac{3}{2}} F\left[\alpha+\tfrac{1}{2},\ \alpha+\tfrac{5}{6},\ \tfrac{3}{2},\ \dfrac{(9-8x)^2 x}{(4x-3)^3}\right]; \end{cases}$$

(125) $F(3\alpha+\tfrac{1}{2},\ \alpha+\tfrac{2}{3},\ \tfrac{3}{2},\ x)$

$$= (1-x)^{\frac{1}{3}-4\alpha} F(1-3\alpha,\ \tfrac{5}{6}-\alpha,\ \tfrac{3}{2},\ x)$$

$$= \begin{cases} \left(1-\dfrac{x}{9}\right)(1-x)^{-2\alpha-1} F\left[\alpha+\tfrac{1}{2},\ \tfrac{2}{3}-\alpha,\ \tfrac{3}{2},\ \dfrac{(x-9)^2 x}{-27(1-x)^2}\right], \\ \left(1-\dfrac{x}{9}\right)\left(1+\dfrac{x}{3}\right)^{-3\alpha-\frac{3}{2}} F\left[\alpha+\tfrac{1}{2},\ \alpha+\tfrac{5}{6},\ \tfrac{3}{2},\ \dfrac{(x-9)^2 x}{(x+3)^3}\right]. \end{cases}$$

Formulæ furnished by the Inverse Transformations of the Other Rational Transformations.

(126) $F(4\alpha,\ 4\alpha+\tfrac{1}{3},\ 6\alpha+\tfrac{1}{2},\ x)$

$$= (1-x)^{\frac{1}{6}-2\alpha} F(2\alpha+\tfrac{1}{6},\ 2\alpha+\tfrac{1}{2},\ 6\alpha+\tfrac{1}{2},\ x)$$

$$= \begin{cases} \left(\dfrac{27-36x+8x^2}{27}\right)^{-2\alpha} F\left[\alpha,\ \alpha+\tfrac{1}{2},\ 2\alpha+\tfrac{5}{6},\ \dfrac{-64x^3(1-x)}{(8x^2-36x+27)^2}\right], \\ \left(1-\dfrac{8x}{9}\right)^{-3\alpha} F\left[\alpha,\ \alpha+\tfrac{1}{3},\ 2\alpha+\tfrac{5}{6},\ \dfrac{64x^3(1-x)}{(9-8x)^3}\right]; \end{cases}$$

(127) $F(4\alpha,\ 4\alpha+\tfrac{1}{3},\ 2\alpha+\tfrac{5}{6},\ x)$

$$= (1-x)^{\frac{1}{2}-6\alpha} F(\tfrac{1}{2}-2\alpha,\ \tfrac{5}{6}-2\alpha,\ 2\alpha+\tfrac{5}{6},\ x)$$

$$= \begin{cases} (1-20x-8x^2)^{-2\alpha} F\left[\alpha,\ \alpha+\tfrac{1}{2},\ 2\alpha+\tfrac{5}{6},\ \dfrac{-64x(1-x)^3}{(1-20x-8x^2)^2}\right], \\ (1+8x)^{-3\alpha} F\left[\alpha,\ \alpha+\tfrac{1}{3},\ 2\alpha+\tfrac{5}{6},\ \dfrac{64x(1-x)^3}{(1+8x)^3}\right]; \end{cases}$$

(128) $F(4\alpha,\ 2\alpha+\tfrac{1}{6},\ 6\alpha+\tfrac{1}{2},\ x)$

$$= (1-x)^{\frac{1}{3}} F(4\alpha+\tfrac{1}{3},\ 2\alpha+\tfrac{1}{2},\ 6\alpha+\tfrac{1}{2},\ x)$$

$$= \begin{cases} \left(\dfrac{27-18x-x^2}{27}\right)^{-2\alpha} F\left[\alpha,\ \alpha+\tfrac{1}{2},\ 2\alpha+\tfrac{5}{6},\ \dfrac{64x^3}{(x^2+18x-27)^2}\right], \\ \left(1-\dfrac{x}{9}\right)^{-3\alpha}(1-x)^{-\alpha} F\left[\alpha,\ \alpha+\tfrac{1}{3},\ 2\alpha+\tfrac{5}{6},\ \dfrac{64x^3}{(x-9)^3(1-x)}\right]; \end{cases}$$

(129) $F(4\alpha,\ \tfrac{1}{2}-2\alpha,\ 2\alpha+\tfrac{5}{6},\ x)$

$$= (1-x)^{\frac{1}{3}} F(4\alpha+\tfrac{1}{3},\ \tfrac{5}{6}-2\alpha,\ 2\alpha+\tfrac{5}{6},\ x)$$

$$= \begin{cases} (1+18x-27x^2)^{-2\alpha} F\left[\alpha,\ \alpha+\tfrac{1}{2},\ 2\alpha+\tfrac{5}{6},\ \dfrac{64x}{(1+18x-27x^2)^2}\right], \\ (1-9x)^{-3\alpha}(1-x)^{-\alpha} F\left[\alpha,\ \alpha+\tfrac{1}{3},\ 2\alpha+\tfrac{5}{6},\ \dfrac{64x}{(9x-1)^3(1-x)}\right]; \end{cases}$$

(130) $F(4\alpha,\ \tfrac{1}{2}-2\alpha,\ \tfrac{2}{3},\ x)$

$$= \begin{cases} (1-x)^{-\alpha} F\left[\alpha,\ \tfrac{1}{6}-\alpha,\ \tfrac{2}{3},\ \dfrac{(8-9x)^3x}{64(1-x)}\right], \\ \left(\dfrac{8-36x+27x^2}{8}\right)^{-2\alpha} F\left[\alpha,\ \alpha+\tfrac{1}{2},\ \tfrac{2}{3},\ \dfrac{-(8-9x)^3x}{(8-36x+27x^2)^2}\right]; \end{cases}$$

(131) $F(4\alpha,\ 2\alpha+\tfrac{1}{6},\ \tfrac{2}{3},\ x)$

$$=\begin{cases}(1-x)^{-3\alpha}F\left[\alpha,\ \tfrac{1}{6}-\alpha,\ \tfrac{2}{3},\ \dfrac{-(x+8)^3x}{64(1-x)^3}\right], \\ \left(\dfrac{8-20x-x^2}{8}\right)^{-2\alpha}F\left[\alpha,\ \alpha+\tfrac{1}{2},\ \tfrac{2}{3},\ \dfrac{(x+8)^3x}{(x^2-20x-8)^2}\right];\end{cases}$$

(132) $F(4\alpha+\tfrac{1}{3},\ \tfrac{5}{6}-2\alpha,\ \tfrac{4}{3},\ x)$

$$=\begin{cases}(1-x)^{-\alpha-\frac{1}{3}}\left(1-\dfrac{9x}{8}\right)F\left[\alpha+\tfrac{1}{3},\ \tfrac{1}{2}-\alpha,\ \tfrac{4}{3},\ \dfrac{(8-9x)^3x}{64(1-x)}\right], \\ \left(\dfrac{8-36x+27x^2}{8}\right)^{-2\alpha-\frac{2}{3}}\left(1-\dfrac{9x}{8}\right) \\ \qquad F\left[\alpha+\tfrac{1}{3},\ \alpha+\tfrac{5}{6},\ \tfrac{4}{3},\ \dfrac{-(8-9x)^3x}{(8-36x+27x^2)^2}\right],\end{cases}$$

(133) $F(4\alpha+\tfrac{1}{3},\ 2\alpha+\tfrac{1}{2},\ \tfrac{4}{3},\ x)$

$$=\begin{cases}(1-x)^{-3\alpha-1}\left(1+\dfrac{x}{8}\right)F\left[\alpha+\tfrac{1}{3},\ \tfrac{1}{2}-\alpha,\ \tfrac{4}{3},\ \dfrac{-(x+8)^3x}{64(1-x)^3}\right], \\ \left(\dfrac{8-20x-x^2}{8}\right)^{-2\alpha-\frac{2}{3}}\left(1+\dfrac{x}{8}\right)F\left[\alpha+\tfrac{1}{3},\ \alpha+\tfrac{5}{6},\ \tfrac{4}{3},\ \dfrac{(x+8)^3x}{(x^2-20x-8)^2}\right];\end{cases}$$

(134) $F(6\alpha,\ 2\alpha+\tfrac{1}{3},\ 4\alpha+\tfrac{2}{3},\ x)$

$=(1-x)^{\frac{1}{3}-4\alpha}F(2\alpha+\tfrac{1}{3},\ \tfrac{2}{3}-2\alpha,\ 4\alpha+\tfrac{2}{3},\ x)$

$$=\begin{cases}\left(\dfrac{2-3x-3x^2+2x^3}{2}\right)^{-2\alpha}F\left[\alpha,\ \alpha+\tfrac{1}{2},\ 2\alpha+\tfrac{5}{6},\ \dfrac{-27x^2(1-x)^2}{(2x^3-3x^2-3x+2)^2}\right] \\ (1-x+x^2)^{-3\alpha}F\left[\alpha,\ \alpha+\tfrac{1}{3},\ 2\alpha+\tfrac{5}{6},\ \dfrac{27x^2(1-x)^2}{4(x^2-x+1)^3}\right];\end{cases}$$

(135) $F(6\alpha,\ \tfrac{2}{3}-2\alpha,\ 2\alpha+\tfrac{5}{6},\ x)$

$=(1-x)^{\frac{1}{6}-2\alpha}F(4\alpha+\tfrac{1}{6},\ \tfrac{5}{6}-4\alpha,\ 2\alpha+\tfrac{5}{6},\ x)$

$$=\begin{cases}1+30x-96x^2+64x^3)^{-2\alpha} \\ F\left[\alpha,\ \alpha+\tfrac{1}{2},\ 2\alpha+\tfrac{5}{6},\ \dfrac{-108x(1-x)}{(64x^3-96x^2+30x+1)^2}\right], \\ (1-16x+16x^2)^{-3\alpha}F\left[\alpha,\ \alpha+\tfrac{1}{3},\ 2\alpha+\tfrac{5}{6},\ \dfrac{108x(1-x)}{(1-16x+16x^2)^3}\right];\end{cases}$$

(136) $F(6\alpha,\, 4\alpha + \frac{1}{6},\, 2\alpha + \frac{5}{6},\, x)$

$$= (1 - x)^{\frac{2}{3} - 8\alpha} F(\tfrac{5}{6} - 4\alpha,\, \tfrac{2}{3} - 2\alpha,\, 2\alpha + \tfrac{5}{6},\, x)$$

$$= \begin{cases} (1 - 33x - 33x^2 + x^3)^{-2\alpha} F\left[\alpha,\, \alpha + \frac{1}{2},\, 2\alpha + \frac{5}{6},\, \dfrac{108x(1-x)^4}{(x^3 - 33x^2 - 33x + 1)^2}\right], \\ (1 + 14x + x^2)^{-3\alpha} F\left[\alpha,\, \alpha + \frac{1}{3},\, 2\alpha + \frac{5}{6},\, \dfrac{-108x(1-x)^4}{(x^2 + 14x + 1)^3}\right]; \end{cases}$$

(137) $F(6\alpha,\, 4\alpha + \frac{1}{6},\, 8\alpha + \frac{1}{3},\, x)$

$$= (1 - x)^{\frac{1}{6} - 2\alpha} F(2\alpha + \tfrac{1}{3},\, 4\alpha + \tfrac{1}{6},\, 8\alpha + \tfrac{1}{3},\, x)$$

$$= \begin{cases} \left(\dfrac{64 - 96x + 30x^2 + x^3}{64}\right)^{-2\alpha} \\ \qquad\qquad F\left[\alpha,\, \alpha + \frac{1}{2},\, 2\alpha + \frac{5}{6},\, \dfrac{108x^4(1-x)}{(64 - 96x + 30x^2 + x^3)^2}\right], \\ \left(\dfrac{16 - 16x + x^2}{16}\right)^{-3\alpha} F\left[\alpha,\, \alpha + \frac{1}{3},\, 2\alpha + \frac{5}{6},\, \dfrac{-108x^4(1-x)}{(16 - 16x + x^2)^3}\right]. \end{cases}$$

The transformations which we can effect when two of the three elements α, β, γ are arbitrary have been completely given by Kummer. In the case where a single element is arbitrary, he has indicated some particular cases of the rational transformation of the fourth degree and a certain number of irrational transformations. The other rational transformations and the greater part of the irrational transformations above seem to be new.

CHAPTER VIII.

IRREDUCIBLE LINEAR DIFFERENTIAL EQUATIONS.

SOME properties of these equations have already been noted in Chapter IV, but we shall study the question in a rather more general manner. Before entering into the study of these equations from the modern point of view which requires a knowledge of the group of substitutions belonging to a given linear differential equation, we will give some general theorems concerning irreducible differential equations taken from the memoir by Frobenius* previously referred to.

In all that follows we will assume $m < n$. Denote by P the operator

$$\frac{d^n}{dx^n} + p_1 \frac{d^{n-1}}{dx^{n-1}} + \ldots + p_n,$$

and by Q the operator

$$\frac{d^m}{dx^m} + q_1 \frac{d^{m-1}}{dx^{m-1}} + \ldots + q_m,$$

where p and q are uniform functions of x.

The differential equation

$$(1) \qquad Py \equiv \frac{d^ny}{dx^n} + p_1 \frac{d^{n-1}y}{dx^{n-1}} + \ldots + p_ny = 0$$

* Frobenius: *Ueber den Begriff der Irreductibilität in der Theorie der linearen Differentialgleichungen.* Crelle, vol. 76, p. 256. Frobenius refers to the memoir by Libri in Crelle, vol. 10, p. 193, and also to the "Note" by Brassinne in the Appendix to Vol. II of Sturm's *Cours d'Analyse.*

is reducible or irreducible according as it has or has not integrals in common with an equation of lower order, say

$$(2) \qquad Qy \equiv \frac{d^m y}{dx^m} + q_1 \frac{d^{m-1} y}{dx^{m-1}} + \ldots + q_m y = 0.$$

Let $n - m = l$, and for brevity write Q instead of Qy; no inconvenience can arise from this abbreviation, as it will always be clear whether we mean by Q the operator or the differential quantic which is the left-hand member of (2). Form now the derivatives

$$\frac{dQ}{dx}, \quad \frac{d^2Q}{dx^2}, \quad \ldots, \quad \frac{d^lQ}{dx^l},$$

and from the equations so obtained find the values of

$$\frac{d^m y}{dx^m}, \quad \frac{d^{m+1} y}{dx^{m+1}}, \quad \ldots, \quad \frac{d^n y}{dx^n}$$

and substitute these in (1); we have then an expression of the form

$$(3) \quad Py = \frac{d^lQ}{dx^l} + r_1 \frac{d^{l-1}Q}{dx^{l-1}} + r_2 \frac{d^{l-2}Q}{dx^{l-2}} + \ldots + r_l Q + r_0 R = 0,$$

where $r_0, r_1, r_2, \ldots, r_l$ are uniform functions of x, and R is a linear function of

$$\frac{d^\lambda y}{dx^\lambda}, \quad \frac{d^{\lambda-1} y}{dx^{\lambda-1}}, \quad \ldots, \quad \frac{dy}{dx}, \quad y \qquad (\lambda \leqq m - 1),$$

having for coefficients uniform functions of x. From this equation it is at once evident that all functions which are at the same time integrals of $P = 0$ and $Q = 0$ must also be integrals of $R = 0$. For convenience we will employ a notation borrowed from the Theory of Numbers and write equation (3) in the form

$$(4) \qquad P \equiv R \mod Q.$$

In this congruence P, Q, and R denote differential expressions in which the coefficient of the highest derivative is unity, and, denoting by n, m, λ the orders of P, Q, and R respectively, we have the inequalities $n \geqq m > \lambda$.

Suppose that *all* the integrals of $Q = 0$ are also integrals of $P = 0$; then it follows from (4) that they must also be integrals of $R = 0$; but the independent integrals of $Q = 0$ are m in number, and the order of $R = 0$ is λ, which is less than m, and as $R = 0$ can only have λ independent integrals, it follows in this case that R is identically zero. Therefore: *If the differential equation $P = 0$ has among its integrals all of the integrals of $Q = 0$, then P can be put in the form*

$$P = \frac{d^lQ}{dx^l} + r_1\frac{d^{l-1}Q}{dx^{l-1}} + \ldots + r_lQ,$$

or

$$P \equiv 0 \mod Q.$$

Again, suppose $Q = 0$ to be an irreducible equation, and suppose the equation $P = 0$ has an integral Y which is also an integral of $Q = 0$; then the order of Q cannot be higher than that of P; if then

$$P \equiv R \mod Q,$$

Y must also be an integral of $R = 0$; but the irreducible equation $Q = 0$ can have no integral in common with an equation of lower order, so that R must be identically zero; that is, we must have

$$P \equiv 0 \mod Q.$$

It follows therefore at once that *If a linear differential equation has among its integrals one which is also an integral of an irreducible linear differential equation, then all the integrals of the latter equation are integrals of the first.*

Suppose we have two equations $P = 0$ of order n and $P_1 = 0$ of order n_1; the integrals, if any, which satisfy these two equations will be integrals of a third equation which can be found by a process quite analogous to that for finding the greatest common divisor. We have, viz.,

$$P \equiv P_2 \quad \text{mod } P_1,$$

$$P_1 \equiv P_3 \quad \text{mod } P_2,$$

$$\cdot \quad \cdot \quad \cdot \quad \cdot$$

$$P_{i-1} \equiv P_{i+1} \text{ mod } P_i.$$

Denoting by n_k the order of P_k, we must have

$$n \overset{=}{>} n_1 > n_2 > n_3 > \ldots,$$

and n_k must vanish at the latest when $k = 1 + n_1$. Suppose $n_{i+1} = 0$ but n_i not zero; then P_{i+1} either reduces to merely y or is zero. In the first case $P = 0$ and $P_1 = 0$ have no integral in common except $y = 0$; that is, they have *no* integral in common. In the second case the integrals common to $P = 0$ and $P_1 = 0$ are integrals of $P_i = 0$, and all the integrals of $P_i = 0$ are integrals of $P = 0$ and $P_1 = 0$. *If*, therefore, *a linear differential equation is reducible, there exists a linear differential equation of lower order all of whose integrals are also integrals of the given equation.*

If $P = 0$ is a reducible linear differential equation, and $Q = 0$ is a linear differential equation of lower order all of whose integrals are integrals of $P = 0$, then, as we have seen, we have

$$P \equiv 0 \quad \text{mod } Q;$$

that is, the left-hand member of a reducible linear differential equation is of the form

$$P = \frac{d^l Q}{dx^l} + r_1 \frac{d^{l-1} Q}{dx^{l-1}} + \ldots + r_l Q.$$

We will revert now to the original definition of P and Q as operators, and add the operator R defined by

$$R = \frac{d^l}{dx^l} + r_1 \frac{d^{l-1}}{dx^{l-1}} + \ldots + r_l.$$

Instead of the congruence

$$P \equiv 0 \mod Q$$

we can now write

$$Py = R(Qy).$$

We recall that the orders of P, Q, R respectively are n, m, l, and $n = m + l$.

Suppose w to be the general integral of the equation $Ry = 0$, and v an integral of $Qy = w$; then v is also an integral of $Py = 0$, and, from what has been assumed concerning this equation, is a regular integral. The function $w = Qv$ is then also a regular function, and consequently $Ry = 0$ is an equation of the same form as $Py = 0$. The differential equation $Qy = w$ containing the l arbitrary constants belonging to the function w is therefore an integral equation for the given reducible equation $Py = 0$, and this last equation is deducible from $Qy = w$ by differentiation and elimination of the l arbitrary constants. If then a linear differential equation of order n has among its integrals all the integrals of an equation $Qy = 0$ of order m, each of the integrals of the given equation satisfies a differential equation $Qy = w$, in which w is an integral of a determinate equation of order $n - m$. Conversely, a linear differential equation is reducible when it has for an integral a differential equation of the form $Qy = w$. This form of the integral equation is the characteristic property of reducible linear differential equations.

So far we have only explicitly considered equations of the type studied by Fuchs in his first two memoirs, viz., equations with uniform coefficients, a finite number of critical points, and having only regular integrals. We may consider more generally linear differential equations of which we assert merely that they have uniform coefficients. An irreducible equation of this sort is defined as one which has no integrals in common with a differential equation of lower order having also uniform coefficients. The theorems already proved can be readily seen to hold for this more extended class of equations; but as the whole matter will be taken up presently from a different point of view, it is not necessary to dwell longer on it here. One general theorem, however, it is desirable to give, viz., that *if of two distinct integrals of a linear differential equation one is equal to a differential expression with uniform coefficients of the other, then the differential equation is reducible.*

Suppose

$$Qy = q_0 \frac{d^m y}{dx^n} + q_1 \frac{d^{m-1} y}{dx^{m-1}} + \ldots + q_m y,$$

where $q_0, q_1, \ldots, q_m$ are uniform functions of x. Denote by y_0 and y_1 two distinct integrals of the equation

$$Py = p_0 \frac{d^n y}{dx^n} + p_1 \frac{d^{n-1} y}{dx^{n-1}} + \ldots + p_n y = 0,$$

where $p_0, p_1, \ldots, p_n$ are uniform functions of x, and suppose

$$y_1 = Qy_0.$$

If now, contrary to the statement in the above theorem, $Py = 0$ is irreducible, then, since one integral of $Py = 0$ is an integral of $P(Qy) = 0$, all of the integrals of $Py = 0$ must be integrals of $P(Qy) = 0$; if then y is an integral of $Py = 0$, Qy is also an integral—that is, $Py = 0$ will be satisfied by the functions

$$y_0, \quad Qy_0, \quad Q^2 y_0, \quad Q^3 y_0 \ldots;$$

but as $Py = 0$ can only have n independent integrals, we must arrive at a function, say $Q^k y$, which is linearly expressible in terms of the preceding ones, or we must have an equation of the form

$$a_0 y_0 + a_1 Qy_0 + \ldots + a_k Q^k y = 0,$$

where a_k is not zero, and between the functions

$$y_0, \quad Qy_0, \ldots Q^{k-1} y_0$$

there is no such linear relation with constant coefficients. The number k must, of course, be greater than unity, since by hypothesis y_0 and y_1, $= Qy_0$, are distinct integrals. If $k < n$, then we have still to find $n - k$ integrals which, together with those written above, will constitute a complete independent system. Write now

$$f(r) = a_0 + a_1 r + \ldots + a_k r^k,$$

and

$$\frac{f(r) - f(s)}{r - s} = f_1(r) + f_2(r)s + \ldots + f_k(r)s^{k-1},$$

and consequently

$$f_1(r) = a_1 + a_2 r + \ldots + a_k r^{k-1},$$
$$f_2(r) = a_2 + a_3 r + \ldots + a_k r^{k-2},$$
$$\cdot \quad \cdot \quad \cdot \quad \cdot \quad \cdot \quad \cdot \quad \cdot$$
$$f_k(r) = a_k .$$

Assume further

$$Ry = f_1(r)y + f_2(r)Qy + \ldots + f_k(r)Q^{k-1}y,$$

then

$$R(Qy) = f_1(r)Qy + \ldots + f_k(r)Q^k y.$$

We have, however,

$$f_k(r)Q^k y_0 = a_k Q^k y_0 = - a_0 y_0 - a_1 Q y_0 - \ldots - a_{k-1} Q^{k-1} y_0 ,$$

and consequently

$$R(Qy_0) = - a_0 y_0 + [f_1(r) - a_1]Qy_0 + \ldots + [f_{k-1}(r) - a_{k-1}]Q^{k-1}y_0$$
$$= r[f_1(r)y_0 + f_2(r)Qy_0 + \ldots + f_k(r)Q^{k-1}y_0] - f(r)y_0 .$$

If therefore r is a root of $f(r) = 0$, we have

$$R(Qy_0) = rRy_0 ,$$

and consequently each integral of the irreducible equation $Py = 0$ must satisfy the equation

$$R(Qy) = rRy.$$

Denote by y_i the integral $Q^i y_0$, $i = 1, 2, \ldots$, and we see that

$$Ry_2 = R(Q^2 y_0) = rR(Qy_0) = r^2 Ry_0 ,$$

and in general, for $\alpha < k$,

$$Ry_\alpha = r^\alpha Ry_0.$$

If now x starts from any non-critical point and describes an arbitrary closed curve returning to the initial point, y_0 will change into an expression of the form

$$c_0y_0 + c_1y_1 + \ldots + c_{k-1}y_{k-1} + c_ky_k + \ldots c_{n-1}y_{n-1},$$

and Ry_0 will become

$$(c_0 + c_1r + \ldots + c_{k-1}r^{k-1})Ry_0 + c_kRy_k + \ldots + c_{n-1}Ry_{n-1},$$

and consequently all the branches of this function are expressible as linear functions (constant coefficient understood) of

$$Ry_0, \quad Ry_k, \quad \ldots, \quad Ry_{n-1}.$$

It follows therefore that Ry_0 satisfies a linear differential equation with uniform coefficients whose order is at the most $n-k+1$. The linear differential equation $Py = 0$ has therefore an integral

$$Ry_0 = f_1(r)y_0 + f_2(r)y_1 + \ldots + f_k(r)y_{k-1}$$

in common with a differential equation of lower order, and in consequence cannot be irreducible.

We will now take up the subject of reducibility of linear differential equations with uniform coefficients from the point of view of the *groups* to which such equations belong. Suppose $P = 0$ to be an equation of order n, and $Q = 0$ an equation of order $m < n$ all of whose integrals are integrals of $P = 0$. Denote by $Y_1 \ldots Y_m$ a system of fundamental integrals of $Q = 0$; then any substitution belonging to the group of $P = 0$ can by hypothesis only change $Y_1 \ldots Y_m$ into linear functions of themselves. Let $y_{m+1} \ldots y_n$ denote the functions which with $Y_1 \ldots Y_m$ form a fundamental system of $P = 0$; then the substitutions of the group of P will change these functions in general into linear functions of themselves and of $Y_1 \ldots Y_m$; that is, all the substitutions belonging to the group of P will be of the form

$$\left| \begin{matrix} Y_1 & \ldots & Y_m; & f_1 & \ldots & f_m \\ y_{m+1} & \ldots & y_n: & f_{m+1} & \ldots & f_n \end{matrix} \right|;$$

where $f \ldots f_m$ are linear functions of $Y_1 \ldots Y_m$ alone, and $f_{m+1} \ldots f_n$ are linear functions of $Y_1 \ldots Y_m, y_{m+1} \ldots y_n$.

Reciprocally, if we can choose $y_1 \ldots y_m$ such that the group G contains only substitutions of this form, then these functions will satisfy the equation

$$\begin{vmatrix} y, & \frac{dy}{dx}, & \ldots, & \frac{d^m y}{dx^m} \\ y_1, & \frac{dy_1}{dx}, & \ldots, & \frac{d^m y_1}{dx^m} \\ \cdot & \cdot & \cdot & \cdot \\ y_m, & \frac{dy^m}{dx}, & \ldots, & \frac{d^m y_m}{dx^m} \end{vmatrix} = 0,$$

where the coefficients of y and its derivatives have for ratios only uniform functions. Suppose x to describe an arbitrary closed contour containing one or more critical points; then $y_1 \ldots y_m$ will change into

$$\begin{array}{c} c_{11}y_1 + \ldots + c_{1m}y_m, \\ \cdot \quad \cdot \quad \cdot \quad \cdot \quad \cdot \\ c_{m1}y_1 + \ldots + c_{mm}y_m, \end{array}$$

and consequently all the coefficients of this equation will be multiplied by one and the same factor, viz., the determinant

$$| c_{ij} |$$

of the substitution corresponding to the path described by x; the ratios of the coefficients are consequently unaltered, and these are therefore uniform functions. The question of the determination of the group of a given linear differential equation will not be taken up here, but we can investigate the following question, viz.: *Having given a group G composed of linear substitutions $S, S_1, S_2, \ldots$ among n variables, required to determine whether or not the linear differential equation which has G for its group is satisfied by the integrals of analogous equations of order lower than n, and to determine the groups of these equations.*

From what precedes this can be stated as follows:

Required to determine in all possible manners a system of linear functions $Y_1, \ldots, Y_m$ *of the variables* $y_1, \ldots, y_n$ *such that each of the substitutions* $S, S_1, \ldots$ *of which* G *is composed shall change* $Y_1, \ldots, Y_m$ *into linear functions of themselves.*

This enunciation may still be slightly modified, and in order to do so it is necessary to introduce a new term which shall replace the French word *faisceau* employed by Jordan. The author has not been able to find any *single* English word for *faisceau* which is not already employed for other purposes, or which would be at all appropriate, and therefore suggests the term *function-group.* We will therefore say that a system of functions forms a function-group when every linear combination of these functions forms a part of the system. If $Y_1, \ldots, Y_m$ forms one of the systems of functions which we have to determine as above described, then each substitution of the group G will transform the different elements of the function-group corresponding to $Y_1, \ldots, Y_m$ into other elements of the same function-group. In analogy with the theory of groups of substitutions, we see that each function-group will contain a certain number of linearly distinct functions in terms of which every other function in the function-group is linearly expressible. We can say then that the function-group is *derived* from these linearly distinct functions. Suppose now that in the function-group corresponding to $Y_1, \ldots, Y_m$ we choose any other m linearly distinct functions; then (using for brevity F. G. to denote *function-group*) every other function of the F. G. is linearly expressible in terms of these new m functions, and these latter will further form a system possessing precisely the same properties as the given system $Y_1, \ldots, Y_m$. If now we consider the F. G. instead of the particular system $Y_1, \ldots, Y_m$, we shall have the advantage of a greater freedom of choice of the functions from which the F. G. is derived. Our problem may therefore be stated as follows:

Required to determine all the function-groups such that each of the transformations $S, S_1, \ldots$ *shall transform any element of one into some other element of the same one.*

There will of course always be *one* such F. G. formed by the aggregate of the linear functions of $y_1, \ldots, y_n$. If there is only one the group G will be said to be *prime*, and the corresponding differential equation will be *irreducible*. If, on the contrary, there should be more than one, then to each of the F. G.'s derived from the linearly distinct functions $Y_1, \ldots, Y_m$ there will correspond a *reduced* differential equation whose group will be formed by the changes which the substitutions of G impose upon $Y_1, \ldots, Y_m$.

We have now first to indicate a means of ascertaining whether the group G is or is not prime, and in the latter case to determine a function-group containing less than n linearly distinct elements. This question will be taken up in Chapter XI. All that immediately precedes concerning the group of an equation, etc., is taken directly from a memoir by Jordan.* Chapter XI. is also taken from this memoir, the only one that the author has knowledge of which deals directly with the subject.

---

* *Mémoire sur une application de la théorie des substitutions à l'étude des équations différentielles linéaires :* Bull. de la Société Mathématique de France, t. 2, p. 100..

CHAPTER IX.

LINEAR DIFFERENTIAL EQUATIONS SOME OF WHOSE INTEGRALS ARE REGULAR.

WE will now resume the direct study of the integrals of linear differential equations with uniform coefficients, and will treat particularly the cases where not all of the integrals are regular, and show how to determine the number of such integrals in each case. The method employed is due to Frobenius and in part to Thomé. The equation is

$$(1) \qquad P = \frac{d^n y}{dx^n} + p_1 \frac{d^{n-1} y}{dx^{n-1}} + p_2 \frac{d^{n-2} y}{dx^{n-2}} + \ldots + p_n y = 0;$$

the critical points of the coefficients $p_1, \ldots, p_n$ are to be poles, and therefore in the development of any p in, say, the region of $x = 0$ we shall have only a finite number of negative powers of x, and so for this region each p may be written in the form

$$(2) \qquad p = \frac{\Pi(x)}{x^k},$$

where $\Pi(x)$ contains only positive powers of x. Denote by ω_1 the degree of x in the denominator of p_1, by ω_2 the degree of x in the denominator of p_2, etc. We will for convenience call ω_i the *order* of the coefficient p_i. Form now the series of numbers ($\omega_0 = 0$)

$$(3) \qquad \omega_0 + n, \quad \omega_1 + n - 1, \quad \omega_2 + n - 2, \quad \ldots, \quad \omega_n,$$

which denote respectively by $\Omega_0, \Omega_1, \Omega_2, \ldots, \Omega_n$. Let g be the greatest of these numbers; obviously, there may be more than one Ω equal to g. Supposing, however, that these numbers are arranged in ascending order of their subscripts,

$$(4) \qquad \Omega_0, \quad \Omega_1, \quad \Omega_2, \quad \ldots, \quad \Omega_n,$$

we shall consider in particular the first Ω, counting from the left, which is equal to g. Suppose this is Ω_i; then

$$(5) \qquad \Omega_i = \omega_i + n - i, = g;$$

Ω_i will be called the *characteristic index* of the given differential equation. This notion of the "characteristic index" is due to Thomé.

In the case studied already, *i.e.* where all the integrals are regular, it is obvious that we have as the greatest values of $\omega_1, \ldots, \omega_n$, $\omega_1 = 1$, $\omega_2 = 2, \ldots$, and the numbers $\Omega_0, \Omega_1, \ldots, \Omega_n$ have n for their maximum value, and so $\Omega_0 = g = n$. The theorem proved by Fuchs in this case may now be stated in the form:

In order that the differential equation $P = 0$ shall have all of its integrals regular in the region of $x = 0$ it is necessary that its coefficients contain in their developments only a finite number of negative powers of x, and, further, that its characteristic index shall be zero.

And the converse theorem is:

If the coefficients of $P = 0$ contain only a finite number of negative powers of x, and if, further, the characteristic index of the equation is zero, then in the region of $x = 0$ all the integrals of the equation are regular.

These integrals are, as we know, each of the form

$$x^\rho[\phi_0 + \phi_1 \log x + \ldots + \phi_\alpha \log^\alpha x],$$

where the functions ϕ are holomorphic in the region of $x = 0$, and therefore contain in their developments only positive powers of x; and further, $\phi_0, \ldots, \phi_\alpha$ do not all vanish for $x = 0$. The exponents $\rho_1, \rho_2, \ldots, \rho_n$ are roots of a certain algebraic equation—the *indicial equation.*

The method of arriving at this equation may be briefly recalled. The differential equation is of the form

$$(6) \qquad \frac{d^n y}{dx^n} + \frac{\Pi_1(x)}{x}\frac{d^{n-1}y}{dx^{n-1}} + \ldots + \frac{\Pi_n(x)}{x^n}y = 0;$$

where the functions Π contain only positive powers of x. Now make $y = x^\rho$, and substitute in this equation; we have

$$(7) \quad x^{\rho-n}\big[\rho(\rho-1)\ldots(\rho-n+1)$$
$$+ \Pi_1(x)\rho(\rho-1)\ldots(\rho-n-2) + \ldots + \Pi_n(x)\big].$$

Multiply this result by $x^{-\rho}$, and equate to zero the coefficient of x^{-n}, and we have the indicial equation

$$(8)\quad \rho(\rho-1)\ldots(\rho-n+1) \\ + \Pi_1(0)\rho(\rho-1)\ldots(\rho-n+2)+\ldots+\Pi_n(0)=0.$$

The following theorems may be easily proved, and the reader is advised to consult Thomé's memoirs in vols. 74 and 75 of Crelle concerning them. They are given here without proof, partly to economize space but mainly because they are in great part included in the more general theorems which follow and of which complete proofs are given.

When the coefficients $p_1, p_2, \ldots, p_s$ of the differential equation $P=0$ contain in their developments only a finite number of negative powers of x, and if the equation has at least $n-s$ regular linearly independent integrals, the remaining coefficients, $p_{s+1}\ldots p_n$, will contain in their developments only a finite number of negative powers of x, and the characteristic index of the equation will be at most equal to s.

As consequences of this we have:

1. *If the $s-1$ first coefficients $p_1 \ldots p_{s-1}$ of $P=0$ contain in their developments only a limited number of negative powers of x, and if p_s does not satisfy this condition, then the equation has at most $n-s$ regular linearly independent integrals.*

2. *If all the coefficients $p_1 \ldots p_n$ contain only a finite number of negative powers of x, the equation has at most $n-i$ regular linearly independent integrals where i is the characteristic index of the equation.*

If $i=0$ (Fuchs's case), we know that the equation always has n regular integrals; if, however, $i>0$, it may be that the equation will not have $n-i$ such integrals. For example: Write

$$Y=\frac{dy}{dx}+ky,$$

and consider the two differential equations

$$(9)\qquad Y=0,$$

$$(9)'\qquad Y+h=0,$$

where h is an arbitrary function of x.

Let us form the equation

(10) $$h\frac{dY}{dx} - Y\frac{dh}{dx} = 0;$$

that is,

(11) $$\frac{d^2y}{dx^2} + p_1\frac{dy}{dx} + p_2 y = 0,$$

where

$$p_1 = k - \frac{d\log h}{dx}, \quad p_2 = \frac{dk}{dx} - k\frac{d\log h}{dx}.$$

All the integrals of (9) and (9)′ satisfy (4); and conversely, as (10) gives by integration

$$Y = Ch$$

if y is a solution of (11) or if y verifies equation (9), $-\frac{y}{C}$ will verify equation (9)′. Therefore, if (9) and (9)′ have no solutions presenting the character of regular integrals, equation (11) will have no regular integrals.

Let p_1 and h be taken arbitrarily,

$$p_1 = \frac{1}{x^4}, \quad h = x,$$

whence

$$k = \frac{1}{x^4} + \frac{1}{x};$$

that is, for $x = 0$, k is infinite of the order 4. This order being superior to 1, equation (9) has no regular integral. Equation (9)′ has now no solution of the nature of regular integrals, for its general integral is

(12) $$y = e^{-\int kdx}(C' - \int he^{\int kdx}dx);$$

or, replacing h and k by their values and effecting the integration within the brackets,

(13) $$y = e^{-\int kdx}[\Theta(x) + C_1 \log x],$$

C_1 being a constant, and the function $\Theta(x)$ containing in its development an unlimited number of powers of x^{-1}, as well as $e^{-\int k dx}$.

Therefore, none of the integrals of (9) and (9)′ being regular, equation (11) has no regular integrals. It is of the form

$$(14) \qquad \frac{d^2y}{dx^2} + \frac{1}{x^4}\frac{dy}{dx} + \frac{f(x)}{x^5}y = 0,$$

where $f(x)$ is a holomorphic function in the region of the point zero, since

$$f(x) = -2x^3 - 5.$$

The characteristic index is 1, and nevertheless the equation, by what precedes, has no regular integral.

The preceding method admits of generalization and enables us to form differential equations of any degree with coefficients infinite of finite order for $x = 0$ and not possessing $n - i$ linearly independent regular integrals, i being their characteristic index.

Start now with the equation

$$P = P(y) = \frac{d^ny}{dx^n} + p_1\frac{d^{n-1}y}{dx^{n-1}} + \ldots + p_ny = 0,$$

in which the uniform coefficients p contain only a finite number of negative powers of x (*i.e.*, we will only consider the region of the critical point $x = 0$). If we substitute for y the value $y = x^\rho$, we obtain a function of x and ρ, viz., $P(x^\rho)$, which Frobenius has called the *characteristic function* of the differential equation $P = 0$, or of the differential quantic P. We find at once

$$(15)\ P(x^\rho) = x^\rho\left[\frac{\rho(\rho-1)\ldots(\rho-n+1)}{x^n} + p_1\frac{\rho(\rho-1)\ldots(\rho-n+2)}{x^{n-1}} + \ldots + p_{n-1}\frac{\rho}{x} + p_m\right].$$

We see at once that the product $x^{-\rho}P(x^\rho)$ can be developed in integer powers of x, and that the development will only contain a finite number of negative powers of x; the coefficients in the development will contain only positive powers of ρ, and ρ^n is the highest power that can occur. When the differential quantic P is given, its characteristic function $P(x^\rho)$ is of course known. Suppose, however, a characteristic function $x^\rho f(x, \rho)$ to be given where $f(x, \rho)$ is an integral function of ρ whose coefficients are functions of x: what is the corresponding differential quantic? Write

$$(16) \qquad f(x, \rho + 1) - f(x, \rho) = \Delta_\rho f(x, \rho)$$

and

$$(17) \qquad U_k = \frac{[\Delta_\rho^{(k)} f(x, \rho)]_{\rho=0}}{1 . 2 \ldots k}.$$

We know now that the integral function $f(x, \rho)$ of ρ can be placed in one way, and in one way only, in the form

$$(18) \quad f(x, \rho) = U_n\rho(\rho - 1) \ldots (\rho - n - 1) + U_{n-1}\rho(\rho - 1) \ldots (\rho - n + 2) + \ldots + U_1\rho + U_0;$$

it follows then that

$$(19) \quad x^\rho f(x, \rho) = x^\rho\left[U_n x^n \frac{\rho(\rho - 1) \ldots (\rho - n + 1)}{x^n} + \ldots + U_1 x \frac{\rho}{x} + U_0 \right]$$

is the characteristic function of the differential quantic

$$U_n x^n \frac{d^n y}{dx^n} + U_{n-1} x^{n-1} \frac{d^{n-1} y}{dx^{n-1}} + \ldots + U_1 x \frac{dy}{dx} + U_0 y.$$

We know that the product

$$(20) \quad x^{-\rho} P(x^\rho) = \frac{\rho(\rho - 1) \ldots (\rho - n + 1)}{x^n} + p_1 \frac{\rho(\rho - 1) \ldots (\rho - n + 2)}{x^{n-1}} + \ldots + p_{n-1} \frac{\rho}{x} + p_n$$

can be developed in a series of ascending powers of x, containing only a finite number of negative powers and in which the coefficients are integral functions of ρ of degrees at most $= n$. We wish now to determine the first term of this series. The exponents of x in the denominators of the expansion of $x^{-\rho}P(x^\rho)$ are obviously the numbers $\Omega_0, \Omega_1, \ldots, \Omega_n$ above defined. If g is the greatest of the numbers Ω, the first term of the series will clearly be of the form $\frac{G(\rho)}{x^g}$, where $G(\rho)$ is an integral function of ρ and does not contain x. If we denote by γ the degree of $G(\rho)$, we have obviously

$$(21) \qquad \gamma = n - i,$$

where i is the characteristic index of the equation, *i.e.*, i is the index of the first Ω which is $= g$. In the particular case of $i = 0$ the equation $G(\rho) = 0$ is the indicial equation already defined in the case where all the integrals of $P = 0$ are regular. In the case of $i > 0$ we can now generalize this notion of the indicial equation and say that $G(\rho) = 0$ is the indicial equation of our present differential equation $P = 0$ or of the differential quantic P. The function $G(\rho)$ will be called the *indicial function.*

Thus to obtain the indicial function of a differential quantic, we form its characteristic function and, after multiplying by $x^{-\rho}$, develop the product in ascending powers of x; the coefficient of the first term is the indicial function. It is to be noted that, knowing g, it is sufficient to multiply the characteristic function by $x^{g-\rho}$ and then make $x = 0$; thus

$$[x^{g-\rho}P(x^\rho)]_{x=0}$$

is the indicial function of $P(y)$.

Some properties of the indicial function of the differential quantic P will now be given.

First: if p_n is identically zero, the characteristic function and consequently the indicial function is divisible by ρ. Effecting this division and changing ρ into $\rho + 1$ in the quotient, we obtain, by equating it to zero, the indicial equation of the differential equation of order $n - 1$ obtained by taking $\frac{dy}{dx}$ for the unknown variable.

Second: if in the equation $P = 0$ we put

$$(22) \qquad y = x^{\rho_0}w,$$

the indicial equation of the differential equation in w thus obtained will have for its roots those of the indicial equation of $P = 0$ diminished by ρ_0. For the characteristic function of $P(x^{\rho_0}w) = 0$ is $P(x^{\rho_0+\rho})$. It is therefore deduced from the characteristic function of the equation in y, by changing ρ to $\rho + \rho_0$; and, consequently, the same is true of the indicial equations.

Finally: if in the equation $P = 0$ we place

$$y = \psi(x)w, \tag{23}$$

$\psi(x)$ being a holomorphic function in the region of the point zero and not vanishing for $x = 0$, the indicial equation of the equation in w thus obtained will be the same as that of the equation in y. For it is easily seen that, the equation in w being of the form

$$\frac{d^n w}{dx^n} + (p_1 + P_1)\frac{d^{n-1}w}{dx^{n-1}} + (p_2 + P_2)\frac{d^{n-2}w}{dx^{n-2}} + \ldots + (p_n + P_n)w = 0, \tag{24}$$

its characteristic function is the sum of two terms. The first term,

$$x^\rho\left[\frac{\rho(\rho-1)\ldots(\rho-n+1)}{x^n} + p_1\frac{\rho(\rho-1)\ldots(\rho-n+2)}{x^{n-1}} + \ldots + p_n\right], \tag{25}$$

is the characteristic function of the equation in y, and in the second term,

$$x^\rho\left[\frac{\rho(\rho-1)\ldots(\rho-n+1)}{x^n} + P_1\frac{\rho(\rho-1)\ldots(\rho-n+2)}{x^{n-1}} + \ldots + P_n\right], \tag{26}$$

the highest exponent of x in the denominator is inferior to the highest power, g, in the denominator of the first term. Whence it follows that upon multiplying by $x^{g-\rho}$ and then making $x = 0$ to obtain the indicial function of the equation in w, the second term will vanish and the result will be the same as if the operation had been performed upon the first term alone, that is, it will be the indicial function of the equation in y.

It follows from these propositions, $\rho_1, \rho_2, \ldots, \rho_\gamma$ being the γ roots of the indicial equation of P:

1. That if in $P = 0$ we put

$$(27) \qquad y = y_1 w,$$

where y_1 is an integral of the form $x^{\rho_0}\psi(x)$, $\psi(x)$ being holomorphic in the region of the point 0 and not zero for $x = 0$, the homogeneous differential equation in w thus obtained will have an indicial equation whose roots are

$$\rho_1 - \rho_0, \quad \rho_2 - \rho_0, \quad \ldots, \quad \rho_\gamma - \rho_0,$$

and, this equation being divisible by ρ, one of the quantities $\rho_1, \rho_2, \ldots, \rho_\gamma$ must be equal to ρ_0; let $\rho_1 = \rho_0$.

2. If in the equation in w we now put

$$w = \int z dx,$$

the indicial function of the equation in z will be of degree $\gamma - 1$, and will have for roots

$$\rho_2 - \rho_0 - 1, \quad \rho_3 - \rho_0 - 1, \quad \ldots, \quad \rho_\gamma - \rho_0 - 1.$$

3. Consequently, if in $P = 0$ we make the substitution

$$y = y_1 \int z dx,$$

the equation in z, of the order $n - 1$, thus obtained will have for indicial equation the equation of degree $\gamma - 1$ which admits the roots

$$\rho_2 - \rho_0 - 1, \quad \ldots, \quad \rho_\gamma - \rho_0 - 1.$$

The simple relation

$$(28) \qquad i + \gamma = n$$

between the order of the differential equation, its characteristic index, and the degree of the indicial equation, enables us to replace the notion of the characteristic index by the more rational consideration of the indicial equation. The characteristic index of an equation is merely the difference between the degree of the equation and that of its indicial equation. Whence it is evident that all propositions concerning the characteristic index and, consequently, relating to equations whose coefficients are infinite of a finite order for $x = 0$ can be expressed by aid of the indicial equation. Thus:

The number of linearly independent regular integrals of the equation $P = 0$ is at most equal to the degree of its indicial equation.

To obtain a clearer idea of the characteristic and indicial functions of $P = 0$, the equation may be put in a certain normal form. Thus we may write

$$(29)\quad P(y) = \frac{1}{x^n} x^n \frac{d^n y}{dx^n} + \frac{p_1}{x^{n-1}} x^{n-1} \frac{d^{n-1} y}{dx^{n-1}} + \ldots + \frac{p_{n-1}}{x} x \frac{dy}{dx} + p_n y = 0.$$

Reducing the fractions $\frac{1}{x^n}, \frac{p_1}{x^{n-1}} \ldots \frac{p_{n-1}}{x}, p_n$ to the least common denominator x^g and multiplying by x^g, the equation becomes

$$(30)\quad T(y) = t_0 x^n \frac{d^n y}{dx^n} + t_1 x^{n-1} \frac{d^{n-1} y}{dx^{n-1}} + \ldots + t_{n-1} x \frac{dy}{dx} + t_n y = 0,$$

where the functions

$$t_0,\ t_1,\ \ldots,\ t_n$$

contain only positive powers of x and do not all vanish for $x = 0$.

This form of the first member of a linear differential equation will be called the *normal form*. The characteristic function of the differential quantic T is

$$(31)\quad T(x^\rho) = x^\rho \big[t_0 \rho(\rho - 1) \ldots (\rho - m + 1) \\ + t_1 \rho(\rho - 1) \ldots (\rho - m + 2) + \ldots + t_{n-1}\rho + t_n\big];$$

hence the product $x^{-\rho} T(x^\rho)$ contains only positive powers of x and does not vanish for $x = 0$; its constant term is the indicial function.

Conversely, it is easy to see that if a linear differential equation has a characteristic function fulfilling these conditions, it is in the normal form. Hence an examination of the characteristic function is sufficient to determine whether the differential equation is or is not in the normal form.

Our next step will be to define a composite differential quantic and prove one of its important properties.

In the differential expression

$$(32)\qquad A(y) = A_0 \frac{d^\alpha y}{dx^\alpha} + A_1 \frac{d^{\alpha-1} y}{dx^{\alpha-1}} + \ldots + A_{\alpha-1} \frac{dy}{dx} + A_\alpha y$$

the letter A will be considered a symbol of operation such that $A(y)$ indicates the definite operation to be performed upon y:

$$(33)\qquad A(y) = \left(A_0 \frac{d^\alpha}{dx^\alpha} + A_1 \frac{d^{\alpha-1}}{dx^{\alpha-1}} + \ldots + A_\alpha\right) y.$$

Also, $A(B)$, or simply AB, will indicate that the same operation is to be performed upon B. If then B is a differential quantic, AB will also represent a differential quantic C, and we shall say that the quantic $AB = C$ *is composed of the quantics A and B taken in that order.* The same definition holds for a quantic composed of more than two quantics.

If the coefficients of the component differential quantics contain only a limited number of negative powers of x as is here supposed, it is clear that it is the same with the composite quantics.

Let $AB = C$, and let

$$(34)\qquad \begin{cases} A(x^\rho) = x^\rho h(x, \rho) = \Sigma_\mu h_\mu(\rho) x^{\rho+\mu}, \\ B(x^\rho) = x^\rho k(x, \rho) = \Sigma_\nu k_\nu(\rho) x^{\rho+\nu}, \\ C(x^\rho) = x^\rho l(x, \rho) = \Sigma_\lambda l_\lambda(\rho) x^{\rho+\lambda}, \end{cases}$$

be the characteristic functions of the quantics A, B, and C.

We have

$$(35)\qquad C(x^\rho) = AB(x^\rho) = A\left[\Sigma_\nu k_\nu(\rho) x^{\rho+\nu}\right] = \Sigma_\nu k_\nu(\rho) A(x^{\rho+\nu});$$

that is,

$$(36)\qquad \Sigma_\lambda l_\lambda(\rho) x^\lambda = \Sigma_\mu \Sigma_\nu h_\mu(\rho+\nu) k_\nu(\rho) x^{\mu+\nu}.$$

From this equality it follows that if A and B have the normal form, since in that case $h_\mu(\rho)$ and $k_\nu(\rho)$ vanish for negative values of μ and ν but not for the value zero, $\Sigma_\lambda l_\lambda(\rho) x^\lambda$ will contain only positive powers of x, and will not vanish for $x = 0$. Therefore, the characteristic function of C divided by x^ρ fulfilling these conditions, C

will itself have the normal form. Moreover, making $x = 0$ in the same equation, we have the following simple relation between the indicial functions of A, B, and C:

$$l_0(\rho) = h_0(\rho)k_0(\rho). \tag{37}$$

We may therefore state the following proposition:

If a differential quantic is composed of several differential quantics each of which is in the normal form, it has itself the normal form, and its indicial function is the product of the indicial functions of the component quantics.

The degree of $l_0(\rho)$ is consequently the sum of the degrees of $h_0(\rho)$ and $k_0(\rho)$.

More generally, we may remark that if *two of the three differential quantics A, B, and C have the normal form, the third has also the normal form.*

The notion of the component factors of a differential quantic leads directly to that of reducibility; but as this subject has been treated elsewhere, it will not be resumed in this connection.

Resuming now the question of regular integrals, we see at once that if the equation $P = 0$ has a regular integral, then $P = 0$ is a reducible equation. For, if $P = 0$ has a regular integral, it necessarily has one of the form $x^\rho \psi(x)$, where $\psi(x)$ is a uniform function which does not vanish for $x = 0$. Now obviously the function

$$y_1 = x^\rho \psi(x) \tag{38}$$

is an integral of the equation of the first order

$$\frac{dy}{dx} + \frac{P_1(x)}{x} y = 0, \tag{39}$$

where

$$\frac{P_1(x)}{x} = -\frac{\rho}{x} + \frac{\psi'(x)}{\psi(x)}, \tag{40}$$

$P_1(x)$ containing only positive powers of x. Therefore $P = 0$ is reducible and has among its integrals all those of

$$B_1 = \frac{dy}{dx} + \frac{P_1(x)}{x} y = 0. \tag{41}$$

We can now place $P = 0$ in the form $Q_1 B_1 = 0$, where Q_1 is of order $n - 1$. If, again, $Q_1 = 0$ has a regular integral, we can write

$$Q_1 = Q_2 B_2, \tag{42}$$

where Q_2 is of order $n - 2$. Continuing this process, we have finally

$$P = QD, \tag{43}$$

where
$$D = B_\beta B_{\beta-1} \ldots B_1,$$

and where each B equated to zero gives a differential equation of the first order having only regular integrals, and $Q = 0$ is a differential equation of order $n - \beta$ which has no regular integral. If P is in the normal form, the same thing may obviously be supposed of Q and D. The regular integrals of $P = 0$ being now those of $D = 0$, suppose $P = 0$ has all of its integrals regular; it follows at once that β, the degree of $D = 0$, is equal to n ($\beta = n$), and so that $P = D$. We have then the theorem:

(α) *If the equation $P = 0$ has all of its integrals regular, it can be composed in one way, and one way only, of differential equations of the first order having each only regular integrals.*

The converse of this is:

(β) *If the differential equation $P = 0$ is composed in one way, and one way only, of equations of the first order each of which has only regular integrals, it will have all of its integrals regular.* The proof of this will be left as an exercise for the reader.

The following theorems are immediately inferrible from what we have just shown, viz.:

(1) *If $P = 0$ admits of regular integrals, there exists an equation $D = 0$ all of whose integrals are regular, and these are the regular integrals of $P = 0$.*

(2) *If $D = 0$ is the equation which gives the regular integrals of $P = 0$, and if we put P in the composite form $P = QD$, then the equation $Q = 0$ has no regular integrals.*

If $A = 0$ and $B = 0$ are two equations having all their integrals regular, then they can be each composed uniquely of equations of the first order each of which admits only regular integrals. The equation $AB = 0$ will then be composed in the same way. And so for any number of equations $A = 0$, $B = 0$, $C = 0$ . . . We have then this generalization of β: If the equation $P = 0$ is composed uniquely of equations of lower orders which have all their integrals regular, then $P = 0$ will have all of its integrals regular. From what precedes, we know that any linear differential quantic, say A, can be put in the form $A = QD$ (where, of course, the order of D may be zero), in which $D = 0$ has only regular integrals and $Q = 0$ has no such integrals. Suppose a new equation $B = 0$ to have only regular integrals; then

$$AB = Q(DB). \tag{44}$$

Now, as already shown, the regular integrals of $AB = 0$ are those of $DB = 0$; but by hypothesis $DB = 0$ has all of its integrals regular; it follows then that the number of linearly independent integrals of $AB = 0$ is equal to the sum of the orders of the differential equations $B = 0$ and $D = 0$, or, what is the same thing, is equal to the sum of the linearly independent integrals of $A = 0$ and $B = 0$. It follows from this that *if the equation $B = 0$ has only regular integrals, the equation $AB = 0$ will have exactly as many regular integrals as the two equations $A = 0$ and $B = 0$ have combined.*

From this we draw at once the conclusion: *If $B = 0$ has only regular integrals, and if $AB = 0$ has s of such integrals which are linearly independent, then $A = 0$ will have $s - \beta$ linearly independent regular integrals, β denoting the order of B.*

Before going farther it will be convenient to make a remark concerning a differential equation of the first order. Suppose the equation is

$$ux\frac{dy}{dx} + wy = 0, \tag{45}$$

and suppose further that it is in the normal form; then, of course, u and w contain only positive powers of x and do not both vanish for $x = 0$. The general integral of this equation is

$$y = e^{-\int \frac{wdx}{ux}}. \tag{46}$$

Now we know that if u does not vanish for $x = 0$, y will be a regular integral of the form $x^\rho \psi(x)$; if, however, for $x = 0$ we have $u = 0$ of order k (in which case of course w is not zero), then

$$y = e^{\frac{C_1}{x} + \frac{C_2}{x^2} + \dots + \frac{C_k}{x^k}} x^\rho \psi(x), \tag{47}$$

which is not a regular integral. The function $\psi(x)$ contains only positive powers of x and does not vanish for $x = 0$.

Suppose u_0 and w_0 to be the values of u and w for $x = 0$. The characteristic function of the considered equation is $x^\rho(u\rho + w)$, and the indicial function is

$$u_0\rho + w_0. \tag{48}$$

In the first case, then, where u_0 is not zero, and consequently where the equation has its integrals regular, the indicial function is an integral function of ρ of the first degree; and in the second case where $u_0 = 0$ and the equation has no regular integral the indicial function is a constant. The converse of this is readily shown.

It is obvious from what precedes that if the equation $P = 0$ has all of its integrals regular, then the degree of its indicial equation is equal to its order; and also that the number of linearly independent regular integrals of $P = 0$ is at most equal to the degree of its indicial equation. These results have, however, been previously arrived at.

We will now obtain the precise condition which the differential equation $P = 0$ must satisfy in order that the number of its linearly independent regular integrals may be *exactly* equal to the degree of its indicial function. We suppose the differential quantics P, Q, D to be in the normal form. If the number of linearly independent regular integrals of $P = 0$ is γ, where γ is the degree of the indicial function of P, then P can be placed in the form

$$P = QD,$$

where Q is of the order $n - \gamma$ and has a constant for its indicial function. For, since $P = 0$ has γ regular integrals, it can be placed in the form $P = QD$, where Q is of order $n - \gamma$ and has no regular integrals, and where D is of order γ and has all of its integrals regular. The indicial function of P is then the product of the indicial functions of Q and D; but since D has all of its integrals regular, its indicial function is of degree γ, and this is also the degree of the indicial function of P; hence the indicial function of Q is a constant.

Conversely, if the differential expression P having an indicial function of degree γ can be placed in the form $P = QD$, where Q is a differential expression of order $n - \gamma$ having a constant as its indicial function, then $P = 0$ will have exactly γ regular integrals. For, since P is of order n and Q of order $n - \gamma$, D is of order γ; now the indicial function of Q being of order zero, that of D is of order γ, and since the indicial function of Q is a constant, Q can have no regular integrals, and so the equation $P = QD = 0$ will have for regular integrals those of $D = 0$; that is, $P = 0$ has exactly γ regular integrals. Therefore, finally: *In order that $P = 0$ having an indicial function of degree γ shall have γ regular integrals, it must be possible to place P in the form $P = QD$, where Q is of order $n - \gamma$ and has a constant for its indicial function.*

Irregular integrals of the form (47) have been called by Thomé *normal integrals.* It is not the intention here to go at all into the theory of these integrals, indeed that theory is in a very imperfect state owing principally to the, in many cases, impossibility of ascertaining anything definite concerning the convergence of the series involved. The reader is referred to Thomé's papers in Crelle, but particularly to the following papers by Poincaré: "*Sur les équations linéaires aux différentielles ordinaires et aux différences finies*" (American Journal of Mathematics, vol. 7, pp. 208–258), "*Sur les intégrals irrégulières des équations linéaires*" (Acta Mathematica, vol. 8, pp. 295–344); see also a note by Poincaré in the Acta, vol. 10, p. 310, entitled "*Remarques sur les intégrales irrégulières des équations linéaires.*" A Thesis by M. E. Fabry (Paris, 1885) entitled "*Sur les intégrales des équations différentielles linéaires à coefficients rationnels*" may also be advantageously consulted. The first paper by Poincaré in the Acta Math. is by far the most important of these references; the author regrets to be unable to give an account of it here, but limits of space prevent.

CHAPTER X.

DECOMPOSITION OF A LINEAR DIFFERENTIAL EQUATION INTO SYMBOLIC PRIME FACTORS.

THE coefficients $p_1, \ldots, p_n$ of the differential equation

$$(1) \qquad P = P(y) = \frac{d^n y}{dx^n} + p_1 \frac{d^{n-1} y}{dx^{n-1}} + \ldots + p_n y = 0$$

are, with the exception of certain isolated critical points, holomorphic functions of x in a portion of the plane which is limited by a simple (*i.e.*, non-crossing) contour.

P can be put in the form

$$(2) \qquad P = A_n A_{n-1} \ldots A_1,$$

where

$$(3) \qquad A_i = \frac{dy}{dx} - a_i y$$

is called a *symbolic prime factor* of P. For convenience the word "symbolic" may be omitted, as it is always understood. We can now show that P is decomposable into prime factors. Let

$$(4) \quad y_1 = v_1, \quad y_2 = v_1 \int v_2 dx, \quad \ldots, \quad y_n = v_1 \int v_2 dx \int v_3 dx \ldots \int v_n dx$$

denote a set of fundamental integrals of $P = 0$. Form now the system of differential equations

$$(5) \qquad A_i = \frac{dy}{dx} - a_i y = 0, \quad i = 1, 2, \ldots, n,$$

having for integrals

$$v_1, \quad v_1v_2, \quad v_1v_2v_3, \quad \ldots, \quad v_1v_2 \ldots v_n;$$

then

(6) $$a_i = \frac{d}{dx} \log (v_1v_2 \ldots v_i).$$

It is obvious now that the composite expression $A_nA_{n-1} \ldots A_1$ is identically equal to P; that is, the coefficients of the derivatives of y of the same order are the same in both expressions, a fact already established.

It is easy to see now that every decomposition of P into prime factors is obtainable by this process, each such decomposition corresponding to a chosen system of fundamental integrals. We see at once an analogy between algebraic equations and linear differential equations. When we know the roots of the algebraic equation, its decomposition into prime factors enables us at once to write out the equation; so, knowing the decomposition into prime factors of a differential equation, we are enabled to form the differential equation which possesses a given system of fundamental integrals. If

$$y_1 = v_1, \quad y_2 = v_1\int v_2dx, \quad \ldots, \quad y_n = v_1\int v_2dx \ldots \int v_ndx$$

are the integrals, the differential equation will be

(7) $$A_nA_{n-1} \ldots A_1 = 0;$$

where

(8) $$a_i = \frac{d}{dx} \log (v_1v_2 \ldots v_i) \quad i = 1, 2, \ldots, n.$$

It is to be noticed, however, that while the arrangement of the factors in the algebraic equation is arbitrary (*i.e.*, the factors are commutative), the arrangement of the symbolic factors in the differential equation is not in general arbitrary.

When we know a system of values of the functions $v_1 \ldots v_n$ we know also all possible systems of fundamental integrals of $P = 0$, and so can in all possible ways decompose P into its prime factors by aid of the formulæ

(9) $$a_i = \frac{d}{dx} \log (v_1, v_2, \ldots, v_n). \quad i = 1, 2, \ldots, n.$$

Conversely, if we know the coefficients a of the prime factors, we can integrate $P = 0$; for we have

$$(10) \quad v_1 = e^{\int a_1 dx}, \quad v_2 = e^{\int (a_2 - a_1) dx}, \quad \ldots, \quad v_n = e^{\int (a_n - a_{n-1}) dx},$$

and a system of fundamental integrals of $P = 0$ is then

$$(11) \qquad y_1 = e^{\int a_1 dx}, \quad y_2 = e^{\int a_1 dx} \int e^{\int (a_2 - a_1) dx}, \quad \ldots .$$

If now we wish to decompose the expression

$$P(y) = \frac{d^n y}{dx^n} + p_1 \frac{d^{n-1} y}{dx^{n-1}} + \ldots + p_n y$$

into its prime factors, we have one of two things to do: either calculate a system of values of the functions $v_1, v_2, \ldots, v_n$, or evaluate directly a system of values of the coefficients $a_1, a_2, \ldots, a_n$. In the first case we should have first to find a particular solution v_1 of the equation $P(z_1) = 0$, then a particular solution v_2 of $P(v_1 \int z_2 dx) = 0$, etc. The equations in $z_1, z_2, \ldots, z_n$ will be linear and of orders $n, n-1, \ldots, 2, 1$ respectively. In the second case we should have to find first a particular solution a_1 of $P(e^{\int u_1 dx}) = 0$, then a particular solution a_2 of $P[e^{\int u_1 dx} \int e^{\int (u_2 - u_1) dx} dx] = 0$, etc. The equations in $u_1, u_2, \ldots, u_n$ will be of orders $n, n-1, \ldots, 2, 1$, but will obviously not be linear.

The formulæ giving the coefficients $p_1 \ldots p_n$ of

$$P(y) = \frac{d^n y}{dx^n} + p_1 \frac{d^{n-1} y}{dx^{n-1}} + \ldots + p_n y = 0$$

in terms of $a_1, a_2, \ldots, a_n$ are easily found. Suppose

$$(12) \qquad P = A_n A_{n-1} \ldots A_1,$$

and further suppose

$$(13) \qquad A_n A_{n-1} \ldots A_2 = \frac{d^{n-1} y}{dx^{n-1}} + q_1 \frac{d^{n-2} y}{dx^{n-2}} + \ldots + q_{n-1} y.$$

In this last equation replace y by $\frac{dy}{dx} - a_1 y$, and so form the expression $A_n A_{n-1} \ldots A_2 A_1$, or $P(y)$; we have then

$$(14)\quad P(y) = \frac{d^n y}{dx^n} + (q_1 - a_1)\frac{d^{n-1}y}{dx^{n-1}} + \left[q_2 - (n-1)\frac{da_1}{dx} - q_1 a_1\right]\frac{d^{n-2}y}{dx^{n-2}}$$

$$+ \ldots + \left[-\frac{d^{n-1}a_1}{dx^{n-1}} - q_1\frac{d^{n-2}a_1}{dx^{n-2}} - \ldots - q_{n-2}\frac{da_1}{dx} - q_{n-1}a_1\right]y;$$

from which we find

$$(15)\quad p_1 = q_1 - a_1,$$

$$p_2 = q_2 - (n-1)\frac{da_1}{dx} - q_1 a_1,$$

$$p_3 = q_3 - \frac{(n-1)(n-2)}{1 \cdot 2}\frac{d^2 a_1}{dx^2} - q_1(n-2)\frac{da_1}{dx} - q_2 a_1,$$

$$\cdot \quad \cdot \quad \cdot \quad \cdot \quad \cdot \quad \cdot \quad \cdot \quad \cdot \quad \cdot \quad \cdot \quad \cdot$$

$$p_n = -\frac{d^{n-1}a}{dx^{n-1}} - q_1\frac{d^{n-2}a_1}{dx^{n-2}} - \ldots - q_{n-2}\frac{da_1}{dx} - q_{n-1}a_1.$$

If now we start from the expression A_n and form successively the expressions

$$A_n,\quad A_n A_{n-1},\quad A_n A_{n-1} A_{n-2},\quad \ldots,\quad A_n A_{n-1} A_{n-2} \ldots A_1,$$

we readily find, by aid of the preceding equations, the following values for $p_1, p_2, \ldots, p_n$, viz.:

$$(16)\quad p_1 = -(a_1 + a_2 + \ldots + a_n),$$

$$p_2 = \Sigma a_i a_j - (n-1)\frac{da_1}{dx} - (n-2)\frac{da_n}{dx} - \ldots - \frac{da_{n-1}}{dx},$$

$$p_3 = -\Sigma a_i a_j a_k - \left[\frac{(n-1)(n-2)}{1 \cdot 2}\frac{d^2 a_1}{dx^2} + \frac{(n-2)(n-3)}{1 \cdot 2}\frac{d^2 a_2}{dx^2} + \ldots\right]$$

$$+ (n-2)(a_2 + a_3 + \ldots + a_n)\frac{da_1}{dx} + (n-3)(a_3 + \ldots + a_n)\frac{da_2}{dx} + \ldots$$

$$+ a_1\left[(n-2)\frac{da_2}{dx} + (n-3)\frac{da_3}{dx} + \ldots\right] + a_2\left[(n-3)\frac{da_3}{dx} + \ldots\right] + \ldots$$

These formulæ give the coefficients p in terms of the coefficients a, and consequently, since the a's are functions of $v_1, v_2, \ldots, v_n$, in terms of $v_1, v_2, \ldots, v_n$. Write, for brevity,

$$(17) \qquad p_i = f_i(a_1, a_2, \ldots, a_n);$$

now, since $p_1, p_2, \ldots, p_n$ have the same values whatever be the chosen system of fundamental integrals of the differential equation, and consequently whatever be the chosen system of the functions v, it follows that the functions f are invariants.

If $a_1, a_2, \ldots, a_n$ are constants, we have

$$(18) \qquad p_1 = -\Sigma a_i, \quad p_2 = \Sigma a_i a_j, \quad p_3 = -\Sigma a_i a_j a_k, \quad \ldots,$$

and therefore $p_1, p_2, \ldots$ are the coefficients of the algebraic equation

$$(t - a_1)(t - a_2) \ldots (t - a_n) = 0,$$

having $a_1, a_2, \ldots, a_n$ for roots.

A transformation which is frequently useful consists in changing $P = 0$ into an equation whose integrals are those of P, each multiplied by the same factor, say N. If z be the dependent variable in the new equation, then

$$(19) \qquad z = Ny \quad \text{and} \quad y = \frac{1}{N}z,$$

and the new equation is therefore

$$(20) \qquad P\left(\frac{1}{N}z\right) = 0.$$

Let

$$(21) \quad y_1 = v_1, \quad y_2 = v_1 \int v_2 dx, \quad \ldots, \quad y_n = v_1 \int v_2 dx \ldots \int v_n dx$$

denote a system of fundamental integrals of $P = 0$, and let

$$P = A_n A_{n-1} \ldots A_1$$

denote the corresponding decomposition of P into prime factors. In order to multiply all these integrals by N it is obviously only necessary to change v_1 into Nv_1; with this change of v_1 the general coefficient a_i becomes

(22) $$a_i + \frac{d}{dx}\log N,$$

and $A_i' = 0$ is

(23) $$\frac{dy}{dx} - \left[a_i + \frac{d}{dx}\log N\right] y = 0.$$

We have thus the identity

(24) $$NP\left(\frac{1}{N}y\right) = A_n' A_{n-1}' \ldots A_1'.$$

Suppose $N = \frac{V}{W}$, then a_i' is

(25) $$a_i' = a_i + \frac{d}{dx}\log V - \frac{d}{dx}\log W.$$

In particular let

(26) $$V = v_1, \quad W = 1,$$

then

(27) $$\frac{d}{dx}\log V = a_1,$$

and we have

(28) $$y_1 P\left(\frac{y}{y_1}\right) = A_n' A_{n-1}' \ldots A_1',$$

where

(29) $$a_i' = a_i + a_1.$$

Other special cases are readily found.

We have, in what precedes, lowered the order of the equation $P = 0$ by unity by assuming that we knew a solution, $y_1 = v_1$, of $A_1 = 0$; that is, a particular solution y_1 of $P = 0$. By aid of this solution we find the equation

(30) $$A_n A_{n-1} \ldots A_2 = \frac{d^{n-1}z}{dx^{n-1}} + q_1 \frac{d^{n-2}z}{dx^{n-2}} + \ldots + q_{n-1} z = 0$$

of order $n-1$, and from equations (15) have its coefficients $q_1 \ldots q_{n-1}$ given as functions of $p_1 \ldots p_n$ and a_1 by

$$(31)\quad q_1 = a_1 + p_1,$$

$$q_2 = q_1 a_1 + p_2 + (n-1)\frac{da_1}{dx},$$

$$q_3 = q_2 a_1 + p_3 + \frac{(n-1)(n-2)}{1\,.\,2}\frac{d^2 a_1}{dx^2} + q_1(n-2)\frac{da_1}{dx},$$

. . . .

Since in A_1, $= \dfrac{dy_1}{dx} - a_1 y_1$, we have from (11)

$$(32)\qquad a_1 = \frac{1}{y_1}\frac{dy_1}{dx},$$

the order of the equation $P = 0$ has been lowered by unity by the transformation

$$(33)\qquad \frac{dy}{dx} - \frac{1}{y_1}\frac{dy_1}{dx}y = z,$$

whereas the ordinary formula for the lowering of the order of the equation is

$$(34)\qquad y = y_1 \int z dx.$$

There is, however, a simple relation connecting these two processes. From the equation $y = y_1 \int z dx$ we derive

$$(35)\qquad z = \frac{d}{dx}\frac{y}{y_1},$$

and from

$$(36)\qquad \frac{dy}{dx} - \frac{1}{y_1}\frac{dy_1}{dx}y = z,$$

we have

$$(37)\qquad z = y_1 \frac{d}{dx}\frac{y}{y_1}.$$

It follows now at once that the integrals of

(38) $$A_n A_{n-1} \ldots A_2 = 0$$

obtained by one of the two methods are equal to those of $P(y_1 \int z\,dx)$ obtained by the other when each of the latter integrals are multiplied by y_1. If we multiply the solutions of

(39) $$P(y_1 \int z\,dx) = 0$$

by y_1, which is done by changing z into $\frac{z}{y_1}$, we have, since the first coefficient of $P\left(y_1 \int \frac{z}{y_1} dx\right)$ is unity, the identity

(40) $$P\left(y_1 \int \frac{y}{y_1} dx\right) = A_n A_{n-1} \ldots A_2.$$

If now we divide the solutions of $A_n A_{n-1} \ldots A_2 = 0$ by y_1, which, by a formula analogous to (29), is done by changing a_i into $a_i - a_1$, $i = 2, 3, \ldots, n$, we have a second identity,

(41) $$\frac{1}{y_1} P(y_1 \int y\,dx) = A_n' A_{n-1}' \ldots A_2',$$

where

(42) $$a_i' = a_i - a_1.$$

Suppose the algebraic equation $F(y) = 0$ has a root $y = y_1$; in order that $y = y_1$ shall be a double root we must have $\frac{dF}{dy} = 0$ for $y = y_1$. A similar property exists for the linear differential equation

$$P(y) = \frac{d^n y}{dx^n} + p_1 \frac{d^{n-1} y}{dx^{n-1}} + \ldots + p_n y = 0.$$

Suppose y_1 a solution of $P = 0$ obtained by equating A_1 to zero in

$$P = A_n A_{n-1} \ldots A_1 = 0.$$

In order that y_1 shall be a solution of

$$A_n A_{n-1} \ldots A_2 = 0$$

where the factor A_1 has been dropped, it is necessary and sufficient that y_1 satisfy the equation

$$(43)\qquad n\frac{d^{n-1}y}{dx^{n-1}} + (n-1)p_1\frac{d^{n-2}y}{dx^{n-2}} + \ldots + p_{n-1}y = 0.$$

The proof of this is very simple. The identity (40) shows that the equation

$$A_n A_{n-1} \ldots A_2 = 0$$

is satisfied by $y = y_1$ if

$$(44)\qquad P\left(y_1 \int \frac{y_1}{y_1}\, dx\right) = 0;$$

that is, if

$$(45)\qquad P(xy_1) = 0.$$

This means that $P = 0$ must have $y = xy_1$ as an integral. Substituting this value of y in $P = 0$, and remembering that $P(y_1) = 0$, we have

$$(46)\qquad n\frac{d^{n-1}y}{dx^{n-1}} + (n-1)p_1\frac{d^{n-2}y}{dx^{n-2}} + \ldots + p_{n-1}y = 0,$$

which proves the proposition.

It is clear now that if the equation

$$(47)\qquad A_n A_{n-1} \ldots A_2 = 0$$

has again the integral y_1, it is necessary and sufficient that we have

$$(48)\qquad P\left(y_1 \int \frac{xy_1}{y_1}\, dx\right) = P(x^2 y_1) = 0;$$

and in general the necessary and sufficient condition that y_1 shall be a solution of $P = 0$, and of the $n - 1$ equations derived from $P = 0$ by the substitution

$$(49)\qquad \frac{dy}{dx} - \frac{1}{y_1}\frac{dy_1}{dx} = z,$$

is that $P = 0$ must admit of the n solutions

(50) $$y_1, \quad xy_1, \quad x^2y_1, \quad \ldots, \quad x^{n-1}y_1.$$

These solutions are called *conjugate solutions* of the differential equation, and they are analogous to the equal roots of an algebraic equation.

Suppose the differential equation $P = 0$ to admit of k conjugate solutions: we can readily show that P can be decomposed into n prime factors of which the last k shall be equal. To prove this, let

(51) $$y_1 = v_1, \quad y_2 = xv_1, \quad y_3 = x^2v_1, \quad \ldots, y_k = x^{k-1}v_1, \quad y_{k+1} \ldots y_n$$

denote a fundamental system of integrals of $P = 0$, and let

(52) $$P = A_nA_{n-1} \ldots A_1$$

be the corresponding decomposition of P. If now we calculate v_2, $v_3, \ldots, v_k$ by the ordinary formulæ, viz.,

(53) $$y_1 = v_1, \quad y_2 = v_1 \int v_2 dx, \quad \ldots, \quad y_k = v_1 \int v_2 dx \ldots \int v_k dx,$$

we find at once

(54) $$v_2 = 1, \quad v_3 = 2, \quad \ldots, \quad v_k = k - 1,$$

and consequently

(55) $$a_i = \frac{d}{dx} \log (v_1v_2 \ldots v_i) = \frac{d}{dx} \log (\underline{i-1}|v_1) = \frac{d}{dx} \log v_1,$$

or

(56) $$a_i = a_1 \quad \text{for} \quad i = 2, 3, \ldots, k.$$

Conversely, if P is decomposable into n prime factors of which the last k are equal, then the equation $P = 0$ admits of k conjugate solutions. If the system of integrals

(57) $$y_1 = v_1, \quad y_2 = v_1 \int v_2 dx, \quad \ldots, \quad y_n = v_1 \int v_2 dx \ldots \int v_n dx$$

corresponds to the decomposition

$$A_nA_{n-1} \ldots A_1$$

of P, we see at once that $v_2, v_3, \ldots, v_k$ given by equations (10) (viz.,

$$v_2 = e^{\int(a_2 - a_1)dx} \quad \ldots \quad v_k = e^{\int(a_k - a_{k-1})dx};$$

where $a_1 = a_2 = \ldots = a_k$) are constants, and therefore $y_1, y_2, \ldots, y_k$ are of the forms

(58) $$y_1 = v_1, \quad y_2 = xv_1, \quad \ldots, \quad y_k = x^{k-1}v_1.$$

It is easy to see what takes place when in the equation

$$A_n A_{n-1} \ldots A_1 = 0$$

we have any k consecutive factors equal. Suppose, for example,

(59) $$A_{k+1} = A_k = \ldots = A_2;$$

the equation

(60) $$A_n A_{n-1} \ldots A_2 = 0$$

admits now the k conjugate solutions

(61) $$v_1 v_2, \quad xv_1 v_2, \quad x^2 v_1 v_2, \quad \ldots, \quad x^{k-1} v_1 v_2,$$

and the proposed equation has the k integrals

(62) $$y_2 = v_1 \int v_2 dx, \quad y_3 = v_1 \int x v_2 dx, \quad \ldots, \quad y_{k+1} = v_1 \int x^{k-1} v_2 dx,$$

satisfying the relations

(63) $$\frac{d}{dx}\frac{y_2}{v_1} = v_2, \quad \frac{d}{dx}\frac{y_3}{v_1} = xv_2, \quad \ldots, \quad \frac{d}{dx}\frac{y_{k+1}}{v_1} = x^{k-1}v_2;$$

that is, the derivatives of the solutions are conjugate, and not the solutions themselves.

We have already remarked that the order of the factors A in the decomposition of

$$P(y) = \frac{d^n y}{dx^n} + p_1 \frac{d^{n-1}y}{dx^{n-1}} + \ldots + p_n y = 0$$

is not arbitrary. Suppose, however, that for any arrangement of these factors we find the same expression P, that is, we find the same values of the coefficients p; the factors are then *commutative*. We have now to find the conditions which must be satisfied in order that P can be decomposed into a system of commutative prime factors.

To do this compare first the two differential expressions

(64) $$A_n A_{n-1} \dots A_2 A_1$$

and

(65) $$A_{n-1} A_n \dots A_2 A_1 .$$

Call for brevity the common part of these expressions A, that is,

(66) $$A_{n-2} A_{n-3} \dots A_2 A_1 = A,$$

and we have at once

(67) $$A_n A_{n-1} A = \frac{d^2 A}{dx^2} - (a_{n-1} + a_n)\frac{dA}{dx} + \left(a_n a_{n-1} - \frac{da_{n-1}}{dx}\right) A,$$

(68) $$A_{n-1} A_n A = \frac{d^2 A}{dx^2} - (a_{n-1} + a_n)\frac{dA}{dx} + \left(a_n a_{n-1} - \frac{da_n}{dx}\right) A.$$

In order that these may be identical, or that (64) and (65) may give the same set of values of the coefficients p, we have

(69) $$\frac{da_n}{dx} - \frac{da_{n-1}}{dx} = 0;$$

that is, $a_n - a_{n-1} =$ constant. We have then that, in order that in a composite differential expression we may be able to change the order of the first two prime factors, the difference of the coefficients a of these factors must be constant. Again, compare

(70) $$A_n A_{n-1} A_{n-2} A_{n-3} A_{n-4} \dots A_1$$

and

(71) $$A_n A_{n-1} A_{n-3} A_{n-2} A_{n-4} \dots A_1 .$$

By aid of the preceding theorem it is clear that if

$$(72) \qquad A_{n-2}A_{n-3}A_{n-4} \ldots A_1, = S,$$

and

$$(73) \qquad A_{n-3}A_{n-2}A_{n-4} \ldots A_1, = T,$$

are identical, we must have

$$(74) \qquad a_{n-2} - a_{n-3} = \text{const.}$$

In this case (70) and (71) are obviously also identical. Conversely, if (70) and (71) are identical, then (72) and (73) are also identical. In fact, we have, if (70) and (71) are identical,

$$(75) \qquad A_nA_{n-1}S - A_nA_{n-1}T = A_nA_{n-1}(S - T)$$

identically zero; but

$$(76) \quad A_nA_{n-1}(S-T) = \frac{d^2(S-T)}{dx^2} - (a_{n-1} + a_n)\frac{d(S-T)}{dx} + \left(a_{n-1}a_n - \frac{da_{n-1}}{dx}\right)(S-T).$$

Suppose this is not identically zero, that is, suppose S and T are not identical, and let $R\dfrac{d^ky}{dx^k}$ be the first term in $S - T$ which does not vanish; then in (76) we shall have the term $R\dfrac{d^{k+2}y}{dx^{k+2}}$, which will not cancel with any other term, which is contrary to hypothesis. Hence S and T are identical. We see at once now that if in a differential expression we can change the order of any two consecutive factors, we must have that the coefficients of these factors differ only by a constant. We are thus led to the general result:

In order that the factors of a differential expression which is composed of prime symbolic factors may be commutative, it is necessary and sufficient that the differences of the coefficients of the factors taken in pairs shall be constants.

Consider now the two equations $A_k = 0$ and $A_l = 0$; that is,

$$\frac{dy}{dx} - a_k y = 0, \quad \frac{dy}{dx} - a_l y = 0; \tag{77}$$

of which y_k and y_l are the general integrals. In the first of these write $y = y_l z$ and we have

$$\frac{dz}{dx} - (a_k - a_l)z = 0; \tag{78}$$

if now $a_k - a_l$ is a constant, we have for z the value

$$z = Ce^{\alpha x}, \tag{79}$$

where C and α are constants; conversely, if z is of this form, $a_k - a_l$ is a constant. The condition, therefore, that the difference $a_k - a_l$ shall be a constant gives for the ratio of the integrals of equations (77) the value

$$\frac{y_k}{y_l} = Ce^{\alpha x}. \tag{80}$$

From this results the following theorem:

If the factors of a differential expression composed of prime symbolic factors are commutative, it is necessary and sufficient that the ratios of pairs of integrals of corresponding factors shall be of the form $Ce^{\alpha x}$, where C and α are constants.

The special forms of this result for equations with constant coefficients are readily found; this will, however, be left as an exercise.

It is evident that a differential expression composed of prime commutative factors vanishes if an expression composed of one or several of its factors vanishes. By aid of this remark we will proceed to determine the form of the differential expression

$$P(y) = \frac{d^n y}{dx^n} + p_1 \frac{d^{n-1}y}{dx^{n-1}} + \cdots + p_n y$$

when it is decomposable into commutative prime factors. We remark first that the coefficients p are symmetric functions of the coefficients a of the commutative factors. These coefficients are found from equations (16) when we write in the latter

$$\frac{da_1}{dx} = \frac{da_2}{dx} = \ldots = \frac{da_n}{dx}, \tag{81}$$

and so for the higher derivatives.

Suppose now P to be decomposable into commutative prime factors; then

$$P = A_n A_{n-1} \ldots A_1, \tag{82}$$

where $\qquad a_i - a_k = \text{const. for } i, k = 1, 2, \ldots, n.$

Consider a solution y_k of $A_k = 0$; y_k is also a solution of $P = 0$. Now we know that

$$\frac{1}{y_k} P(y_k y) = A'_n A'_{n-1} \ldots A_1' \tag{83}$$

where

$$a_i' = a_i - a_k, \qquad i = 1, 2, \ldots, n. \tag{84}$$

It follows then that the expression

$$\frac{1}{y_k} P(y_k y)$$

has its coefficients constants, and so

$$\frac{1}{y_k} P(y_k y), = Q(y), = \frac{d^n y}{dx^n} + q_1 \frac{d^{n-1} y}{dx^{n-1}} + \ldots + q_{n-1} \frac{dy}{dx}, \tag{85}$$

where $q_1, \ldots, q_{n-1}$ are constants. Changing y into $\dfrac{y}{y_k}$, we have

$$P(y) = y_k Q\left(\frac{y}{y_k}\right). \tag{86}$$

Therefore: *Every differential expression $P(y)$ which is decomposable into commutative prime factors is of the form*

$$P(y) = y_k Q\left(\frac{y}{y_k}\right); \tag{87}$$

where y_k is a function of x and $Q(y)$ is a linear differential expression with constant coefficients.

Conversely: *Every linear differential expression of the form*

$$P(y) = y_k Q\left(\frac{y}{y_k}\right)$$

is decomposable into commutative prime factors.

We have in fact

$$(88) \qquad Q(y) = B_n B_{n-1} \ . \ . \ . \ B_1,$$

where the coefficients b are constants; now we know that

$$(89) \qquad y_k Q\left(\frac{y}{y_k}\right) = B_n' B_{n-1}' \ . \ . \ . \ B_1',$$

where the coefficients b' have the form

$$(90) \qquad \begin{cases} b_i' = b_i + \dfrac{d}{dx} \log y_k, \\ \\ b_j' = b_j + \dfrac{d}{dx} \log y_k; \end{cases}$$

and so the differences $b_i' - b_j'$, $i, j = 1, 2, \ldots, n$ are constants, and consequently the factors B' are commutative. It is therefore necessary and sufficient that $P(y)$ should have the form $y_k Q\left(\frac{y}{y_k}\right)$ in order that it may be decomposed into commutative prime factors. It is now easy to see what in the supposed case is the form of the integrals of $P = 0$. Let $w_1, w_2, \ldots, w_n$ denote a system of integrals of $Q = 0$; one of these is of course a constant, since Q contains no term in y, and the others are of the form $x^{\alpha} e^{\beta x}$. The equation $Q\left(\frac{y}{y_k}\right) = 0$, or $P(y) = 0$, will then have

$$y_k w_1, \quad y_k w_2, \quad \ldots, \quad y_k w_n$$

as a system of linearly independent integrals. Finally, then, if P is decomposable into a system of commutative prime factors, and if y_k is a function which causes one of these factors to vanish, $P = 0$

has a system of linearly independent integrals of the form $y_k x^\alpha e^{\beta x}$, the integral y_k being included in this system. As an example, let us find the condition to be satisfied in order that the differential expression

(91) $$P(y) = \frac{d^2y}{dx^2} + p_1 \frac{dy}{dx} + p_2 y$$

of the second order may be decomposed into commutative factors, say

(92) $$P = A_2 A_1 = A_1 A_2.$$

We have directly

(93) $$A_2 A_1 = \frac{d^2y}{dx^2} - (a_1 + a_2)\frac{dy}{dx} + \left(a_1 a_2 - \frac{da_1}{dx}\right) y,$$

giving

(94) $$p_1 = -(a_1 + a_2), \quad p_2 = a_1 a_2 - \frac{da_1}{dx},$$

with the condition

(95) $$\frac{da_1}{dx} = \frac{da_2}{dx}.$$

Now

(96) $$\frac{da_1}{dx} = \frac{da_2}{dx} = \frac{1}{2}\frac{d}{dx}(a_1 + a_2) = -\frac{1}{2}\frac{dp_1}{dx},$$

and

(97) $$a_1 a_2 = p_2 - \frac{1}{2}\frac{dp_1}{dx};$$

consequently

(98) $$\frac{(a_1 - a_2)^2}{4} = -p_2 + \frac{1}{2}\frac{dp_1}{dx} + \frac{p_1^2}{4};$$

but $a_1 - a_2 = \text{const.}$, therefore

(99) $$p_2 - \frac{1}{2}\frac{dp_1}{dx} - \frac{p_1^2}{4} = \text{const.},$$

which is the required necessary and sufficient condition. If the two commutative factors are equal, then

$$(100) \qquad p_2 - \frac{1}{2}\frac{dp_1}{dx} - \frac{p_1^2}{4} = 0.$$

If we transform the equation

$$(101) \qquad \frac{d^2y}{dx^2} + p_1\frac{dy}{dx} + p_2 y = 0$$

by removing in the usual way its second term, we have

$$(102) \qquad \frac{d^2y}{dx^2} + Iy = 0,$$

where

$$(103) \qquad I = p_2 - \frac{1}{2}\frac{dp_1}{dx} - \frac{p_1^2}{4};$$

then if the seminvariant I is a constant we can always decompose

$$\frac{d^2y}{dx^2} + p_1\frac{dy}{dx} + p_2 y = 0$$

into commutative prime factors.

We will proceed now to apply the results arrived at concerning decomposition into prime factors to the subject of the regular integrals of a linear differential equation. We will recall first the properties of the integrals of the equation of the first order,

$$(104) \qquad \frac{dy}{dx} - ay = 0,$$

where a is a uniform function of x containing both negative and positive powers of x, and where the double series so formed is convergent in the region of $x = 0$. In order that this equation shall have only regular integrals it is necessary and sufficient that a shall be of the form $\frac{\alpha}{x}$, where α is a uniform function of x containing only positive powers of x, and which *may* contain a power of x as a factor, that is, may vanish for $x = 0$. If these conditions are satisfied, the integrals of (104) are of the form $y = x^\rho \psi(x)$, where ρ is the single root of the indicial equation, and $\psi(x)$ is a uniform function of x which does not vanish for $x = 0$. If α does not contain only positive powers of x, and if a becomes infinite of order $\nu + 1$ for $x = 0$, the integrals are of the form

$$y = e^{\frac{c_1}{x} + \frac{c_2}{x^2} + \cdots + \frac{c_\nu}{x^\nu}} x^\rho \psi(x),$$

ρ and $\psi(x)$ being characterized as before.

When α contains only positive powers of x we will call

$$\frac{dy}{dx} - \frac{\alpha}{x} y$$

a regular prime factor of a differential expression. It is easy to see from equations (16) that a differential expression P which is made up of n regular prime factors is of the form

$$(105) \qquad \frac{d^n y}{dx^n} + \frac{P_1(x)}{x}\frac{d^{n-1} y}{dx^{n-1}} + \cdots + \frac{P_n(x)}{x^n} y,$$

where $P_1(x), \ldots, P_n(x)$ are holomorphic functions of x in the region of $x = 0$, and may vanish for $x = 0$; and further, that such an expression is put into its normal form when we multiply it by x^n. We have shown that if a given differential expression is composed of differential expressions which are in the normal form, the given differential expression will itself be in the normal form, and its indicial function will be the product of the indicial functions of its components. Bearing this result in mind, we will proceed to the consideration of expressions of the form

$$\frac{d^n y}{dx^n} + p_1 \frac{d^{n-1} y}{dx^{n-1}} + \cdots + p_n y,$$

which is not in the normal form, but where the coefficients p are developable in series going according to integral powers of x and containing only a finite number of negative powers of x. Suppose, first, that a differential expression P is composed uniquely of regular factors, say

$$(106) \qquad P = A_n A_{n-1} \ldots A_1,$$

where the factors

$$(107) \qquad A_i = \frac{dy}{dx} - \frac{\alpha_i}{x} y \quad (i = 1, 2, \ldots, n)$$

are all regular, and let

$$y_1 = v_1, \quad y_2 = v_1 \int v_2 dx, \quad \ldots, \quad y_n = v_1 \int v_2 dx \ldots \int v_n dx$$

denote the fundamental system of integrals corresponding to this decomposition of P. How shall the factors A of (106) be modified in order that their new expressions shall be in the normal form, and that at the same time they will give P in its normal form, viz. $x^n P$? This question is readily answered. Multiply each factor A by x, and let A'' denote the product; A'' is of course in the normal form, and since the equations

$$(108) \qquad A_1'' = 0, \quad A_1'' = v_1 v_2, \quad A_2'' A_1'' = v_1 v_2 v_3, \quad \ldots$$

admit the integrals

$$(109) \qquad v_1, \quad v_1 \int \frac{v_1}{x} dx, \quad v_1 \int \frac{v_2}{x} dx \int \frac{v_3}{x} dx, \quad \ldots$$

we see that

$$(110) \qquad P'' = A_n'' A_n''_{-1} \ldots A_1'' = 0$$

has the same integrals for a fundamental system. Suppose then we add to the coefficients $\frac{\alpha_i}{x}$ the quantities

$$(111) \qquad \frac{d}{dx} \log x^{i-1}, \quad = \frac{i-1}{x},$$

or, what is the same thing, increase the functions α_i by the numbers $i - 1$; then, as already shown, the functions $v_2, v_3, \ldots, v_n$ will be each multiplied by x; the factors A'' are now changed into factors A' of the form

$$(112) \qquad A_i' = x\frac{dy}{dx} - (\alpha_i + i - 1)y,$$

and these, on being combined in the usual way, give rise to the expression

$$(113) \qquad P' = A_n'A_{n-1}' \ldots A_1',$$

which equated to zero has for a fundamental system of integrals

$$y_1 = v_1, \quad y_2 = v_1 \int v_2 dx, \quad \ldots, \quad y_n = v_1 \int v_2 dx \ldots \int v_n dx;$$

but these constitute a fundamental system of integrals of $P = 0$, and consequently, since the first coefficient of P is unity and that of P' is x^n, we have

$$(114) \qquad P = x^n P'.$$

We have thus the theorem: *Having given a differential expression composed uniquely of regular factors of the form*

$$\frac{dy}{dx} - \frac{\alpha_i}{x}y,$$

it can be at once thrown into the normal form by adding the numbers $i - 1$ *to each of the quantities* α_i, *and then putting each of the regular factors into the normal form by multiplying them by* x.

Consider now the composite expression

$$(115) \qquad P = QD,$$

where Q and D are linear homogeneous differential quantics of orders $n - s$ and s respectively, each of which has unity for its first coefficient and rational functions of x for the remaining coefficients. P is then of the same form as its components Q and D, and is of order n. Let x^g, x^k, x^h denote the powers of x by which we must multiply the differential expressions P, Q, D, respectively, in order to put them into their normal forms. We will now find the relation connecting the indicial functions $g(\rho)$, $k(\rho)$, $h(\rho)$ of P, Q, D. Make in D the substitution

$$y = x^h z;$$

the resultant expression D' has the normal form, a fact which is seen at once when we recall the law of formation of the coefficients in D'. The expression $P(y)$ now becomes QD', or $P(x^h z)$. Now put Q into its normal form Q', by multiplying it by x^k, and we have

(116) $$x^k P(x^h y) = Q'D'.$$

The expression $x^k P(x^h y)$ is now in the normal form, and, denoting by $g'(\rho)$, $k'(\rho)$, $h'(\rho)$ the indicial functions of $x^k P(x^h y)$, Q', D' respectively, we have

(117) $$g'(\rho) = k'(\rho) h'(\rho).$$

We have of course the relation $g = k + h$ among these exponents. We have already shown that the indicial functions of

$$x^k P(x^h y), \quad \text{or} \quad P(x^h y), \quad \text{and} \quad D(x^h y)$$

are immediately deducible from those of $P(y)$ and $D(y)$ by changing ρ into $\rho + h$; consequently

(118) $$g'(\rho) = g(\rho + h), \quad h'(\rho) = h(\rho + h),$$

and, since $k'(\rho) = k(\rho)$,

(119) $$g(\rho + h) = k(\rho) h(\rho + h).$$

Changing now ρ into $\rho - h$, we have the identity

(120) $$g(\rho) = h(\rho) k(\rho - h).$$

In particular, if D is of the form

(121) $$\frac{d^s y}{dx^s} + \frac{P_1(x)}{x} \frac{d^{s-1} y}{dx^{s-1}} + \ldots + \frac{P_s(x)}{x^s} y,$$

we have

(122) $$g(\rho) = h(\rho) k(\rho - s),$$

since h is now equal to s. By proceeding to the case where P is of the form $P = QDE$, then to the case where $P = QDEF$, etc., we will easily arrive at the following general theorem: *If P denote a differential expression of order n composed of θ differential expressions, viz.,*

$$P = D_\theta D_{\theta-1} \ldots D_1,$$

where the components D are linear homogeneous differential expressions whose first coefficients are each unity and whose remaining coefficients are rational functions of x, and if x^{h_i} is the power of x by which it is necessary to multiply D_i in order to reduce it to the normal form: then, denoting by $g(\rho)$ the indicial function of P, and by $h_i(\rho)$ the indicial function of D_i, we have the identity

$$(123) \qquad g(\rho) = h_1(\rho)h_2(\rho - h_1)h_3(\rho - h_1 - h_2) \ldots h_n(\rho - h_1 - h_2 - \ldots - h_n),$$

and also

$$h_1 + h_2 + \ldots + h_n = g,$$

where g is the exponent of the factor x^g, by which we have to multiply P in order that the product may have the normal form.

If all the D's have all their factors regular, we have

$$(124) \qquad h_1 = h_2 = \ldots = h_n = 1,$$

and consequently

$$(125) \qquad g(\rho) = h_1(\rho)h_2(\rho - 1) \ldots h_n(\rho - n + 1).$$

We deduce at once from this general theorem the corollary: *The degree of the indicial function of P is equal to the sum of the degrees of the component's functions D.*

We will now take up the general question of the decomposition of the differential expression

$$P = \frac{d^n y}{dx^n} + p_1 \frac{d^{n-1} y}{dx^{n-1}} + \ldots + p_n y,$$

where the coefficients p are developable in double series proceeding according to positive and negative integer powers of x. The restriction to a finite number of negative powers of x will not be here applied.

We know that $P = 0$ always admits an integral of the form $y_1 = x^r\phi(x)$, where $\phi(x)$ is a double series, similar therefore to the coefficients p and $r = \frac{\log s}{2\pi i}$, where s is a root of the characteristic equation for the point $x = 0$.

Suppose now $P = 0$ to admit an integral of the form $y_1 = x^\rho\psi(x)$ (that is, a regular integral), where $\psi(x)$ is a holomorphic function of x and $\psi(0)$ is not zero. Make the substitution

$$y = y_1 \int z dx,$$

and we have

$$(126) \qquad \frac{d^{n-1}z}{dx^{n-1}} + q_1 \frac{d^{n-2}z}{dx^{n-2}} + \dots + q_{n-1}z = 0,$$

where

$$(127) \qquad \begin{cases} q_1 = \dfrac{1}{y_1}\left[n\dfrac{dy_1}{dx} + p_1 y_1\right], \\ q_2 = \dfrac{1}{y_1}\left[\dfrac{n(n-1)}{1 \cdot 2}\dfrac{d^2y_1}{dx^2} + (n-1)p_1\dfrac{dy_1}{dx} + p_2 y_1\right], \\ \cdot \quad \cdot \quad \cdot \quad \cdot \quad \cdot \quad \cdot \quad \cdot \quad \cdot \quad \cdot \end{cases}$$

The forms of the remaining coefficients are easily found. It is easy to see that these coefficients q are, like the coefficients p, uniform in the region of $x = 0$. In fact the coefficients q are made up from the p's and sums of the form

$$\sum_{\alpha=0}^{\alpha=h} c_\alpha x^{-\alpha} \frac{\psi^{(h-\alpha)}_{(x)}}{\psi(x)},$$

where α is an integer. Each of these sums is then uniform and continuous in the region of $x = 0$, but of course having $x = 0$ as a critical point. The same conclusion then holds for the coefficients q. It is important to remark here that though this result has been arrived at by assuming $\psi(x)$ to contain only positive powers of x, this restriction is not necessary: it suffices that $\psi(x)$ be developable in a double series proceeding according to positive and negative powers of x and convergent in the region of $x = 0$, and further that $\psi(x)$ shall have no zeros infinitely near $x = 0$. The necessity for this last remark is easily shown as follows: Suppose $f(x)$ to be a uniform function of x in the region of $x = 0$, and suppose the point $x = 0$ to be an isolated essential singular point for $f(x)$; *i.e.*, there must exist no pole or essential singular point infinitely near the point $x = 0$. In this case the function $f(x)$ is developable in a double series proceeding according to positive and negative powers of x and convergent in the region of $x = 0$. The derivatives $f', f'', \ldots$ are of course developable in the same way. We require for our purpose, however, that this shall also be true for the logarithmic derivative $\frac{f'}{f}$ and for the functions $\frac{f^{(i)}}{f}$. If $f(x)$ has no zeros infinitely near $x = 0$, then $\frac{f^{i}(x)}{f(x)}$ can have no point of discontinuity infinitely near $x = 0$, and is consequently developable in a convergent double series in this region. If, however, $f(x) = 0$ has an infinite number of solutions in any region however small of $x = 0$, then $\frac{f^{(i)}_{(x)}}{f(x)}$ will have in this region an infinite number of poles, and consequently will not be developable in the same way as $f(x)$. This shows us at once the necessity for the above-mentioned restriction upon the function $\psi(x)$. Suppose now

$$(128) \qquad P = A_n A_{n-1} \ldots A_1,$$

and let

$$(129) \qquad y_1 = v_1, \; y_2 = v_1 \int v_2 dx, \; \ldots, \; y_n = v_1 \int v_2 dx \ldots \int v_n dx$$

denote the system of fundamental integrals corresponding to this decomposition of P. As the auxiliary equations of which $v_1, v_2, \ldots, v_n$ are integrals are of the same form as P, we can, and will, choose $v_1, v_2, \ldots, v_n$ of the forms

$$(130) \qquad v_1 = x^{r_1} \phi_1(x), \quad v_2 = x^{r_2} \phi_2(x), \quad \ldots, \quad v_n = x^{r_n} \phi_n(x).$$

We have now

$$(131) \qquad a_i = \frac{d}{dx} \log (v_1 v_2 \ldots v_i) = \frac{d}{dx} \log (x^{r_1 + r_2 + \ldots + r_i} \phi_1 \phi_2 \ldots \phi_i).$$

If now the functions ϕ have no zeros infinitely near the point $x = 0$, it follows that the coefficients a are, in the region of $x = 0$, continuous and monogenic functions having this point as a critical point; they are also uniform: for, when the variable turns round the point $x = 0$, the quantity under the logarithmic sign is multiplied by $e^{2\pi i(r_1+r_2+\ldots+r_n)}$, and therefore its logarithmic derivative is unaltered. The coefficients a have therefore the same properties as the coefficients p. We have thus the result: *If the functions $\phi(x)$ admit of no zeros infinitely near the point $x = 0$, it is possible to decompose the differential expression P into prime factors having uniform coefficients of the form*

$$\sum_{-\infty}^{+\infty} C_i x^i,$$

where i is an integer.

Resume now the particular case where the coefficients p contain only a finite number of negative powers of x. The decomposition of P into prime factors of the form

$$\frac{dy}{dx} - y \sum_{-\infty}^{+\infty} C_i x_i$$

leads to two important propositions:

(A) *If P is decomposable into prime factors of the form*

$$\frac{dy}{dx} - y \sum_{-\infty}^{+\infty} C_i x^i,$$

the degree γ of its indicial function is equal to the total number j of regular factors which enter into this decomposition.

To prove this, we will consider the decomposition

$$P = A_n A_{n-1} \ldots A_1,$$

which contains j regular factors. Let us neglect for a moment, in factors which contain an infinite number of negative powers of x, all of the powers of x^{-1} which, in absolute value, are greater than a given arbitrary number k, and let

$$P'_{,} = B_n B_{n-1} \ldots B_1,$$

denote the modified form of P. The expression P' is of the same character as P, and the regular factors in P' of course coincide with the j regular factors in P. If now γ' denote the degree of the indicial function of P', we know that γ' is equal to the sum of the degrees of the indicial functions of the factors B. Now the irregular factors B have constants as their indicial functions; the regular factors B are all of their first order, and so their indicial functions are of the first degree, consequently we have $\gamma' = j$. This equality evidently exists however great k may be; we may then increase k indefinitely, and since γ' has γ for its limit we have, finally, $\gamma = j$. The proposition A is thus proved. As a corollary it may be remarked that the number of regular factors which enter into a decomposition of P into prime factors of the form

$$\frac{dy}{dx} - y\sum_{-\infty}^{+\infty} C_i x^i$$

is constant whatever be the chosen method of decomposition. In a decomposition $P = A_n A_{n-1} \ldots A_1$ let σ denote the greatest possible number of consecutive regular factors which such a decomposition can have when we count back from the factor A_1; that is,

$$A_\sigma, \quad A_{\sigma-1}, \quad \ldots, \quad A_1$$

are all the possible consecutive regular factors which can appear at the right-hand end of the decomposition $A_n A_{n-1} \ldots A_1$. Our second proposition is now:

(B) *The number, s, of linearly independent regular integrals of the equation $P = 0$ is equal to the greatest number, σ, of consecutive regular factors which can terminate a decomposition of P into symbolic factors of the form*

$$\frac{dy}{dx} - y\sum_{-\infty}^{+\infty} C_i x^i.$$

This is readily proved. Since $P = 0$ has s regular integrals, it has at least one integral of the form $y_1 = v_1 = x^{\rho_1}\psi_1(x)$, where $\psi_1(x)$ is in the region of $x = 0$ a holomorphic function of x, and $\psi(0)$ is not equal to zero. Make now the substitution $y = v_1 \int z dx$, and we have a differential equation $Q(z) = 0$ of order $n - 1$ having $s - 1$ regular integrals. $Q(z) = 0$ has then at least one integral of the form

$$z_2 = v_2 = x^{\rho_2}\psi_2(x).$$

Make now a substitution similar to the preceding one, and continue this process until we arrive at an equation, say $H(u) = 0$, which has one regular integral of the form $v_s = x^{\rho_s}\psi_s(x)$. Now make the substitution $u = v_s \int v ds$; the reduced equation will have at least one integral of the form $x^r\phi(x)$ which contains no logarithms. By an analogous series of reductions we will obtain the functions

$$v_{s+2}, \quad v_{s+3}, \quad \ldots, \quad v_n,$$

and can then give the decomposition of P into prime factors corresponding to the system of fundamental integrals

$$y_1 = v_1, \quad y_2 = v_1 \int v_2 dx, \quad \ldots, \quad y_n = v_1 \int v_2 dx \ldots \int v_n dx.$$

Let

$$P = A_n A_{n-1} \ldots A_1$$

denote this decomposition, where

$$A = \frac{dy}{dx} - y \sum_{-\infty}^{+\infty} C_i x^i,$$

and where the equations

$$A_1 = 0, \quad A_2 = 0, \quad \ldots, \quad A_s = 0$$

admit the regular integrals

$$v_1 = x^{\rho_1}\psi_1, \quad v_1 v_2 = x^{\rho_1+\rho_2}\psi_1\psi_2, \quad \ldots,$$
$$v_1 v_2 \ldots v_s = x^{\rho_1+\rho_2+\ldots+\rho_s}\psi_1\psi_2 \ldots \psi_s.$$

Since these last s factors are regular, we have $\sigma \geqq s$. It is easy to see, however, that we will not have $\sigma > s$. Suppose in fact that it is possible to so decompose P into prime factors of the form

$$\frac{dy}{dx} - y \sum_{-\infty}^{+\infty} C_i x^i$$

that the last $s + t$ factors shall be regular,—say

$$P = B_n B_{n-1} \ldots B_{s+t} \ldots B_2 B_1.$$

Denote by

$$w_1, \quad w_1 w_2, \quad w_1 w_2 w_3, \quad \ldots, \quad w_1 w_2 \ldots w_{s+t}$$

the respective solutions of the equations

$$B_1 = 0, \quad B_2 = 0, \quad B_3 = 0, \quad \ldots, \quad B_{s+t} = 0.$$

These solutions are regular and of the form $x^\rho\psi(x)$, where $\psi(x)$ is a holomorphic function of x in the region of $x = 0$ and $\psi(0)$ does not vanish. It follows then that the ratios,

$$\frac{w_1 w_2 \ldots w_{i-1} w_i}{w_1 w_2 \ldots w_{i-1}},$$

that is, the integrals,

$$w_1, \; w_2, \; \ldots, \; w_{s+t},$$

of the auxiliary equations, deduced the one from the other by the ordinary substitutions, are regular integrals of the form $x^\rho\psi(x)$. Now since the equation which gives w_{s+t} has at least one regular integral, that which gives w_{s+t-1} must have at least two and that which gives w_{s+t-2} must have at least three regular integrals, etc., finally, then, the equation which gives w_1, that is, $P = 0$, must have at least $s + t$ regular integrals; therefore, since this is contrary to our hypothesis, we must have exactly $\sigma = s$. It is easy to see that proposition (B) is still true when the prime factors are not restricted to be of the form

$$\frac{dy}{dx} - y\sum_{-\infty}^{+\infty} C_i x^i.$$

From (A) and (B) we derive at once the following theorems:

(*a*) *If $P = 0$ has all of its integrals regular, P is decomposable into n regular prime factors.*

(*b*) *If P is decomposable into n regular factors, $P = 0$ has all of its integrals regular.*

(*c*) *If $P = 0$ has all of its integrals regular, the degree of its indicial equation is equal to its order.*

(*d*) *If the degree of the indicial equation of P is equal to the order of P, then the equation $P = 0$ has all of its integrals regular.*

The following theorem is easily established:

(*c*) *If $P = 0$ has all of its integrals regular, it has a fundamental system of integrals belonging to exponents which are roots of its indicial equation.*

If $P = 0$ has all of its integrals regular, P is decomposable into n regular factors, viz.,

$$P = A_n A_{n-1} \ . \ . \ . \ A_1 ,$$

and from (125) we have the relation

$$g(\rho) = h_1(\rho) h_2(\rho - 1) \ . \ . \ . \ h_n(\rho - n + 1),$$

connecting the indicial functions g and h of P and its factors A. If now we write

$$A_1 = 0, \quad A_2 = 0, \quad . \ . \ . \ , \quad A_n = 0,$$

we see that the integrals

$$v_1 , \quad v_1 v_2 , \quad . \ . \ . \ , \quad v_1 v_2 \ . \ . \ . \ v_n$$

of these equations belong respectively to the exponents

$$\rho_1 , \quad \rho_2 - 1, \quad \rho_3 - 2, \quad . \ . \ . \ , \quad \rho_n - n + 1,$$

where

$$\rho_1 , \quad \rho_2 , \quad . \ . \ . \ , \quad \rho_n$$

are the roots of the indicial equation $g(\rho) = 0$. It is obvious then that the integrals

$$y_1 = v_1 , \quad y_2 = v_1 \int v_2 dx, \quad . \ . \ . \ , \quad y_n = v_1 \int v_2 dx \ . \ . \ . \int v_n dx$$

of $P = 0$ belong to the exponents $\rho_1 , \rho_2 , \ . \ . \ . \ , \rho_n$ respectively. In the case where the integrals of the differential equation are not all regular, propositions (A) and (B) give the following theorem:

The number of linearly independent integrals of $P = 0$ is at most equal to the degree of its indicial equation.

The form which the expression P must have in order that the equation $P = 0$ shall have s regular integrals is given in the following theorem:

In order that $P = 0$ shall have s linearly independent regular integrals it is necessary and sufficient that P can be put in the form

$$P = QD,$$

where Q and D are respectively of the orders $n - s$ and s, and are expressions of the same kind as P (that is, the first coefficient in each is unity and the remaining coefficients are rational functions of x), and where $D = 0$ has all of its integrals regular while $Q = 0$ has no such integral.

We will show first that this condition is necessary. We have shown that if $P = 0$ has s regular integrals, it admits of a decomposition

$$P = A_n A_{n-1} \ldots A_{s+1} A_s \ldots A_1$$

into prime factors of the form

$$A = \frac{dy}{dx} - y \sum_{-\infty}^{+\infty} C_i x^i,$$

where the last s factors are regular. Write now

$$A_s \ldots A_2 A_1 = D, \quad A_n A_{n-1} \ldots A_{s+1} = Q;$$

then

$$P = QD.$$

Since D is of order s and contains only regular factors, we have

$$D = \frac{d^s y}{dx^s} + \frac{P_1(x)}{x} \frac{d^{s-1} y}{dx^{s-1}} + \ldots + \frac{P_s(x)}{x^s} y_s,$$

the form of the P's being known, and $D = 0$ has then all of its integrals regular. The expression Q is of order $n - s$. Effecting now the operation QD and identifying the result with P, we obtain a system of equations which show (under the above-mentioned restriction as to the functions ϕ) that the coefficients of Q are, like those of P and D, rational functions. Further, Q has no regular integral; for, if it had, we could terminate a decomposition of Q by at least one regular factor, and so $P, = QD$, would have more than s regular integrals, which is impossible by proposition (B). Second, this condition is necessary. Suppose $P = QD$, where $D = 0$ has all of its integrals regular and $Q = 0$ has no regular integral. Every solution of $P = 0$ which satisfies $D = 0$ is regular. Every solution of $P=0$ which does not satisfy $D=0$ will satisfy one of the equations $D=u$, where u denotes some one of the integrals of $Q=0$. Now by hypothesis none of the integrals of $Q=0$ are regular, and if in D we replace y by a regular function, the result is a regular expression, and therefore no regular value of y can make D equal to u. It follows then that $D=u$ can have no regular solution, and consequently that the only regular integrals of $P=0$ are the s regular integrals of $D=0$.

It is easy to see the cause of the difference $\gamma-s$ between the degree γ of the indicial equation of $P=0$, and the number s of linearly independent regular integrals of this equation. Suppose in fact that P is decomposed into prime factors of the form

$$\frac{dy}{dx}-y\sum_{-\infty}^{+\infty}C_i x^i.$$

This decomposition, whatever it may be, will, as we know, contain γ regular factors; of these regular factors a group of s at most can be placed at the right-hand end of the decomposition, and the remaining $\gamma-s$ factors cannot come into this group (of course in particular cases we will have $\gamma=s$). The difference $\gamma-s$ then arises from the presence of regular factors in Q which cannot be placed as consecutive factors at the right-hand end of the decomposition of Q.

Observing that the indicial functions of D and Q are respectively of the degrees s and $\gamma-s$, we can enunciate the two following theorems:

(α) *In order that the linear differential equation $P=0$ of order n possessing an indicial equation of order γ shall have $\gamma-\mu$ linearly independent regular integrals, it is necessary and sufficient that P shall have the form $P=QD$, Q and D being of the same form as P, and that $Q=0$, being of order $n-\gamma+\mu$, shall have no regular integrals and shall have an indicial equation of order μ.*

(β) *In order that the linear differential equation $P=0$ of order n and having an indicial equation of degree γ shall have γ linearly independent regular integrals, it is necessary and sufficient that P shall be of the form $P=QD$, Q and D being of the same form as P, Q being of order $n-\gamma$ and having a constant as its indicial function.*

We can further show again the truth of the theorem:

(γ) *The s linearly independent regular integrals of $P = 0$ belong to exponents which are certain s of the roots of the indicial equation for $P = 0$.*

This is immediately seen, since if we put P in the form

$$P = QD,$$

we have the relation

$$g(\rho) = h(\rho)k(\rho - s)$$

between the indicial functions. Now the s regular integrals of $D = 0$, that is, all the linearly independent regular integrals of $P = 0$, belong, as we have seen, to exponents which are the roots of $h(\rho) = 0$. It follows then, from the above relation connecting the indicial functions, that the regular integrals of $P = 0$ belong to exponents which are certain s of the roots of $g(\rho) = 0$.

The following theorems concerning the adjoint equation are easily seen to be true; their verification will, however, be left as exercises.

I. *The indicial functions $g(\rho)$ and $\mathfrak{g}(\rho)$ of the adjoint expressions P and $\mathfrak{P}$ are derived the one from the other by changing ρ into $-\rho + \beta - 1$, where x^β is the power of x by which we must multiply P in order to put it in the normal form.*

II. *If the linear differential equation $P = 0$ has all of its integrals regular, the same is true of its adjoint equation $\mathfrak{P} = 0$.*

The theorem (γ) above is here changed into the following:

III. *In order that the linear differential equation $P = 0$ of order n having an indicial equation of degree γ shall have exactly γ linearly independent regular integrals, it is necessary and sufficient that its adjoint equation $\mathfrak{P} = 0$ shall have among its integrals all the integrals of a linear differential equation of order $n - \gamma$ which has a constant for its indicial function.*

It is easy to see from the preceding investigations that all the properties of the regular integrals of linear differential equations can be arrived at by aid of the decomposition into prime symbolic factors.

CHAPTER XI.

APPLICATION OF THE THEORY OF SUBSTITUTIONS TO LINEAR DIFFERENTIAL EQUATIONS.

I.

We may recall the definition of the product of two substitutions; viz., the product SS_1 of the two substitutions S and S_1 is the new substitution which produces the same effect as if the substitutions S and S_1 were applied successively. A similar definition applies to the product of any number of substitutions. In particular, we speak of the power of a substitution, say S^α, which means that the substitution S has been made α times.

In Chapter III, equation (2), we have the general type of the substitution which corresponds to any critical point of the linear differential equation with uniform coefficients; viz., if $y_1, \ldots, y_n$ denote a system of fundamental integrals in the region of the assumed critical point, we have

$$(1) \qquad S = \begin{vmatrix} y_1; & c_{11}y'_1 + c_{12}y'_2 + \ldots + c_{1n}y_n \\ y_2; & c_{21}y'_1 + c_{22}y'_2 + \ldots + c_{2n}y_n \\ \cdot & \cdot \quad \cdot \quad \cdot \quad \cdot \quad \cdot \quad \cdot \\ y_n; & c_{n1}y'_1 + c_{n2}y'_2 + \ldots + c_{nn}y_n \end{vmatrix}$$

It was also shown how by aid of the characteristic equation

$$(2) \qquad \begin{vmatrix} c_{11} - s, & c_{21} & \ldots & c_{n1} \\ c_{12}, & c_{22} - s & \ldots & c_{n2} \\ \cdot & \cdot & \cdot & \cdot \\ c_{1n}, & c_{2n} & \ldots & c_{nn} - s \end{vmatrix} = 0$$

this could be reduced to the canonical form given in equation (80′) of the same chapter. The equation (80′) may, by a slight change in notation, be written in the form

$$(3) \qquad S = \begin{vmatrix} y_1' & \dots & y_1^{k_1}; & s_1 y_1' & \dots & s_1 y_1^{k_1} \\ y_2' & \dots & y_2^{k_2}; & s_1(y_2' + y_1') & \dots & s_1(y_2^{k_2} + y_1^{k_2}) \\ \cdot & \cdot & \cdot & \cdot & \cdot & \cdot \\ y_p' & \dots & y_p^{k_p}; & s_1(y_p' + y_{p-1}') & \dots & s_1(y_p^{k_p} + y_{p-1}^{k_p}) \\ z_1' & \dots & z_1^{l_1}; & s_2 z_1' & \dots & s_2 z_1^{l_1} \\ z_2' & \dots & z_2^{l_2}; & s_2(z_2' + z_1') & \dots & s_2(z_2^{l_2} + z_1^{l_2}) \\ \cdot & \cdot & \cdot & \cdot & \cdot & \cdot \end{vmatrix};$$

where the integers $k, l, \dots$ satisfy the inequalities

$$k_1 \overline{\geqq} k_2 \dots \overline{\geqq} k_k, \quad l_1 \overline{\geqq} l_2 \quad \dots,$$

and where $k_1 + k_2 + \dots + k_p, l_1 + l_2 + \dots, \dots$ are the respective degrees of multiplicity of the roots $s_1, s_2, \dots$ of equation (2).

We shall speak of the functions $y, z, \dots$ as forming classes corresponding to the roots $s_1, s_2, \dots$, respectively, of the characteristic equation; thus, we shall speak of the functions y as being of the first class, the functions z as being of the second class, etc.

The group of the differential equation has been already defined, but it will be convenient to repeat the definition here. Suppose a, $b, c, \dots$ to be the critical points of the equation; when the variable

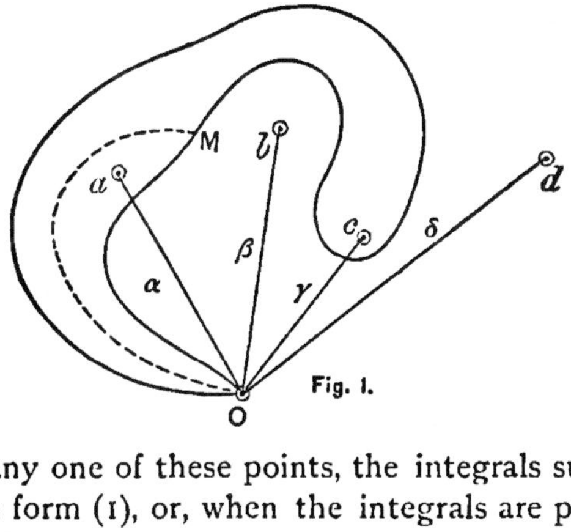

Fig. 1.

travels round any one of these points, the integrals submit to a substitution of the form (1), or, when the integrals are properly chosen, of the canonical form (3).

Any path going round only one critical point can of course be replaced by the *loop* (*lacet*) belonging to this point, and any path going round more than one of these points may be replaced by the loops, taken in proper order, belonging to these points.

Corresponding to each of these paths there is a certain substitution; denote by A the substitution corresponding to the critical point a, by B the substitution corresponding to b, etc. If in Fig. 1 we start from O and travel along the curved line back to O, we shall have gone round the two points a and c. Now we can in the usual manner (by aid of the dotted line OM) see that this path is equivalent to the two loops α and γ; we therefore apply to the functions first the substitution A and then the substitution C, or simply the substitution AC.

Again, if in Fig. 2 we start from O and go round the curved

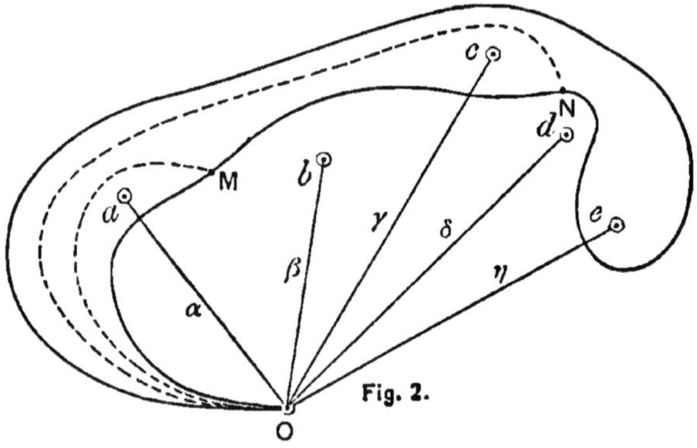

Fig. 2.

line and come back to O, it is easy to see that we have applied the substitution ACE. All possible substitutions of this sort that we can form from the simple substitutions A, B, C, . . . , their powers and products, form the group G of the equation.

Denote by F a function-group of the equation such that all the substitutions of G transform the functions y, z, u, . . . among themselves. It is clear that this function-group can be regarded as *containing* certain other function-groups. For example, suppose we consider the single function y and its transforms by A and the powers of A; these will form a function-group which is obviously *contained* in F; and so in many different ways we may form function-groups which will all be *contained* in F. We will call such function-groups *minors* of F.

We will show now that F can be considered as derived from a certain number of distinct functions, each of which contains only variables of a single class. Any function whatever in F is made up of the sum of a certain linear function of the variables y (say Y), a linear function of the variables z (say Z), a linear function of the variables u (say U), etc. If then f denote a function contained in F, we can write

(4) $$f = Y + Z + U + \ldots .$$

Let Φ denote the function-group derived from f and its transforms f', f'', . . . which are obtained by operating on f by the different powers of A; again, let Φ' denote the function-group formed from the partial functions Y, Z, . . . , and their transforms by A, A^2, ... We may write

(5) $$\Phi' = \Phi'_y + \Phi'_z + \ldots ;$$

Φ'_y being made up of y and its transforms, Φ'_z of Z and its transforms, etc. It is obvious that the function-group Φ is a minor of Φ'. Let now ρ denote the greatest of the indices k of those of the functions y_i^k which enter in Y; σ the greatest of the indices l of those of the functions z_i^l which enter in Z; etc. We can now establish the following theorem:

The function-group Φ contains each of the partial functions Y, Z, . . . and is derived from $\rho + \sigma + \ldots$ distinct functions, of which ρ are formed exclusively with the variables y_i^k, σ are formed exclusively with the variables z_i^l, etc.

We have to show first that the number of distinct functions in Φ is not greater than $\rho + \sigma + \ldots$

We will determine first how many distinct functions will be given by Y and its transforms. Denote by $Y_{\rho-i}$ a linear function of the y's whose index is less than or equal to $\rho - i$. Now it is obvious that Φ' contains the function

(6) $$Y = \lambda' y_\rho' + \lambda'' y_\rho'' + \ldots + Y_{\rho-1},$$

and it also contains the transform AY, and consequently it contains the function

(7) $$Y' = \frac{1}{s_1} AY - Y,$$

which has obviously the form

(8) $$Y = \lambda' y'_{\rho-1} + \lambda'' y''_{\rho-1} + \ldots + Y_{\rho-2}.$$

The functions Y and Y' are by definition elements of the sub-function-group Φ'_y. By a continuation of the above process we see that Φ'_y contains the elements

$$(9) \qquad Y'' = \frac{1}{s_1} A Y' - Y' = \lambda' y'_{\rho-2} + \lambda'' y''_{\rho-2} + \ldots + Y_{\rho-3};$$

. . . .

$$(10) \qquad Y^{\rho-1} = \frac{1}{s_1} A Y^{\rho-2} - Y^{\rho-2} = \lambda' y_1' + \lambda'' y_1'' + \ldots .$$

This last function gives identically

$$(11) \qquad \frac{1}{s_1} A Y^{\rho-1} - Y^{\rho-1} = 0.$$

The functions $Y, Y', \ldots, Y^{\rho-1}$ are evidently linearly independent, as each of them contains certain of the variables y which are not contained in the others. Again, we have from the above equations

$$(12) \quad A Y = s_1(Y + Y'), \ A Y' = s(Y' + Y''), \ldots, A Y^{\rho-1} = s_1 Y^{\rho-1};$$

that is, the substitution A changes these functions into linear functions of themselves alone. The different powers of A will obviously do the same thing, and consequently the number of linearly independent functions in Φ'_y is exactly equal to ρ. In the same way we see that the number of distinct functions in Φ'_z is equal to σ, and so for the other sub-function-groups. It follows then that Φ' contains exactly $\rho + \sigma + \ldots$ distinct functions, and consequently that Φ, which is a minor of Φ', can contain no more than $\rho + \sigma + \ldots$ distinct functions. To complete the proof of the above theorem, it is now necessary to show that Φ contains at least $\rho + \sigma + \ldots$ linearly independent functions.

It is obvious that among the functions composing Φ we have

$$(13) \qquad f = \lambda y_\rho + \lambda' y'_\rho + \ldots + Y_{\rho-1} + Z + U + \ldots;$$

Z being a linear function of the variables z, U a linear function of the variables u, etc. From the definition of Φ we see that this function-group contains $Af, A^2f, \ldots$ It follows then that Φ contains also the function

$$(14) \quad f' = \frac{1}{s_1 - s_2}(Af - s_2 f) = \lambda y_\rho + \lambda' y'_\rho + \ldots + Y'_{\rho-1} + Z' + \ldots$$

$Y'_{\rho-1}$ is analogous to $Y_{\rho-1}$, and Z' is analogous to Z, but differs from Z in that the highest index of the variables z is now equal to or less than $\sigma - 1$. In like manner we have in Φ the function

$$(15)\quad f'' = \frac{1}{s_1 - s_2}(Af' - s_2 f') = \lambda y_\rho + \lambda' y'_\rho + \ldots + Y''_{\rho-1} + Z'' + \ldots;$$

the highest index of z in Z'' being equal to or less than $\sigma - 2$, etc.

We can obviously continue this process until we arrive at a function contained in Φ and containing none of the variables z. Suppose now that the variables u belong to the root s_3 of the characteristic equation; then by a process entirely similar to the preceding we can find a function contained in Φ which contains none of the variables u, and so for all the other variables except those of the first class. We have then finally, as one of the functions in Φ, the expression

$$(16)\qquad \phi = \lambda y_\rho + \lambda' y_\rho' + \ldots + \mathfrak{Y}_{\rho-1};$$

$\mathfrak{Y}_{\rho-1}$ being a linear function of the variables y whose index is less than ρ.

The functions $A\phi$, $A^2\phi$, ... are all contained in Φ, and of the set ϕ, $A\phi$, $A^2\phi$, ... there are evidently ρ distinct functions. In the same way we can show that Φ contains σ distinct functions involving only the variables z, etc.—that is, Φ contains in all $\rho + \sigma + \ldots$ distinct functions; and since Φ is a minor of Φ', which contains the same number of distinct functions, it follows that Φ and Φ' are identical, and therefore Φ contains the functions $Y, Z, U, \ldots$ above defined. Now Φ is a minor of the function-group F; therefore F contains $Y, Z, \ldots$ among its elements. We have then, finally, that if the function-group F contains the function

$$(4)\qquad f = Y + Z + U + \ldots,$$

it contains also the functions $Y, Z, U, \ldots$

Let now $Y_1, Y_2, \ldots$ denote the linearly independent functions contained in F which contain only the variables y; $Z_1, Z_2, \ldots$ the analogous functions of the variables z; etc. The function-group F is *derived* from the elements $Y_1, Y_2, \ldots, Z_1, Z_2, \ldots$, etc. For, denoting by

$$f_1 = Y + Z + \ldots,$$

any one of the functions contained in F, we know that Y, which depends only on the variables y, is a linear function of $Y_1, Y_2, \ldots$, and that Z, containing only the variables z, is a linear function of $Z_1, Z_2, \ldots$, etc. We have thus established the above proposition concerning the function-group F, viz.: *that F can be considered as derived from a certain number of distinct functions $Y_1, Y_2, \ldots, Z_1, Z_2, \ldots$, etc., each of which contains only variables of a single class.*

Suppose that in F we have the function Y of the variables y, say

$$Y_1 = f_1 = \lambda' y_1' + \lambda'' y_1'' + \ldots \tag{17}$$

The different transforms of f are also contained in F. If we perform the substitution A any number of times in Y, we obtain functions which only contain the variables y; but if we perform the substitutions B, C, etc., we get new functions which are linear on the one hand with respect to the variables $y, z, u, \ldots$, and on the other hand with respect to the constants $\lambda', \lambda'', \ldots$ Denote by $P_1, P_2, \ldots$ the different products that can be formed by multiplying one of the constants λ by one of the variables $y, z, u, \ldots$; as both variables and constants are finite in number, there will be only a finite number of the products P. Let $\eta_1, \eta_2, \ldots$ denote linear functions of the y's alone, $\zeta_1, \zeta_2, \ldots$ linear functions of the z's alone, etc.; then we can write

$$Bf = \eta_1 + \zeta_1 + \ldots, \tag{18}$$

$$Cf = \eta_2 + \zeta_2 + \ldots, \tag{19}$$

$$\cdot \quad \cdot \quad \cdot \quad \cdot \quad \cdot$$

From what we have above proved it is clear now that F contains $\eta_1, \eta_2, \ldots, \zeta_1, \zeta_2, \ldots$, and also

$$\left\{\begin{array}{ll} \eta_1' = \frac{1}{s_1} A\eta_1 - \eta_1, & \eta_1'' = \frac{1}{s_1} A\eta_1' - \eta_1' \ldots, \\ \eta_2' = \frac{1}{s_1} A\eta_2 - \eta_2, & \eta_2'' = \frac{1}{s_1} A\eta_2' - \eta_2' \ldots, \\ \cdot \quad \cdot \quad \cdot & \cdot \quad \cdot \quad \cdot \\ \zeta_1' = \frac{1}{s_2} A\zeta_1 - \zeta_1, & \zeta_1'' = \frac{1}{s_2} A\zeta_1' - \zeta_1' \ldots, \\ \cdot \quad \cdot \quad \cdot & \cdot \quad \cdot \quad \cdot \end{array}\right. \tag{20}$$

each of which contains only variables of a single class. These different functions $\eta_1, \eta_1', \eta_1'', \ldots, \zeta_1, \zeta_1', \ldots$ are of course linear functions of $P_1, P_2, \ldots$ It may be, however, that some of these functions $\eta, \zeta, \ldots, \eta', \zeta', \ldots$ are linearly expressible in terms of the remaining ones and f; if such functions exist we will discard them, and so have finally a certain set of distinct functions, say $f, f', f'', \ldots, f^r$, of which $f' \ldots f^r$ have been derived by operating upon f by B, $C, \ldots$ and by powers of A. We may take now each of these new functions $f', f'', \ldots, f^r$, and by precisely similar operations arrive at still other functions which will be contained in F and each of which will contain only the variables of a single class. These new functions are again linear functions of $P_1, P_2, \ldots$; neglect, as before, all of them that are linear functions of the remaining ones and of $f, f', \ldots, f^r$, and say the new set of distinct functions is

$$f, f' \ldots f^r, f^{r+1} \ldots f^s.$$

Treat the functions $f^{r+1} \ldots f^s$ as before, and neglect all the new functions which are linearly expressible in terms of the remaining ones and of $f, f', \ldots, f^s$. We will continue this process until we can find no new function which is not linearly expressible in terms of those already found. That the series of operations is limited is clear from the fact that each of the functions which we form is a linear function of the quantities $P_1, P_2, \ldots$ which are limited in number. Suppose the functions finally obtained are $f, f', \ldots, f^t$, each of which contains only the variables of a single class (of course there will in general be several functions which contain the variables of the same class). Each of the functions $f, f', \ldots, f^t$ is transformed by each of the substitutions of the group G into a linear function of $f, f', \ldots, f^t$. In order to prove this it is only necessary to show its truth for each of the substitutions $A, B, C, \ldots$ from which G is derived. Take, for example, the function $f', = \eta_1'$. The transform of this by the substitution A is

$$A\eta_1' = s_1(\eta_1'' + \eta_1'); \tag{21}$$

operating with $B, C, \ldots$, we have

$$B\eta_1' = \bar{\eta}_1 + \bar{\zeta}_1 + \ldots, \tag{22}$$

$$C\eta_1' = \bar{\eta}_2 + \bar{\zeta}_2 + \ldots; \tag{23}$$

$\bar{\eta}$, $\bar{\zeta}$, . . . containing respectively only the variables y, z, . . . ; but each of these partial functions $\bar{\eta}$, $\bar{\zeta}$, . . . is a linear function of f, f', . . . , f^t. From the functions $f, f', \ldots, f^t$ we can form a function-group Ψ which possesses the property of having its different functions transformed into one another by the substitutions of the group G.

Suppose $f, f_1, f_2, \ldots$ to be the functions which depend only on the variables y, and let Ψ_y be the function-group derived from f, f_1, $f_2, \ldots$, or, what amounts to the same thing, derived from f, ϕ_1, $\phi_2, \ldots$, where

$$(24) \qquad \phi_1 = f_1 + \omega_1 f, \quad \phi_2 = f_2 + \omega_2 f, \quad \ldots ;$$

$\omega_1, \omega_2, \ldots$ being constants each of which is subject to only one condition, viz., that it shall not be a root of a certain algebraic equation. The function f is of the form

$$(25) \qquad f = \lambda' y_1' + \lambda'' y_1'' + \ldots ,$$

and f_1 of the form

$$(26) \quad f_1 = [ay_1' + by_1'' + \ldots + dy_2' + \ldots]\lambda' \\ + [a'y_1' + b'y_1'' + \ldots + d'y_2' + \ldots]\lambda'' + \ldots ;$$

therefore ϕ_1 is of the form

$$(27) \quad \phi_1 = [(a + \omega_1)y_1' + by_1'' + \ldots + dy_2' + \ldots]\lambda' \\ + [a'y_1' + (b' + \omega_1)y_1'' + \ldots]\lambda'' + \ldots .$$

We will now determine ω_1 in such a manner that the determinant

$$(28) \qquad \Delta_1 = \begin{vmatrix} a + \omega_1, & b & , d & \ldots \\ a' & , b' + \omega_1, & d' & \ldots \\ a'' & , b'' & , d'' + \omega_1 & \ldots \\ \cdot & \cdot & \cdot & \cdot \end{vmatrix}$$

shall not vanish. It is obviously only necessary to choose for ω_1 any value that is not a root of the equation $\Delta_1 = 0$. The constants ω_2, $\omega_3, \ldots$ are to be determined in a similar manner; that is, ω_2 must not be a root of a certain equation $\Delta_2 = 0$, etc. The function ϕ_1 is therefore obtained by operating upon

$$f = \lambda' y_1' + \lambda'' y_1'' + \ldots$$

with the substitution

$$\Sigma_1 = \begin{vmatrix} y_1' ; & (a + \omega_1) y_1' + & b y_1'' & + \ldots \\ y_1'' ; & a' y_1' & + (b' + \omega_1) y_1' + & \ldots \\ \vdots & \vdots & \vdots & \vdots \end{vmatrix}.$$

The functions ϕ_2, ϕ_3, . . . will be obtained by operating upon f with similar substitutions Σ_2, Σ_3, ... It follows, therefore, that the function-group Ψ_y will be identical with the function-group derived from the transforms of f by the substitutions of a substitution-group Γ which is derived from Σ_1, Σ_2, . . . Let us assume now that we are able to ascertain whether the group Γ is prime or not, and in this second case we will suppose that we can determine a function-group Θ which is derived from a number of distinct functions, this number being less than the number of the variables y; further, Θ is to be such that the substitutions Σ_1, Σ_2, . . . shall replace its functions the one by another. Suppose, first, that Γ is not prime, and denote by ξ any one of the functions in Θ. Choose the constants λ', λ'', . . . , such that

(29) $$\xi = \lambda' y_1' + \lambda'' y_1'' + \ldots .$$

The function ξ and all of its transforms by the substitutions Σ_1, Σ_2, . . . will form a function-group contained in Θ, and consequently depending upon a number, say ρ, of distinct functions, at most equal to the number of distinct functions in Θ. Now among the distinct functions from which Ψ is derived, and each of which contains only one class of variables, there will be only ρ containing the variables y, and the number of these variables is by hypothesis greater than ρ. In the case of each of the other classes there can of course be no more functions than the number of variables of the corresponding class. The function-group Ψ will then contain less than n distinct functions, which will be obtained by substituting in $f, f, \ldots, f^t$ the special values of λ', λ'', . . . and, after effecting the substitutions, neglecting those functions which are linearly expressible in terms of the others.

In this case, then, the group G is not prime, and we have determined a function-group containing less than n distinct functions.

In the second case, suppose Γ to be prime. Whatever be the system of values chosen for $\lambda', \lambda'', \ldots$, the function-group Ψ_y obtained from $\lambda' y_1' + \lambda'' y_1'' + \ldots$ by the substitutions of the group Γ will depend upon a number of distinct functions equal to the number of the variables y. It follows then that Ψ_y will contain among its functions all those that can be formed from these variables. Suppose y_1' one of these functions; then y_1' is necessarily contained in F. Since $\lambda', \lambda'', \ldots$ are no longer restricted in value, we may write $\lambda' = 1$, $\lambda'' = \lambda''' = \ldots = 0$, and ascertain whether or not the function-group Ψ formed under this hypothesis contains n or less than n distinct functions. If Ψ contains n distinct functions, then F, of which Ψ is a minor, will also contain n distinct functions, and G will therefore be prime. If Ψ contains less than n distinct functions, G will not be prime, and the system of distinct functions upon which Ψ depends will have to be determined.

Suppose now that F contains no function of the variables y of the first class, but contains a function Z of the variables z of the second class. By a procedure identically the same as the one just described we can determine a function-group Ψ (if such exists) which shall be a minor of F depending upon less than n distinct functions. And so in like manner we can proceed if F contains no function of the variables y and no function of the variables z, but a function U of the variables u of the third class, etc. By continuing this process we see that we can always determine a function-group Ψ containing less than n distinct functions, unless indeed F is known to contain always n distinct functions, in which case G is prime by definition.

The question proposed in Chapter VIII concerning the group G was, "*Having given a group G composed of linear substitutions A, B, C, . . . among n variables, required to determine whether or not the linear differential equation which has G for its group is satisfied by the integrals of analogous differential equations of orders lower than n, and to determine the groups of these equations.*"

In what precedes we have supposed that we knew how to answer this question for the group Γ and all similar groups. The number of variables in Γ is less than those in G (since Γ only contains the variables of a single class), and the problem is therefore reduced to a simpler form, and may be considered as solved when we examine the case so far excluded; viz., the case when the characteristic equation corresponding to the substitution S has all of its roots equal—or, say, has but one root.

We will suppose (changing very slightly the previous notation) that the characteristic equation corresponding to the substitution A has the single root a. In order that we may have a new problem to solve it is necessary that the characteristic equation corresponding to each and every substitution in G shall have but one root; if, for example, there existed in G a substitution, say S, corresponding to which there was more than one root, then we might reason with S as we have already with A and effect the required reduction by the processes described above. We must therefore suppose that there is no substitution in G whose equation has more than one root. We will, in this case, first show that G cannot be prime, and will give the canonical form of its substitutions which shall bring this fact into evidence.

This being done, we will show how to ascertain whether or not the assumed hypothesis is admissible, and, if it is, give the canonical form of the substitutions of G. In case the hypothesis is not correct we will show either how to obtain directly a function-group containing less than n distinct functions, or a substitution, say S, corresponding to which the characteristic equation has several distinct roots, and so be conducted to the case already investigated.

Denote by $a, b, \ldots$ the single roots of the characteristic equations corresponding to the substitutions $A, B, \ldots$ We may write

$$A = a\mathfrak{A}, \quad B = b\mathfrak{B}, \quad \ldots$$

Here $a, b, \ldots$ represent substitutions which multiply all the variables by $a, b, \ldots$; and $\mathfrak{A}, \mathfrak{B}, \ldots$ substitutions whose determinant is unity and corresponding to each of which we have the characteristic equation

$$(s - 1)^n = 0.$$

We have now a group $\mathfrak{G}$ derived from $\mathfrak{A}, \mathfrak{B}, \ldots$ as G is derived from $A, B, \ldots$ Consider any substitution, say T, of G, and let

$$T = a\, b^\beta \ldots \mathfrak{A}^\alpha \mathfrak{B}^\beta \ldots;$$

then

$$\mathfrak{T} = \mathfrak{A}^\alpha \mathfrak{B}^\beta \ldots,$$

and this is a substitution contained in $\mathfrak{G}$. Let

$$(30) \qquad \varDelta = \begin{vmatrix} \lambda - s, & \mu \ldots\ldots \\ \lambda_1, & \mu_1 - s \ldots \\ . & . \quad . \quad . \quad . \end{vmatrix} = 0$$

be the characteristic equation corresponding to $\mathfrak{T}$. The roots of the characteristic equation corresponding to T are evidently the roots of $\varDelta = 0$ multiplied by the constant $(a^\alpha b^\beta \ldots)^n$; but by hypothesis this last equation has only one root, therefore $\varDelta = 0$ has only one root.

We will now show that *the single root of $\varDelta = 0$ is unity.* In order to do this, we will show that if this statement is true for any two substitutions $\mathfrak{S}$ and $\mathfrak{T}$, it is true for their product $\mathfrak{S}\mathfrak{T}$; then as we know that the proposition is true for the substitutions $\mathfrak{A}, \mathfrak{B}, \ldots,$ from which $\mathfrak{G}$ is derived, it must also be true for their products taken two and two, three and three, etc., and consequently for all the substitutions of $\mathfrak{G}$.

We may evidently suppose the variables $y, z, \ldots$ so chosen that the substitution $\mathfrak{S}$ shall be in the canonical form; viz., by the above hypothesis,

$$(31) \qquad \mathfrak{S} = \begin{vmatrix} y_1 \ldots y_r; & y_1 \ldots y_r \\ z_1 \ldots z_s; & z_1 + y_1 \ldots z_s + y_s \\ u_1 \ldots u_t; & u_1 + z_1 \ldots u_t + z_t \\ . \quad . \quad . & . \quad . \quad . \quad . \end{vmatrix}.$$

The substitution $\mathfrak{T}$ cannot, of course (with these variables), be supposed to be in its canonical form, so we may write

$$(32) \qquad \mathfrak{T} = \begin{vmatrix} y_1; & a_1 y_1 + \ldots + b_1 z_1 + \ldots + c_1 u_1 + \ldots \\ . & . \quad . \quad . \quad . \quad . \quad . \quad . \quad . \\ u_1; & d_1 y_1 + \ldots + e_1 z_1 + \ldots + f_1 u_1 + \ldots \\ . & . \quad . \quad . \quad . \quad . \quad . \quad . \quad . \end{vmatrix}.$$

From these last two equations we have (λ denoting an integer)

$$(33)\qquad \mathfrak{S}^\lambda\mathfrak{T} = \left| \begin{array}{l} y_1; \quad \left[a_1 + \lambda b_1 + \frac{\lambda(\lambda - 1)}{1 . 2} c_1 + \ldots\right] y_1 + \ldots \\ \quad +(b_1 + \lambda c_1 + \ldots) z_1 + \ldots + (c_1 + \ldots) u_1 + \ldots \\ \cdot \quad \cdot \quad \cdot \quad \cdot \quad \cdot \quad \cdot \quad \cdot \quad \cdot \quad \cdot \quad \cdot \\ u_1; \quad \left[d_1 + \lambda e_1 + \frac{\lambda(\lambda - 1)}{1 . 2} f_1 + \ldots\right] y_1 + \ldots \\ \quad +(e_1 + \lambda f_1 + \ldots) z_1 + \ldots + (f_1 + \ldots) u_1 + \ldots \\ \cdot \quad \cdot \quad \cdot \quad \cdot \quad \cdot \quad \cdot \quad \cdot \quad \cdot \quad \cdot \quad \cdot \end{array} \right| .$$

The characteristic equation corresponding to this substitution may be written in the form

$$(34)\qquad \varDelta = s^n + \phi s^{n-1} + \phi_1 s^{n-2} + \ldots + (-1)^n D = 0,$$

where D is the determinant of $\mathfrak{S}\mathfrak{T}$, and where $\phi, \phi_1, \ldots$ are integral functions; but since $\mathfrak{S}$ and $\mathfrak{T}$ are products formed with the substitutions $\mathfrak{A}, \mathfrak{B}, \ldots$ each of whose determinants is unity, it follows that $D = 1$. The single root of $\varDelta = 0$ will then be an nth root of unity, since the nth power of this root is (to sign près) the last term of equation (32)—*i.e.*, unity. We have then

$$(35)\qquad \varDelta = (s - \theta)^n = 0,$$

where θ is an nth root of unity. By comparing (34) and (35), we see that we must have

$$(36)\qquad \phi = -n\theta^{n-1}, \quad \phi_1 = \frac{n(n-1)}{1 . 2}\theta^{n-2}, \ \ldots,$$

and consequently

$$(37)\qquad \phi^n = (-n)^n, \quad \phi_1^n = \left[\frac{n(n-1)}{1 . 2}\right]^n, \ \ldots .$$

These equations are of finite degree in λ and are satisfied by hypothesis for all values of λ; they are therefore identities. The coefficients $\phi, \phi_1, \ldots$ are therefore constants, and $\varDelta = 0$ is independent of λ. Suppose now $\lambda = 0$; then by hypothesis again we have $\varDelta = (s - 1)^n$. It follows then that all the substitutions

$\mathfrak{S}\mathfrak{T} \ldots \mathfrak{S}^\lambda\mathfrak{T}$ have for the corresponding characteristic equation simply $(s-1)^n = 0$, and so finally every substitution in $\mathfrak{G}$ has for its characteristic equation $(s-1)^n = 0$.

We will now suppose the variables so chosen that the substitution $\mathfrak{A}$ is in its canonical form, and for simplicity we will assume

$$(38) \qquad \mathfrak{A} = \left| \begin{array}{l} y'_1,\ y'_2,\ y'_3,\ y'_4;\ \ y_1,\ y_2,\ y_3,\ y_4 \\ z_1,\ z_2,\ z_3 \qquad\quad ;\ \ z_1+y'_1,\ z_2+y'_2,\ z_3+y'_3 \end{array} \right|.$$

Any other substitution, say $\mathfrak{T}$, of $\mathfrak{G}$ may be written in the form

$$(39) \quad \mathfrak{T} = \left| \begin{array}{l} \cdot \quad \cdot \quad \cdot \quad \cdot \quad \cdot \quad \cdot \quad \cdot \quad \cdot \quad \cdot \\ y'_\rho;\ \ a_{\rho 1}y'_1 + \ldots + a_{\rho 4}y'_4 + b_{\rho 1}z_1 + \ldots + b_{\rho 3}z_3 \\ \cdot \quad \cdot \quad \cdot \quad \cdot \quad \cdot \quad \cdot \quad \cdot \quad \cdot \quad \cdot \\ z_\rho;\ \ c_{\rho 1}y_1 + \ldots + c_{\rho 4}y_4 + d_{\rho 1}z_1 + \ldots + d_{\rho 3}z_3 \\ \cdot \quad \cdot \quad \cdot \quad \cdot \quad \cdot \quad \cdot \quad \cdot \quad \cdot \quad \cdot \end{array} \right|.$$

We have now

$$(40) \quad \mathfrak{A}_\lambda\mathfrak{T} =$$

$$\left| \begin{array}{l} \cdot \quad \cdot \quad \cdot \quad \cdot \quad \cdot \quad \cdot \quad \cdot \quad \cdot \quad \cdot \quad \cdot \quad \cdot \quad \cdot \\ y_\rho;\ \ (a_{\rho 1}+\lambda b_{\rho 1})y_1 + \ldots + (a_{\rho 3}+\lambda b_{\rho 3})y_3 + a_{\rho 4}y_4 + b_{\rho 1}z_1 + \ldots \\ \cdot \quad \cdot \quad \cdot \quad \cdot \quad \cdot \quad \cdot \quad \cdot \quad \cdot \quad \cdot \quad \cdot \quad \cdot \quad \cdot \\ z_\rho;\ \ (c_{\rho 1}+\lambda d_{\rho 1})y_1 + \ldots + (c_{\rho 3}+\lambda d_{\rho 3})y_3 + c_{\rho 4}y_4 + d_{\rho 1}z_1 + \ldots \\ \cdot \quad \cdot \quad \cdot \quad \cdot \quad \cdot \quad \cdot \quad \cdot \quad \cdot \quad \cdot \quad \cdot \quad \cdot \quad \cdot \end{array} \right|.$$

The characteristic equation corresponding to this is obviously

$$(41) \qquad \left| \begin{array}{ll} a_{11}+\lambda b_{11}-s, & a_{12}+\lambda b_{12} \ldots\ldots \\ a_{21}+\lambda b_{21}, & a_{22}+\lambda b_{22}-s \ldots \\ \cdot \quad \cdot \quad \cdot \quad \cdot & \cdot \quad \cdot \quad \cdot \quad \cdot \end{array} \right| = 0,$$

which, as already shown, is independent of λ. In expanding this we find certain terms, viz., those contained in the expression

$$(42) \qquad \left| \begin{array}{lll} \lambda b_{11}-s, & \lambda b_{12}, & \lambda b_{13} \\ \lambda b_{21}, & \lambda b_{22}-s, & \lambda b_{23} \\ \lambda b_{31}, & \lambda b_{32}, & \lambda b_{33}-s \end{array} \right| s^4,$$

involving only the constants $b_{11}, b_{12}, \ldots, b_{33}$ and not cancelling with any other terms. The terms containing these constants b must therefore vanish identically, and so, equating to zero the coefficients of λs^6, $\lambda^2 s^5$, $\lambda^3 s^4$, we must have

$$b_{11} + b_{22} + b_{33} = 0; \tag{43}$$

$$\begin{vmatrix} b_{11}, & b_{12} \\ b_{21}, & b_{22} \end{vmatrix} + \begin{vmatrix} b_{22}, & b_{23} \\ b_{32}, & b_{33} \end{vmatrix} + \begin{vmatrix} b_{33}, & b_{31} \\ b_{13}, & b_{11} \end{vmatrix} = 0; \tag{44}$$

$$\begin{vmatrix} b_{11}, & b_{12}, & b_{13} \\ b_{21}, & b_{22}, & b_{23} \\ b_{31}, & b_{32}, & b_{33} \end{vmatrix} = 0. \tag{45}$$

We may remark here that since we have only the three independent variables z_1, z_2, z_3, we may, without altering the substitution $\mathfrak{A}$, replace y_1, y_2, y_3 by arbitrary linear functions of themselves, pro vided that we make an analogous change in z_1, z_2, z_3. By aid of this very obvious remark we may obtain a simpler form for the substitution $\mathfrak{T}$.

From (45) we can clearly determine three constants, say l_1, l_2, l_3, such that

$$\begin{aligned} & l_1(b_{11}z_1 + b_{12}z_2 + b_{13}z_3) \\ & + l_2(b_{21}z_1 + b_{22}z_2 + b_{23}z_3) \\ & + l_3(b_{31}z_1 + b_{32}z_2 + b_{33}z_3) = 0. \end{aligned} \tag{46}$$

Supposing then that l_1, l_2, l_3 are determined so as to satisfy (46); if now we replace one of the variables y_1, y_2, y_3 by

$$l_1y_1 + l_2y_2 + l_3y_3,$$

it is easy to see that the corresponding transform of this variable in the substitution $\mathfrak{T}$ will not contain z; if y_3 be the variable so replaced, the effect in $\mathfrak{T}$ will be the same as if the coefficients b_{31}, b_{32}, b_{33} were made zero. Suppose then that this operation has been performed and that we have

$$b_{31} = b_{32} = b_{33} = 0; \tag{47}$$

it follows now that (44) reduces to

(48) $$\begin{vmatrix} b_{11}, & b_{12} \\ b_{21}, & b_{22} \end{vmatrix} = 0.$$

By aid of this equation we can determine two constants m_1, m_2, such that

(49) $$m_1(b_{11}z_1 + b_{12}z_2) + m_2(b_{21}z_1 + b_{22}z_2) = 0;$$

if now we replace y_2 by $m_1y_1 + m_2y_2$, we will not affect the substitution but $\mathfrak{A}$, we will produce in $\mathfrak{T}$ the same result as if we made $b_{21} = b_{22} = 0$. This change effected, reduces (43) to $b_{11} = 0$. We see then, finally, that we may write

(50) $$b_{11} = b_{21} = b_{22} = b_{31} = b_{32} = b_{33} = 0.$$

Further, it is easy to show that if one of the coefficients b_{12}, b_{23} is zero, we may also make the other zero.

(1) Let $b_{12} = 0$: we can obviously find l_1 and l_2, such that $l_1b_{13} + l_2b_{23} = 0$. If b_{13} is not zero, then l_2 cannot be zero, and we may replace y_2 and z_2 by $l_1y_1 + l_2y_2$ and $l_1z_1 + l_2z_2$, which will not affect $\mathfrak{A}$, but will make the resulting coefficient to $\mathfrak{T}$ which is analogous to the original b_{23} equal to zero. If $b_{13} = 0$, we shall arrive at the same result by permuting the variables y_1, z_1 with y_2, z_2.

(2) Suppose $b_{23} = 0$: if b_{13} is not zero, we can cause the new coefficient in $\mathfrak{T}$ which is analogous to the original b_{12} to vanish by taking for new variables $b_{12}y_2 + b_{13}y_3$, $b_{12}z_2 + b_{13}z_3$ in place of y_3 and z_3; if $b_{13} = 0$, the same result will be obtained by permuting y_2, z_2 with y_3, z_3.

Denote now by

(51) $$\mathfrak{U}, = \begin{vmatrix} \cdot & \cdot & \cdot & \cdot & \cdot & \cdot & \cdot & \cdot & \cdot & \cdot \\ y_\rho; & a'_{\rho 1}y_1 + & \dots & + a'_{\rho 4}y_4 + b'_{\rho 1}z_1 + & \dots & + b'_{\rho 3}z_3 \\ \cdot & \cdot & \cdot & \cdot & \cdot & \cdot & \cdot & \cdot & \cdot & \cdot \\ z_\rho; & c'_{\rho 1}y_1 + & \dots & + c'_{\rho 4}y_4 + d'_{\rho 1}z_1 + & \dots & + d'_{\rho 3}z_3 \\ \cdot & \cdot & \cdot & \cdot & \cdot & \cdot & \cdot & \cdot & \cdot & \cdot \end{vmatrix},$$

any substitution of the group $\mathfrak{G}$; write, for brevity,

(52) $$b''_{\rho\sigma} = (a'_{\rho 1} + \mu b'_{\rho 1})b_{1\sigma} + (a'_{\rho 2} + \mu b'_{\rho 2})b_{2\sigma} + (a'_{\rho 3} + \mu b'_{\rho 3})b_{3\sigma} + a'_{\rho 4}b_{4\sigma} + b'_{\rho 1}d_{1\sigma} + b'_{\rho 2}d_{2\sigma} + b'_{\rho 3}d_{3\sigma}, \text{ etc.},$$

where μ is an indeterminate constant. The substitution $\mathfrak{T}\mathfrak{A}^\mu\mathfrak{T}$, $= \mathfrak{V}$, will be of the form

$$(53)\quad \mathfrak{V} = \left| \begin{array}{l} \cdot \quad \cdot \quad \cdot \quad \cdot \quad \cdot \quad \cdot \quad \cdot \quad \cdot \quad \cdot \quad \cdot \\ y_\rho ;\ a''_{\rho 1}y_1 + \ldots + a''_{\rho 4}y_4 + b''_{\rho 1}z_1 + \ldots + b''_{\rho 3}z_3 \\ \cdot \quad \cdot \quad \cdot \quad \cdot \quad \cdot \quad \cdot \quad \cdot \quad \cdot \quad \cdot \quad \cdot \\ z_\rho ;\ c''_{\rho 1}y_1 + \ldots + c''_{\rho 4}y_4 + d''_{\rho 1}z_1 + \ldots + d''_{\rho 3}z_3 \\ \cdot \quad \cdot \quad \cdot \quad \cdot \quad \cdot \quad \cdot \quad \cdot \quad \cdot \quad \cdot \quad \cdot \end{array} \right| .$$

As already shown, the coefficients b_{11}'', . . . , b_{33}'' must satisfy equations of the same form as (43), (44), and (45), and in particular

$$(54)\qquad b_{11}'' + b_{22}'' + b_{33}'' = 0.$$

Substituting here the values of b_{11}'', b_{22}'', and b_{33}'' from (52), and equating to zero the coefficients of μ, we have, on taking account of equations (50),

$$(55)\qquad b'_{21}b_{12} + b'_{31}b_{13} + b'_{32}b_{23} = 0.$$

This equation must necessarily be satisfied whatever be the substitution $\mathfrak{T}$ of the group $\mathfrak{G}$; it must therefore be satisfied when we replace the coefficients b'_{21}, b'_{31}, b'_{32} of $\mathfrak{T}$ by the corresponding coefficients b_{21}'', b_{31}'', b_{32}'' of $\mathfrak{V}$; consequently, whatever may be the value of μ, we must have

$$(56)\qquad b_{21}''b_{12} + b_{31}''b_{13} + b_{32}''b_{23} = 0.$$

If now in this equation we replace b_{21}'', b_{31}'', b_{32}'' by their values and equate to zero the terms multiplied by μ, we shall have, by aid of (50),

$$(57)\qquad b_{12}b_{23}b'_{31} = 0.$$

This equation can be satisfied in two different ways: first, by making

$$(58)\qquad b'_{31} = 0;$$

second, by making $b_{12}b_{23} = 0$. But we have shown that if either b_{12} or b_{23} is zero, the other may also be made zero, and so the equation $b_{12}b_{23} = 0$ is equivalent to

$$(59)\qquad b_{12} = 0, \quad b_{23} = 0,$$

and finally, from (55),

(60) $$b_{13} = 0.$$

It is now easy to establish the first of our theorems, viz., that under the assumed hypothesis as to the roots of the characteristic equations corresponding to the substitutions of the group $\mathfrak{G}$, this group cannot be prime. There are two cases to be considered according as b'_{31} is or is not equal to zero. Suppose first b'_{31} not zero. We know now that each of the quantities

$$\begin{matrix} b_{11}, & b_{12}, & b_{13}, \\ b_{21}, & b_{22}, & b_{23}, \\ b_{31}, & b_{32}, & b_{33} \end{matrix}$$

is zero, and consequently that the substitution $\mathfrak{T}$, which is any substitution whatever of the group $\mathfrak{G}$, gives as the transforms of y_1, y_2, y_3 linear functions of $y_1, \ldots, y_4$ only. Denote by $\mathfrak{T}, \mathfrak{T}' \ldots$ the different substitutions of $\mathfrak{G}$; by $Y, Y', \ldots$ the corresponding transforms of y_1. Each of these functions depends only on y_1, y_2, y_3, y_4, and consequently the function-group formed by the linear functions

$$\alpha Y + \alpha' Y' + \ldots$$

can contain at the most but 4 distinct functions. Let $Y_1, \ldots, Y_k$ denote these functions, where $k \leqq 4$.

Further, we remark that substitution in $\mathfrak{G}$ changes the functions of the series $Y, Y', \ldots$ among themselves. Suppose $\mathfrak{T}_1$ changes Y into Y_1; we know that $\mathfrak{T}$ changes y_1 into Y, and consequently $\mathfrak{T}\mathfrak{T}_1$ changes y_1 in Y_1; and since $\mathfrak{T}\mathfrak{T}_1$ is a substitution of the group $\mathfrak{G}$, it follows that Y_1 belongs to the series $Y, Y', \ldots$. The same reasoning of course applies to the general case, and in this case we see that the substitutions of $\mathfrak{G}$ transform into one another the functions of the function-group $\alpha Y + \alpha' Y' + \ldots$. The number of distinct functions $Y_1, \ldots, Y_k$ contained in this function-group being less than the number of variables, it follows that $\mathfrak{G}$ cannot be prime. The substitutions of $\mathfrak{G}$ can obviously be put in the form

(61) $$\left| \begin{matrix} Y_1 \ldots Y_k; & \alpha_1 Y_1 + \ldots + \alpha_k Y_k, \ldots, \beta_1 Y_1 + \ldots + \beta_k Y_k \\ Y_{k+1} \ldots & ; \gamma_1 Y_1 + \ldots + \gamma_k Y_k + \gamma_{k+1} Y_{k+1} + \ldots, \quad \ldots \end{matrix} \right|.$$

In the second case, suppose $b'_{31} = 0$, and let $Y, Y', \ldots$ be the transforms of y_3 by the substitutions of $\mathfrak{G}$. None of these can contain z_3, and consequently the function-group formed from $Y, Y', \ldots$ must contain a number of distinct functions $Y_1, \ldots, Y_k$ less in number than the variables $y_1, y_2, y_3, y_4, z_1, z_2, z_3$. The same sort of reasoning as above will conduct us to the form (61) of the substitutions in $\mathfrak{G}$.

We will suppose now that the substitutions of $\mathfrak{G}$ have all been brought to the form (61). Consider the characteristic equation corresponding to a substitution of $\mathfrak{G}$ of this form; the first member of this equation is evidently divisible by the first member of the characteristic equation corresponding to the substitution

$$(62) \quad | Y_1 \ldots Y_k;\ \alpha_1 Y_1 + \ldots + \alpha_k Y_k,\ \ldots,\ \beta_1 Y_1 + \ldots + \beta_k Y_k |.$$

But by hypothesis all the substitutions of $\mathfrak{G}$ have for the right-hand side of their characteristic equations a power of $s - 1$; it follows then that the right-hand side of the equation corresponding to (62) must also be a power of $s - 1$. The group $\mathfrak{G}_k$ formed by the partial substitutions (62) cannot therefore be prime if $k > 1$, and so, by a proper choice of the independent variables, the substitutions of this group can be put in the form

$$(63) \quad \left| \begin{matrix} Y_1 \ldots Y_l & ; & \phi_1(Y_1 \ldots Y_l), \ldots, & \phi_l(Y_1 \ldots Y_l) \\ Y_{l+1} \ldots Y_k; & & \phi_{l+1}(Y_1 \ldots Y_k), \ldots, & \phi_k(Y_1 \ldots Y_k) \end{matrix} \right|,$$

where $l < k$ and ϕ denotes a linear function of the quantities in parenthesis.

If $l > 1$, we can again so choose the variables $Y_1 \ldots Y_l$ that the partial substitutions

$$(64) \quad | Y_1 \ldots Y_l;\ \phi_1(Y_1 \ldots Y_l),\ \ldots,\ \phi_l(Y_1 \ldots Y_l) |$$

shall take the simpler form

$$(65) \quad \left| \begin{matrix} Y_1 \ldots Y_m & ; & \psi_1(Y_1 \ldots Y_m), \ldots, & \psi_m(Y_1 \ldots Y_m) \\ Y_{m+1} \ldots Y_l; & & \psi_{m+1}(Y_1 \ldots Y_l), \ldots, & \psi_l(Y_1 \ldots Y_l) \end{matrix} \right|,$$

where $m < l$. By continuing this process we can obviously so choose the independent variables Y that the first one, Y_1, shall, for each substitution in $\mathfrak{G}$, be changed into itself multiplied by a constant:

and since the characteristic equation corresponding to each substitution in $\mathfrak{G}$ is a power of $s-1$, it is obvious that this factor reduces to unity. We have then the theorem:

If each substitution in $\mathfrak{G}$ has $(s-1)^n = 0$ for its characteristic equation, there must exist at least one function of the variables $y_1 \ldots y_n$ which is not altered by any substitution.

There may obviously be more than one such function. Suppose $Y_1 \ldots Y_p$ are p distinct functions possessing this property; then the substitutions of $\mathfrak{G}$ can be put in the form

$$(66) \left| \begin{array}{l} Y_1 \ldots Y_p \quad ; \quad Y_1 \ldots Y_p \\ Y_{p+1} \ldots Y_n ; \alpha_1 Y_1 + \ldots + \alpha_n Y_n, \ldots, \beta_1 Y_1 + \ldots + \beta_n Y_n \end{array} \right| .$$

Now the first member of the characteristic equation of such a substitution is equal to the product of $(s-1)^p$ by the first member of the characteristic equation belonging to

$$(67) \mid Y_{p+1} \ldots Y_n ; \alpha_{p+1} Y_{p+1} \ldots + \alpha_n Y_n, \ldots, \beta_{p+1} Y_{p+1} \ldots + \beta_n Y_n \mid .$$

Substitutions of this form corresponding to the different substitutions of $\mathfrak{G}$ will then have $(s-1)^{n-p} = 0$ for their characteristic equation. We conclude, then, that we may so choose the variables as to make these substitutions of the form

$$(68) \left| \begin{array}{l} Y_{p+1} \ldots Y_{p+q} \quad ; \quad Y_{p+1} \ldots Y_{p+q} \\ Y_{p+q+1} \ldots Y_n ; \alpha'_{p+1} Y_{p+1} \ldots + \alpha'_n Y_n \ldots \beta'_{p+1} Y_{p+1} + \ldots + \beta'_n Y_n \end{array} \right| .$$

By a continuation of this process we see that all of the substitutions of $\mathfrak{G}$ can be put in the form

$$(69) \qquad \left| \begin{array}{l} Y_1 \ldots Y_p \quad ; \quad Y_1 \ldots Y_p \\ Y_{p+1} \ldots Y_q : \ Y_{p+1} + f_{p+1} \ldots Y_q + f_q \\ Y_{q+1} \ldots Y_r ; \ Y_{q+1} + f_{q+1} \ldots Y_r + f_r \\ \cdot \quad \cdot \quad \cdot \quad \cdot \quad \cdot \quad \cdot \quad \cdot \quad \cdot \end{array} \right| ,$$

where $f_{p+1} \ldots f_q$ are linear functions of $Y_1, \ldots, Y_p$; $f_{q+1} \ldots f_r$ are linear functions of $Y_1, \ldots, Y_q$; etc.

As every substitution in $\mathfrak{G}$ has now this form, the elementary substitutions $\mathfrak{A}, \mathfrak{B}, \ldots$, from which $\mathfrak{G}$ is derived must also have the form of (69).

We arrive now at the second part of our problem. So far we have assumed that all the substitutions in $\mathfrak{G}$ have unity as the single root of all their characteristic equations, and on this hypothesis have found the form (69) for each of the substitutions. We have now to show how, for any group $\mathfrak{G}$, we can ascertain whether or not the substitutions $\mathfrak{A}, \mathfrak{B}, \ldots$ are of this form. To do this we will endeavor by the method of indeterminate coefficients to find out whether or not there exist linear functions of the independent variables $y_1, \ldots, y_n$ which are unaltered by the substitutions $\mathfrak{A}, \mathfrak{B}, \ldots$ We will assume certain linear functions of $y_1, \ldots, y_n$ of the form, say,

$$\alpha_1 y_1 + \ldots + \alpha_n y_n,$$

and operate upon them by the substitutions $\mathfrak{A}, \mathfrak{B}, \ldots$ If the transforms of such functions are the same as the functions themselves, we will arrive at a series of linear homogeneous equations for the determination of the constants $\alpha_1, \ldots, \alpha_n$, the number of these equations being, of course, in general greater than n. It may happen that these equations can all be satisfied (as certain determinants may vanish) and still leave a number, say p, of the coefficients α arbitrary. If, however, we have p arbitrary constants, we will also have p functions $Y_1, \ldots, Y_p$ possessing the required property; viz., the transforms of $Y_1, \ldots, Y_p$ by each substitution in $\mathfrak{G}$ will be again $Y_1, \ldots, Y_p$. The next step is to ascertain whether or not there exist functions $Y_{p+1}, \ldots, Y_q$ whose transforms by the substitutions of $\mathfrak{G}$ are equal to themselves increased by linear functions of $Y_1, \ldots, Y_p$. We thus arrive at another system of linear homogeneous equations of the first degree, etc. If in continuing this process we never arrive at an incompatible system of linear homogeneous equations we may take $Y_1, \ldots, Y_p, Y_{p+1}, \ldots, Y_q, \ldots$ for the system of variables which shall throw the substitutions $\mathfrak{A}, \mathfrak{B}, \ldots$, and consequently all the substitutions of $\mathfrak{G}$, in the form (69). The transition to the substitutions $A = a\mathfrak{A}$, $B = b\mathfrak{B}, \ldots$ of the group G is at once effected by multiplying the transforms in $\mathfrak{G}$ by $a, b, \ldots$ Since now each of the substitutions of G multiplies each of the functions $Y_1, \ldots, Y_p$ by a constant factor, the problem proposed concerning this group is solved.

We will now take up the case where, in following the above indicated process, we arrive at a system of incompatible linear equations for the determination of the coefficients α; if this case presents itself, we know at once that $\mathfrak{G}$ contains at least one substitution whose characteristic equation has several distinct roots. To solve the problem which thus presents itself, we have to do one of two things; viz., we may either determine a function-group Ψ containing less than n distinct functions, or we may determine a substitution, say S, whose characteristic equation contains several distinct roots.

We will assume for our independent variables those which shall give the substitution $\mathfrak{A}$ in its canonical form, and, for simplicity, will take the special case above considered, where

$$(70) \qquad \mathfrak{A} = \left| \begin{matrix} y_1, y_2, y_3, y_4; & y_1, y_2, y_3, y_4 \\ z_1, z_2, z_3, \quad ; & z_1 + y_1, z_2 + y_2, z_3 + y_3 \end{matrix} \right| .$$

We will now determine the transforms of y_1 by $\mathfrak{A}$, $\mathfrak{B}$, . . . ; if among these transforms there are any which are linear functions of the others and of y_1, we will discard them; suppose the remaining distinct functions to be y_1', . . . , y_1^r. Operate now on y_1', . . . , y_1^r by $\mathfrak{A}$, $\mathfrak{B}$, . . . and obtain new functions, from which discard again all those that are linear functions of the others and of y_1, y_1', . . . , y_1^r; if, again, there remain new independent functions y_1^{r+1}, . . . , y_1^s, we will transform them by $\mathfrak{A}$, $\mathfrak{B}$, . . . and discard as before. By continuing this process, which is necessarily limited, since the total number of distinct functions cannot exceed the number of the independent variables y_1, y_2, y_3, y_4, we shall finally come to a point where every new function obtained by the operation of the substitutions $\mathfrak{A}$, $\mathfrak{B}$, . . . is a linear function of the preceding ones. The linearly distinct functions so obtained, say y_1, y_1', . . . are evidently changed into linear functions of themselves by $\mathfrak{A}$, $\mathfrak{B}$, . . . (and consequently by all the substitutions of $\mathfrak{G}$). The problem now divides into the two cases above mentioned. First: Suppose that none of the functions y_1, y_1', . . . contain any of the variables z_1, z_2, z_3; then the number of these functions is at most equal to the number of the functions y_1, y_2, y_3, y_4. We can at once proceed to the general case of n independent variables y, and so see at once that we can obtain a system of less than n functions such that the substitutions of $\mathfrak{G}$, and consequently those of G, transform these functions into linear functions of themselves without introducing any new functions. The function-group Ψ formed from y_1, y_1', . . . gives us the solution of our problem.

Secondly: Suppose that among the functions y_1, y_1', . . . there is one, say y_1^ρ, which contains z_1, z_2, z_3; we will then stop the preceding series of operations when we arrive at y_1^ρ, and by retracing our steps find the substitution, say $\mathfrak{T}$, which changes y_1 into y_1^ρ. This substitution will of course be of the form (39), and the coefficients b_{11}, . . . , b_{33} may or may not satisfy equations (43), (44), and (45). Suppose the coefficients b_{ij} do not satisfy these equations; we can then determine a substitution $\mathfrak{A}^\lambda\mathfrak{T}$ whose characteristic equation shall have several distinct roots. The first member of this equation will be the same as the first member of (41), and on being developed will be of the form

$$s^n + \phi s^{n-1} + \phi_1 s^{n-2} + \ldots = 0.$$

It is only necessary here to assign a value to λ which shall not satisfy equations (37). This can be done by a number of trials, at most equal to $n\tau$, where τ is the degree in λ of one of the coefficients ϕ, ϕ_1, . . .

If equations (43), (44), and (45) are satisfied, we can choose the independent variables so that equations (50) shall also be satisfied. If, further, we have $b_{12}b_{23} = 0$, we can, as shown above, so choose the variables that both b_{12} and b_{23} shall be zero, but b_{13} shall not be zero. We now proceed to determine the transforms of y_3 by the substitutions $\mathfrak{A}$, $\mathfrak{B}$, . . . , neglecting all transforms which are linear functions of the remaining ones and of y_3. Repeat this process on the new functions so obtained, and continue in the manner already several times described. We will finally arrive at a, necessarily limited, series of functions y_3, y_3', . . . , which the transformations of G will transform into linear functions of themselves. Suppose none of the functions y_3, y_3', . . . contain z_1; then their number is at most $= n - 1$, this being the number of independent variables when z_1 is omitted. We will thus have a system of less than n functions forming a function-group and such that the transformations of G simply interchange these functions among themselves.

Suppose now that one of the functions, say y_3^σ, contains z_1, and let $\mathfrak{U}$ be the transformation which changes y_3 into y_3^σ; then $\mathfrak{U}$ will be of the form (50), and obviously the coefficient b'_{31} will not be zero. Since neither b_{13} nor b'_{31} are zero, it follows that equations (55) and (57) cannot be simultaneously satisfied. If $\mathfrak{U}$ does not satisfy (55), then $\mathfrak{V} = \mathfrak{T}\mathfrak{A}^\mu\mathfrak{U}$ will be of the form (53), and we can determine μ so that (54) shall not be satisfied, and then determine λ so that the substitution $\mathfrak{A}^\lambda\mathfrak{V}$ shall have a characteristic equation other than $(s-1)^n = 0$. If $\mathfrak{U}$ does not satisfy (57), we will determine μ in such a way that (56), which is analogous to (55), shall not be satisfied, and then reason with $\mathfrak{V}$ as we have with $\mathfrak{U}$ in the previous case.

In what precedes we have shown how to determine whether or not the group G is prime, and, if it is not, how we can choose its variables so as to throw its substitutions in the form

$$(71)\quad \left| \begin{matrix} Y_1 \ldots Y_k & ; \phi_1(Y_1 \ldots Y_k), & \ldots, & \phi_k(Y_1 \ldots Y_k) \\ Y_{k+1} \ldots Y_n; & \phi_{k+1}(Y_1 \ldots Y_n), & \ldots, & \phi_n(Y_1 \ldots Y_n) \end{matrix} \right| .$$

If the group

$$(72)\quad | Y_1 \ldots Y_k; \ \phi_1(Y_1 \ldots Y_k), \ \ldots, \ \phi_k(Y_1 \ldots Y_k) |$$

formed by these partial substitutions is not prime, we can again so choose its variables as to throw its substitutions in the form $l < k$:

$$(73)\quad \left| \begin{matrix} Y_1 \ldots Y_l & ; \psi_1(Y_1 \ldots Y_l), & \ldots, & \psi_l(Y_1 \ldots Y_l) \\ Y_{l+1} \ldots Y_k; & \psi_{l+1}(Y_1 \ldots Y_k), & \ldots, & \psi_k(Y_1 \ldots Y_k) \end{matrix} \right| .$$

We can continue this process and finally get for the form of the substitutions in G

$$(74)\quad \left| \begin{matrix} Y_1 \ldots Y_m & ; \alpha_1 Y_1 + \ldots + \alpha_m Y_m, \ldots, \beta_1 Y_1 + \ldots + \beta_m Y_m \\ Y_{m+1} \ldots Y_n; & \gamma_1 Y_1 + \ldots + \gamma_n Y_n, \ldots, \delta_1 Y_1 + \ldots + \delta_n Y_n \end{matrix} \right| ,$$

the group formed by the partial substitutions

$$(75)\quad | Y_1 \ldots Y_m; \ \alpha_1 Y_1 + \ldots + \alpha_m Y_m, \ \ldots, \ \beta_1 Y_1 + \ldots + \beta_m Y_m |$$

being prime. If now the group formed by the substitutions

$$(76)\quad \left| \begin{matrix} Y_{m+1} \ldots Y_n; & Y_{m+1} Y_{m+1} + \ldots + \gamma_n Y_n, \ \ldots, \\ & \delta_{m+1} Y_{m+1} + \ldots + \delta_n Y_n \end{matrix} \right|$$

is not prime, we can choose the variables $Y_{m+1}, \ldots, Y_n$ in such a way as to throw these substitutions in the form

$$(77)\quad \left| \begin{array}{ll} Y_{m+1} \ldots Y_{m'}; & \alpha'_{m+1}Y_{m+1} + \ldots + \alpha'_{m'}Y_{m'}, \quad \ldots, \\ & \beta'_{m+1}Y_{m+1} + \ldots + \beta_{m'}Y_{m'} \\ Y_{m'+1} \ldots Y_n\ ; & \gamma'_{m+1}Y_{m+1} + \ldots + \gamma'_nY_n, \quad . \quad , \\ & \delta'_{m+1}Y_{m+1} + \ldots + \delta'_nY_n \end{array} \right|,$$

the partial substitutions

$$(78)\quad \left| \begin{array}{ll} Y_{m+1} \ldots Y_{m'}; & \alpha'_{m+1}Y_{m+1} + \ldots + \alpha'_{m'}Y_{m'}, \quad \ldots, \\ & \beta'_{m+1}Y_{m+1} + \ldots + \beta'_{m'}Y_{m'} \end{array} \right|$$

forming a prime group. By continuing this process we can finally throw all the substitutions of G in the form

$$(79)\quad \left| \begin{array}{l} Y_1 \ldots Y_m \quad ; \quad \alpha_1Y_1 + \ldots + \alpha_mY_m, \ \ldots \\ \qquad\qquad\qquad\quad \beta_1Y_1 + \ldots + \beta_mY_m \\ Y_{m+1} \ldots Y_{m'}; \quad \alpha'_{m+1}Y_{m+1} + \ldots \\ + \alpha'_{m'}Y_{m'} + f_{m+1}(Y_1 \ldots Y_m), \ \ldots, \ \beta'_{m+1}Y_{m+1} + \ldots \\ + \beta'_{m'}Y_{m'} + f_{m'}(Y_1 \ldots Y_m) \\ Y_{m'+1} \ldots Y_{m''}; \quad \alpha''_{m'+1}Y_{m'+1} + \ldots \\ + \alpha''_{m''}Y_{m''} + f_{m'+1}(Y_1 \ldots Y_{m'}), \ \ldots, \ \beta''_{m'+1}Y_{m'+1} + \ldots \\ + \beta''_{m''}Y_{m''} + f_{m''}(Y_1 \ldots Y_{m'}) \\ . \quad . \quad . \quad . \quad . \quad . \quad . \quad . \quad . \quad . \quad . \quad . \end{array} \right|,$$

the partial substitutions

$$(80)\quad \left| \begin{array}{ll} Y_1 \ldots Y_m \quad ; & \alpha_1Y_1 + \ldots + \alpha_mY_m, \quad \ldots, \\ & \beta_1Y_1 + \ldots + \beta_mY_m \end{array} \right|$$

$$(81)\quad \left| \begin{array}{ll} Y_{m+1} \ldots Y_{m'}; & \alpha'_{m+1}Y_{m+1} + \ldots + \alpha'_{m'}Y_{m'}, \quad \ldots, \\ & \beta'_{m+1}Y_{m+1} + \ldots + \beta'_{m'}Y_{m'} \end{array} \right|$$

. . . .

forming prime groups.

We can now show how to determine all possible function-groups which are such that the substitutions of G simply change among themselves the functions of a given function-group. Denote by F such a function-group. We will say that F is of the first class if the functions of which it is composed depend only on the variables $Y_1, \ldots, Y_m$; of the second class if some of its functions depend on the variables $Y_{m+1}, \ldots, Y_{m'}$ without containing any of the following variables, etc. Since the group formed by the substitutions (80) is prime, there can exist but one function-group of the first class which will be formed by all the linear combinations of $Y_1 \ldots Y_m$.

We will now give the means of determining the function-groups of class k when we know all those of classes inferior to k. Knowing the function-group of the first class, we can then proceed to build up all of the function-groups of higher classes. To fix the ideas, suppose $k = 3$, and denote by $F, F', \ldots$ the different function-groups of the first and second classes, and by Φ one of the unknown function-groups of the third class. Denote by $P, P_1, \ldots$ linear functions of $Y_{m'+1}, \ldots, Y_{m''}$; by $Q, Q_1, \ldots$, linear functions of $Y_1, \ldots, Y_{m'}$; then the different functions of Φ will be of the form $P + Q$, $P_1 + Q_1$, ... Now the substitutions of G (when G is in the form (79)) transform among themselves the functions $P + Q$, $P_1 + Q_1, \ldots$. In order that this may be so it is evidently necessary that the substitutions

$$(82) \quad \left| \begin{array}{l} Y_{m'+1} \ldots Y_{m''}; \quad \alpha''_{m'+1} Y_{m'+1} + \ldots + \alpha''_{m''} Y_{m''}, \\ \qquad \ldots, \quad \beta''_{m'+1} Y_{m'+1} + \ldots + \beta''_{m''} Y_{m''} \end{array} \right|$$

transform among themselves the partial functions $P, P_1, \ldots$. But by hypothesis these substitutions form a prime group, say Γ, and consequently the function-group formed by $P, P_1, \ldots$ will contain all the linear functions of $Y_{m'+1}, \ldots, Y_{m''}$, and in particular will contain $Y_{m'+1}$. It follows then that Φ will contain a function of the form

$$(83) \qquad f = Y_{m'+1} + Q = Y_{m'+1} + \lambda_1 Y_1 + \ldots + \lambda_{m'} Y_{m'}.$$

Form now the transforms $f', f'', \ldots$ of this function by the substitutions $A, B, \ldots$, considering $\lambda_1, \ldots, \lambda_{m'}$ as indeterminates, and discard all transforms which are linear functions of the remaining ones and of f. We will continue this process in the manner which has already been several times described, and finally arrive at a series of independent functions $f, f', \ldots, f^t$. That the series of operations which give rise to new functions which are linearly independent of the preceding ones is limited is easy to see; for the transforms which are obtained by the successive substitutions are linear functions of $Y_{m'+1}, \ldots, Y_{m''}$ and the m'^2 products of $\lambda_1 Y_1, \ldots, \lambda_{m'} Y_{m'}$. The number of the functions $f, f', \ldots, f^t$ is thus at most $= m'' - m' + m'^2$. By combining $f, f', \ldots, f^t$ linearly we shall obtain a function-group whose functions are transformed the one into the other by all the substitutions of G. These functions have the forms $P + Q, P' + Q', \ldots, P^t + Q^t$, where $P, P', \ldots, P^t$ are linear functions of $Y_{m'+1}, \ldots, Y_{m''}$, and $Q, Q', \ldots, Q^t$ are similar functions of $Y_1, \ldots, Y_{m'}$. The substitutions of Γ permute among themselves the functions of the function-group formed by the linear combinations of $P, P', \ldots, P^t$; but Γ is prime by hypothesis. Therefore, among the functions $P, P', \ldots, P^t$ which are formed by aid of the $m'' - m'$ variables $Y_{m'+1}, \ldots, Y_{m''}$, there will be $m'' - m'$ distinct functions.

Suppose that $P, P', \ldots, P^{m''-m'-1}$ are these distinct functions, and let $\phi = \mathfrak{P} + \mathfrak{Q}$ be any one of the functions of the function-group Φ. The function $\mathfrak{P}$ of the variables $Y_{m'+1}, \ldots, Y_{m''}$ can be written as a linear function of the $m'' - m'$ distinct functions $P, P', \ldots, P^{m''-m'-1}$.

For example, let

$$\mathfrak{P} = dP + \ldots + d^{m''-m'-1} P^{m''-m'-1};$$

then evidently

$$\phi = df + \ldots + d^{m''-m'-1} f^{m''-m'-1} + \mathfrak{Q}_1,$$

where $\mathfrak{Q}_1$ depends only on the variables $Y_1, \ldots, Y_{m'}$. The function-group Φ will now be obtained by combining the $m'' - m'$ functions $P, \ldots, P^{m''-m'-1}$ with the functions $\mathfrak{Q}_1, \mathfrak{Q}_2, \ldots$ which depend only on the variables $Y_1, \ldots, Y_{m'}$. It may happen that all the functions $\mathfrak{Q}_1, \mathfrak{Q}_2, \ldots$ reduce to zero; but if they do not, they will form a function-group whose elements are permuted among themselves by the substitutions of G. The function-group so formed will then be one of the assumed function-groups $F, F', \ldots$ Among the functions $\mathfrak{Q}_1, \mathfrak{Q}_2, \ldots$ it is easy to determine those which belong to the function-group derived from $f, f', \ldots, f^t$, this function-group being obviously a minor of Φ. In order to obtain the required functions, we remark that by subtracting properly chosen linear functions of $f, f', \ldots, f^{m''-m'-1}$ from $f^{m''-m'}, \ldots, f^t$ we obtain functions $\mathfrak{Q}^{m''-m'}, \ldots, \mathfrak{Q}^t$ which contain only the variables $Y_1, \ldots, Y_{m'}$; and further, the coefficients of the variables in these functions will be linear in $\lambda_1, \ldots, \lambda_{m'}$. If we wish to have $\mathfrak{Q}_1, \mathfrak{Q}_2, \ldots$ reduce to zero, it will be *à fortiori* necessary that the functions $\mathfrak{Q}^{m''-m'}, \ldots, \mathfrak{Q}^t$ which form a part of the series $\mathfrak{Q}_1, \mathfrak{Q}_2, \ldots$ shall vanish. In order that this may happen, it is obviously necessary that all of the coefficients of $Y_1, \ldots, Y_{m'}$ shall vanish in each of the functions $\mathfrak{Q}^{m''-m'}, \ldots, \mathfrak{Q}^t$; we will then have for $\lambda_1, \ldots, \lambda_{m'}$ a system of linear equations in general greater in number than m'. If these equations are not incompatible, then to each system of values of the λ's which satisfy them there will correspond a function-group Φ derived from the functions $f, f', \ldots, f^{m''-m'-1}$.

If we require the functions $\mathfrak{Q}_1, \mathfrak{Q}_2, \ldots$ to form by their linear combinations a function-group included among those already known, *i.e.*, $F, F', \ldots$, it is necessary that $\mathfrak{Q}^{m''-m'-1}, \ldots, \mathfrak{Q}^t$ shall belong to this function-group, say F. Now let $\chi, \chi', \ldots$ be the distinct functions in F; all the functions of F will then be of the form

$$\mu\chi + \mu'\chi' + \ldots,$$

and we must have

$$\begin{aligned} \mathfrak{Q}^{m''-m'} &= \mu\chi + \mu'\chi' + \ldots, \\ \mathfrak{Q}^{m''-m'+1} &= \mu_1\chi + \mu_1'\chi' + \ldots, \\ &\cdot \quad \cdot \quad \cdot \quad \cdot \quad \cdot \end{aligned}$$

Replacing in these equations $\mathfrak{Q}^{m''-m'}, \ldots, \mathfrak{Q}^t, \chi, \chi', \ldots$ by their values in $Y_1, \ldots, Y_{m'}$ and equating to zero the coefficients of each variable, we shall have a system of linear equations for the determination of $\lambda_1, \ldots, \lambda_{m'}, \mu, \mu', \ldots, \mu_1, \mu_1', \ldots, \ldots$ If these equations are incompatible, there will be no function-group Φ of the third class containing F; if the equations are compatible, then to each system of solutions there will correspond a function-group Φ. We may remark, however, that if in the expression

$$f = Y_{m'+1} + \lambda_1 Y_1 + \ldots + \lambda_{m'} Y_{m'}$$

$\lambda_1, \ldots, \lambda_{m'}$ have had values given them which permit us to determine a function-group Φ of which F is a minor, we can obtain an infinite number of such systems of values of these constants each of which will give rise to the same function-group by simply considering its functions in the form

$$f + \nu\chi + \nu'\chi' + \nu''\chi'' + \ldots,$$

where $\nu, \nu', \nu'', \ldots$ are arbitrary constants.

CHAPTER XII.

EQUATIONS WHOSE GENERAL INTEGRALS ARE RATIONAL. HALPHEN'S EQUATIONS.

A SPECIAL class of Fuchs's regular equations, that is, equations all of whose integrals are regular, is the class of equations all of whose integrals are algebraic, and a still more special class is that in which the general integral is rational. The investigation of equations whose general integrals are algebraic is reserved for Vol. II, but a brief account will be given here of the equation whose general integral is rational. Write the equation in the form

$$(1) \qquad P(y) = \frac{d^n y}{dx^n} + P_1 \frac{d^{n-1} y}{dx^{n-1}} + P_2 \frac{d^{n-2} y}{dx^{n-2}} + \ldots + P_n y = 0,$$

and let $x_1, x_2, \ldots, x_\rho$ denote its finite critical points; then, since the equation is to be in Fuchs's form, we have

$$P_k = \frac{F_k(x)}{[\psi(x)]^k},$$

where F_k is a polynomial in x of degree $k(\rho - 1)$ at most, and

$$\psi(x) = (x - x_1)(x - x_2) \ldots (x - x_\rho).$$

As by hypothesis the integrals of (1) can only have poles as critical points, it follows that the roots of the indicial equation for each of the critical points $x_1, x_2, \ldots, x_\rho$ must all be integers, and, further, that no logarithms can appear in the expressions for the integrals.

For any given equation we can find at once whether the first of these conditions is or is not satisfied by simply forming the indicial equation corresponding to each critical point and obtaining its roots. That the second condition may also be satisfied the equation must be such that equations (47) of Chapter IV are satisfied for all the groups of integrals belonging to each finite singular point.

Supposing all these conditions fulfilled, we see that the general integral is uniform throughout the plane since its singular points at a finite distance are all poles. To show that it is not only uniform but also rational, it is only necessary to show that the point infinity is also a pole. To show this we form the indicial equation for the point $x = \infty$. Divide the roots of this indicial equation into groups, the roots in each group differing from each other only by integers. Suppose the smallest root in each group to be denoted by α, α', . . . respectively, and the number of roots in each group to be denoted by β, β', . . . respectively. For very large values of x the general integral will be of the form

$$\frac{1}{x^{\alpha}}\left[\phi_0 + \phi_1 \log\frac{1}{x} + \phi_2 \log^2\frac{1}{x} + \ldots + \phi_{\alpha-1} \log^{\alpha-1}\frac{1}{x}\right]$$

$$+\frac{1}{x^{\alpha'}}\left[\phi_0' + \phi_1' \log\frac{1}{x} + \phi_2' \log^2\frac{1}{x} + \ldots + \phi'_{\alpha'-1} \log^{\alpha'-1}\frac{1}{x}\right]$$

$$+ \ldots \ldots \ldots \ldots,$$

the functions ϕ, ϕ', ϕ'', . . . being series going according to ascending integral powers of $\frac{1}{x}$. As already seen, however, this expression must be uniform, and so the logarithms must disappear and the exponents α, α', . . . must be integers. It follows, therefore, that the general integral has the point infinity as a pole, and so is a rational function of the form $\frac{P}{Q}$, P and Q being polynomials in x. These polynomials are very readily found; the denominator Q is known at once, since from the differential equation and the various indicial equations we know the finite poles of the general integral and their respective orders of multiplicity. Suppose the poles $x_1, x_2, \ldots, x_\rho$ to be of orders of multiplicity $\alpha_1, \alpha_2, \ldots, \alpha_\rho$ respectively; then obviously

$$Q = A(x - x_1)^{\alpha_1}(x - x_2)^{\alpha_2} \ldots (x - x_\rho)^{\alpha_\rho},$$

where A is a constant. The order of multiplicity of $x = \infty$ is known from the development according to powers of $\frac{1}{x}$, and so the degree of the numerator P is known; to find its coefficients we have simply to identify the development of $\frac{P}{Q}$ according to powers of $\frac{1}{x}$ with the corresponding development furnished by the differential equation itself. In connection with the preceding the reader is referred to a note by Mittag-Leffler in the Comptes Rendus, vol. xc. p. 218.

HALPHEN'S EQUATIONS.

A rather more general class of equations than the preceding has been studied by Halphen. Consider those equations whose integrals are regular and *uniform* in every region of the plane that does not contain the point x_1. The other critical points, $x_2, x_3, \ldots, x_\rho$, must now be merely poles of the integrals, and so the roots of the indicial equations corresponding to these points must all be integers; and further, no logarithms can appear in the developed forms of the corresponding integrals. Let us suppose these conditions all satisfied for the points $x_2, x_3, \ldots, x_\rho$; we can now find the general integral. Consider the region of the point x_1, and let $y_0, y_1, \ldots, y_k$ denote one of the groups of the system of fundamental integrals belonging to this point. If r denote the corresponding root of the indicial equation, we have

$$y_0 = (x - x_1)^r u_0,$$

$$y_1 = (x - x_1)^r [\theta_1 u_0 + u_1],$$

$$\cdot \quad \cdot \quad \cdot \quad \cdot \quad \cdot \quad \cdot \quad \cdot \quad \cdot \quad \cdot$$

$$y_k = (x - x_1)^r [\theta_k u_0 + \theta_{k-1} u_1 + \ldots + u_k],$$

where $u_0, u_1, \ldots, u_k$ are uniform in the region of x_1, and

$$\theta_1 = \frac{\log(x - x_1)}{2\pi i}, \ldots, \theta_k = \frac{\theta_1(\theta_1 - 1) \ldots (\theta_1 - k + 1)}{\underline{k|}}.$$

We can make a further assertion concerning the functions u, viz., they are *rational* functions; in fact, the points $x_2, x_3, \ldots, x_\rho$ being ordinary points for the functions

$$(x - x_1)^{-r} \quad \text{and} \quad \log(x - x_1),$$

and poles for $y_0, y_1, \ldots, y_k$, must also be poles for $u_0, u_1, \ldots, u_k$; these functions are therefore uniform not only in the region of x_1, but throughout the plane. Again, the functions

$$(x - x_1)^{-r}, = \frac{1}{x^r}\left(1 - \frac{x_1}{x}\right)^{-r},$$

and

$$\log(x - x_1), = -\log\frac{1}{x} + \log\left(1 - \frac{x_1}{x}\right),$$

are regular expressions for $x = \infty$; the same is true for $y_0, y_1, \ldots, y_k$, and consequently for $u_0, u_1, \ldots, u_k$. These two properties of the functions u taken together show that they are rational functions, say

$$u_0 = \frac{P_0}{Q}, \quad u_1 = \frac{P_1}{Q}, \quad \ldots, \quad u_k = \frac{P_k}{Q}.$$

The polynomials P and Q are found just as in the preceding case; we know the poles $x_2, x_3, \ldots, x_\rho$ and their respective orders of multiplicity, and so the denominator Q is formed at once; a development of the general integral for very great values of x will give us the degrees of the numerators $P_0, P_1, \ldots, P_k$; to obtain their coefficients we have only to substitute the preceding expressions in the differential equation and identify the result with zero.

Another very interesting class of equations also due to Halphen * is the class where the general integral is of the form

$$y = c_1 e^{a_1 x} f_1(x) + c_2 e^{a_2 x} f_2(x) + \ldots + c_n e^{a_n x} x f_n(x),$$

where $f_1, f_2, \ldots, f_n$ are rational fractions. Halphen's investigation involves certain properties of differential equations whose coefficients are doubly periodic functions of x. The following investigation is due to Jordan:

Consider the differential equation

$$(2) \qquad P(y) = P_0\frac{d^n y}{dx^n} + P_1\frac{d^{n-1}y}{dx^{n-1}} + P_2\frac{d^{n-2}y}{dx^{n-2}} + \ldots + P_n y = 0,$$

* Comptes Rendus, vol. 101, p. 1238.

where P_0, P_1, . . . , P_n are polynomials the degree of any one of which is at most equal to the degree of the first one, P_0; this is of course equivalent to saying that the developments of

$$\frac{P_1}{P_0}, \ \frac{P_2}{P_0}, \ \ldots, \ \frac{P_n}{P_0}$$

according to decreasing powers of x contain no positive powers of x. We will suppose that the integral of (2) contains as critical points at a finite distance only poles; the preceding considerations will of course enable us to make sure of this fact in any particular case.

Write now

$$(3) \qquad y = Rt,$$

where R is a rational function of x. Equation (2) now takes the form

$$(4) \quad P_0R\frac{d^nt}{dx^n} + \left.\begin{matrix} nP_0R' \\ + P_1R \end{matrix}\right| \frac{d^{n-1}t}{dx^{n-1}} + \left.\begin{matrix} \frac{n(n-1)}{2}P_0R'' \\ +(n-1)\ P_1R' \\ + \qquad P_2R \end{matrix}\right| \frac{d^{n-2}t}{dx^{n-2}} + \ldots = 0,$$

and this is of the same form as (2); for, after clearing of fractions, its coefficients will be rational polynomials in x, and its integrals possess, by (3), only polar singularities; finally, if we admit that in the development of R according to descending powers of x the first term is Ax^p, then the first terms in the developments of R', R'', . . . will be, to a constant factor près, x to the powers $p-1$, $p-2$, . . . respectively. It is easy to see, then, that after dividing by P_0R the coefficients of (4) can contain in their developments no positive powers of x; for example, take the coefficient of $\frac{d^{n-1}t}{dx^{n-1}}$, viz.,

$$\frac{nP_0R' + P_1R}{P_0R}, \quad = n\frac{R'}{R} + \frac{P_1}{P_0};$$

from what has been said it is clear that no positive power of x can appear in the development of this.

A particular form of the preceding transformation will now be applied to equation (2); we can of course determine in advance the poles, say $x_1, x_2, \ldots, x_\rho$, of the general integral of this equation, and also their respective degrees of multiplicity; say these are $\mu_1, \mu_2, \ldots, \mu_\rho$. Now transform (2) by the relation

$$y = \frac{t}{(x-x_1)^{\mu_1}(x-x_2)^{\mu_2}\ldots(x-x_\rho)^{\mu_\rho}};$$

the result of the transformation, say

$$(5) \qquad Q_0\frac{d^nt}{dx^n} + Q_1\frac{d^{n-1}t}{dx^{n-1}} + \ldots + Q_nt = 0,$$

is of the same type as (2), but its integrals have no poles. Again, make

$$(6) \qquad t = e^{\lambda x}v;$$

the new transformed equation, viz.,

$$(7) \qquad Q_0\frac{d^nv}{dx^n} + \left.\begin{matrix} n\lambda Q_0 \\ + \quad Q_1 \end{matrix}\right| \frac{d^{n-1}v}{dx^{n-1}} + \ldots + \left.\begin{matrix} \lambda^n Q_0 \\ + \lambda^{n-1}Q_1 \\ + \ldots \\ + Q_n \end{matrix}\right| v = 0,$$

or

$$(8) \qquad R_0\frac{d^nv}{dx^n} + R_1\frac{d^{n-1}v}{dx^{n-1}} + \ldots + R_nv = 0,$$

is obviously of the same type as the original; we will suppose λ so determined that the coefficient of the term of highest degree in R_n is made to vanish; R_n will then be a polynomial of lower degree than R_0.

Denote by $\xi_1, \xi_2, \ldots$ the roots of $R_0 = 0$; by decomposition into partial fractions we have, remembering that the degree of R_k is at most equal to the degree of R_0,

$$(9) \qquad \frac{R_k}{R_0} = A_k + \sum_{k,l}\frac{B_{ikl}}{(x-\xi_i)^l},$$

where A and B are constants, and in particular $A_n = 0$. The points ξ being ordinary points, in the region of which the integrals are regular, the index l can, in the enumeration, only take the values 1, 2, . . . , k.

The indicial equation relative to ξ_i is

$$(10)\quad r(r-1)\ldots(r-n+1)+B_{i11}r(r-1)\ldots(r-n+2) + B_{i22}r(r-1)\ldots(r-n+3)+\ldots=0;$$

the sum of the roots of this is

$$(11)\qquad \frac{n(n-1)}{2}-B_{i11}.$$

Now since ξ_i is an ordinary point for the integrals, these roots are necessarily unequal non-negative integers; the least values which they can have, therefore, are given by the series 0, 1, 2, . . . , $n-1$; their sum is, therefore, at least equal to

$$0+1+2+\ldots+n-1=\frac{n(n-1)}{2}.$$

It follows from this and (11) that B_{i11} is either zero or a negative integer, and *à fortiori* the sum

$$(12)\qquad S=\Sigma B_{i11}$$

taken over all the points ξ is zero or a negative integer.

Let us suppose first that R_n is not zero, and take $v'=\frac{dv}{dx}$ for a new variable. The equation in v thus becomes

$$(13)\qquad R_0\frac{d^{n-1}v'}{dx^{n-1}}+R_1\frac{d^{n-2}v'}{dx^{n-2}}+\ldots+R_nv=0;$$

differentiating this, then eliminating v between (13) and the new equation, we have

$$(14)\quad R_0R_n\frac{d^nv'}{dx^n}+[(R_0'+R_1)R_n-R_0R_n']\frac{d^{n-1}v'}{dx^{n-1}}+\ldots + [(R'_{n-1}+R_n)R_n-R_{n-1}R_n']v'=0.$$

This is obviously of the same type as the equation in v; the ratio of the first two coefficients, which in the v-equation is $\frac{R_1}{R_0}$, is now

$$\frac{R_1}{R_0} + \frac{R_0'}{R_0} - \frac{R_n'}{R_n}.$$

Now let

$$R_0 = (x - \xi_1)^{\alpha_1}(x - \xi_2)^{\alpha_2} \ldots,$$

$$R_n = (x - \eta_1)^{\beta_1}(x - \eta_2)^{\beta_2} \ldots;$$

we have then

$$\frac{R_0'}{R_0} = \frac{\alpha_1}{x - \xi_1} + \frac{\alpha_2}{x - \xi_2} + \ldots,$$

$$\frac{R_n'}{R_n} = \frac{\beta_1}{x - \eta_1} + \frac{\beta_2}{x - \eta_2} + \ldots;$$

the sum S' formed for the v'-equation in the same way that S was formed for the v-equation is now obviously

(15) $$S' = S + \Sigma\alpha - \Sigma\beta.$$

Since $\Sigma\alpha$, the degree of R_0, is by hypothesis greater than $\Sigma\beta$, the degree of R_n, it follows that S' is greater than S. In like manner, if we form a v''-equation in the same way as we formed the v'-equation, we should find $S'' > S'$. If we could continue this process indefinitely, we would form an unlimited series of increasing integers $S, S', S'', \ldots$, none of which, however, are positive, which is absurd. It must be then that there is an equation in the series the coefficient of whose last term is zero. Suppose this to be the case for the v'''-equation; this equation admits a constant as one of its integrals, and so the equation in v admits a polynomial $\pi(x)$ of degree m as an integral, and finally the t-equation has a particular integral $e^{\lambda x}\pi(x)$. Make now

$$t = e^{\lambda x}\pi \int t_1 dx;$$

t_1 satisfies an equation of order $n - 1$ which is of the same type as the t-equation. This new equation therefore admits a particular integral of the form $e^{\lambda_1 x}\pi_1(x)$, $\pi_1(x)$ being a polynomial. Again, write

$$t_1 = e^{\lambda_1 x}\pi_1 \int t_2 dx,$$

and continue the above process. We of course come at last to an equation of the first order whose integral is of the form

$$t_{n-1} = c_n e^{\lambda_{n-1} x} \pi_{n-1}(x),$$

where c_n is an arbitrary constant. The general value of t is now immediately seen; we have merely to retrace our steps from this last equation, performing successively the indicated integrations. As we know how to effect the integrations, we can see immediately what the final form of t is; it is, viz.,

$$t = \Sigma c_k e^{\alpha_k x} \Psi_k,$$

the c's being arbitrary constants and the Ψ's polynomials. Now, since

$$y = \frac{t}{(x - x_1)^{\mu_1}(x - x_2)^{\mu_2} \ldots (x - x_\rho)^{\mu_\rho}},$$

we have finally, for the general integral of the equation in y,

(16) $$y = c_1 e^{\alpha_1 x} f_1(x) + c_2 e^{\alpha_2 x} f_2(x) + \ldots + c_n e^{\alpha_n x} f_n(x),$$

$f_1, f_2, \ldots, f_n$ being rational fractions.

Reciprocally, every differential equation whose general integral is of this form belongs to the type considered. To show this eliminate the constants c_k between (16) and its derivatives, and suppress the common exponential factors; we will thus obtain an equation with rational coefficients which, by clearing of fractions, can be made integral. Suppose the equation is then

$$P_0 \frac{d^n y}{dx^n} + P_1 \frac{d^{n-1} y}{dx^{n-1}} + \ldots + P_n y = 0.$$

As we know, its integral has, at a finite distance, only polar singularities. It remains now to show that the degrees of $P_1, P_2, \ldots, P_n$ are not greater than that of P_0. This last, however, will be left as an exercise for the student. Halphen, in his celebrated "*Mémoire sur la réduction des équations différentielles linéaires aux formes intégrables*,"* and also in his paper in the Comptes Rendus, gives the

* *Savants Étrangère*, t. xxviii. pp. 111, 180, 273.

following three examples of this class of equations, viz. (n is an integer throughout):

$$\text{(I)} \qquad \frac{d^2y}{dx^2} - \left[\frac{n(n+1)}{x^2} + \alpha\right] = 0,$$

which is a very well known equation;

$$\text{(II)} \qquad \frac{d^3y}{dx^3} + \frac{1-n^2}{x^2}\frac{dy}{dx} - \left[\frac{1-n^2}{x^3} + \alpha\right] y = 0,$$

—in this n must be prime to 3;

$$\text{(III)} \quad \frac{d^4y}{dx^4} - \frac{2n(n+1)}{x^2}\frac{d^2y}{dx^2} + \frac{4n(n+1)}{x^3}\frac{dy}{dx} + \left[\frac{n(n+1)(n+3)(n-2)}{x^4} + \alpha\right] y = 0.$$

In the Comptes Rendus paper the following examples are also given, viz.:

$$\text{(IV)} \qquad \frac{d^3y}{dx^3} - \frac{2(n+1)}{x}\frac{d^2y}{dx^2} + \left(\frac{6n}{x^2} - \alpha\right)\frac{dy}{dx} + \frac{2\alpha}{x} y = 0;$$

this has one solution $y = \alpha x^2 - 2(2n - 1)$, and two solutions of the form

$$e^{\pm\sqrt{\alpha}x} f(x),$$

where f is rational.

$$\text{(V)} \quad \frac{d^2y}{dx^2} - \frac{2x}{x^2-1}\frac{dy}{dx} - \left[\frac{2a}{x^2-1} + \frac{n(n+1)}{x^2} + (a-n)(a+n+1)\right] y = 0;$$

for this, supposing n positive, there are two solutions of the form

$$\frac{e^{ax}}{x^n} f(x),$$

where $f(x)$ is a polynomial of degree $n + 1$.

When the general equation is restricted to have the single critical point $x = 0$ in the region of which the integrals belong to the exponents 0, 1, 2, . . . , $n-2$, n, Halphen shows that the form of the equation is

$$\frac{d^n y}{dx^n}+\left(A_1-\alpha-\frac{1}{x}\right)\frac{d^{n-1}y}{dx^{n-1}}+\left(A_2-A_1\alpha-\frac{A_1}{x}\right)\frac{d^{n-2}y}{dx^{n-2}}+\ldots$$
$$+\left(A_{n-1}-A_{n-2}\alpha-\frac{A_{n-2}}{x}\right)\frac{dy}{dx}-\left(A_{n-1}\alpha+\frac{A_{n-1}}{x}\right)y=0.$$

The solutions of this are exponentials e^{ax}, where a is a root of the equation

$$f(a)=a^{n-1}+A_1a^{n-2}+\ldots+A_{n-1}=0;$$

and the remaining solution is

$$y=e^{\alpha x}\left[x-\frac{f'(\alpha)}{f(\alpha)}\right].$$

CHAPTER XIII.

TRANSFORMATION OF A LINEAR DIFFERENTIAL EQUATION. FORSYTH'S CANONICAL FORM. ASSOCIATE EQUATIONS.

ALTHOUGH, as previously stated, it is not intended to go into the theory of Invariants in the present volume, it is nevertheless desirable to give an account of the transformation of the differential equation and its reduction to the canonical form adopted by Forsyth.*

The differential equation may be written in the form

$$R_0 \frac{d^n Y}{dx^n} + nR_1 \frac{d^{n-1} Y}{dx^{n-1}} + \frac{n(n-1)}{1 \cdot 2} R_2 \frac{d^{n-2} Y}{dx^{n-2}} + \ldots + R_n Y = 0.$$

This, by the familiar transformation

$$Y = ye^{-\int \frac{R_1}{R_0} dx}$$

and subsequent division throughout by R_0, is changed into

$$\text{(i)} \quad \frac{d^n y}{dx^n} + \frac{\underline{n|}}{\underline{2|}\,\underline{n-2|}} P_2 \frac{d^{n-2} y}{dx^{n-2}} + \frac{\underline{n|}}{\underline{3|}\,\underline{n-3|}} P_3 \frac{d^{n-3} y}{dx^{n-3}} + \ldots + P_n y = 0.$$

To the letters $P_2, P_3, \ldots$ we may add P_0 and P_1, understanding, of course, that

$$P_0 = 1, \quad P_1 = 0.$$

The numerical coefficients are obviously equal to

$$\frac{n(n-1)}{\underline{2|}}, \quad \frac{n(n-1)(n-2)}{\underline{3|}}, \quad \frac{n(n-1)(n-2)(n-3)}{\underline{4|}} \ldots,$$

and are written in these forms by Laguerre,† but it is desirable for

* Invariants, Covariants, and Quotient-derivatives associated with Linear Differential Equations, by A. R. Forsyth. Phil. Trans. of the Royal Society, vol. 179 (1888), pp. 377–489.

† *Sur quelques invariants des équations différentielles linéaires.* Comptes Rendus, t. 88 (1879), pp. 224–227.

the present purpose to write them in the form chosen by Forsyth. We find readily,

$$P_2 = R_0R_2 - R_1^2 - (R_0R_1' - R_0'R_1),$$
$$P_3 = R_0^2R_3 - 3R_0R_1R_2 + 2R_1^3 - (R_0R_1'' - R_0''R_1),$$
$$\cdot \quad \cdot \quad \cdot \quad \cdot \quad \cdot \quad \cdot \quad \cdot \quad \cdot \quad \cdot \quad \cdot$$

From these equations we see that the functions P are independent of any particular choice of the dependent variable y, and are therefore *seminvariants* of the original equation. Suppose now that

$$(1) \qquad y = u\lambda,$$

where λ is a function of x, and suppose that u satisfies the equation

$$\text{(ii)} \qquad \frac{d^nu}{dz^n} + \frac{\underline{n|}}{\underline{2|}\,\underline{n-2|}} Q_2 \frac{d^{n-2}u}{dz^{n-2}} + \frac{\underline{n|}}{\underline{3|}\,\underline{n-3|}} Q_3 \frac{d^{n-3}u}{dz^{n-3}} + \ldots + Q_nu = 0;$$

in order that (i) may be transformable into (ii), z must be some function of x, and when this is the case there will be n equations connecting λ, z, x, and the two sets of coefficients P and Q. These equations are obtained in the following manner: Making the substitution $y = u\lambda$ in (i), we have

$$(\alpha) \qquad \frac{d^nu}{dx^n}\lambda + n\frac{d\lambda}{dx}\frac{d^{n-1}u}{dx^{n-1}} + \frac{\underline{n|}}{\underline{2|}\,\underline{n-2|}}\frac{d^2\lambda}{dx^2}\frac{d^{n-2}u}{dx^{n-2}} + \ldots$$
$$+ \frac{\underline{n|}}{\underline{2|}\,\underline{n-2|}} P_2 \left\{ \frac{d^{n-2}u}{dx^{n-2}}\lambda + (n-2)\frac{d\lambda}{dx}\frac{d^{n-3}u}{dx^{n-3}} + \ldots \right\} + \ldots = 0.$$

In this equation we have to change the independent variable from x to z in order to compare it with (ii). Write

$$(2) \qquad z = \phi(x);$$

then, by a formula due to Schlömilch, we have

$$\frac{d^mu}{dx^m} = \sum_{s=1}^{s=m} \frac{A_{m,s}}{\underline{s|}} \frac{d^su}{dz^s}$$

where

$$A_{m,s} = \text{Limit, when } \rho = 0, \text{ of } \frac{d^m}{d\rho^m}\{\phi(x+\rho) - \phi(x)\}^s.$$

Writing $\underline{m|}\, C_{m,s} = A_{m,s}$, it follows from this last that

$$C_{m,s} = \text{coefficient of } \rho^m \text{ in } \left(\rho\phi' + \frac{1}{\underline{2|}}\rho^2\phi'' + \frac{1}{\underline{3|}}\rho^3\phi''' + \ldots\right)^s.$$

Substituting now in the semi-transformed equation (α), the coefficient of $\frac{1}{\underline{s|}}\frac{d^s u}{dz^s}$ is readily found to be

$$= \lambda A_{n,s} + n\frac{d\lambda}{dx}A_{n-1,s} + \frac{\underline{n|}}{\underline{2|}\,\underline{n-2|}}\frac{d^2\lambda}{dx^2}A_{n-2,s} + \ldots$$

$$+ \frac{\underline{n|}}{\underline{s|}\,\underline{n-s|}}\frac{d^{n-s}\lambda}{dx^{n-s}}A_{s,s} + \frac{\underline{n|}}{\underline{2|}\,\underline{n-2|}}P_2\left\{\lambda A_{n-2,s} + (n-2)\frac{d\lambda}{dx}A_{n-3,s}\right.$$

$$\left. + \ldots + \frac{\underline{n-2|}}{\underline{s|}\,\underline{n-s-2|}}\frac{d^{n-s-2}\lambda}{dx^{n-s-2}}A_{s,s}\right\} + \ldots$$

$$= \sum_{r=0}^{r=n-s}\frac{\underline{n|}}{\underline{r|}\,\underline{n-r|}}P_r\left\{\sum_{t=0}^{t=n-r-s}\frac{\underline{n-r|}}{\underline{t|}\,\underline{n-r-t|}}\frac{d^t\lambda}{dx^t}A_{n-r-t,s}\right\}$$

$$= \sum_{r=0}^{r=n-s}\sum_{t=0}^{t=n-r-s}\left\{\frac{\underline{n|}}{\underline{r|}\,\underline{t|}\,\underline{n-r-t|}}P_r A_{n-r-t,s}\frac{d^t\lambda}{dx^t}\right\}.$$

In this it is of course understood that $P_0 = 1$ and $P_1 = 0$. But the present form of the equation must be effectively the same as (ii), and the coefficients of corresponding derivatives of u must therefore be proportional to one another. In the transformed equation the coefficient of $\frac{1}{\underline{n|}}\frac{d^n u}{dz^n}$ is (here $s = n$, so that $n = 0$, $t = 0$ are the only values for terms in the summation) $A_{n,n}\lambda$, or the coefficient of $\frac{d^n u}{dz^n}$ is $\frac{A_{n,n}\lambda}{\underline{n|}}$; that is, it is $\lambda z'^n$.

Hence we have

$$\frac{\underline{n|}}{\underline{s|}\,\underline{n-s|}}Q_{n-s}\lambda z'^n = \frac{1}{\underline{s|}}\sum_{r=0}^{r=n-s}\sum_{t=0}^{t=n-r-s}\left\{\frac{\underline{n|}}{\underline{r|}\,\underline{t|}\,\underline{n-r-t|}}P_r A_{n-r-t,s}\frac{d^t\lambda}{dx^t}\right\},$$

or, what is the same thing,

$$\frac{\lambda Q_{n-s}}{\underline{|n-s}} z'^n = \sum_{r=0}^{r=n-s} \sum_{t=0}^{t=n-r-s} \frac{P_r A_{n-r-t,s}}{\underline{|r}\,\underline{|t}\,\underline{|n-r-t}} \frac{d^t\lambda}{dx^t}.$$

Write now

$$W_\theta = \frac{d^\theta\lambda}{dx^\theta} + \frac{\underline{|\theta}}{\underline{|2}\,\underline{|\theta-2}} P_2 \frac{d^{\theta-2}\lambda}{dx^{\theta-2}} + \ldots + P_\theta\lambda,$$

or, symbolically,

(4) $$W_\theta = \left(1, 0, P_2 \ldots P_\theta \between \frac{d}{dx}, 1\right)^\theta \lambda;$$

then the coefficient of

$$\frac{A_{n-r-t,s}}{\underline{|n-r-t}}$$

in the foregoing expression is

$$\Sigma \frac{P_r}{\underline{|r}\,\underline{|t}} \frac{d^t\lambda}{dx^t},$$

the summation extending to those values of r and t that leave the sum $r+t$ unchanged throughout—that is, the coefficient is

$$\frac{W_{r+t}}{\underline{|r+t}};$$

and therefore

$$\frac{\lambda Q_{n-s}}{\underline{|n-s}} z'^n = \sum_{\theta=0}^{\theta=n-s} \frac{A_{n-\theta,s}}{\underline{|n-\theta}} \frac{W_\theta}{\underline{|\theta}}.$$

Changing s into $n-s$ and introducing the quantities C from (3), and this becomes

(iii) $$\frac{\lambda Q_s}{\underline{|s}} z'^n = \sum_{\theta=0}^{\theta=s} C_{n-\theta,n-s} \frac{W_\theta}{\underline{|\theta}}.$$

If it were desirable, the summation in this might extend to the value $\theta = n$, for $C_{m, m'}$ vanishes if $m < m'$. Writing (iii) in detail for the lowest values of s and giving P_1 and Q_1 their zero values, we have in succession

$$(5)' \qquad 0 = W_0 C_{n, n-1} + W_1 C_{n-1, n-1},$$

$$(6)' \quad \frac{\lambda z'^n}{\underline{2|}} Q_2 = W_0 C_{n, n-2} + W_1 C_{n-1, n-2} + \frac{1}{\underline{2|}} W_2 C_{n-2, n-2},$$

$$(7)' \quad \frac{\lambda z'^n}{\underline{3|}} Q_3 = W_0 C_{n, n-3} + W_1 C_{n-1, n-3} + \frac{1}{\underline{2|}} W_2 C_{n-2, n-3} + \frac{1}{\underline{3|}} W_3 C_{n-3, n-3},$$

. . . .,

n equations in all.

There is a *first invariant* of the differential equation which is readily derived from these last equations. The invariant might properly be called Brioschi's invariant, as he first showed that it existed in the case of differential equations of the third and fourth orders. Forsyth's deduction of this invariant (which follows) from equations (5)′, (6)′, (7)′, involves an important modification of the forms of these equations. We have from (4)

$$W_0 = \lambda, \quad W_1 = \lambda',$$

and from (3)

$$C_{n, n-1} = \tfrac{1}{2}(n-1) z'^{n-2} z'', \quad C_{n-1, n-1} = z'^{n-1}$$

while generally

$$C_{m, m} = z'^m;$$

so that (5)′ is now

$$(5) \qquad z'\lambda' + \tfrac{1}{2}(n-1)\lambda z'' = 0.$$

Writing

$$(8) \qquad \frac{z''}{z'} = Z = -\frac{2}{n-1}\frac{\lambda'}{\lambda},$$

we have

$$z'' = z'Z,$$
$$z''' = z'(Z' + Z^2),$$
$$z^{iv} = z'(Z'' + 3ZZ' + Z^3);$$
$$\lambda' = -\tfrac{1}{2}(n-1)\lambda Z,$$
$$\lambda'' = -\tfrac{1}{2}(n-1)\lambda\{Z' - \tfrac{1}{2}(n-1)Z^2\},$$
$$\lambda''' = -\tfrac{1}{2}(n-1)\lambda\{Z'' - \tfrac{3}{2}(n-1)ZZ' + \tfrac{1}{4}(n-1)^2Z^3\}.$$

Again,

$$C_{n,n-2} = \tfrac{1}{24}(n-2)z'^{n-4}\{4z'z''' + 3(n-3)z''^2\}$$
$$= \tfrac{1}{24}(n-2)z'^{n-2}\{4Z' + (3n-5)Z^2\},$$
$$C_{n-1,n-2} = \tfrac{1}{2}(n-2)z'^{n-3}z'' = \tfrac{1}{2}(n-2)z'^{n-2}Z.$$

Introducing these and the values of λ', λ'' in (6)′, it becomes

$$(6) \qquad 2Z' - Z^2 = \frac{12}{n+1}(P_2 - Q_2z'^2).$$

The values of $C_{n,n-3}$, $C_{n-1,n-3}$, $C_{n-2,n-3}$ are similarly found to be

$$C_{n,n-3} = \tfrac{1}{48}(n-3)z'^{n-3}\{2Z'' + (4n-10)ZZ' + (n-2)(n-3)Z^3\},$$
$$C_{n-1,n-3} = \tfrac{1}{24}(n-3)z'^{n-3}\{4Z' + (3n-8)Z^2\},$$
$$C_{n-2,n-3} = \tfrac{1}{2}(n-3)z'^{n-3}Z.$$

By means of these and the above values of λ', λ'', λ''', (7)′ changes into

$$(7) \qquad Z'' - 3Z'Z + Z^3 = \frac{4}{n+1}(P_3 - Q_3z'^3) - \frac{12}{n+1}P_2Z.$$

Differentiating (6) with respect to x, and remembering that Q_2 is a function of z, and that $z'z'' = z'^2Z$, we have

$$Z'' - ZZ' = \frac{6}{n+1}\left(\frac{dP_2}{dx} - z'^3\frac{dQ_2}{dz}\right) - \frac{12}{n+1}z'^2ZQ_2;$$

subtracting (7) from this gives

$$2ZZ' - Z^2 = \frac{6}{n+1}\left(\frac{dP_2}{dx} - z'^3\frac{dQ_2}{dz}\right) - \frac{12}{n+1}z'^2 ZQ_2$$

$$- \frac{4}{n+1}(P_3 - z'^3 Q_3) + \frac{12}{n+1}P_2 Z;$$

multiply (6) by Z and subtract from this last, and we have

$$\frac{6}{n+1}\left(\frac{dP_2}{dx} - z'^3\frac{dQ_2}{dx}\right) - \frac{4}{n+1}(P_3 - z'^3 Q_3) = 0,$$

or, finally,

$$\left(3\frac{dQ_2}{dz} - 2Q_3\right)z'^3 = 3\frac{dP_2}{dx} - 2P_3,$$

the result of the elimination of Z between (6) and (7). This last result is Brioschi's invariant.

It appears from the preceding investigation that there exist rational integral functions of the coefficients of a linear differential equation and their derivatives such that, when the same function is formed for the transformed equation, the two functions are equal to a factor près, which factor is a positive integral power of z'. These functions are the *invariants* of the differential equation, and the exponent of z' is the *index* of the invariant.

Reverting now to equation (5) and integrating it, we have

$$\lambda z'^{\frac{1}{2}(n-1)} = \text{constant};$$

since equations (iii) are homogeneous in the dimensions of λ, this constant may have any arbitrary value other than zero; taking then the value unity, we have

(iv) $$\lambda = z'^{-\frac{1}{2}(n-1)}.$$

This establishes only one relation between the quantities z and λ, and we may suppose z arbitrary so far. We may now impose any other condition we please upon z and λ which does not violate (iv) (or (5)), say such a condition as will make $Q_2 = 0$. By (6) we must then have

$$2Z' = Z^2 + \frac{12}{n+1} P_2,$$

where, by (8),

$$\frac{z''}{z'} = Z = -\frac{2}{n-1} \frac{\lambda'}{\lambda}.$$

If we write

$$Z = -2 \frac{\theta'}{\theta}$$

the equation which determines Z becomes

$$(9) \qquad \frac{d^2\theta}{dx^2} + \frac{3}{n+1} P_2 \theta = 0,$$

and so we may write

$$(10) \qquad \lambda = \theta^{n-1}, \quad z' = \theta^{-2}.$$

Hence by a solution of a linear differential equation of the second order we can remove the second and third terms of the general linear differential equation of any order. This result was first given by Laguerre.* The modified form of the equation to which we have now come is Forsyth's *canonical form*,† and for the future we shall speak of a linear differential equation of order n as being in its canonical form when the terms involving the derivatives of order $(n-1)$ and $(n-2)$ of the dependent variable are lacking.

Since $\theta = z'^{-\frac{1}{2}}$, we have $\frac{d^2\theta}{dx^2} = \frac{3}{4} z'^{-\frac{5}{2}} z''^2 - \frac{1}{2} z'^{-\frac{3}{2}} z'''$; substituting in (9), we get

$$\frac{3}{4} \frac{z''^2}{z'^{\frac{5}{2}}} - \frac{z'''}{2z'^{\frac{3}{2}}} + \frac{3P_2}{n+1} \frac{1}{z'^{\frac{1}{2}}} = 0,$$

---

* Comptes Rendus, t. 88 (1879), p. 226.

† Forsyth's canonical form differs very little from Halphen's, but is more convenient in computing the invariants.

or, on reducing,

$$\{z, x\}, \quad = \frac{z'''}{z'} - \frac{3}{2}\left(\frac{z''}{z'}\right)^2, \quad = \frac{6P_2}{n+1},$$

$\{z, x\}$ being the Schwarzian derivative $\dfrac{z'''}{z'} - \dfrac{3}{2}\left(\dfrac{z''}{z'}\right)^2$.

ASSOCIATE EQUATIONS.

In Chapter I, equation (42), we defined Lagrange's "*équation adjointe*," and there spoke of it as the "adjunct equation;" for the future, however, we shall use the term *associate equation.** The differential equation being given in the form

$$(11) \qquad \frac{d^n y}{dx^n} + R_2 \frac{d^{n-2}y}{dx^{n-2}} + R_3 \frac{d^{n-3}y}{dx^{n-3}} + \ldots + R_n y = 0,$$

its Lagrangian associate is

$$(12) \quad \frac{d^n v}{dx^n} + \frac{d^{n-2}}{dx^{n-2}}(vR_2) - \frac{d^{n-3}}{dx^{n-3}}(vR_3) + \ldots + (-1)^n vR_n = 0.$$

Let $y_1, y_2, \ldots, y_n$ be a system of fundamental integrals of (11); a selection of any $n-1$ of them will suffice to determine the $n-1$ coefficients R. Say we take $y_1, y_2, \ldots, y_{n-1}$; substituting these in (11) and solving, we have

$$R_i = \begin{vmatrix} \dfrac{d^{n-2}y_1}{dx^{n-2}} & \ldots & \dfrac{d^{n-i-1}y_1}{dx^{n-i-1}} & \dfrac{d^n y_1}{dx^n} & \dfrac{d^{n-i+1}y_1}{dx^{n-i+1}} & \ldots y_1 \\ \dfrac{d^{n-2}y_2}{dx^{n-2}} & \ldots & \dfrac{d^{n-i-1}y_2}{dx^{n-i-1}} & \dfrac{d^n y_2}{dx^n} & \dfrac{d^{n-i+1}y_2}{dx^{n-i+1}} & \ldots y_2 \\ \vdots & & \vdots & \vdots & \vdots & \vdots \\ \dfrac{d^{n-2}y^{n-1}}{dx^{n-2}} & \ldots & \dfrac{d^{n-i-1}y_{n-1}}{dx^{n-i-1}} & \dfrac{d^n y_{n-1}}{dx^n} & \dfrac{d^{n-i+1}y_{n-1}}{dx^{n-i+1}} & \ldots y_{n-1} \end{vmatrix} \div v_n,$$

*When Chapter I was written, and indeed when an earlier form of the present chapter was written, I had not seen Forsyth's memoir, and had not been able to find an adopted English term for Lagrange's "*équation adjointe*," so I used the word *adjunct*, suggested by the German "*adjungirte*," and not unlike the French "*adjointe*." It seems better now, however, to employ the word *associate*, or, when speaking simply of Lagrange's "*équation adjointe*," the word *adjoint*. It is unfortunately too late to make this change in Chapter I.

where the subscript n in v_n indicates that the function y_n is absent, and where

$$v_n = \begin{vmatrix} \frac{d^{n-2}y_1}{dx^{n-2}} & \frac{d^{n-3}y_1}{dx^{n-3}} & \cdots & y_1 \\ \frac{d^{n-2}y_2}{dx^{n-2}} & \frac{d^{n-3}y_2}{dx^{n-3}} & \cdots & y_2 \\ \vdots & \vdots & & \\ \frac{d^{n-2}y_{n-1}}{dx^{n-2}} & \frac{d^{n-3}y_{n-1}}{dx^{n-3}} & \cdots & y_3 \end{vmatrix}.$$

Substituting these values in (11) and multiplying through by v_n, this equation becomes

$$\begin{vmatrix} \frac{d^n y}{dx^n} & \frac{d^{n-2}y}{dx^{n-2}} & \cdots & \frac{dy}{dx} & y \\ \frac{d^n y_1}{dx^n} & \frac{d^{n-2}y_1}{dx^{n-2}} & \cdots & \frac{dy}{dx} & y_1 \\ \frac{d^n y_2}{dx^n} & \frac{d^{n-2}y_2}{dx^{n-2}} & \cdots & \frac{dy_2}{dx} & y_2 \\ \cdot & \cdot & \cdot & \cdot & \cdot \\ \frac{d^n y_{n-1}}{dx^n} & \frac{d^{n-2}y_{n-1}}{dx^{n-2}} & \cdots & \frac{dy_{n-1}}{dx^{n-1}} & y_{n-1} \end{vmatrix} = 0,$$

or

$$\frac{d}{dx}\begin{vmatrix} \frac{d^{n-1}y}{dx^{n-1}} & \frac{d^{n-2}y}{dx^{n-2}} & \cdots & \frac{dy}{dx} & y \\ \frac{d^{n-1}y_1}{dx^{n-1}} & \frac{d^{n-2}y_1}{dx^{n-2}} & \cdots & \frac{dy_1}{dx} & y_1 \\ \frac{d^{n-1}y_2}{dx^{n-1}} & \frac{d^{n-2}y_2}{dx^{n-2}} & \cdots & \frac{dy_2}{dx} & y_2 \\ \cdot & \cdot & \cdot & \cdot & \cdot \\ \frac{d^{n-1}y_{n-1}}{dx^{n-1}} & \frac{d^{n-2}y_{n-1}}{dx^{n-2}} & \cdots & \frac{dy_{n-1}}{dx} & y_{n-1} \end{vmatrix} = 0.$$

It follows therefore that v_n is an integrating factor for (11), and consequently that $v_1, v_2, \ldots, v_n$ are integrals of (12); that is, the integrals of (12) are the n determinants of

$$\left\| \begin{array}{ccccc} \frac{d^{n-2}y_n}{dx^{n-2}}, & \frac{d^{n-2}y_{n-1}}{dx^{n-2}}, & \ldots, & \frac{d^{n-2}y_2}{dx^{n-2}}, & \frac{d^{n-2}y_1}{dx^{n-2}} \\ \frac{d^{n-3}y_n}{dx^{n-3}}, & \frac{d^{n-3}y_{n-1}}{dx^{n-3}}, & \ldots, & \frac{d^{n-3}y_2}{dx^{n-3}}, & \frac{d^{n-3}y_1}{dx^{n-3}} \\ \vdots & \vdots & & \vdots & \vdots \\ y_n, & y_{n-1}, & & y_2, & y_1 \end{array} \right\| .$$

[It is clear from this that if all of the integrals of the given differential equation are regular, then all the integrals of the adjoint equation are also regular. This remark applies to all the associate equations.]

It is known that if (12) is the Lagrangian associate of (11), then reciprocally (11) is the Lagrangian associate of (12); and it is evident that if either be in its canonical form, the other will also be in its canonical form. Forsyth shows now that the dependent variable v of Lagrange's associate equation is merely the last one of a set of dependent variables associated with the dependent variable of the given equation. These variables are all transformable by a substitution similar to that which transforms the original dependent variable, viz., multiplication by some power of $\frac{dz}{dx}$; and they possess the property that all combinations of them, similar to those by which they are constructed, are expressible explicitly in terms of the variables of the set. The following is Forsyth's account of these new variables:

Let $y_1, y_2, \ldots, y_n$ be a set of fundamental integrals of the given equation; they are of course linearly independent, and we will further assume concerning them (an assumption justifiable in the general case, but not necessarily so in a particular case) that there exists no linear function of them with constant coefficients which is equal to a polynomial of degree less than $n - 1$. Then of course the linear independence of the functions will hold for their derivatives up to the $(n-1)^{\text{th}}$ inclusive. Any other set of fundamental integrals $Y_1, Y_2, \ldots, Y_n$ are linear functions with constant coefficients of $y_1, y_2, \ldots, y_n$. We may write

$$(Y_1, Y_2, \ldots, Y_n) = S(y_1, y_2, \ldots, y_n),$$

where as before S denotes a substitution with non-vanishing determinant. We have also

$$\left(\frac{d^r Y_1}{dx^r}, \frac{d^r Y_2}{dx^r}, \ldots, \frac{d^r Y_n}{dx^r}\right) = S\left(\frac{d^r y_1}{dx^r}, \frac{d^r y_2}{dr^r}, \ldots, \frac{d^r y_n}{dx^r}\right)$$

for any value of r. If we retain this last equation for values of r equal $0, 1, 2, \ldots, n-1$, we shall have n sets of variables subject to the same linear transformation; and these variables are linearly independent of one another, since for the satisfaction of the differential equation we need the nth differential coefficients of the functions y, but these have been specially excluded. Since the n quantities y are linearly independent they may be looked upon as the co-ordinates of a point in a manifoldness of $n-1$ dimensions; similarly, under the hypothesis made concerning the derivatives up to order $n-1$, each of the $n-1$ sets of derivatives, each set being made up of derivatives of the same order, may be looked upon as representing the co-ordinates of a point in a manifoldness of $n-1$ dimensions. And, since the law of linear transformation is the same for all the sets, all these points may be taken as belonging to the same manifoldness. There are thus n different and independent sets of cogredient variables connected with the single manifoldness of $n-1$ dimensions.

In the theory of the concomitants of algebraical quantities of any order in the variables of a manifoldness of $n-1$ dimensions, it is necessary to consider all the possible classes of variables which can enter into the expressions of these concomitants. Clebsch* has proved that there are in all $n-1$ different classes of variables which thus need to be considered, and that if $x_1, x_2, \ldots, x_n$; $y_1, y_2, \ldots, y_n$; $z_1, z_2, \ldots, z_n$; ... be n sets of cogredient variables, the several classes are constituted by minors of varying orders of the determinant (itself an identical covariant)

* "*Ueber eine Fundamentalaufgabe der Invariantentheorie*," *Göttingen, Abhandlungen*, vol. 17, 1872.

$$\begin{vmatrix} x_1, x_2, & \dots, & x_n \\ y_1, y_2, & \dots, & y_n \\ z_1, z_2, & \dots, & z_n \\ \cdot & \cdot & \cdot \\ \cdot & \cdot & \cdot \end{vmatrix},$$

those of one class being minors of one and the same order. The variables of any class are linearly, but not algebraically, independent of one another, except in the case of the first class, constituted by minors of order unity, and the last class, constituted by minors of order $n - 1$ (the complementaries of those of the first class), in each of which classes the n variables are quite independent of one another. And all similar combinations of variables are expressible in terms of variables actually included in the classes.

In connection with our differential equation we have obtained n different and algebraically independent sets of cogredient variables; the functional derivation of the sets, one from another in succession, by the process of differentiation has been excluded from any interference with their algebraical independence. We already have one class of variables, viz., $y_1, y_2, \dots, y_n$, analogous to the first class of algebraical variables, and another class of variables, viz., $v_1, v_2, \dots, v_n$, analogous to the $(n-1)^{\text{th}}$ class of algebraical variables; and the relation

$$y_1v_1 + y_2v_2 + \dots + y_nv_n = 0,$$

which is satisfied, is precisely the same as the corresponding relation between the similar variables helping to define the higher class (Clebsch, *l. c.*, p. 4). Hence, from the point of view of purely algebraical forms, we infer that the suitable algebraical combinations of the sets of variables, which have arisen in connection with the differential equation, are the minors of varying orders of the determinant

$$\Delta = \begin{vmatrix} y_1, & y_2, & \dots, & y_n \\ \dfrac{dy_1}{dx}, & \dfrac{dy_2}{dx}, & \dots, & \dfrac{dy_n}{dx} \\ \cdot & \cdot & \cdot & \cdot \\ \dfrac{d^{n-1}y_1}{dx^{n-1}}, & \dfrac{d^{n-1}y_2}{dx^{n-1}}, & \dots, & \dfrac{d^{n-1}y_n}{dx^{n-1}} \end{vmatrix},$$

which, since $R_1 = 0$, is a non-evanescent constant. These variables may be arranged in classes which may be called linear, bilinear, trilinear, and so on. In the case of the algebraical quantities it is a matter of indifference which set of minors of a given order be taken to constitute the variables of a class corresponding to that order. Thus for the second class the same kind of variable is obtained by taking the (x, y) minors, the (x, z) minors, the (y, z) minors, and so on. A difference, however, arises in the case of the variables connected with the differential equation. There are n sets of linear variables distinct in character from one another, as the variables of any one set, say $y_1', y_2', \ldots, y_n'$, though submitting to the same substitution as $y_1, y_2, \ldots, y_n$, satisfy an entirely different differential equation. There are $\dfrac{n(n-1)}{2}$ sets of bilinear variables distinct in character; thus

$$\begin{vmatrix} y_1', & y_1 \\ y_2', & y_2 \end{vmatrix}, \quad \begin{vmatrix} y_1'', & y_1' \\ y_2'', & y_2' \end{vmatrix}, \quad \begin{vmatrix} y_1'', & y_1 \\ y_2'', & y_2 \end{vmatrix}$$

are three distinct variables of this class, subject to the same law of linear transformation; and so on for the higher classes.

Most of these must, however, be excluded, and of the foregoing algebraical combinations we must for our purpose select only those which possess, what we may call, the *functional invariantive property*, that is, those which have the invariantive property of reproducing themselves, save as to a power of $\dfrac{dz}{dx}$, after the transformation.

Of the n sets of linear variables constituted by the several sets of n quantities y, n quantities y', and so on, only the first set possesses the functional invariantive property, and we already know by (iv), if u be the new dependent variable, that we have the relation

$$(13) \qquad y = uz'^{-\frac{1}{2}(n-1)}, = u\lambda.$$

Of the $\frac{1}{2}n(n-1)$ sets of bilinear variables, each set containing $\frac{1}{2}n(n-1)$ variables, only one possesses the functional invariantive property, viz., the set constituted by the $\frac{1}{2}n(n-1)$ variables of the type

$$\begin{vmatrix} y_a', & y_a \\ y_\beta', & y_\beta \end{vmatrix}.$$

This statement is readily verified by applying the substitution (13). Suppose t_2 denotes the original bilinear variable, say the one just written, and v_2 the transformed bilinear variable,

$$\begin{vmatrix} \dfrac{du_a}{dz}, & u_a \\ \dfrac{du_\beta}{dz}, & u_\beta \end{vmatrix}.$$

We have, since $\lambda = z'^{-\frac{1}{2}(n-1)}$,

$$y_a = u_a\lambda, \quad y_\beta = u_\beta\lambda,$$

$$y_a' = \frac{du_a}{dz}z'\lambda - \tfrac{1}{2}(n-1)u_a\lambda z'^{-1}z'', \quad y_\beta' = \frac{du_\beta}{dz}z'\lambda - \tfrac{1}{2}(n-1)u_\beta\lambda z'^{-1}z'',$$

giving

$$\begin{vmatrix} y_a', & y_a \\ y_\beta', & y_\beta \end{vmatrix} = \begin{vmatrix} \dfrac{du_a}{dz}z'\lambda - \tfrac{1}{2}(n-1)u_a\lambda z'^{-1}z'', & u_a\lambda \\ \dfrac{du_\beta}{dz}z'\lambda - \tfrac{1}{2}(n-1)u_\beta\lambda z'^{-1}z'', & u_\beta\lambda \end{vmatrix},$$

or

$$\begin{vmatrix} y_a', & y_a \\ y_\beta', & y_\beta \end{vmatrix} = \lambda^2 z' \begin{vmatrix} \dfrac{du_a}{dz}, & u_a \\ \dfrac{du_\beta}{dz}, & u_\beta \end{vmatrix},$$

or, finally,

(14) $$t_2 = \lambda^2 z' v_2 = v_2 z'^{-(n-2)} = v_2 z'^{-\frac{1}{2}2(n-2)}.$$

It is obvious from this that the functional invariantive property does not hold for any bilinear variable of the form

$$\begin{vmatrix} y_a', & y_\gamma \\ y_\beta', & y_\delta \end{vmatrix},$$

where γ and δ are respectively different from α and β.

In precisely the same way we can show that of the $\frac{1}{6}n(n-1)(n-2)$ sets of trilinear variables, each set being constituted by corresponding minors of the third order, there is only one set of which each variable possesses the functional invariantive property, viz., the set of which the typical variable is

$$t_3 = \begin{vmatrix} y_\alpha'', & y_\alpha', & y_\alpha \\ y_\beta'', & y_\beta', & y_\beta \\ y_\gamma'', & y_\gamma', & y_\gamma \end{vmatrix}.$$

The relation of transformation in this case is easily seen to be

$$(15) \qquad t_3 = v_3 z'^{-3\frac{1}{2}(n-3)},$$

where v_3 is the corresponding transformed trilinear variable. In general, of the

$$\frac{\underline{|n}}{\underline{|p}\ \underline{|n-p}}$$

sets of p-linear variables, each set being constituted by minors of the pth order, there is only one set which has variables possessed of the functional invariantive property, viz., the set of which

$$t_p = \begin{vmatrix} y_1, & y_1', & y_1'', & \dots, & y_1^{(p-1)} \\ y_2, & y_2', & y_2'', & \dots, & y_2^{(p-1)} \\ \cdot & \cdot & \cdot & \cdot & \cdot \\ y_p, & y_p', & y_p'', & \dots, & y_p^{(p-1)} \end{vmatrix}$$

is a typical variable. If v_p denote the same p-linear variable associated with the transformed equation, the law of transformation is

$$(16) \qquad \begin{aligned} t_p &= v_p \lambda^p z'^{1+2+\dots+p-1} \\ &= v_p z'^{-\frac{1}{2}p(n-1)+\frac{1}{2}p(p-1)} \\ &= v_p z'^{-\frac{1}{2}p(n-p)}. \end{aligned}$$

The last set of variables is that for which $p = n - 1$; and the typical variable of the set is the variable of the Lagrangian adjoint equation.

We see now that there are in all $n-1$ sets of variables; all the variables in any one set are particular and linearly independent solutions of a differential equation the dependent variable of which is a typical variable of the set. Hence, connected with the given differential equation, there are $n-2$ other differential equations; these are the *associate equations*. The $n-2$ new dependent variables, derived by definite laws of formation, may be called the associate dependent variables; and, calling them in turns the associate variables of the first, second, . . . , $(n-2)^{\text{th}}$ rank, the differential equation of which the dependent variable is the associate of the $(p-1)^{\text{th}}$ rank is linear and of order $\dfrac{\underline{|n}}{\underline{|p}\,\underline{|n-p}}$. For the functional transformation of the original dependent variable given by (13) the law of transformation of the associate variable of the $(p-1)^{\text{th}}$ rank is given by (16); and if we call two ranks complementary when the sum of their orders is $n-2$, their associate variables of complementary rank are transformed by the same relation, since for such variables the index of the factor power of z' has the same value. For example: in t_2 the power of z' is $-(n-2)$, and in t_{n-2} the power of z' is $-(n-2)$, and so t_2 and t_{n-2}, whose ranks (or orders) are respectively 1 and $n-3$, are complementary. The associate variables may therefore be arranged in pairs of complementary rank; in the case of n even there is one dependent variable of self-complementary rank. Each pair has the index of the factor power of z' different from that for any other pair. The simplest case of this arrangement is that which combines in a pair the original variable y and the variable t_{n-1} of Lagrange's adjoint equation; and the two dependent variables have the same functional transformation.

Leaving for the present the general subject of associate equations, we will derive some properties of the Lagrangian associate, or adjoint, equation, due in part to Thomé and in part to Floquet. Let the given differential equation be

$$(17)\qquad P(y)=\frac{d^ny}{dx^n}+P_1\frac{d^{n-1}y}{dx^{n-1}}+P_2\frac{d^{n-2}y}{dx^{n-2}}+\ldots+P_ny=0.$$

This can be placed in a determinate composite form in the following manner: Writing

$$A_j=\frac{dy}{dx}-K_jy=0,$$

construct the series of linear differential equations of the first order

$$A_1 = 0, \quad A_2 = 0, \quad \ldots, \quad A_n = 0,$$

admitting respectively as integrals

$$v_1, \quad v_1v_2, \quad \ldots, \quad v_1v_2 \ldots v_j, \quad \ldots, \quad v_1v_2 \ldots v_n,$$

where, if $y_1, y_2, \ldots, y_n$ are fundamental integrals of (17), we have

$$y_1 = v_1, \quad y_2 = v_1\int v_2dx, \quad \ldots, \quad y_n = v_1\int v_2dx\int v_3dx \ldots \int v_ndx.$$

For brevity write

$$V_j = v_1v_2 \ldots v_j, \quad j = 1, 2, \ldots, n.$$

The value of K_j is then given by

$$K_j = \frac{1}{V_j}\frac{dV_j}{dx} = \frac{d}{dx}\log V_j.$$

We have identically

$$P = A_nA_{n-1} \ldots A_1. \tag{18}$$

In fact, the expression on the right-hand side of this equation is annulled by the general integrals of the equations

$$A_1 = 0, A_1 = v_1v_2, A_2A_1 = v_1v_2v_3, \ldots, A_{n-1}A_{n-2} \ldots A_1 = v_1v_2 \ldots v_n.$$

Now $A_1 = 0$ is satisfied for $y = v_1$, $A_1 = v_1v_2$ is satisfied by $y = v_1\int v_2dx$, $A_2A_1 = v_1v_2v_3$ is satisfied by $y = v_1\int v_2dx\int v_3dx$, and so on; thus the two equations

$$A_nA_{n-1} \ldots A_1 = 0 \quad \text{and} \quad P = 0,$$

each of order n, have a system of fundamental integrals in common, and consequently, as the coefficient of $\frac{d^ny}{dx^n}$ is 1 in both cases, the first members of these equations must be identical, for, if they were not, their difference, which is at most of order $n - 1$, would be annulled by n linearly independent functions, which is impossible. It follows then that (18) is an identity. In the same way we see that the equation of order $n - 1$,

$$A(y) = A_{n-1}A_{n-2} \ldots A_1 = 0,$$

admits the $n - 1$ linearly independent integrals $y_1, y_2, \ldots, y_{n-1}$, and consequently has

$$C_1y_1 + C_2y_2 + \ldots + C_{n-1}y_{n-1}$$

as its general integral. Again, the equation

$$A(y) = V_n$$

admits the particular solution y_n, and consequently the equation

$$A(y) = C_nV_n, \tag{19}$$

where C_n is an arbitrary constant, admits the particular solution C_ny_n, and so the general integral of this last equation of order $n - 1$ is

$$C_1y_1 + C_2y_2 + \ldots + C_ny_n,$$

which is the same as that of the equation $P = 0$ of order n. It follows therefore that (19) is a first integral of $P = 0$. Writing (19) in the form

$$V_n^{-1}A(y) = C_n$$

and differentiating, we have

$$\frac{d}{dx}[V_n^{-1}A(y)] = 0.$$

Make the coefficient of $\frac{d^ny}{dx^n}$ in this unity by multiplying by V_n, and we have the differential equation

$$V_n\frac{d}{dx}[V_n^{-1}A(y)] = 0, \tag{20}$$

whose first member is identical with the first member of $P = 0$. Writing $V_n^{-1} = v$, this identity is

$$vP(y) = \frac{d}{dx}[vA(y)], \tag{21}$$

and we see that

(22) $$v = V_n^{-1} = \frac{1}{v_1 v_2 \ldots v_n}$$

is an integrating factor of $P = 0$, and consequently is the dependent variable of the adjoint equation

(23) $$\mathfrak{P}(v) = \frac{d^n v}{dx^n} - \frac{d^{n-1}}{dx^{n-1}}(P_1 v) + \frac{d^{n-2}}{dx^{n-2}}(P_2 v) + \ldots + (-1)^n P_n v = 0.$$

Consider in general two differential quantics

(24) $$\begin{cases} S(y) = S_0 \dfrac{d^\sigma y}{dx^\sigma} + S_1 \dfrac{d^{-1} y}{dx^{\sigma-1}} + \ldots + S_\sigma y, \\ \mathfrak{S}(y) = \mathfrak{S}_0 \dfrac{d^\sigma y}{dx^\sigma} + \mathfrak{S}_1 \dfrac{d_{\sigma-1} y}{dx^{\sigma-1}} + \ldots + \mathfrak{S}_\sigma y; \end{cases}$$

these will be said to be adjoint differential quantics, or simply adjoint quantics, when the following relations exist, viz.,

(25) $$\begin{cases} S(y) = (-1)^\sigma \dfrac{d^\sigma(\mathfrak{S}_0 y)}{dx^\sigma} + (-1)^{\sigma-1} \dfrac{d^{\sigma-1}(\mathfrak{S}_1 y)}{dx^{\sigma-1}} + \ldots - \dfrac{d(\mathfrak{S}_{\sigma-1} y)}{dx} \\ \qquad + \mathfrak{S}_\sigma y, \\ \mathfrak{S}(y) = (-1)^\sigma \dfrac{d^\sigma(S_0 y)}{dx^\sigma} + (-1)^{\sigma-1} \dfrac{d^{\sigma-1}(S_1 y)}{dx^{\sigma-1}} + \ldots - \dfrac{d(S_{\sigma-1} y)}{dx} \\ \qquad + S_\sigma y. \end{cases}$$

We proceed now to establish an interesting relation existing be tween a differential quantic, when in its composite form, and its adjoint quantic. Suppose the quantics in (24) to be adjoint, and let λ denote a function of x; we will show in the first place that the adjoint quantic of $S(\lambda y)$ is $\lambda\mathfrak{S}(y)$. In the first of equations (25) replace y by λy, and we have

$$S(\lambda y) = (-1)^\sigma \frac{d^\sigma(\mathfrak{S}_0 \lambda y)}{dx^\sigma} + (-1)^{\sigma-1} \frac{d^{\sigma-1}(\mathfrak{S}_1 \lambda y)}{dx^{\sigma-1}} + \ldots + \mathfrak{S}_\sigma \lambda y;$$

the adjoint quantic to $S(\lambda y)$ is therefore

$$\mathfrak{S}_0 \lambda \frac{d^\sigma y}{dx^\sigma} + \mathfrak{S}_1 \lambda \frac{d^{\sigma 1-} y}{dx^{\sigma-1}} + \ldots + \mathfrak{S}_\sigma \lambda y,$$

that is, $\lambda\mathfrak{S}(y)$. Q. E. D.

In connection with this result we can make the following remark: We have supposed that the coefficient of $\frac{d^n y}{dx^n}$ in P was unity, and so have found that the adjoint quantic to P, which may be written in the form

$$\frac{d^n y}{dx^n} + \frac{P_1}{P_0}\frac{d^{n-1}y}{dx^{n-1}} + \frac{P_2}{P_0}\frac{d^{n-2}y}{dx^{n-2}} + \ldots + \frac{P_n}{P_0}y,$$

is

$$(-1)^n \frac{d^n v}{dx^n} + (-1)^{n-1}\frac{d^{n-1}}{dx^{n-1}}\left(\frac{P_1}{P_0}v\right) + \ldots + \frac{P_n}{P_0}v;$$

it follows therefore that the adjoint to

$$P_0\frac{d^n y}{dx^n} + P_1\frac{d^{n-1}y}{dx^{n-1}} + \ldots + P_n y$$

is

$$(-1)^n \frac{d^n (P_0 v)}{dx^n} + (-1)^{n-1}\frac{d^{n-1}(P_1 v)}{dx^{n-1}} + \ldots + P_n v$$

a result of course immediately obtained by Lagrange's method.

Letting λ still denote a function of x, it is easily shown that the adjoint of the quantic $S\left(\lambda\frac{d^k y}{dx^k}\right)$ is

$$(-1)^k \frac{d^k}{dx^k}[\lambda(\mathfrak{S}y)].$$

The differential quantic

$$S\left(\frac{d^k y}{dx^k}\right) = S_0\frac{d^{\sigma+k}y}{dx^{\sigma+k}} + S_1\frac{d^{\sigma+k-1}y}{dx^{\sigma+k-1}} + \ldots + S_\sigma\frac{d^k y}{dx^k}$$

has

$$(-1)^{\sigma+k}\frac{d^{\sigma+k}(S_0 y)}{dx^{\sigma+k}} + (-1)^{\sigma+k-1}\frac{d^{\sigma+k-1}(S_1 y)}{dx^{\sigma+k-1}} + \ldots + (-1)^k\frac{d^k(S_\sigma y)}{dx^k},$$

or

$$(-1)^k\frac{d^k\mathfrak{S}(y)}{dx^k},$$

as its adjoint, and consequently, if $T(y)$ is the quantic which has for its adjoint $\lambda\mathfrak{S}(y)$, $(-1)^k \frac{d^k}{dx^k}[\lambda\mathfrak{S}(y)]$ will be the adjoint to $T\left(\frac{d^k y}{dx^k}\right)$. Now $T(y)$ is given by the equation

$$T(y) = S(\lambda y),$$

and therefore $(-1)^k \frac{d^k}{dx^k}[\lambda\mathfrak{S}(y)]$ is adjoint to $S\left(\lambda \frac{d^k y}{dx^k}\right)$.

Suppose, finally, that $(S, \mathfrak{S})$, $(R, \mathfrak{R})$ are two pairs of adjoint differential quantics, R and $\mathfrak{R}$ being formed in the same way as S and $\mathfrak{S}$; we have now

$$RS = R\left[S_0 \frac{d^\sigma y}{dx^\sigma}\right] + R\left[S_1 \frac{d^{\sigma-1} y}{dx^{\sigma-1}}\right] + \ldots + R[S_\sigma y],$$

and as it is evident that the adjoint of a sum is the sum of the respective adjoints, we have at once as the adjoint of RS the differential quantic

$$(-1)^\sigma \frac{d^\sigma}{dx^\sigma}[S_0 \mathfrak{R}(y)] + (-1)^{\sigma-1} \frac{d^{\sigma-1}}{dx^{\sigma-1}}[S_1 \mathfrak{R}(y)] + \ldots + S_\sigma \mathfrak{R}(y),$$

that is, $\mathfrak{S}\mathfrak{R}$.

This gives the following theorem:

The adjoint quantic of RS is $\mathfrak{R}\mathfrak{S}$.

In the same way we can show that the adjoint quantic of QRS is $\mathfrak{S}\mathfrak{R}\mathfrak{Q}$, and so on for any number of differential quantics. We can therefore generalize the preceding theorem and say that—*If a differential quantic is composed of any number of differential quantics arranged in a given order, the adjoint quantic to the given one is composed of the corresponding component adjoint quantics arranged in the inverse order.*

The relation between the integrals of the adjoint equations

$$P(y) = 0 \qquad \text{and} \qquad \mathfrak{P}(y) = 0$$

is now easily found. Observe in the first place that the equation of the first order

$$\frac{dy}{dx} - hy = 0$$

has as its adjoint equation

$$\frac{dv}{dx} + hv = 0;$$

if the first of these has a solution $y = u$, the second has the solution $v = u^{-1}$. Now write $P(y)$ in the composite form

$$P(y) = A_n A_{n-1} \ldots A_1,$$

where the equations

$$A_1 = 0, \quad A_2 = 0, \quad \ldots, \quad A_n = 0,$$

of the first order, have respectively the solutions

$$v_1, \quad v_1 v_2, \quad \ldots, \quad v_1 v_2 \ldots v_n;$$

the equation $P(y) = 0$ then has, as already shown, the following system of fundamental integrals, viz.:

$$y_1 = v_1, \quad y_2 = v_1 \int v_2 dx, \quad \ldots, \quad y_n = v_1 \int v_2 dx \ldots \int v_n dx.$$

From the above theorem and the remark on page 483, the equations

$$\mathfrak{A}_n = 0, \quad \mathfrak{A}_{n-1} = 0, \quad \ldots, \quad \mathfrak{A}_1 = 0$$

admit the respective solutions

$$V_n^{-1}, \quad V_{n-1}^{-1}, \quad \ldots, \quad V_1^{-1},$$

and so $\mathfrak{P}(w) = 0$, the adjoint equation, admits the fundamental integrals

$$(26) \quad \begin{cases} w_1 = V_n^{-1}, \\ w_2 = V_n^{-1} \int v_n dx, \\ w_3 = V_n^{-1} \int v_n dx \int v_{n-1} dx, \\ \cdot \quad \cdot \quad \cdot \quad \cdot \quad \cdot \quad \cdot \quad \cdot \quad \cdot \\ w_n = V_n^{-1} \int v_n dx \int v_{n-1} dx \ldots \int v_2 dx, \end{cases}$$

which show the relation between the integrals of the two adjoint equations $P(y) = 0$ and $\mathfrak{P}(w) = 0$ when P_1 is not equal zero.

A few other properties of adjoint equations, or adjoint quantics, may be mentioned. Writing

$$A(y) = A_{n-1}A_{n-2}\ldots A_1,$$

we have the identity

(27) $$vP(y) = \frac{d}{dx}[vA(y)];$$

if $A(y)$ be expanded, it is of the form

$$A(y) = \frac{d^{n-1}y}{dx^{n-1}} + L_1\frac{d^{n-2}y}{dx^{n-2}} + \ldots + L_{n-1}y.$$

Equating the coefficients of the same derivatives in (27), we have for the determination of $L_1, L_2, \ldots, L_{n-1}$ the equations

(28) $$\begin{cases} \dfrac{dv}{dx} + vL_1 = P_1v, \quad \dfrac{d(vL_1)}{dx} + vL_2 = P_2v, \quad \ldots, \\ \dfrac{d(vL_{n-2})}{dx} + vL_{n-1} = P_{n-1}v, \quad \dfrac{d(vL_{n-1})}{dx} = P_nv. \end{cases}$$

The elimination of the quantities L between these gives the adjoint equation

$$\mathfrak{P}(v) = \frac{d_nv}{dx^n} - \frac{d^{n-1}(P_1v)}{dx^{n-1}} + \frac{d^{n-2}(P_2v)}{dx^{n-2}} - \ldots + (-1)^nP_nv = 0.$$

Suppose the expression $\mathfrak{A}(v)$ to be formed for the adjoint equation in the same way that $A(y)$ is formed for the given equation; we have then the two identities

$$y\mathfrak{P}(v) = \frac{d}{dx}[y\mathfrak{A}(v)],$$

$$vP(y) = \frac{d}{dx}[vA(y)],$$

giving

(29) $$vP(y) - y\mathfrak{P}(v) = \frac{d}{dx}[vA(y) - y\mathfrak{A}(v)];$$

that is, the difference

$$vP(y) - y\mathfrak{P}(v)$$

is the derivative of a differential quantic which is linear and homogeneous in y and v and whose coefficients are linear homogeneous functions of $P_1, P_2, \ldots, P_n$ and their derivatives. We may write (29) in the form

$$(30) \qquad vP(y) - y\mathfrak{P}(v) = \frac{d}{dx} B(y, v).$$

The differential quantic $B(y, v)$ is called by Frobenius * the *begleitende bilineare Differentialausdruck* to $P(y)$; we shall call it the associate bilinear differential quantic, or, when there can be no ambiguity, simply the associate quantic. It is easy to find now the condition to be satisfied in order that a given differential quantic shall be self-adjoint. Let $P(y)$ be the given quantic and $\mathfrak{P}(v)$ its adjoint; if the coefficient of the highest derivative of y in $P(y)$ is P_0, then the coefficient of the highest derivative of v in $\mathfrak{P}(v)$ will be $(-1)^n P_0$; in order then that the quantics $P(y)$ and $\mathfrak{P}(v)$ shall be the same (save of course as to the letter which is used to denote the dependent variable in each) we must first have that the order n of the quantic is even, say $n = 2\nu$. This being granted, equation (30) must have the form

$$(31) \qquad vP(y) - yP(v) = \frac{d}{dx} B(y, v).$$

If now we interchange y and v, the left-hand member of this equation changes sign, and so we have

$$\frac{d}{dx} B(y, v) = - \frac{d}{dx} B(v, y),$$

or

$$(32) \qquad B(y, v) = - B(v, y).$$

---

* *Ueber adjungirte lineare Differentialausdrücke*, G. Frobenius, Crelle, vol. 85, p. 185.

Suppose $P(y)$ is the negative of $\mathfrak{P}(v)$, then we have

$$(33) \qquad vP(y) + yP(v) = \frac{d}{dx}B(y, v);$$

interchanging y and v does not alter the left-hand member of this equation, and therefore

$$B(y, v) = B(v, y).$$

In this case, of course, n is odd. We can state the general result as follows:

(1) If a differential quantic is self-adjoint it must be of even order, and its associate quantic must change sign when its two dependent variables are interchanged. (2) If the adjoint of a given differential quantic is the negative of that quantic, then the common order of the two must be odd, and the associate quantic must be symmetric in its two dependent variables. It is obvious that the restriction of evenness is unnecessary in the case of a differential *equation*.

The following remarks on adjoint expressions, due to Halphen,* though reproducing to some extent what has already been said, will be useful for future reference. Halphen's notation is retained, at a slight sacrifice of uniformity of notation in this chapter, in order to facilitate reference to his important memoir.

Denote by $g_0, g_1, \ldots, \gamma_0, \gamma_1, \ldots$ given functions of the independent variable x; by y and η indeterminate functions of x; consider the differential quantics

$$G(y) = g_0 y^{(n)} + ng_1 y^{(n-1)} + \frac{n(n-1)}{1\cdot 2} g_2 y^{(n-2)} + \ldots + ng_{n-1}y' + g_n y,$$

$$\Gamma(\eta) = \gamma_0 \eta^{(n)} + n\gamma_1 \eta^{(n-1)} + \frac{n(n-1)}{1\cdot 2} \gamma_2 \eta^{(n-2)} + \ldots + n\gamma_{n-1}\eta' + \gamma_n \eta.$$

These two linear quantics are said to be adjoint to one another if there exists a third bilinear quantic $B(y, \eta)$ which is linear and homogeneous both with respect to y and its derivatives up to the

* *Sur un problème concernant les équations différentielles linéaires.* Par M. G.-H. Halphen. Journal de Mathématiques pures et appliquées, 4me Série, t. i. p. 11.

order $n-1$, and with respect to η and its derivatives up to the same order, and which is such that we have *identically*, that is, whatever be the functions y and η,

$$\eta G(y) + (-1)^{n-1} y \Gamma(\eta) = B'(y, \eta), \tag{34}$$

accents as usual denoting differentiation with respect to x. This relation completely determines one of the quantics $G(y)$, $\Gamma(\eta)$ when the other is arbitrarily given. The relation between these quantics is expressed by either of the following systems of equations:

$$(35)\quad \begin{cases} \gamma_0 = g_0 & g_0 = \gamma_0 \\ \gamma_1 = -g_1 + g_0' & g_1 = -\gamma_1 + \gamma_0' \\ \gamma_2 = g_2 - 2g_1' + g_0'' & g_2 = \gamma_2 - 2\gamma_1' + \gamma_0'' \\ \gamma_3 = -g_3 + 3g_2' - 3g_1'' + g_0''' & g_3 = -\gamma_3 + 3\gamma_2' - 3\gamma_1'' + \gamma_0''' \\ \cdot\quad\cdot\quad\cdot\quad\cdot\quad\cdot & \cdot\quad\cdot\quad\cdot\quad\cdot\quad\cdot \end{cases}$$

The law of these equations is obvious, and the two adjoint quantics have the forms

$$(36)\quad \begin{cases} G(y) = (\gamma_0 y)^{(n)} - n(\gamma_1 y)^{(n-1)} + \dfrac{n(n-1)}{1\cdot 2}(\gamma_2 y)^{(n-2)} + \dots \\ \qquad\qquad + (-1)^{n+1}\gamma_n y, \\ \Gamma(\eta) = (g_0 \eta)^{(n)} - n(g_1 \eta)^{(n-1)} + \dfrac{n(n-1)}{1\cdot 2}(g_2 \eta)^{(n-2)} + \dots \\ \qquad\qquad + (-1)^{n+1} g_n \eta. \end{cases}$$

The two quantics $G(y)$ and $\Gamma(\eta)$ being formed in this manner, (34) is satisfied when the associate quantic $B(y, \eta)$ has either of the forms

$$(37)\quad \begin{cases} B = \beta_0 y^{(n-1)} - \beta_1 y^{(n-2)} + \beta_2 y^{(n-3)} + \dots \\ \qquad\qquad + (-1)^n \beta_{n-2} y' + (-1)^{n+1}\beta_{n-1} y, \\ (-v)^{n-1} B = b_0 \eta^{(n-1)} - b_1 \eta^{(n-2)} + b_3 \eta^{(n-3)} + \dots \\ \qquad\qquad + (-1)^n b_{n-2}\eta' + (-1)^{n+1} b_{n-1}\eta, \end{cases}$$

where

$$
(38)\quad \begin{cases} \beta_0 = g_0\eta, \\ \beta_1 = (g_0\eta)' - ng_1\eta, \\ \beta_2 = (g_0\eta)'' - n(g_1\eta)' + \dfrac{n(n-1)}{1\cdot 2}g_2\eta, \\ \beta_3 = (g_0\eta)''' - n(g_1\eta)'' + \dfrac{n(n-1)}{1\cdot 2}(g_2\eta)' - \dfrac{n(n-1)(n-2)}{1\cdot 2\cdot 3}g_3\eta, \\ \cdot\quad\cdot\quad\cdot\quad\cdot\quad\cdot\quad\cdot\quad\cdot\quad\cdot\quad\cdot\quad\cdot\quad\cdot\quad\cdot, \end{cases}
$$

$$
(39)\quad \begin{cases} b_0 = \gamma_0 y, \\ b_1 = (\gamma_0 y)' - n\gamma_1 y, \\ b_2 = (\gamma_0 y)'' - n(\gamma_1 y)' + \dfrac{n(n-1)}{1\cdot 2}\gamma_2 y, \\ b_3 = (\gamma_0 y)''' - n(\gamma_1 y)'' + \dfrac{n(n-1)}{1\cdot 2}(\gamma_2 y)' - \dfrac{n(n-1)(n-2)}{1\cdot 2\cdot 3}\gamma_3 y, \\ \cdot\quad\cdot\quad\cdot\quad\cdot\quad\cdot\quad\cdot\quad\cdot\quad\cdot\quad\cdot\quad\cdot\quad\cdot\quad\cdot\quad\cdot \end{cases}
$$

In order that two quantics, say $P(y)$ and $\mathfrak{P}(v)$, may be self-adjoint, we have seen that they must be of even order, and that on interchanging y and v, $B(y, v)$ must change sign. As illustrations take the cases of $n = 4$ and $n = 6$, and use (37). For $n = 4$ the quantic P is

$$
(\alpha)\qquad \frac{d^4y}{dx^4} + P_1\frac{d^3y}{dx^3} + P_2\frac{d^2y}{dx^2} + P_3\frac{dy}{dx} + P_4 y,
$$

and its adjoint expanded is

$$
(\alpha)'\quad \frac{d^4v}{dx^4} - P_1\frac{d^3v}{dx^3} + \left(P_2 - 3\frac{d^2P_1}{dx^2}\right)\frac{d^2v}{dx^2} - \left(P_3 - 2\frac{dP_2}{dx} + 3\frac{d^2P_1}{dx^2}\right)\frac{dv}{dx} + \left(P_4 - \frac{dP_3}{dx} + \frac{d^2P_2}{dx^2} - \frac{d^3P_1}{dx^3}\right)v.
$$

The conditions that (α) and $(\alpha)'$ shall be the same are at once found, by comparing coefficients, to be

$$
P_1 = 0,\quad P_2 = P_2,\quad P_3 = \frac{dP_2}{dx},\quad P_4 = P_4.
$$

The associate quantic is

$$(\beta)\quad B(y, v) = vy''' - [v' - P_1v]y'' + [v'' - P_1v' + (P_2 - P_1')v]y' \\ - [v''' - P_1v'' + (P_2 - 2P_1')v' - (P_3 - P_2' + P_1'')v]y;$$

arranged according to v this is

$$(\beta)'\quad -v'''y + v''[P_1y + y'] - v'[(P_2 - 2P_1')y + P_1y' + y''] \\ + v[(P_3 - P_2' + P_1'')y + (P_2 - P_1')y' + P_1y'' + y'''].$$

In (β) interchange y and v and we have

$$(\beta)''\quad v'''y - v''[y' - P_1y] + v'[y'' - P_1y' + (P_2 - P_1')y] \\ - v[y''' - P_1y'' + (P_2 - 2P_1')y' - (P_3 - P_2' + P_1'')y].$$

That $(\beta)'$ and $(\beta)''$ may have opposite signs we must obviously have

$$y' - P_1y = P_1y + y', \quad \therefore P_1 = 0;$$

using this, we have next

$$y'' + P_2y = P_2y + y'', \quad \therefore P_2 = P_2;$$

again,

$$-y''' - P_2y' + (P_3 - P_2')y = -y''' - P_2y' - (P_3 - P_2')y,$$

giving

$$P_3 = \frac{dP_2}{dx},$$

and finally $P_4 = P_4$. These are the same results as found by direct comparison of (α) and $(\alpha)'$.

Take the case now of $n = 6$. From the general form of the adjoint equation it is obvious *à priori* that the coefficient of the second highest derivative must be zero, *i.e.*, $P_1 = 0$; the sextic can then be written

$$(\gamma)\qquad \frac{d^6y}{dx^6} + P_2\frac{d^4y}{dx^4} + P_3\frac{d^3y}{dx^3} + P_4\frac{d^2y}{dx^2} + P_5\frac{dy}{dx} + P_6y;$$

the adjoint sextic is

$$(\gamma)' \quad \frac{d^6v}{dx^6}+P_2\frac{d^4v}{dx^4}-\left[P_3-4\frac{dP_2}{dx}\right]\frac{d^3v}{dx^3}+\left[P_4-3\frac{dP_3}{dx}+6\frac{dP_2}{dx}\right]\frac{d^2v}{dx^2}$$
$$-\left[P_5-2\frac{dP_4}{dx}+3\frac{d^2P_3}{dx^2}-4\frac{d^3P_2}{dx^3}\right]\frac{dv}{dx}$$
$$+\left[P_6-\frac{dP_5}{dx}+\frac{d^2P_4}{dx^2}-\frac{d^3P_3}{dx^3}+\frac{d^4P_2}{dx^4}\right]v.$$

In order that (γ) and $(\gamma)'$ shall be the same we must have, as is easily seen by comparing coefficients,

$$P_3=2\frac{dP_2}{dx},\quad P_5=\frac{dP_4}{dx}-\frac{d^3P_2}{dx^3},$$

P_2, P_4, and P_6 remaining arbitrary. The same results of course must be obtained by considering the associate quantic $B(y, v)$; this, which is easily found, is

$$(\delta)\quad vy^{v}-v'y^{iv}+[v''+P_2v]y'''-[v'''+P_2v'-(P_3-P_2')v]y''$$
$$+[v^{iv}+P_2v''-(P_3-2P_2')v'+(P_4-P_3'+P_2'')v]y'$$
$$-[v^{v}+P_2v'''-(P_3-3P_2')v''+(P_4-2P_3'+3P_2'')v'$$
$$-(P_5-P_4'+P_3''-P_2''')v]y;$$

arranged according to v this is

$$(\delta)'\quad -v^{v}y+v^{iv}y'-v'''[P_2y+y'']+v''[(P_3-3P_2')y+P_2y'+y''']$$
$$-v'[(P_4-2P_3'+3P_2'')y+(P_3-2P_2')y'+P_2y''+y^{iv}]$$
$$+v[(P_5-P_4'+P_3''-P_2''')y+(P_4-P_3'+P_2'')y'$$
$$+(P_3-P_2')y''+P_2y'''+y^{v}].$$

In (δ) interchange y and v and we have

$$(\delta)''\quad v^{v}y-v^{iv}y'+v'''[y''+P_2y]-v''[y'''+P_2y'-(P_3-P_2')y]$$
$$+v'[y^{iv}+P_2y''-(P_3-2P_2')y'+(P_4-P_3'+P_2'')y]$$
$$-v[y^{v}+P_2y'''-(P_3-3P_2')y''+(P_4-2P_3'+3P_2'')y'$$
$$-(P_5-P_4'+P_3''-P_2''')y].$$

In order that $(\delta)'$ and $(\delta)''$ shall have opposite signs we must have, as is readily seen,

$$P_3 = 2\frac{dP_2}{dx}, \quad P_5 = \frac{dP_4}{dx} - \frac{d^3P_2}{dx^3}.$$

In the case of the quartic (assuming always $P_1 = 0$) replace P_2, P_3 by $6P_2, 4P_3$, and in the case of the sextic replace P_2, P_3, P_4, P_5 by $15P_2, 20P_3, 15P_4, 6P_5$; that is, multiply the coefficients of the two quantics by the corresponding binomial coefficients. The conditions that these quantics shall each be self-adjoint are, for the quartic,

$$(40) \qquad P_3 - \frac{3}{2}\frac{dP_2}{dx} = 0;$$

for the sextic,

$$(41) \qquad \begin{cases} P_3 - \dfrac{3}{2}\dfrac{dP_2}{dx} = 0, \\ P_5 - \dfrac{5}{2}\dfrac{dP_4}{dx} - \dfrac{5}{3}\dfrac{d^2P_3}{dx^2} = 0. \end{cases}$$

Forsyth in his memoir already referred to gives the forms of certain invariants connected with the general linear differential equation; among these are two, denoted by Θ_3 and Θ_5 respectively, which have the forms

$$(42) \qquad \begin{cases} \Theta_3 = P_3 - \dfrac{3}{2}\dfrac{dP_2}{dx}, \\ \Theta_5 = P_5 - \dfrac{5}{2}\dfrac{dP_4}{dx} + \dfrac{15}{7}\dfrac{d^2P_3}{dx^2} - \dfrac{5}{7}\dfrac{d^3P_2}{dx^3} - \dfrac{10}{7}\dfrac{7n+13}{n+1}P_2\Theta. \end{cases}$$

Comparing these with (40) and (41), it is seen at once that the conditions to be satisfied in order that a quartic or sextic linear differential quantic shall be self-adjoint are, for the quartic,

$$\Theta_3 = 0;$$

for the sextic,

$$\Theta_3 = 0, \quad \Theta_5 = 0.$$

In the case of the octic

$$\left(1,\, 0,\, P_2,\, P_3 \,.\,.\,.\, P_8\right)\!\left(\frac{d}{dx},\, 1\right)y$$

the conditions for self-adjointness are

$$P_3 - \frac{3}{2}\frac{dP_2}{dx} = 0,$$

$$P_5 - \frac{5}{2}\frac{dP_4}{dx} + \frac{5}{3}\frac{d^2P_3}{dx^2} = 0,$$

or

$$P_5 - \frac{5}{2}\frac{dP_4}{dx} + \frac{5}{2}\frac{d^3P_2}{dx^3} = 0,$$

and

$$P_7 - \frac{7}{2}\frac{dP_6}{dx} + \frac{35}{4}\frac{d^3P_4}{dx^3} - \frac{21}{2}\frac{d^5P_2}{dx^5} = 0.$$

In terms of Forsyth's invariants these are

$$\Theta_3 = 0, \quad \Theta_5 = 0, \quad \Theta_7 = 0.$$

The general form of Θ_7 is

$$\Theta_7 = P_7 - \frac{7}{2}\frac{dP_6}{dx} + \frac{105}{22}\frac{d^2P_5}{dx^2} - \frac{35}{11}\frac{d^3P_4}{dx^3} + \frac{35}{33}\frac{d^4P_3}{dx^4} - \frac{7}{44}\frac{d^5P_2}{dx^5}$$

$$- \frac{7}{11}\frac{P_2}{n+1}\left\{\frac{3}{2}(11n+31)\left(2P_5 - 5\frac{dP_4}{dx}\right) + 5(15n+41)\frac{d^2P_3}{dx^2} - 15(2n+5)\frac{d^3P_2}{dx^3}\right\}$$

$$- \frac{7}{11}\frac{3n+4}{n+1}\left\{3\frac{d^2P_2}{dx^2}\left(P_3 + \frac{dP_2}{dx}\right) - 5\frac{dP_2}{dx}\frac{dP_3}{dx}\right\}$$

$$+ P_2^2\,\Theta_3\,\frac{1155n^2 + 6048n + 6909}{22(n+1)^2}.$$

Mr. G. F. Metzler has put this invariant in the following form, which, for the present purpose at least, is rather more desirable than the above:

$$\Theta_7 = P_7 - \frac{7}{2}\frac{dP_6}{dx} + \frac{105}{22}\frac{d^2P_5}{dx^2} - \frac{35}{11}\frac{d^3P_4}{dx^3} + \frac{35}{33}\frac{d^4P_3}{dx^4} - \frac{7}{44}\frac{d^5P_2}{dx^5}$$

$$- \frac{7}{11}\frac{P_2}{n+1}(33n+93)\Theta_5 - \frac{9}{22}P_2^2\Theta_3\frac{385n^2+1728n+1919}{(n+1)^2}$$

$$- \frac{(3n+4)}{11(n+1)}\left[10P_2\frac{d^2\Theta_3}{dx^2} - 35\frac{dP_2}{dx}\frac{d\Theta_3}{dx} + 21\frac{d^2P_2}{dx^2}\Theta_3\right].$$

The subject of adjoint quantics will be taken up again after an account of the invariantive theory has been given; enough has already been said, however, to show that a differential quantic is self-adjoint when certain of its invariants vanish, and it is easy to see that in any case three of these invariants must be Θ_3, Θ_5, and Θ_7.*

An interesting theorem due to Appell, and closely related to the invariantive theory, may be merely stated here without proof; the reader is referred to Appell's paper in vol. 90 of the Comptes Rendus, p. 1477. Given the equation

$$\frac{d^ny}{dx^n} + P_1\frac{d^{n-1}y}{dx^{n-1}} + P_2\frac{d^{n-2}y}{dx^{n-2}} + \ldots + P_ny = 0;$$

let $y_1, y_2, \ldots, y_n$ be a set of fundamental integrals; Appell's theorem is:

Every integral algebraic function, F, of $y_1, y_2, \ldots, y_n$ and the derivatives of these functions, which reproduces itself multiplied by a constant factor other than zero when we replace $y_1, y_2, \ldots, y_n$ by the elements of another fundamental system, is equal to an integral algebraic function of the coefficients of the differential equation and their derivatives multiplied by a power of $e^{-\int P_1 dx}$.

---

* Mr. Metzler has proved that the condition for self-adjointness in general is that the invariants Θ with odd suffixes must all vanish. The proof is rather too long and complicated to give here.

CHAPTER XIV.

LINEAR DIFFERENTIAL EQUATIONS WITH UNIFORM DOUBLY-PERIODIC COEFFICIENTS.

BEFORE taking up the subject of the integrals of these equations it will be convenient to recall a few points in the theory of doubly-periodic functions. These functions are all constructed by aid of the element function $\theta_1(x)$ (or, if a be a constant, $\theta_1(x-a)$). If ω and ω' denote the periods of the doubly-periodic function, then we know that

$$\theta_1(x+\omega) = -\theta_1(x), \quad \theta_1(x+\omega') = -e^{-\frac{2\pi i x}{\omega} - \frac{\pi i \omega'}{\omega}} \theta_1(x).$$

Another element function which plays an important part in the theory is the logarithmic derivative of θ_1, viz.,

$$Z(x) = \frac{\theta_1'(x)}{\theta_1(x)}.$$

For $x = 0$, $\theta_1(x)$ has a simple zero, and so $Z(x)$ has a simple pole. Inside each parallelogram of periods the doubly-periodic function will have a certain number of zeros and a certain number of poles (account being taken of the orders of multiplicity). Now it is a known theorem that the number of zeros inside a parallelogram of periods is equal to the number of poles. Let this number be m, and let $a_1, \ldots, a_m$ denote the zeros and $\alpha_1, \ldots, \alpha_m$ the poles inside a parallelogram of periods; then by another theorem we have

$$(1) \qquad \sum_1^m a_i - \sum_1^m \alpha_i = \mu\omega + \mu'\omega',$$

where μ and μ' are integers. Now a doubly-periodic function having ω and ω' as periods is given by

$$(2) \qquad e^{-\frac{2\mu'\pi x i}{\omega}} \frac{\theta_1(x-a_1) \ldots \theta_1(x-a_m)}{\theta_1(x-\alpha_1) \ldots \theta_1(x-\alpha_m)}.$$

This obviously has $a_1, \ldots, a_m$ as zeros and $\alpha_1, \ldots, \alpha_m$ as poles. Further, ω is a period; in fact the exponential is unaltered by the change of x into $x+\omega$, and each function θ_1 merely changes its sign to minus—there are, however, an even number of these functions, and so the sign remains unaltered. Again, ω' is a period: changing x into $x+\omega'$ reproduces the exponential multiplied by the factor

$$e^{-\frac{2\mu'\pi i\omega'}{\omega}};$$

$\theta_1(x-a_1)$ is changed into $-q^{-1}e^{-\frac{2\pi i}{\omega}(x-a_1)}\theta_1(x-a_1)$, and so for all of the other functions θ_1. The whole function is then multiplied by

$$e^{\frac{2\pi i}{\omega}(-\mu'\omega+a_1+\ldots+a_m-\alpha_1-\ldots-\alpha_m)} = e^{2\pi i\mu} = 1.$$

Another representation of the doubly-periodic function is due to M. Hermite. Suppose $a, b, \ldots$ to be the poles of the function inside a given parallelogram of periods, and $\alpha, \beta, \ldots$ to be their respective orders of multiplicity; the doubly-periodic function $f(x)$ is then given by the formula

$$\begin{aligned}(3)\quad f(x) = {} & A_1Z(x-a) - A_2Z'(x-a) + \ldots \\ & \qquad + \frac{(-)^{\alpha-1}}{1.2\ldots(\alpha-1)}Z^{(\alpha-1)}(x-a) \\ & + B_1Z(x-b) - B_2Z'(x-b) + \ldots \\ & \qquad + \frac{(-)^{\beta-1}}{1.2\ldots(\beta-1)}Z^{(\beta-1)}(x-b) \\ & + \ldots\ldots\ldots\ldots \\ & + C,\end{aligned}$$

where C is an arbitrary constant, and where the sum of the *residues* $A_1, B_1, \ldots$ is zero, viz.,

$$(4)\qquad A_1 + B_1 + \ldots = 0.$$

This need not be verified here. Another and very important class of doubly-periodic functions are the *doubly-periodic functions of the second kind*, as they are called by Hermite. Denote by Θ such a function with periods ω and ω' as before. Then Θ satisfies the relations

$$\Theta(x+\omega) = s_1\Theta(x), \quad \Theta(x+\omega') = s_1'\Theta(x),$$

where s_1 and s_1' are constants. It is easy to form such a function as this by aid of the function $\theta_1(x)$, and at the same time to let the constants s_1 and s_1' take any value we please. Write, viz.,

(5) $$\Theta(x) = e^{px}\frac{\theta_1(x-q)}{\theta_1(x)},$$

where p and q are arbitrary constants. We have now

(6) $$\Theta(x+\omega) = e^{p\omega}\Theta(x), \quad \Theta(x+\omega') = e^{p\omega'+\frac{2\pi i}{\omega}q}\,\Theta(x).$$

Now since p and q are arbitrary constants, we may write

(7) $$e^{p\omega} = s_1, \quad e^{p\omega'+\frac{2\pi i}{\omega}q} = s_1',$$

where s_1 and s_1' are equally arbitrary constants. Suppose now that among the infinite number of values of the logarithms of s_1 and s_1' we choose arbitrarily two values which we will denote by $\mathrm{Lg}s_1$ and $\mathrm{Lg}s_1'$; denoting now by m and m' two arbitrary integers, we have, from (7),

(8) $$\begin{cases} p\omega = \mathrm{Lg}s_1 + 2m'\pi i, \\ p\omega' + \dfrac{2\pi i}{\omega}q = \mathrm{Lg}s_1' + 2m\pi i, \end{cases}$$

from which follow

(9) $$\begin{cases} p = \dfrac{\mathrm{Lg}s_1}{\omega} + \dfrac{2m'\pi i}{\omega}, \\ q = \dfrac{1}{2\pi i}[\omega\mathrm{Lg}s_1' - \omega'\mathrm{Lg}s_1] + m\omega - m'\omega'. \end{cases}$$

Among the systems of values of p and q given by these equations there is one only for which the point q lies inside the parallelogram of periods having the origin as one vertex and otherwise determined by the periods ω and ω'; this system of values of p and q will be the one employed in all that follows.

The function $\Theta(x)$ so formed admits the multipliers s_1 and s_1' with respect to the periods ω and ω', and inside the parallelogram of periods considered has the simple zero $x = q$ and the simple pole $x = 0$. The case of $q = 0$ must of course be excepted, as in this case $\Theta(x)$ reduces to the exponential e^{px} and has no zero or pole in the parallelogram. A doubly-periodic function, which is to be considered presently, has periods ω and ω' and poles $a+q$, a, b, . . . of orders of multiplicity respectively equal to 1, $\alpha+i-2$, β, . . . Denoting such a function by Φ_{ik}, we have, from (3),

$$(10)\quad \Phi_{ik}(x) = A_1 Z(x-a) + A_2 Z'(x-a) + \dots + A_{\alpha+i-2} Z^{(\alpha+i-3)}(x-a) + B_1 Z(x-b) + B_2 Z'(x-b) + \dots + B_\beta Z^{\beta-1}(x-b) + \dots + MZ(x-a-q) + C,$$

where A, B, . . . C are constants and

$$(11)\qquad A_1 + B_1 + \dots + M = 0.$$

This function can be given in terms of elliptic functions by aid of the relation

$$(12)\quad Z(x-a) - Z(x) = \frac{\operatorname{sn} a}{\operatorname{sn} x \operatorname{sn}(x-a)} + Z\left(\frac{\omega'}{2} - a\right) - Z\left(\frac{\omega'}{2}\right).$$

The truth of (12) is easily seen: in fact both members of the equation have the same periods ω and ω', the same simple poles 0 and a, and the same residues; finally, they are identical for $x = \frac{\omega'}{2}$. We can obtain from (12) (or can verify directly) the relation

$$(13)\qquad Z'(x-a) = -\frac{1}{\operatorname{sn}^2(x-a)} + Z'\left(\frac{\omega'}{2}\right).$$

From this last equation by differentiation we get the values of

$$Z''(x-a), \quad Z'''(x-a), \quad \dots$$

Form in the same way the values of

$$Z(x-b), \quad Z'(x-b), \quad \dots$$

Eliminating now all of these quantities from (10), and denoting by C' the new additive constant, we have

$$(14)\quad \Phi_{ik}(x) = A_1 \frac{\operatorname{sn} a}{\operatorname{sn} x \operatorname{sn}(x-a)} - \frac{A_2}{\operatorname{sn}^2(x-a)} - \ldots$$

$$- A_{a+i-2} \frac{d^{a+i-4}}{dx^{a+i-4}} \frac{1}{\operatorname{sn}^2(x-a)}$$

$$+ B_1 \frac{\operatorname{sn} b}{\operatorname{sn} x \operatorname{sn}(x-b)} - \ldots + M \frac{\operatorname{sn}(a+q)}{\operatorname{sn} x \operatorname{sn}(x-a-q)} + C'.$$

For the complete determination of the function $\Phi_{ik}(x)$ in any particular case it is only necessary to determine the constants $A, B, \ldots, M, C'$.

One more preliminary remark may be made before proceeding directly to the investigation of the integrals of the given differential equation. The function $Z(x)$ satisfies the relations

$$(15)\qquad Z(x+\omega) = Z(x), \quad Z(x+\omega') = -\frac{2\pi i}{\omega} + Z(x).$$

Form now the function

$$(16)\qquad mx + m'Z(x-a), = \phi(x),$$

where m, m', and a are constants. If we change x into $x+\omega$ and $x+\omega'$ successively, this function $\phi(x)$ will receive increments $\Delta\phi(x)$ and $\Delta'\phi(x)$ defined by the equations

$$(17)\qquad \begin{cases} \Delta\phi(x) = m\omega, \\ \Delta'\phi(x) = m\omega' - \dfrac{2\pi i}{\omega} m'. \end{cases}$$

We can clearly choose the constants m and m' so that these increments shall take any values we please. Suppose then we construct a function μ of the form (16) such that

$$(18)\qquad \Delta\mu = 1, \quad \Delta'\mu = 0;$$

also a function μ' of the same form such that

$$(19)\qquad \Delta\mu' = 0, \quad \Delta'\mu' = 1.$$

These functions μ and μ' will have inside the parallelogram of periods the same simple pole $x = a$ as that possessed by the function $Z(x - a)$ from which they are derived.

If now we write

$$(20)\quad \begin{cases} \mu_0 = 1, \quad \mu_1 = \mu, \quad \mu_2 = \dfrac{\mu(\mu - 1)}{1\,.\,2}, \quad \ldots, \\ \qquad\qquad \mu_n = \dfrac{\mu(\mu - 1)\ldots(\mu - n + 1)}{1\,.\,2\ldots n}, \\ \mu_0' = 1, \quad \mu_1' = \mu', \quad \mu_2' = \dfrac{\mu'(\mu' - 1)}{1\,.\,2}, \quad \ldots, \\ \qquad\qquad \mu_n' = \dfrac{\mu'(\mu' - 1)\ldots(\mu' - n + 1)}{1\,.\,2\ldots n}, \end{cases}$$

we have obviously

$$(21)\quad \begin{cases} \Delta\mu_n = \dfrac{(\mu + 1)\mu(\mu - 1)\ldots(\mu - n + 2)}{1\,.\,2\ldots n} \\ \qquad\qquad - \dfrac{\mu(\mu - 1)\ldots(\mu - n + 1)}{1\,.\,2\ldots n} = \mu_{n-1}, \\ \Delta'\mu_n = 0, \\ \Delta\mu_n' = 0, \qquad \Delta'\mu'_n = \mu'_{n-1}, \end{cases}$$

a series of relations similar to those obtained in Chapter III (equations 84–86) for the functions $\theta_1, \theta_2, \ldots, \theta_k, \ldots$. Every polynomial, integral in both μ and μ', can, considered as a function of μ, be written in one way only in the form

$$A_0\mu_0 + A_1\mu_1 + \ldots,$$

where the coefficients A are polynomials in μ' having the form

$$A_i = B_{i0}\mu_0' + B_{i1}\mu_1' + \ldots,$$

$B_{i0}, B_{i1}, \ldots$ being constants. Any such polynomial then, say Π, can in one way only be placed in the form

$$(22)\qquad \Pi = \sum_{\alpha,\alpha'} B_{\alpha\alpha'}\mu_\alpha \mu'_{\alpha'}.$$

By aid of (21) we derive at once

$$(23) \qquad \begin{cases} \Delta\Pi = \sum\limits_{a,\, a'} B_{aa'}\mu_{a-1}\mu'_{a'}\,, \\ \Delta'\Pi = \sum\limits_{a,\, a'} B_{aa'}\mu_a\,\mu'_{a'-1} \end{cases}$$

Consider for a moment the polynomials

$$(24) \qquad \left.\begin{cases} \Xi^{lm}_{\lambda k} = \sum\limits_{a,\, a'} \chi_{aa'}\,\mu_a\,\mu'_{a'}, \\ \Xi'^{lm}_{\lambda k} = \sum\limits_{a,\, a'} \chi'_{aa'}\,\mu_a\,\mu'_{a'}, \end{cases}\right\} \qquad \alpha + \alpha' = \lambda - 1 - l,$$

which are of the same form as Π. We have at once

$$(25) \qquad \begin{cases} \Delta'\,\Xi^{lm}_{\lambda k} = \sum\limits_{a,\, a'} \chi_{aa'}\,\mu_a\,\mu'_{a'-1}\,, \\ \Delta\Xi'^{lm}_{\lambda k} = \sum\limits_{a,\, a'} \chi'_{aa'}\,\mu_{a-1}\mu'_{a'}. \end{cases}$$

The expressions for $\Delta\,\Xi^{lm}_{\lambda k}$ and $\Delta'\,\Xi'^{lm}_{\lambda k}$ are not needed: their forms are, however, obvious. The expressions in (25) are again polynomials in μ and μ'. Suppose the condition

$$(26) \qquad \Delta'\,\Xi^{lm}_{\lambda k} = \Delta\,\Xi'^{lm}_{\lambda k}$$

is to be satisfied; from (25) we must then have

$$(27) \qquad \chi_{a-1,\, a'} = \chi'_{a,\, a'-1}\,.$$

We can now determine a polynomial of order $\lambda - l$ in μ and μ', say

$$(28) \qquad \Pi^{lm}_{\lambda k} = \sum\limits_{a,\, a'} B_{aa'}\,\mu_a\,\mu'_{a'}, \qquad \alpha + \alpha' \leqq l,$$

such that its variations

$$(29) \qquad \begin{cases} \Delta\Pi^{lm}_{\lambda k}, = \sum\limits_{a,\, a'} B_{aa'}\,\mu_{a-1}\,\mu'_{a'}\,, \\ \Delta'\Pi^{lm}_{\lambda k}, = \sum\limits_{a,\, a'} B_{aa'}\,\mu_a\,\mu'_{a'-1}\,, \end{cases}$$

shall be respectively equal to $\Xi^{lm}_{\lambda k}$, $\Xi'^{lm}_{\lambda k}$; for, in order that these relations may hold we must have

$$(30) \qquad B_{aa'} = \chi_{a-1,\,a'}, \quad B_{aa'} = \chi'_{a,\,a'-1}.$$

Now if the functions Ξ, Ξ' have been determined in such a way that equation (26) is satisfied, it follows immediately from equation (27) that equations (30) can also be satisfied.

We will take up now the investigation of the integrals of the linear differential equation with uniform doubly-periodic coefficients. We have seen in the case of the linear differential equations already studied that there always exists at least one integral of the equation which, when the variable travels round a critical point, is changed into itself multiplied by a constant factor, the factor being a root of the characteristic equation corresponding to the particular critical point; or, in other words, the effect of imposing upon this integral the *substitution* corresponding to the critical point considered is to multiply the integral by a constant which is a root of the characteristic equation of the substitution. The uniform coefficients of the equation are unaltered when the variable turns round the critical point. In the equation with uniform doubly-periodic coefficients these coefficients are equally unaltered when the variable x is changed into $x + \omega$ or into $x + \omega'$. The question now naturally arises in this case as to whether or not the equation possesses an integral which is multiplied by a certain constant, say s, when x changes into $x + \omega$, and by a certain constant, say s', when x changes into $x + \omega'$. Corresponding to the changes of x into $x + \omega$ and $x + \omega'$ respectively, we have the substitutions S and S'; and so, if such an integral as we have described exists, we might expect by analogy to find the multipliers s and s' as roots of the characteristic equations corresponding to the substitutions S and S'. Picard has shown that every linear differential equation with uniform doubly-periodic coefficients and possessing only uniform integrals has always at least one integral which is a doubly-periodic function of the second kind whose multipliers s and s' are roots of the characteristic equations corresponding to the substitutions S and S'. That is, the equation always has at least one integral such that when x is changed into $x + \omega$ the integral is multiplied by s, and when x is changed into

$x + \omega'$ the integral is multiplied by s'.* It has been supposed that the integrals of the differential equation are all uniform; whether or not the integrals possess this property can always be ascertained by developing them in the form of series. In what follows we will assume that the integrals always satisfy the condition of being uniform. We have seen in Chapter III that it is possible to determine a system of fundamental integrals

$$(31) \begin{cases} y_{11}, & \ldots, & y_{1l_1}, & y_{21}, & \ldots, & y_{2l_2}, & \ldots, & y_{\lambda 1}, & \ldots, & y_{\lambda l_\lambda}, \\ z_{11}, & \ldots, & z_{1m_1}, & z_{21}, & \ldots, & z_{2m_2}, & \ldots, & z_{\mu 1}, & \ldots, & z_{\mu m_\mu} \\ \cdot & \cdot & \cdot & \cdot & \cdot & \cdot & \cdot & \cdot & \cdot & \cdot \;, \end{cases}$$

such that the substitutions S and S' shall take the canonical forms

$$(32)\quad S = \begin{vmatrix} y_{1k}, & \ldots, & y_{ik}, & \ldots; & s_1 y_{1k}, & \ldots, & s_1(y_{ik} + Y_{ik}), & \ldots \\ z_{1k}, & \ldots, & z_{ik}, & \ldots; & s_2 z_{1k}, & \ldots, & s_2(z_{ik} + Z_{ik}), & \ldots \\ \cdot & \cdot & \cdot & \cdot & \cdot & \cdot & \cdot & \cdot \end{vmatrix},$$

$$(33)\quad S' = \begin{vmatrix} y_{1k}, & \ldots, & y_{ik}, & \ldots; & s_1' y_{1k}, & \ldots, & s_1'(y_{ik} + Y'_{ik}), & \ldots \\ z_{1k}, & \ldots, & z_{ik}, & \ldots; & s_2' z_{1k}, & \ldots, & s_2'(z_{ik} + Z'_{ik}), & \ldots \\ \cdot & \cdot & \cdot & \cdot & \cdot & \cdot & \cdot & \cdot \end{vmatrix},$$

where $s_1, s_2, \ldots$ are distinct roots of the characteristic equation corresponding to the substitution S, and $s_1', s_2', \ldots$ are distinct roots of the characteristic equation corresponding to S', and where Y_{ik}, Y'_{ik} are linear functions of the integrals y whose first suffix is less than i, etc. The integrals now being supposed to be chosen so that (32) and (33) are satisfied, let us consider the class $y_{11}, \ldots, y_{ik}, \ldots$ If now we change x into $x + \omega$ and $x + \omega'$, we have the partial substitutions

$$(34)\quad \sigma = |\, y_{1k}, \ \ldots, \ y_{ik}, \ \ldots; \ s_1 y_{1k}, \ \ldots, \ s_1(y_{ik} + Y_{ik}), \ \ldots \,|,$$

$$(35)\quad \sigma' = |\, y_{1k}, \ \ldots, \ y_{ik}, \ \ldots; \ s_1' y_{1k}, \ \ldots, \ s_1'(y_{ik} + Y'_{ik}), \ \ldots \,|,$$

---

* Picard: *Sur les équations différentielles linéaires à coefficients doublement périodiques,* Crelle, vol. 90, p. 281. See also Floquet: *Sur les équations différentielles linéaires à coefficients doublement périodiques.* Annales de l'École Normale Supérieure, 1884. The student should also consult Hermite: *Sur quelques applications des Fonctions Elliptiques.* Gauthier-Villars, Paris, 1885.

where, by hypothesis,

$$(36)\qquad Y_{ik} = \sum_{l,m} a_{ik}^{lm}\, y_{lm}, \quad Y'_{ik} = \sum_{l,m} b_{ik}^{lm}\, y_{lm},$$

where the summations extend to all values of the first index l which are less than i, and to the corresponding values of the second index m. Suppose, for example, $l = \nu$, where ν is less than i; then the corresponding values of m will be, from (31), $m = 1, 2, \ldots l_\nu$.

Since we have the relation $SS' = S'S$, we must also have, from (34) and (35),

$$(37)\qquad \sigma\sigma' = \sigma'\sigma.$$

This relation can be expressed in another form by aid of equations (36). Consider the function y_{ik}; the substitution σ changes it into

$$(38)\qquad \sigma y_{ik} = s_1(y_{ik} + Y_{ik}) = s_1\Big[y_{ik} + \sum_{l,m} a_{ik}^{lm}\, y_{lm}\Big],$$

and the substitution σ' gives

$$(39)\qquad \sigma' y_{ik} = s_1'(y_{ik} + Y'_{ik}) = s_1'\Big[y_{ik} + \sum_{l,m} b_{ik}^{lm}\, y_{lm}\Big].$$

If now we first make the substitution σ and then follow it by σ', *i.e.*, if we make the substitution $\sigma\sigma'$, we find at once

$$(40)\qquad \sigma\sigma' y_{ik} = s_1 s_1'\Big[y_{ik} + Y_{ik} + Y'_{ik} + \sum_{l,m}\Big\{a_{ik}^{lm} \sum_{l',m'} b_{lm}^{l'm'}\, y_{l'm'}\Big\}\Big],$$

where the summation with respect to l' and m' extends over all values of l' which are less than l, and over the corresponding values of m'. The substitution $\sigma'\sigma$ applied to the same function y_{ik} will clearly only have the effect of interchanging the letters a and b in (40). Now since

$$\sigma\sigma' y_{ik} = \sigma'\sigma y_{ik},$$

we have (since the functions y are linearly independent), on equating the coefficients of $y_{l'm'}$, the system of relations

$$(41)\qquad \sum_{l,m}\left(a_{ik}^{lm}\, b_{lm}^{l'm'} - b_{ik}^{lm}\, a_{lm}^{l'm'}\right) = 0,$$

the summations extending over all values of l which are less than i and greater than l', and over the corresponding values of m.

We will now proceed to form functions

$$y_{11}, \ldots, y_{ik}, \ldots$$

which submit to the substitutions (34) and (35), satisfying the condition (37) or, what is the same thing, the system of conditions (41). Denoting by a a constant, and recalling the definition given above of the functions Θ, we have

$$(42) \qquad \begin{cases} \Theta(x - a + \omega) = s_1 \Theta(x - a), \\ \Theta(x - a + \omega') = s_1' \Theta(x - a). \end{cases}$$

Write now

$$(43) \qquad \begin{cases} y_{1k} = \Theta(x - a)\Phi_{1k}, \\ \cdot \quad \cdot \quad \cdot \quad \cdot \quad \cdot \quad \cdot \quad \cdot \\ y_{ik} = \Theta(x - a)\left[\Phi_{ik} + \sum_{l, m} \Pi_{ik}^{lm} \Phi_{lm}\right]. \end{cases}$$

In these equations the functions Π_{ik}^{lm} are determinate polynomials of degree $i - l$ in μ and μ'; the functions Φ are ordinary doubly-periodic functions with periods ω and ω', *i.e.*,

$$\Phi(x + \omega) = \Phi(x), \quad \Phi(x + \omega') = \Phi(x);$$

and the summations extend over all values of l less than i and over the corresponding values of m. We have now to show that the functions y defined by these equations satisfy all the prescribed conditions and are integrals of the given differential equation

$$\frac{d^n y}{dx^n} + p_1 \frac{d^{n-1} y}{dx^{n-1}} + \ldots + p_n y = 0,$$

where the coefficients p are uniform and doubly periodic, having ω and ω' for periods. From equations (42) we see that all of the functions y in (43) whose first suffix is unity satisfy the conditions of (34) and (35). To investigate the remaining ones we will write

$$(44) \qquad y_{ik} = \Theta(x - a) z_{ik}.$$

Now since the change of x into $x+\omega$ multiplies $\Theta(x-a)$ by s_1, and the change of x into $x+\omega'$ multiplies this same function by s_1', it is clear, from (34) and (35), that these changes must cause the functions z to submit to the substitutions

$$(45)\qquad \tau = | z_{1k}, \ \ldots, \ z_{ik}, \ \ldots; \ z_{1k}, \ \ldots, \ z_{ik} + Z_{ik}, \ \ldots |,$$

$$(46)\qquad \tau' = | z_{1k}, \ \ldots, \ z_{ik}, \ \ldots; \ z_{1k}, \ \ldots, \ z_{ik} + Z'_{ik}, \ \ldots |,$$

where the functions Z_{ik}, Z'_{ik} are deduced from the functions Y_{ik}, Y'_{ik} of (34) and (35) by replacing in the latter the functions y by the new functions z. We have of course, from (45) and (46), the relation

$$(47)\qquad \tau\tau' = \tau'\tau.$$

As all the functions z whose first suffix is unity remain unaltered by the substitutions τ and τ', they are ordinary doubly-periodic functions, and we can therefore write

$$(48)\qquad z_{1k} = \Phi_{1k};$$

and consequently, so far as the functions y_{1k} are concerned, all the required conditions are satisfied. Let us assume that we have succeeded step by step in constructing all of the functions z whose first suffix is less than λ, and that the general form of these functions is given by

$$(49)\qquad z_{ik} = \Phi_{ik} + \sum_{a,\,a'} \Pi_{ik}^{lm} \Phi_{lm} \quad (i < \lambda).$$

We have now to construct the functions $z_{\lambda k}$ such that

$$(50)\qquad z_{\lambda k}(x+\omega) = z_{\lambda k} + Z_{\lambda k}, \quad z_{\lambda k}(x+\omega') = z_{\lambda k} + Z'_{\lambda k}.$$

The functions Z and Z' being linear functions of the z's whose first suffix is less than λ, we have, by hypothesis,

$$(51)\qquad Z_{\lambda k} = \sum_{l,\,m} \Xi_{\lambda k}^{lm} \phi_{lm}, \quad Z'_{\lambda k} = \sum_{l,\,m} \Xi'^{lm}_{\lambda k} \Phi_{lm},$$

where $\Xi_{\lambda k}^{lm}$ and $\Xi'^{lm}_{\lambda k}$ are polynomials of order $\lambda - 1 - l$ in μ and μ' and depend linearly on the coefficients of $Z_{\lambda k}$ and $Z'_{\lambda k}$, and where the summations extend to all systems of values of l and m for which $l < \lambda$.

The substitutions $\tau\tau'$ and $\tau'\tau$ applied to $z_{\lambda k}$ give

$$(52)\qquad \begin{cases} \tau\tau' z_{\lambda k} = z_{\lambda k} + Z_{\lambda k} + Z'_{\lambda k} + \delta' Z_{\lambda k}, \\ \tau'\tau z_{\lambda k} = z_{\lambda k} + Z_{\lambda k} + Z'_{\lambda k} + \delta Z'_{\lambda k}; \end{cases}$$

$\delta' Z_{\lambda k}$ denoting the increment received by $Z_{\lambda k}$ from the substitution τ', and $\delta Z'_{\lambda k}$ denoting the increment received by $Z'_{\lambda k}$ from the substitution τ. Now since $\tau\tau' = \tau'\tau$, we must have

$$(53)\qquad \delta' Z_{\lambda k} = \delta Z'_{\lambda k}.$$

We have to bear in mind now that we have to form functions in general which are such that the results of the substitutions τ and τ' shall respectively be equal to the results arising from changing x into $x + \omega$ and $x + \omega'$. Now by hypothesis these equalities hold for the $z_{ik} (i < \lambda)$ already constructed, and consequently they hold for $Z_{\lambda k}$, $Z'_{\lambda k}$, which are linear functions of the same. It follows then that

$$(54)\qquad \begin{cases} \delta' Z_{\lambda k} = \Delta' Z_{\lambda k} = \sum\limits_{l,m} \Delta' \Xi^{lm}_{\lambda k} \Phi_{lm}, \\ \delta Z'_{\lambda k} = \Delta Z'_{\lambda k} = \sum\limits_{l,m} \Delta \Xi'^{lm}_{\lambda k} \Phi_{lm}. \end{cases}$$

(Since Φ is an ordinary doubly-periodic function, we have $\Delta\Phi = \Delta'\Phi = 0$.) From (53) we must now have, since the functions Φ_{lm} are arbitrary,

$$(55)\qquad \Delta' \Xi^{lm}_{\lambda k} = \Delta \Xi'^{lm}_{\lambda k}.$$

The conditions that these equations may be satisfied are given in equations (27), and in equations (28), (29), and (30) we have seen that a polynomial $\Pi^{lm}_{\lambda k}$ can be constructed such that

$$(56)\qquad \begin{cases} \Delta \Pi^{lm}_{\lambda k} = \Xi^{lm}_{\lambda k}, \\ \Delta' \Pi^{lm}_{\lambda k} = \Xi'^{lm}_{\lambda k}. \end{cases}$$

Supposing the polynomials Π to be so constructed, write

$$(57)\qquad z_{\lambda k} = U_{\lambda k} + \sum_{l,m} \Pi^{lm}_{\lambda k} \Phi_{lm};$$

in this change x into $x + \omega$, and the resulting increment is

$$(58)\quad \Delta U_{\lambda k} + \sum_{l,m} \Delta \Pi_{\lambda k}^{lm} \Phi_{lm} = \Delta U_{\lambda k} + \sum_{l,m} \Xi_{\lambda k}^{lm} \Phi_{lm} = \Delta U_{\lambda k} + Z_{\lambda k}.$$

Again changing x into $x + \omega$, we have as the corresponding increment

$$(59)\quad \Delta' U_{\lambda k} + Z'_{\lambda k}.$$

From equations (50) to (55), inclusive, we see that $z_{\lambda k}$ will satisfy all the required conditions if we have

$$(60)\quad \Delta U_{\lambda k} = 0 \quad \text{and} \quad \Delta' U_{\lambda k} = 0;$$

that is, if $U_{\lambda k}$ is a doubly-periodic function with periods ω and ω'. We have thus established the fact that the general form of the integrals y_{ik} is that given by equations (43). In order to satisfy the given differential equation, however, we must still determine the doubly-periodic functions Φ_{ik} and find the values of all the constants which have entered into the preceding analysis. We know in the first place that the critical points of the integrals are those of the coefficients $p_1, \ldots, p_n$ of the differential equation, and therefore, since these points are known, we know all the critical points (in our case these are poles) of the integrals. Suppose $a, b, \ldots$ to denote the poles of the integrals. By the test which we suppose to have been made in order to determine whether or not the general integral of the equation is uniform (as in our case it is supposed to be) and which consists in developing the integral according to increasing powers of $x - a$, $x - b, \ldots$, we have also discovered the various orders of multiplicity, say $\alpha, \beta, \ldots$, of these poles. We know then that the orders of multiplicity of the poles $a, b, \ldots$ in the particular integrals cannot respectively exceed $\alpha, \beta, \ldots$ Consider now the function y_{ik}, that is,

$$(61)\quad y_{ik} = \Theta(x - a)\Big[\Phi_{ik} + \sum_{l,m} \Pi_{ik}^{lm} \Phi^{lm}\Big] = \Theta(x - a) z_{ik}.$$

The function $\Theta(x - a)$ already defined admits only a simple pole a* and a simple zero $a + q$; the constant a, which has so far

* $\Theta(x - a)$ has of course, in addition to a, the poles $a + m\omega + m'\omega'$, where m and m' are integers; but we confine ourselves to the fundamental or primary pole a.

been arbitrary, will now be taken as one of the poles of the coefficients p. The functions z_{ik} as defined by (61) can then only have as poles the points $a+q$, a, b, . . . , with orders of multiplicity respectively equal at most to 1, $\alpha - 1$, β, . . . ; further, the functions μ and μ' have only the simple pole of the function $Z(x-a)$, viz., $x = a$, and therefore the polynomials Π_{ik}^{lm} have this point as their only pole, and its order of multiplicity is $i - l$. From the relations

$$(62) \qquad z_{ik} = \Phi_{ik} + \sum_{l,\,m} \Pi_{ik}^{lm} \Phi_{lm}$$

we now see at once that the functions Φ_{ik} can have no other poles than the points

$$a+q, \quad a, \quad b, \quad . \; . \; . \; ,$$

and that these have respectively orders of multiplicity at most equal to

$$1, \quad \alpha + i - 2, \quad \beta, \quad . \; . \; .$$

The forms of these functions are given by equations (10) and (14). The *forms* of the integrals y_{ik} are now known, but we have still to determine the parameters p and q and the constant coefficients which enter into the expressions of the functions Φ_{ik} and the linear functions Y_{ik}. The necessity for determining the constants in the functions Y_{ik} arises from the fact that the polynomials Π depend upon these functions. These constants will be determined by substituting in the differential equation the expressions above found for the integrals y_{ik}.

Consider first the integrals which are doubly-periodic functions of the second kind, *i.e.*, the integrals

$$y_{11}, \; y_{12} \; . \; . \; . \; y_{1k} \; . \; . \; . \; , \quad \text{(multipliers } s_1 \text{ and } s_1'\text{,)}$$
$$z_{11}, \; z_{12} \; . \; . \; . \; z_{1k} \; . \; . \; . \; , \quad \text{(multipliers } s_2 \text{ and } s_2'\text{,)}$$
$$. \quad . \quad . \quad . \quad , \qquad . \quad . \quad . \quad .$$

Take first the integral y_{11} given by the equation

$$(63) \qquad y_{11} = \Theta(x-a)\Phi_{11} = \Theta\Phi,$$

where for brevity we have written

$$(64) \qquad \Phi = \Phi_{11} \quad \text{and} \quad \Theta = \Theta(x-a), = e^{p(x-a)}\frac{\theta_1(x-a-q)}{\theta_1(x-a)}.$$

The logarithmic derivative of Θ is

$$(65) \qquad \frac{\Theta'}{\Theta} = p + Z(x-a-q) - Z(x-a),$$

or, from equation (12),

$$(66) \qquad \frac{\Theta'}{\Theta} = p + \frac{\operatorname{sn} q}{\operatorname{sn}(x-a)\operatorname{sn}(x-a-q)} + Z\left(\frac{\omega'}{2} - q\right) - Z\left(\frac{\omega'}{2}\right).$$

Since ω and ω' are the periods of the elliptic function $\operatorname{sn} t$, the second member of this equation is a doubly-periodic function which we will denote by Ω, that is,

$$(67) \qquad p + \frac{\operatorname{sn} q}{\operatorname{sn}(x-a)\operatorname{sn}(x-a-q)} + Z\left(\frac{\omega'}{2} - q\right) - Z\left(\frac{\omega'}{2}\right) = \Omega,$$

or

$$(68) \qquad \frac{\Theta'}{\Theta} = \Omega.$$

From this last equation we have

$$(69) \qquad \Theta' = \Theta\Omega,$$

and consequently

$$(70) \qquad \begin{cases} \dfrac{d}{dx}\Theta\Phi = \Theta'\Phi + \Theta\Phi' = \Theta\{\Omega\Phi + \Phi'\}, \\ \dfrac{d^2}{dx^2}\Theta\Phi = \Theta\{\Omega[\Omega\Phi + \Phi'] + [\Omega\Phi + \Phi']'\} \\ \qquad\qquad = \Theta\{[\Omega^2 + \Omega']\Phi + 2\Omega\Phi' + \Phi''\}, \\ \cdot \quad \cdot \quad \cdot \quad \cdot \quad \cdot \quad \cdot \quad \cdot \end{cases}$$

Substituting these values in the first member of the differential equation

$$\frac{d^n y}{dx^n} + p_1\frac{d^{n-1}y}{dx^{n-1}} + \ldots + p_n y = 0,$$

we have as the result the expression ΘA, denoting by A the doubly-periodic function

$$(71) \quad A = p_n\Phi + p_{n-1}\{\Omega\Phi + \Phi'\} + p_{n-2}\{[\Omega^2 + \Omega']\Phi + 2\Omega\Phi' + \Phi''\} + \dots .$$

The coefficients p_l are doubly-periodic functions which by hypothesis have the points $a, b, \dots$ as poles of orders of multiplicity at most $= l$, and so from (14) they are of the form

$$(72) \quad p_l = A_1^l \frac{\operatorname{sn} a}{\operatorname{sn} x \operatorname{sn}(x-a)} - A_2^l \frac{1}{\operatorname{sn}^2(x-a)} - \dots - A_l^l \frac{d^{l-2}}{dx^{l-2}} \frac{1}{\operatorname{sn}^2(x-a)}$$
$$+ B_1^l \frac{\operatorname{sn} b}{\operatorname{sn} x \operatorname{sn}(x-b)} - B_2^l \frac{1}{\operatorname{sn}^2(x-a)} - \dots + C^l.$$

In equation (14) we have a similar form for Φ which is linear and homogeneous in the as yet unknown constant coefficients

$$A_1, A_2, \dots, B_1, B_2, \dots, M, C'.$$

From (14) we deduce by differentiation the values of Φ', Φ'', ... Finally, if we write, for brevity,

$$(73) \qquad p' = p + Z\left(\frac{\omega'}{2} - q\right) - Z\left(\frac{\omega'}{2}\right),$$

we have

$$(74) \quad \begin{cases} \Omega = p' + \dfrac{\operatorname{sn} q}{\operatorname{sn}(x-a)\operatorname{sn}(x-a-q)}, \\ \Omega' = Z'(x-a-q) - Z'(x-a), \\ \quad = -\dfrac{1}{\operatorname{sn}^2(x-a-q)} + \dfrac{1}{\operatorname{sn}^2(x-a)} \end{cases}$$

and from this last equation we derive at once the values of

$$\Omega'', \quad \Omega''', \quad \dots .$$

Substituting these values in (71), we obtain the final expression for the doubly-periodic function A. If now y_{11} is an integral of the equation, A must vanish, and by a known theorem A will vanish if we can show that it has more zeros than it has poles.

The function y_{11} has the points $a, b, \ldots$ as poles with orders of multiplicity at most equal to $\alpha, \beta, \ldots$, and consequently $\dfrac{d^i y_{11}}{dx^i}$ has these same points as poles of orders of multiplicity at most equal to $\alpha + i, \beta + i, \ldots$. If now we multiply $\dfrac{d^i y_{11}}{dx^i}$ by the coefficient p_{n-i}, we obtain as the product an expression which has the points $a, b, \ldots$ as poles of orders of multiplicity at most equal to

$$\alpha + n, \quad \beta + n, \quad \ldots.$$

It follows then that ΘA, which is a sum of such expressions, possesses the same property. Further, Θ never has more than one pole and one zero, and so the total number of poles of A, when we take account of their orders of multiplicity, is at most equal to

$$\alpha + n + \beta + n + \ldots.$$

In order then that A may vanish identically it is only necessary to show that it admits of arbitrarily chosen zeros the sum of whose orders of multiplicity is greater than

$$\alpha + n + \beta + n + \ldots;$$

or, in other words, if the sum of the orders of multiplicity of the poles of A is less than δ, there must exist

$$\alpha + n + \beta + n + \ldots - \delta + 1 \text{ zeros.}$$

The system of equations so obtained is generally superabundant, but we nevertheless know *à priori* that they admit of solutions. These equations are linear and homogeneous with respect to the constants

$$A_1, A_2, \ldots, \quad B_1, B_2, \ldots, \quad M, \quad C'$$

of Φ. If now we eliminate these constants, which may be done in different ways, we obtain algebraic relations connecting p', $\operatorname{sn} q$ and its derivative $\operatorname{cn} q \operatorname{dn} q$. To each such system of values of p' and q corresponds for p a value

$$(75) \qquad p = p' + Z\left(\frac{\omega'}{2}\right) - Z\left(\frac{\omega'}{2} - q\right),$$

and for s_1 and s_1' the values

$$(76) \qquad s_1 = e^{p\omega}, \quad s_1' = e^{p\omega' + \frac{2\pi i q}{\omega}}.$$

Substituting the values of p' and q in the equations of condition, these will determine certain of the coefficients

$$A_1, \ A_2, \ \ldots, \ B_1, \ B_2, \ \ldots, \quad M, \quad C'$$

in terms of the remaining ones, and if these remaining ones are ν in number, we shall have ν particular integrals $y_{11}, y_{12}, \ldots, y_{1\nu}$, which are doubly-periodic functions of the second kind admitting s_1 and s_1' as multipliers. We will thus have obtained all of the integrals which are doubly-periodic functions of the second kind. To another system of values of p' and q correspond other values of the multipliers, say s_2 and s_2', etc. We determine then in the above manner all the pairs

$$s_1, \ s_1'; \quad s_2, \ s_2'; \quad s_3, \ s_3'; \quad \ldots$$

of multipliers and the corresponding integrals

$$\begin{array}{ll} y_{11}, \ \ldots, & y_{1k}, \ldots, \\ z_{11}, \ \ldots, & z_{1k}, \ldots, \\ \cdot \quad \cdot \quad \cdot & \cdot \end{array}$$

which are doubly periodic of the second kind. If the number of these integrals is equal to n (which is generally the case), their linear combination will give the general integral; if this number is less than n, we have still to find some integrals. Let us suppose that we have constructed all of the integrals $y_{ik}, z_{ik}, \ldots$ whose first index is less than λ, and that we have determined the corresponding linear functions $Y_{ik}, Y_{ik}', \ldots$. We seek now to determine the integrals $y_{\lambda k}$ (if such exist) and the corresponding functions $Y_{\lambda k}, Y_{\lambda k}', \ldots$. Now we know that

$$(77) \qquad y_{\lambda k} = \Theta\left[\Phi_{\lambda k} + \Sigma \Pi_{\lambda k}^{lm} \Phi_{lm}\right],$$

where, remembering the limits of the summation, everything is known save the indeterminate coefficients

$$A_1^{\lambda k}, \ A_2^{\lambda k}, \ \ldots, \quad M^{\lambda k}, \quad C'^{\lambda k}$$

of $\Phi_{\lambda k}$ and the coefficients of $Y_{\lambda k}$, $Y'_{\lambda k}$ which enter linearly into the polynomials $\Pi^{lm}_{\lambda k}$.

Substituting the above expression for $y_{\lambda k}$ in the given differential equation, we obtain as the result $\Theta\Lambda_{\lambda k}$, where $\Lambda_{\lambda k}$ is a function such that the sum of the orders of multiplicity of its poles is not greater than

$$\alpha + n + \beta + n + \ldots.$$

Further, $\Lambda_{\lambda k}$ is doubly periodic: to show this it is only necessary to change x into $x + \omega$; the result of this change is obviously the same whether it be made in $y_{\lambda k}$ before the substitution in the differential equation or be made in $\Theta\Lambda_{\lambda k}$ after the substitution. The result of changing x into $x + \omega$ is to change $y_{\lambda k}$ into $s_1(y_{\lambda k} + Y_{\lambda k})$; and since $Y_{\lambda k}$ is a linear function of the integrals y already found, it follows that on substituting $s_1(y_{\lambda k} + Y_{\lambda k})$ in the differential equation the result will be

$$s_1\Theta\Lambda_{\lambda k}.$$

But $\Theta(x + \omega) = s_1\Theta(x)$, and consequently

$$\Lambda_{\lambda k}(x + \omega) = \Lambda_{\lambda k}(x).$$

A similar result will obviously be obtained if we change x into $x + \omega'$; that is, we should find

$$\Lambda_{\lambda k}(x + \omega') = \Lambda_{\lambda k}(x),$$

and therefore $\Lambda_{\lambda k}$ is a doubly-periodic function. In order now that $\Lambda_{\gamma k}$ may vanish (as it must if $y_{\lambda k}$ is an integral) it is only necessary to show, in the manner already described, that the sum of the orders of multiplicity of its zeros is greater than

$$\alpha + n + \beta + n + \ldots.$$

We will thus determine a system of linear homogeneous equations for the determination of the unknown coefficients. If this system is compatible, there will still exist, and we can determine, new integrals y_{ik}; if it be not compatible, we will know that the integrals y'_{ik} of the first class have been all determined, and in order to obtain new integrals we shall have to start with those of the second class, viz., the integrals z_{ik}, etc. This process being continued, we will finally arrive at a system of n linearly independent integrals.

Lamé's equation

$$\frac{d^2y}{dx^2} - [m(m-1)k^2\text{sn}^2x + h]y = 0$$

(m a positive integer, h an arbitrary constant, and k the modulus of the elliptic function sn x) is probably the most important known equation of the type above considered. It is not possible, however, in the limits of this treatise to investigate this equation, and so the reader is referred to the various papers on the subject by M. Hermite in the Comptes Rendus of the Academy of Sciences (Paris) in Crelle's Journal, and particularly to M. Hermite's treatise "*Sur quelques applications des Fonctions Elliptiques*" (Paris: Gauthier-Villars, 1885). Two memoirs by Count de Sparre in Vol. III of the Acta Mathematica should also be consulted. The common title of de Sparre's two memoirs is

"*Sur l'Équation*

$$\frac{d^2y}{dx^2} + \left[2\nu\frac{k^2 \text{ sn } x \text{ cn } x}{\text{dn } x} + 2\nu_1\frac{\text{sn } x \text{ dn } x}{\text{cn } x} - 2\nu_2\frac{\text{cn } x \text{ dn } x}{\text{cn } x}\right]\frac{dy}{dx}$$

$$= \left[\frac{1}{\text{sn}^2 x}(n_3 - \nu_2)(n_3 + \nu_2 + 1) + \frac{\text{dn}^2 x}{\text{cn}^2 x}(n_2 - \nu_1)(n_2 + \nu_1 + 1)\right.$$

$$+ k^2\frac{\text{cn}^2 x}{\text{dn}^2 x}(n_1 - \nu)(n_1 + \nu + 1) + k^2 \text{ sn}^2 x(n + \nu + \nu_1 + \nu_2)$$

$$\left.(n - \nu - \nu_1 - \nu_2 + 1) + h\right]y.$$

Équation où ν, ν_1, ν_2 désignent des nombres quelconques, n, n_1, n_2, n_3 des nombres entiers positive ou négative, et h une constante arbitraire."

The subject of equations with doubly-periodic coefficients will be resumed later in connection with the invariantive theory; the reader is here, however, advised to consult Halphen's "*Mémoire sur la réduction des équations différentielles linéaires aux Formes intégrables*" (Savants Etrangères, vol. xxviii), and also Chapter XIII of the second volume of his "*Traité des Fonctions Elliptiques.*"

END OF VOL. I.